# Advances in Intelligent Systems and Computing

Volume 980

The series "Advances in Intelligent Systems and Computing" contains publications on theory, applications, and design methods of Intelligent Systems and Intelligent Computing. Virtually all disciplines such as engineering, natural sciences, computer and information science, ICT, economics, business, e-commerce, environment, healthcare, life science are covered. The list of topics spans all the areas of modern intelligent systems and computing such as: computational intelligence, soft computing including neural networks, fuzzy systems, evolutionary computing and the fusion of these paradigms, social intelligence, ambient intelligence, computational neuroscience, artificial life, virtual worlds and society, cognitive science and systems, Perception and Vision, DNA and immune based systems, self-organizing and adaptive systems, e-Learning and teaching, human-centered and human-centric computing, recommender systems, intelligent control, robotics and mechatronics including human-machine teaming, knowledge-based paradigms, learning paradigms, machine ethics, intelligent data analysis, knowledge management, intelligent agents, intelligent decision making and support, intelligent network security, trust management, interactive entertainment, Web intelligence and multimedia.

The publications within "Advances in Intelligent Systems and Computing" are primarily proceedings of important conferences, symposia and congresses. They cover significant recent developments in the field, both of a foundational and applicable character. An important characteristic feature of the series is the short publication time and world-wide distribution. This permits a rapid and broad dissemination of research results.

**** Indexing: The books of this series are submitted to ISI Proceedings, EI-Compendex, DBLP, SCOPUS, Google Scholar and Springerlink ****

More information about this series at http://www.springer.com/series/11156

Karsten Berns · Daniel Görges
Editors

# Advances in Service and Industrial Robotics

## Proceedings of the 28th International Conference on Robotics in Alpe-Adria-Danube Region (RAAD 2019)

Springer

*Editors*
Karsten Berns
Department of Computer Science
Robotics Research Lab
Technische Universität Kaiserslautern
Kaiserslautern, Germany

Daniel Görges
Department of Electrical and Computer Engineering
Electromobility Group
Technische Universität Kaiserslautern
Kaiserslautern, Germany

ISSN 2194-5357 ISSN 2194-5365 (electronic)
Advances in Intelligent Systems and Computing
ISBN 978-3-030-19647-9 ISBN 978-3-030-19648-6 (eBook)
https://doi.org/10.1007/978-3-030-19648-6



This Springer imprint is published by the registered company Springer Nature Switzerland AG
The registered company address is: Gewerbestrasse 11, 6330 Cham, Switzerland

# Preface

This book is the collection of the accepted papers presented at the 28th International Conference on Robotics in Alpe-Adria-Danube Region, RAAD 2019. RAAD 2019 is held in Kaiserslautern at the Fraunhofer Zentrum and the Technische Universität in the state Rhineland-Palatinate of Germany, during June 19–21, 2019.

The conference offers an international and cooperative platform to exchange and discuss recent research results and future ideas about robotic technology, engineering, and science. Researchers from 20 countries mostly affiliated to the Alpe-Adria-Danube Region participate.

Continuing its tradition, RAAD 2019 covers all major areas of robotic research, development, and innovation. Among others, new applications and current trends such as Industry 4.0, human–robot interaction, medical and rehabilitation robotics, robot control, and machine learning are covered. In relation to these topics, we are very proud that three renowned keynote speakers present the current state as well as future innovations in their main research areas. They are Giovanni Muscato (navigation of mobile robots in rough outdoor environments), Andreas Müller (geometric modeling of robotic manipulators—a short overview for practitioners), and J. Marius Zöllner (from deep learning to autonomous driving).

In addition, three special sessions offer deep insight into today's hot research topics with a strong influence on application areas. For proposing, promoting, and collecting the related papers, a special acknowledgment is given to: Med Amine Laribi and Giuseppe Carbone (Safety-Related Devices and Applications), Achim Wagner (Flexible Manufacturing in Industry 4.0 Environments), and Kosta Jovanović and Tadej Petrič (Advances in Human–Robot Interaction).

Keeping the overall quality of the conference contributions high, a thorough review process was held for paper acceptance. This includes three detailed reviews per paper covering its thematic relevance, quality, originality, significance, and the content presentation. At this point, we would like to thank all the reviewers for their valuable time and effort they spent on reviewing the assigned papers. The review process resulted in the acceptance of 44 regular papers and 19 special session contributions out of 71 original submissions. A special acknowledgment goes to all authors for their excellent work and effort to submit and improve their papers.

According to conference sessions, this book is organized in 9 chapters each covering one major research area of the conference:

- Mechanical Design and Kinematics
- Robot Control
- Industry Robotics
- Path Planning and Navigation
- Advances in Human–Robot Interaction
- Medical and Rehabilitation Robotics
- Safety-Related Devices and Applications
- Robot Vision
- Multi-robot and Communication Systems

There are many bodies sponsoring the conference. First of all, we would like to thank our main sponsor, the Deutsche Forschungsgemeinschaft (DFG). Further, we would like to thank the Fraunhofer Zentrum Kaiserslautern and the Deutsches Forschungszentrum für Künstliche Intelligenz (DFKI) and the Technische Universität Kaiserslautern (TUK) for providing the conference locations, technical demonstrations, and administration. A special acknowledgment goes to the RAAD Scientific Committee for support and conference promotion.

June 2019

K. Berns
D. Görges

# Organization

## General Chair

| | |
|---|---|
| Karsten Berns | Technische Universität Kaiserslautern, Germany |

## Co-chair

| | |
|---|---|
| Daniel Görges | Technische Universität Kaiserslautern, Germany |

## International Scientific Committee

| | |
|---|---|
| Nikos A. Aspragathos | University of Patras, Greece |
| Karsten Berns | Technische Universität Kaiserslautern, Germany |
| Theodor Borangiu | Polytehnica University of Bucharest, Romania |
| Ivana Budinská | Slovak Academy of Sciences, Bratislava, Slovakia |
| Marco Ceccarelli | University of Cassino, Italy |
| Karol Dobrovodský | Slovak Academy of Sciences, Slovakia |
| Carlo Ferraresi | Polytechnic University of Turin, Italy |
| Alessandro Gasparetto | University of Udine, Italy |
| Nick Andrei Ivanescu | Polytehnica University of Bucharest, Romania |
| Roman Kamnik | University of Ljubljana, Slovenia |
| Gernot Kronreif | ACMIT GmbH, Austria |
| Andreas Mueller | Johannes Kepler University Linz, Austria |
| Tadej Petrič | Jožef Stefan Institute, Ljubljana, Slovenia |
| Ivan Petrovic | University of Zagreb, Croatia |
| Doina Pîsla | Technical University of Cluj-Napoca, Romania |
| Alexander Rodic | Institute Mihailo Pupin, Belgrade, Serbia |

# Contents

# Mechanical Design and Kinematics

# An Integrated Taxonomy and Critical Review of Module Designs for Serial Reconfigurable Manipulators

N. A. Stravopodis(✉), C. Valsamos, and V. C. Moulianitis

Department of Product and Systems Design Engineering,
University of the Aegean, Syros, Greece
{n.stravopodis,balsamos,moulianitis}@syros.aegean.gr

**Abstract.** Modular, reconfigurable, self-reconfigurable and metamorphic robots have already been in research since the last quarter of the previous century however and quite several different module designs have been presented in the relevant literature. As such, these modules need to be taxonomized and critically reviewed in order to form the basis for the design of new modules. The present paper introduces such a taxonomy based on the governing aspects of the modules in the literature and a weighted grading system for their critical review. Fifteen different module designs are examined and graded to identify their suitability for serving as the basis for the design of new modules for a self-metamorphic robotic manipulator system.

**Keywords:** Modular robots · Reconfigurable robots self-reconfigurable robots · Metamorphic robots · Self metamorphic robots

## 1 Introduction

Modularity and reconfiguration have been under increasing research efforts in the field of robotics since the last two decades of the previous century since they were defined as the driving concepts that would allow the improvement of robots abilities and performance. Modularity allowed for a robot to be constructed of a varying number of different basic components (connectors, joints, end effectors etc.), in an effort to allow the end user to be able to construct a robot structure with respect to their needs and also mitigate the relative equipment and maintenance costs.

Reconfiguration was identified as the next step in the evolution of robots, making use of modularity and allowing the use of interchangeable basic components that were further evolved to be self-sufficient mechatronic devices (equipped with actuation, computational power and sensors). The end user was provided with the ability to match their robot structure to task in order to achieve the highest possible performance during task execution, albeit performance was in the respective literature defined in a multitude of ways (kinematic, dynamic, energy usage, cycle time etc.) and to same respect application space within the robot's workspace (local, task-based or global).


K. Berns and D. Görges (Eds.): RAAD 2019, AISC 980, pp. 3–11, 2020.
https://doi.org/10.1007/978-3-030-19648-6_1

From then on, the evolution of these concepts moved on in two distinctive paths: Self-reconfiguration and Metamorphic systems. Although the distinction between the two is usually quite difficult, since they usually share several similar aspects, authors characterized their proposed systems respectively, based on their characteristics ([1, 2]). However, it is safe to define that the main common characteristic of the two concepts was the *achievement of different physical forms (anatomies) from a given structure.*

In retrospect, there is a common and standing feature that is common to almost all proposed designs and systems in the relevant literature of the aforementioned robotic evolution concepts: Modularity was and is at the core of any of the following paradigms whether they were reconfiguration, self- reconfiguration or metamorphosis.

This paper presents a comprehensive state of the art review of various module types that were presented in the relative literature regarding the design of modules to be used as building blocks for the structuring of robots that adhere to the aforementioned paradigms. The proposed module designs for serial modular and modular reconfigurable robotic systems are critically reviewed under different aspects and are subsequently grated regarding these aspects. The aim of the presented work is to define the driving characteristics/aspects for the design of modules that will be used for the structuring of self-metamorphic open chain robotic manipulator systems of Technology Readiness Level (TRL) 4 and 5 in accordance to the definitions described in the Multi-Annual Roadmap (MAR) For Robotics in Europe [3].

The paper is structured as follows. First a module taxonomy is introduced, with the relative classes and taxes described in detail. In the next session, a grading system on the suitability of a module design as basis for the design of new modules for the structuring of self-metamorphic serial manipulators is introduced, and 15 modules from the relevant literature are subsequently graded accordingly. The paper closes with the discussion of the results and the description of future work on the subject of module design.

## 2 Module Categories – Module Taxonomy and Typology

When viewing the relevant literature on modular, reconfigurable, self-reconfigurable and metamorphic robots, the numbers of different proposed systems and approaches in module design is quite large. The quite large level of variance between these approaches provides increased levels in complexity when considering their categorization aiming at providing a concise state of the art review on modules for the considered robot types. As such, in this paragraph an attempt is presented to provide a taxonomy of the different module types, based on certain defining and distinct aspects of their design. Although the presented taxes may overlap to a certain degree, such an effort is deemed necessary for the better understanding of the presented module designs. The selected aspects are those that adhere to the capability of a given module to facilitate the structuring of metamorphic manipulator and it's metamorphosis (or self metamorphosis).

## 2.1 First Aspect: Operation – Active vs Passive

A short definition of the terms "autonomous" and "passive" used for this characterization are the following; an autonomous module is one capable of operating individually, i.e. has computational power and censoring capabilities, actuation and is capable of some sort of movement, whereas a passive module is a "dumb" device, unable to operate individually.

Passive modules are quite often met in the proposed designs for modules used to structure either modular or reconfigurable manipulators. Their most common usage is to provide the necessary lengths for the developed system and to provide the main load bearing structure of the system. In most cases [4–7] they are totally rigid with regards to this aspect and they are manually connected to the rest of the modules making up the robot's lattice. However in some later designs, like the modular metamorphic manipulator system proposed in [8] and the reconfigurable manipulator system proposed in [9] the passive modules used were given abilities allowing the end user to mechanically configure some governing aspect (namely the angle between the module two parts in [8] and the module's overall length in [9]) so as to allow a level of metamorphosis or reconfiguration respectively. The alteration of the governing aspect allowed the changing of the structured manipulators D-H parameters, leading to new anatomies.

On the other hand, active modules are evident in all cases of proposed robotic systems of the examined types, since their existence in the robot's lattice is necessary as they usually form the robot's joints. In the case of robotic manipulators, such modules are usually self-contained devices built around an actuator, and as such providing both computational power and 1 degree of freedom (DOF) actuation [4–9]. However, this is not the case for self-reconfigurable and metamorphic robots that are usually built for mobility. Active modules in such cases are complex mechatronic intelligent devices that are quite capable of achieving not only actuation but a certain level of motion autonomously [10–13].

## 2.2 Second Aspect: Connectivity – Automatic vs User Operated and Multiplicity

Automatic connectivity may best be described as the ability of several modules to rearrange their connections to form different structures or anatomies without the necessity of the end user intervening by either mechanically making the connections between the modules or assisting with the (possibly) necessary motion of one module with respect to others to achieve the envisaged connection.

This aspect can be deemed as a governing one separating modular and reconfigurable robots from self-reconfigurable and metamorphic robots. Almost all the systems of the first type are not autonomous and require user help for the structuring of a form or the reconfiguration to a new one [4–9]. The level of the required effort by the end user may range from the complete or partial assembly/disassembly of the modules [4–9] to the mechanical alteration of module properties that allow the transition to the new form [8, 9]. Additionally, in almost all cases, the connections available for the formation are limited to either two single points of connection (fore and aft of the module) [4–6, 8, 9] or on the modules sides therefore providing up to four additional connection surfaces [7, 8].

In the case of self-reconfigurable or metamorphic robots however, modules are semi- or completely autonomous units and can connect/disconnect without any interference by the end user [10–14]. The modules of these systems also provide various alternative connecting surfaces allowing different connections between two adjoint modules, for example the M-Tran [10] and Polybot [11] modules that allow the connection of modules in four different orientations using their connection surfaces. The SMORES module [14] is able to rotate it's docking port to any angle along the docking axis to effectively connect active docking ports to neighboring modules in arbitrary orientations, as long as the docking ports have aligned the docking axis to be parallel and coincident.

### 2.3 Third Aspect: Homogeneity vs Heterogeneity

Homogeneity and Heterogeneity are a direct function of the different module types require to structure a modular (whether reconfigurable, self-reconfigurable or metamorphic) robotic system. Homogenous systems are composed of a single type of module while heterogenous systems make use of different types of modules to construct a robotic form. In the basis of the distinction lies the task type that the structured robotic forms will be utilized to perform, their operability and scalability. Homogenous systems provide significant levels of scalability, diagnosable and cost effectiveness. As such this design aspect is utilized almost extensively for self-reconfigurable and metamorphic systems [10–14] due to their mobility aspect. Most artificial and natural swarms consist of structurally homogeneous elements, capable of behavioral specialization, e.g. [15]. However, due to this fact, limitations do arise in terms of power, weight and locomotion capabilities [16]. As such, most modules for reconfigurable and modular manipulator systems are heterogenous in order to provide the necessary diversity in specialization. For example, in [17] two distinct types of modules are used to structure a reconfigurable modular manipulator, an active module that provides actuation and computational power and a passive connection module that provides the necessary geometric characteristics as well as forming the load bearing structure of the robot's form.

## 3 State of the Art Module Designs as a Basis for the Design of Modules for a Self-metamorphic Manipulator System – Identification of Major Design Perquisites

A total of 15 different module designs for use in the structuring of different serial modular and reconfigurable robotic systems from the relative literature were selected for review in this work. The module designs were subsequently graded based on six equally weighted criteria, which validate their design regarding the possibility for its usage as basis for the design of a novel self-contained autonomous module for the structuring of a serial self-metamorphic serial manipulator system.

Equally weighted criteria were used since all are considered equally important for the design and development of robotic modules for the metamorphic robotic system. The criteria selected are based on the specific application considered and its constraints. Next, the selected criteria are justified and their relative grades, are presented.

- C1: *Self contained module:* Independence and completeness of each module is important since they enhance the ability to develop a high-integrated robotic system. (graded 0 if not and 1 if yes).
- C2: The *weight* of the module: A lightweight design is of high importance since the developed robotic system should be able to be moves effortlessly between different workstations. (graded as: 1 for modules >6 kg, 2 for modules 5–6 kg, 3 for modules of 4 kg, 4 for modules of 3 kg and 5 for modules <2 kg).
- C3: *Scalability:* The number of DoF of the system can be altered by adding or removing modules. Additionally, due to static and dynamic concerns, different module sizes in the robot's lattice will be required (i.e. a larger and more robust module for the base of the manipulator and smaller ones in the wrist) (graded as: 0 for not scalable systems, 1 for scalable systems).
- C4: *Facilitation of metamorphosis* of the constructed manipulator: Altering the configuration of the robot leads to a high level of configurability. The metamorphosis procedure for the envisaged system should present limited down-time. (graded as: 1 for not modular systems, 2 for requirement of manual disassembly/reassembly, 3 for manual metamorphosis for structure change, 4 for system assisted manual metamorphosis of the system, 5 for autonomous metamorphosis of the system).
- C5: *Estimated cost:* The overall cost of the robotic system is highly connected to the cost of each module. Due to the aim of the metamorphic system as a cost effective alternative as a tool for SME's in different sectors the overall cost of the robotic system should be lower than fixed-anatomy robots for similar applications. As such the cost of each module is of high importance. A grade of 0-1 is given based on the following attributes: Level of technology equipment used (0.2) Number of different sensors used (0.3) Custom made parts (0.4) Founding reported (0.1).
- C6: *Connectivity:* The versatility of the robotic system depends on the simplicity and the amount of the functional capabilities performed via the interconnection between the different modules. A grade of 0-1 is given based on the following attributes: (0.2) For simple mechanical only, for manual assembly-disassembly, (0.4) For fast-plug mechanical, (0.6) For fast-plug mechanical and electronics, (0.8) For fast-plug mechanical, electronics, communication and (1.0) for fast-plug mechanical, electronics, communication and power bus.

The finalized score board for the selected modules is presented in Table 1.

From the results presented in Table 1, based on the selected criteria, the T&I type joint modules, Modular Joints for industrial applications and the WALK-MAN actuation system are proposed as the best basis for the design of a novel module for a serial self-metamorphic manipulator. All three are scalable and able to form a n-DoF reconfigurable robotic system. In T&I type joint modules the kinematic topology can be altered, since different relative positions between adjacent modules can be autonomously achieved. In Modular Joints for industrial applications metamorphosis is

**Table 1.** Total score for the reviewed modules based on the selected criteria

| Robotic system or Joint module reviewed | C1 | C2 | C3 | C4 | C5 | C6 | Total score |
|---|---|---|---|---|---|---|---|
| Contactless active robotic joint using active magnetic bearing [18] | 1 | 0.2 | 0 | 0.2 | 0.435 | 0.2 | 2.035 |
| DEXARM [19] | 1 | 0.6 | 1 | 0.4 | 0.295 | 1 | 4.295 |
| T&I type joint modules [20] | 1 | 0.6 | 1 | 1 | 0.73 | 1 | 5.330 |
| Spring-Assisted MRR [21] | 1 | 0.4 | 1 | 1 | 0.47 | 0.6 | 4.470 |
| LCT-Robot [22] | 1 | 0.2 | 1 | 1 | 0.42 | 0.4 | 4.020 |
| MARS manipulator [23] | 1 | 0.8 | 0 | 1 | 0.275 | 0.2 | 3.275 |
| MODMAN [17] | 1 | 1 | 1 | 0.6 | 0.265 | 1 | 4.865 |
| Novel task-based manipulator [24] | 0 | 1 | 0 | 0.6 | 0.92 | 0.2 | 2.720 |
| Versatile passive connector [8] | 0 | 0.8 | 1 | 0.6 | 0.92 | 0.4 | 3.720 |
| TOMMS [25] | 0 | 1 | 1 | 0.6 | 0.45 | 0.6 | 3.650 |
| RMMS [4] | 1 | 0.4 | 1 | 0.4 | 0.57 | 1 | 4.370 |
| Drive module based on dynamic load characteristics [26] | 1 | 0.8 | 1 | 0.4 | 0.44 | 0.2 | 3.840 |
| Modular joint for industrial applications [27] | 1 | 1 | 1 | 0.4 | 0.7 | 1 | 5.100 |
| WALK-MAN actuation system [28] | 1 | 1 | 1 | 1 | 0.225 | 0.8 | 5.025 |
| THOR modular mobile manipulator [29] | 1 | 0.6 | 1 | 0.4 | 0.83 | 0.4 | 4.230 |

manually implemented limiting its full potential. WALK-MAN is used in a humanoid robot, capable of autonomous reconfiguration. All modules are self-contained, with fast mechanical connection of adjacent modules with power and communication buses also integrated. Finally a key aspect in all three systems is their lightweight design.

Other modules with lower grade, but worth mentioning are the MARS Manipulator and LCT-Robot modules. Low connectivity scores and increased weight per module respectively, contribute to the downgraded scoring. MODMAN modules and DEXARM despite high connectivity abilities, scalability, independence and completeness require manual metamorphosis.

The proposed robotic systems comply with key system abilities as specified in [3]. Configurability Level 3 (Run-time self configuration) is achieved, since set of alternative configurations are pre-determined and the robotic systems are programmed to alter their structural topology depending on user-specified demands. To achieve that, the proposed systems integrate some of the most important ability targets specified such as Plug and play standards, User configuration interfaces, mechatronic and distributed configuration. A key issue for future work is the integration of the different aspects and system abilities from the proposed modules in a single module (Fig. 1)

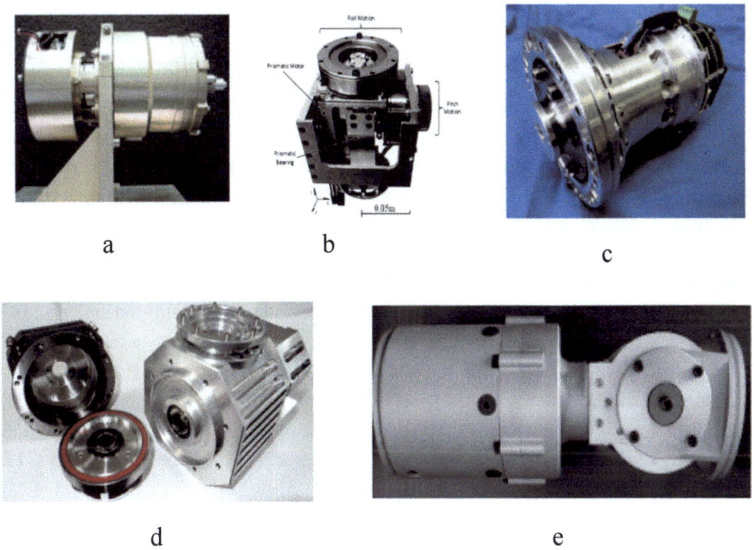


**Fig. 1.** Images of some of the highest scoring modules, (a) DEXARM [19], (b) MARS Manipulator [23], (c) WALK-MAN actuation system [28], (d) MODMAN [17], (e) T&I type joint modules [20]

## 4 Discussion and Future Work

In this paper a consisted State of the Art and critical review of different designs for modules proposed or presented in the relevant literature for the structuring of serial modular and/or reconfigurable robotic systems was presented. A taxonomy based on the identification of key governing aspects of the modules was proposed. Fifteen different designs of modules for serial modular and/or reconfigurable robotic systems were critically reviewed, categorized by the proposed taxonomy system and graded based on selected criteria, to determine which would present the most promising basis for the design of a module for the development of a serial self-metamorphic robotic manipulator.

Future work, includes the widening of the critical review to all types of systems modules (parallel, Serial, Hybrid, self-reconfigurable) and subsequent grading to determine the governing aspects and specifications for the design of the aforementioned system's modules.

**Acknowledgement.** This research has been financially supported by General Secretariat for Research and Technology (GSRT) and the Hellenic Foundation for Research and Innovation (HFRI) (Code: 1184)

## References


1. Chirikjian, G., Pamecha, A.: Bounds for self-reconfiguration of metamorphic robots. In: 1996 IEEE International Conference on Robotics and Automation. Proceedings, vol. 2. IEEE (1996)
2. Stoy, K., et al.: Self-Reconfigurable Robots: An Introduction. MIT Press, Cambridge (2010)
3. Multi-Annual Roadmap (MAR) For Robotics in Europe. https://ec.europa.eu/digital-single-market/en/news/multi-annual-roadmap-call-ict-24-robotics-now-available. Accessed Dec 2018
4. Paredis, C., Brown, B., Khosla, P.: A rapidly deployable manipulator system. In: Proceedings of the International Conference on Robotics and Automation, pp. 1434–1439 (1996)
5. US Department of Energy (DOE): Modular Manipulator for Robotic Applications, DOE/EM-0641 (2002)
6. Matsumaru, T.: Design and control of the modular robot system: TOMMS. In: 1995 IEEE International Conference on Robotics and Automation. Proceedings, vol. 2, pp. 2125–2131, 21–27 May 1995
7. Chen, I.-M.: Rapid response manufacturing through a rapidly reconfigurable robotic workcell. Robot. Comput. Integr. Manuf. **17**, 199–213M (2001)
8. Valsamos, C., Moulianitis, V., Aspragathos, N.: Kinematic synthesis of structures for metamorphic serial manipulators. J. Mech. Robot. **6**, 041005 (2014). 14 pages. https://doi.org/10.1115/1.4027741
9. Aghili, F., Parsa, K.: Design of a reconfigurable space robot with lockable telescopic joints. In: Proceedings of the 2005 IEEE/RSJ International Conference on Intelligent Robots and Systems, Beijing, China, 9–15 October 2005
10. Kurokawa, H., et al.: Self-reconfigurable M-TRAN structures and walker generation. Robot. Auton. Syst. **54**(2), 142–149 (2006)
11. Yim, M., et al.: Connecting and disconnecting for chain self-reconfiguration with PolyBot. IEEE/ASME Trans. Mechatron. **7**(4), 442–451 (2002)
12. Rubenstein, M., et al.: Docking among independent and autonomous CONRO self-reconfigurable robots. In: 2004 IEEE International Conference on Robotics and Automation. Proceedings, ICRA 2004, vol. 3. IEEE (2004)
13. Yim, M., Shen, W.M., Salemi, B., Rus, D., Moll, M., Lipson, H., Klavins, E., Chirikjian, G. S.: Modular self-reconfigurable robot systems [grand challenges of robotics]. IEEE Robot. Autom. Mag. **14**(1), 43–5 (2007)
14. Davey, J., Kwok, N., Yim, M.: Emulating self-reconfigurable robots-design of the SMORES system. In: 2012 IEEE/RSJ International Conference on Intelligent Robots and Systems (IROS). IEEE (2012)
15. Kernbach, S., Kernbach, O.: Collective energy homeostasis in a large-scale micro-robotic swarm. Robot. Auton. Syst. (2011). https://doi.org/10.1016/j.robot.2011.08.001
16. Kernbach, S., et al.: Heterogeneity for increasing performance and reliability of self-reconfigurable multi-robot organisms. arXiv preprint arXiv:1109.2288 (2011)
17. Hong, S., et al.: Design of manually reconfigurable modular manipulator with three revolute joints and links. In: 2016 IEEE International Conference on Robotics and Automation (ICRA). IEEE (2016)
18. Selmy, M., Fanni, M., Mohamed, A.M.: Design and control of a novel contactless active robotic joint using AMB. In: 2015 IEEE International Conference on Autonomous Robot Systems and Competitions (ICARSC). IEEE (2015)

19. Schuler, S., et al.: Design and development of a joint for the dexterous robot arm. In: Ninth ESA Workshop on Advanced Space Technologies for Robotics and Automation, Noordwijk, The Netherlands, November 2006
20. Guan, Y., et al.: 1-dof robotic joint modules and their applications in new robotic systems. In: IEEE International Conference on Robotics and Biomimetics, ROBIO 2008. IEEE (2009)
21. Liu, G., Liu, Y., Goldenberg, A.A.: Design, analysis, and control of a spring-assisted modular and reconfigurable robot. IEEE/ASME Trans. Mechatron. **16**(4), 695–706 (2011)
22. Wang, P., Cao, Q., Gu, K.: The modular robot joint design and experience verification for small openings and enclosed space. In: 2014 IEEE International Conference on Information and Automation (ICIA). IEEE (2014)
23. Kereluk, J.A., Reza Emami, M.: A new modular, autonomously reconfigurable manipulator platform. Int. J. Adv. Robot. Syst. **12**(6), 71 (2015)
24. Singla, E., Singh, S., Gupta, S.: "A." Proceedings of the 1st International and 16th National Conference on Machines and Mechanisms (iNaCoMM 2013) (2013)
25. Matsumaru, T.: Design and control of the modular robot system: TOMMS. In: 1995 IEEE International Conference on Robotics and Automation. Proceedings, vol. 2. IEEE (1995)
26. Seo, J.-M., et al.: Selection and design of drive modules for robot joint based on dynamic load characteristics. Int. J. Control. Autom. Syst. **15**(2), 790–801 (2017)
27. Liu, Y., et al.: A modular manipulator for industrial applications: design and implement. In: 2017 2nd International Conference on Robotics and Automation Engineering (ICRAE). IEEE (2017)
28. Negrello, F., et al.: A modular compliant actuator for emerging high performance and fall-resilient humanoids. In: 2015 IEEE-RAS 15th International Conference on Humanoid Robots (Humanoids). IEEE (2015)
29. Lyder, A., Garcia, R.F.M., Stoy, K.: Genderless connection mechanism for modular robots introducing torque transmission between modules. In: Proceedings of the ICRA Workshop on Modular Robots, State of the Art (2010)

# Geometry and Inverse Kinematics of 3-PRRS Type Parallel Manipulator

Zhumadil Baigunchekov[1,2(✉)], M. A. Laribi[3], Azamat Mustafa[2], Rustem Kaiyrov[1], Bekzat Amanov[1], and Abzal Kassinov[1]

[1] Al-Farabi Kazakh National University, Almaty, Kazakhstan
bzh47@mail.ru
[2] Satbayev University, Almaty, Kazakhstan
[3] Université de Poitiers, Poitiers, France
med.amine.laribi@univ-poitiers.fr

**Abstract.** In this paper the methods of structural synthesis and inverse kinematics of 3-PRRS type parallel manipulator (PM) are developed. This PM is formed by connecting of a moving platform with a fixed base by three passive closing kinematic chains of PRRS type. Constant and variable parameters characterizing the geometry of links and relative motions of elements of joints are defined. Inverse kinematics of the PM is solved of the basis of solution of the loop-closure matrix equations of the legs.

**Keywords:** Parallel manipulator · Tripod · Geometry · Inverse kinematics

## 1 Introduction

Parallel manipulators with six DOF and three legs or tripods have a large workspace and less complex singular configurations compared to the parallel manipulators with six legs. The following URS [1], ESR [2], RRPS [3], RES [4–6], PPSP [7], PRPS [8], CRS [9], CCC [10] of tripods are known.

In this paper the methods for determining the geometry and solving the inverse kinematics of 3-PRRS type tripod with six DOF are presented. According to the developed principle of forming of parallel manipulators [11], the considered parallel manipulator is formed by connecting a moving platform with a fixed base by three passive closing kinematic chains of PRRS type with zero DOF. To describe the geometry and study the kinematics of this parallel manipulator, two Cartesian coordinate systems *UVW* and *XYZ* fixed at two elements of each kinematic pair are used [12]. The *W* and *Z* axes of these coordinate systems are directed along the axis of rotational and translational motions of the kinematic pairs elements, and the *U* and *X* axes are directed along the direction of the perpendicular *t*, drawn from the *W* axis to the *Z* axis.


K. Berns and D. Görges (Eds.): RAAD 2019, AISC 980, pp. 12–18, 2020.
https://doi.org/10.1007/978-3-030-19648-6_2

The transformation matrix between the coordinate systems $U_jV_jW_j$ and $X_kY_kZ_k$ (Fig. 1) has the form

$$\mathbf{T}_{jk} = \mathbf{T}_{jk}\left(a_{jk}, \alpha_{jk}, b_{jk}, \beta_{jk}, c_{jk}, \gamma_{jk}\right) =$$
$$= \left[\begin{array}{c|c|c|c} 1 & 0 & 0 & 0 \\ \hline \begin{matrix} a_{jk}\cdot c\gamma_{jk} + \\ +b_{jk}\cdot s\gamma_{jk}\cdot s\alpha_{jk} \end{matrix} & \begin{matrix} c\gamma_{jk}\cdot c\beta_{jk} - \\ -s\gamma_{jk}\cdot c\alpha_{jk}\cdot s\beta_{jk} \end{matrix} & \begin{matrix} -c\gamma_{jk}\cdot s\beta_{jk} - \\ -s\gamma_{jk}\cdot c\alpha_{jk}\cdot c\beta_{jk} \end{matrix} & s\gamma_{jk}\cdot s\alpha_{jk} \\ \hline \begin{matrix} a_{jk}\cdot s\gamma_{jk} - \\ -b_{jk}\cdot c\gamma_{jk}\cdot s\alpha_{jk} \end{matrix} & \begin{matrix} s\gamma_{jk}\cdot c\beta_{jk} + \\ c\gamma_{jk}\cdot c\alpha_{jk}\cdot s\beta_{jk} \end{matrix} & \begin{matrix} c\gamma_{jk}\cdot c\alpha_{jk}\cdot c\beta_{jk} - \\ -s\gamma_{jk}\cdot s\beta_{jk} \end{matrix} & -c\gamma_{jk}\cdot s\alpha_{jk} \\ \hline c_{jk} + b_{jk}\cdot c\alpha_{jk} & s\alpha_{jk}\cdot s\beta_{jk} & s\alpha_{jk}\cdot c\beta_{jk} & c\alpha_{jk} \end{array}\right], \quad (1)$$

where $a_{jk}$: distance between the $W_j$ and $Z_k$ axes measured in the direction of the common perpendicular $t_{jk}$ between these axes; $\alpha_{jk}$: angle between the positive directions of the $W_j$ and $Z_k$ axes measured about the positive direction $t_{jk}$ according to the right-hand rule; $b_{jk}$: distance from the direction $t_{jk}$ to the direction of the $X_k$ axis measured along the positive $Z_k$ axis direction; $\beta_{jk}$: angle between the positive directions of the $t_{jk}$ and $X_k$ axis measured according to the right-hand rule; $c_{jk}$: distance from the positive direction of the $U_j$ axis to the direction $t_{jk}$ measured in the positive direction of the $W_j$ axis; $\gamma_{jk}$: angle between the positive directions of the $U_j$ axis and $t_{jk}$ measured according to the right-hand rule.

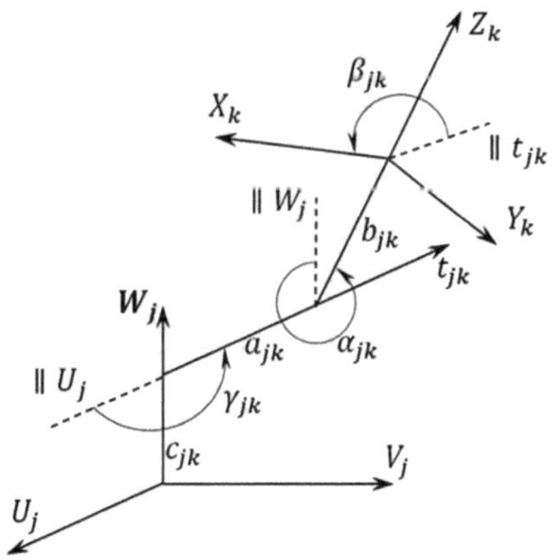


**Fig. 1.** Cartesian coordinate systems $U_jV_jW_j$ and $X_kY_kZ_k$

The transformation matrix of type (1) of the coordinate systems fixed at the ends of binary link has constant parameters, and it is called the matrix of binary link. The transformation matrix between the coordinate systems fixed at the two elements of the kinematic pair is called the kinematic pair matrix, which contains variable parameters characterizing the relative positions of the kinematic pair's elements.

The transformation matrix (1) contains six parameters, but not four parameters as the Denavit-Hartenberg transformation matrix [13]. This allows to fully characterize the geometry of the parallel manipulators. In fact, the position of a solid body in space is determined by six independent parameters.

## 2 Geometry of the Parallel Manipulator

Figure 2 shows the 3D model of the parallel manipulator, where the moving platform 4 is connected to the fixed base 0 by three identical legs of type <u>P</u>RRS. Let consider one of the legs *OABCD* of the parallel manipulator (Fig. 3), where the rod 1 of hydraulic cylinder together with the moving carriage $1'$ on which the motor $1''$ with gearbox is mounted, makes translational motion relative to the fixed base 0, forming the prismatic kinematic pair *A*. Link 2 is connected with the shaft of the gearbox by revolute kinematic pair *B*, it is connected with the link 3 by revolute kinematic pair *C*. Link 3 is connected with the moving platform 4 by spherical kinematic pair *D*. The hydraulic cylinder rod and the gearbox shaft are input links or the kinematic pairs *A* and *B* are active kinematic pairs.

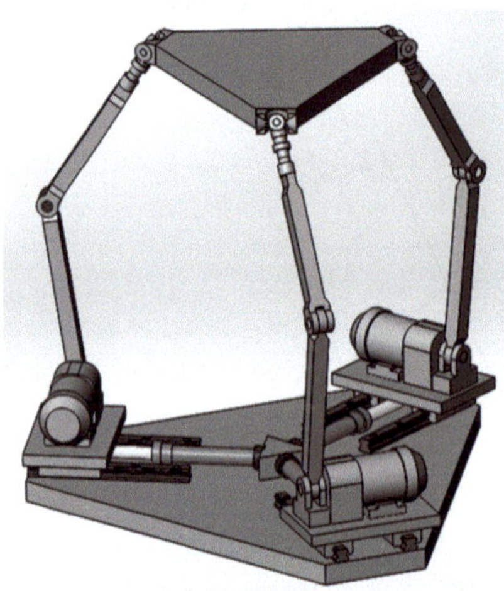

**Fig. 2.** 3D model of the 3-<u>P</u>RRS type parallel manipulator

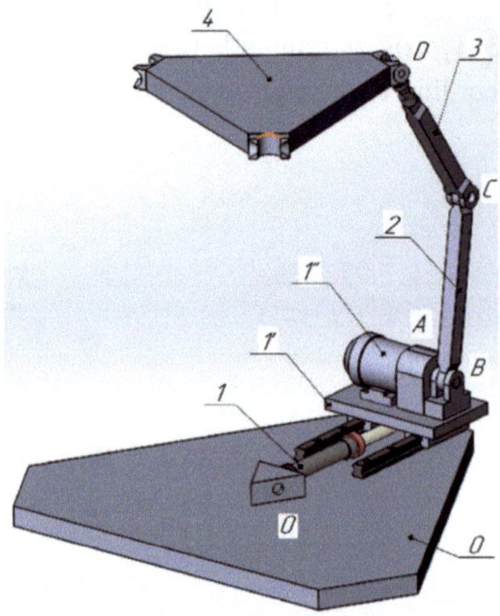


**Fig. 3.** Leg *OABCD* of the 3-PRRS type parallel manipulator

Let connect the absolute coordinate system $U_0V_0W_0$ with the fixed base, and the coordinate system $X_PY_PZ_P$ with the moving platform, where the $W_0$ and $Z_P$ axes are perpendicular to the planes of the fixed base and the moving platform, respectively. The coordinate systems $U_jV_jW_j$ and $X_kY_kZ_k$ are fixed at the elements of the kinematic pairs, where $j, k = 1, 2, 3, 4$. Here the kinematic pairs *A, B, C, D* are designated by numbers 1, 2, 3, 4, respectively. As can be seen from Fig. 4, the matrices of the links 1, 2, 3 and the kinematic pairs *A, B, C* have the following elements: $\mathbf{T}_{01} = \mathbf{T}_{01}(0, \alpha_{01}, 0, 0, c_{01}, \gamma_{01})$, $\mathbf{P}_1^P = \mathbf{P}_1^P(0, 0, 0, 0, c_1, 0)$, $\mathbf{T}_{12} = \mathbf{T}_{12}(a_{12}, \alpha_{12}, b_{12}, \beta_{12}, 0, 0)$, $\mathbf{P}_2^R = \mathbf{P}_2^R(0, 0, 0, \beta_{22}, 0, 0)$, $\mathbf{T}_{23} = \mathbf{T}_{23}(a_{23}, 0, 0, 0, 0, 0)$, $\mathbf{P}_3^R = \mathbf{P}_3^R(0, 0, 0, \beta_{33}, 0, 0)$, $\mathbf{T}_{34} = \mathbf{T}_{34}(a_{34}, 0, 0, 0, 0, 0)$, where $\alpha_{01} = 90^0$, $c_{01} = OO_1$, $c_{11} = O_1O_1' = s_1$, $a_{12} = O_1'O_2$, $\alpha_{12} = -90^0$, $b_{12} = O_2O_2'\beta_{12} = 90^0$, $\beta_{22} = \theta_2$, $a_{23} = O_2O_3$, $\beta_{33} = \theta_3$, $a_{34} = O_3', O_4$.

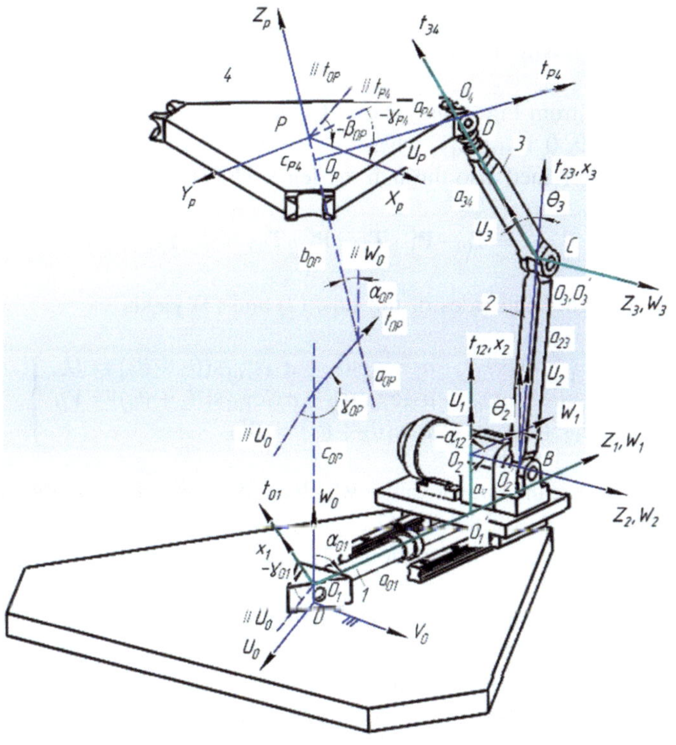


**Fig. 4.** Parameters of the leg *OABCD*

Of these the parameters $\alpha_{01}$, $c_{01}$, $\gamma_{01}$, $a_{12}$, $\alpha_{12}$, $b_{12}$, $\beta_{12}$, $a_{23}$, $a_{34}$ are constants, which characterize the geometry of the links, and the parameters $c_{11}$, $\beta_{22}$, $\beta_{33}$ are variables which characterize the relative movements of the elements of kinematic pairs, where the variable parameters $c_{11} = s_1$ and $\beta_{22} = \theta_2$ are input and $\beta_{33} = \theta_3$ is output parameters. The remaining two legs of the considered parallel manipulator have similar parameters.

## 3 Inverse Kinematics of the Parallel Manipulator

Position of the coordinate system $X_P\,Y_P\,Z_P$ fixed at the moving platform 4 relative to the absolute coordinate system $U_0\,V_0\,W_0$ is determined by the transformation matrix $\mathbf{T}_{04}$, which has the following constant parameters $\mathbf{T}_{04} = \mathbf{T}_{04}(a_{04},\ \alpha_{04},\ b_{04},\ \beta_{04},\ c_{04},\ \gamma_{04})$ as shown in Fig. 4. Given the position of the coordinate system $X_P\,Y_P\,Z_P$, it is necessary to determine the input variable parameters $s_1$ and $\theta_2$.

Coordinates of the spherical pair center $D$ can be determined through the coordinate system $X_P\,Y_P\,Z_P$

$$[U_D,\ V_D,\ W_D]^T = \mathbf{T}_{OP} \cdot \mathbf{T}_{P4} \cdot [1,\ 0,\ 0,\ 0]^T. \tag{2}$$

As can be seen from Fig. 4 the matrix $\mathbf{T}_{P4}$ has the following constant parameters $\mathbf{T}_{P4} = \mathbf{T}_{P4}(a_{P4},\ 0,\ 0,\ 0,\ c_{P4},\ \gamma_{P4})$, where $a_{P4} = O_P D$, $c_{P4} = PO_P$. Coordinates of the point $D$ can be determined also through the leg $OABCD$

$$[U_D,\ V_D,\ W_D]^T = \mathbf{T}_{01} \cdot \mathbf{P}_1^P \cdot \mathbf{T}_{12} \cdot \mathbf{P}_2^R \cdot \mathbf{T}_{23} \cdot \mathbf{P}_3^R \cdot \mathbf{T}_{34} \cdot [1,\ 0,\ 0,\ 0]^T. \tag{3}$$

Equating the right-hand sides of the Eqs. (2) and (3) yields

$$\left.\begin{array}{l} -b_{12}\mathrm{c}\gamma_{01} + s_1\mathrm{s}\gamma_{01} - a_{23}\mathrm{s}\gamma_{01}\mathrm{s}\theta_2 - a_{34}\mathrm{s}\gamma_{01}\mathrm{s}(\theta_2+\theta_3) = U_D \\ -b_{12}\mathrm{s}\gamma_{01} - s_1\mathrm{c}\gamma_{01} + a_{23}\mathrm{c}\gamma_{01}\mathrm{s}\theta_2 + a_{34}\mathrm{c}\gamma_{01}\mathrm{s}(\theta_2+\theta_3) = V_D \\ c_{01} + a_{12} + a_{23}\mathrm{c}\theta_2 + a_{34}\mathrm{c}(\theta_2+\theta_3) = W_D \end{array}\right\}, \tag{4}$$

where $U_D$, $V_D$, $W_D$ are the known values which are determined through the elements of the matrices $\mathbf{T}_{0P}$ and $\mathbf{T}_{P4}$ of the Eq. (2).

Analysis of the system of equations (4) shows that the first two equations are linearly dependent. Therefore, one of the parameters $s_1$, $\theta_2$ and $\theta_3$ can be chosen free. Indeed, as can be seen from the Fig. 4, the given position of the point $D$ of the moving platform 4 is not changed at different positions of the rod of the hydraulic cylinder 1, i.e. the $s_1$ is a free parameter. Its values are in the interval satisfying the condition $\left|l_{O_2'D}\right|_{\max} \geq a_{23} + a_{34}$.

After transforming the first and third equations of the system (4) and the following notations

$$K = \frac{U_D + b_{12} \cdot \mathrm{c}\gamma_{01}}{\mathrm{s}\gamma_{01}},\ M = W_D - c_{01} - a_{12},$$

yields

$$E \cdot \mathrm{s}^2\theta_2 - F \cdot \mathrm{s}\theta_2 + G = 0, \tag{5}$$

where

$$E = (s_1 - K)^2 + M^2,\ F = 2 \cdot (a_{23} + a_{34} \cdot c\theta_3) \cdot (s_1 - K),$$
$$G = (a_{23} + a_{34} \cdot c\theta_3)^2 - M^2,\ c\theta_3 = \frac{(s_1 - K)^2 + M^2 - a_{23}^2 - a_{34}^2}{2 \cdot a_{23} \cdot a_{34}}.$$

From the Eq. (5) we obtain

$$\theta_2 = \text{arcs}\frac{F \pm \sqrt{F^2 - 4EG}}{2E}, \tag{6}$$

which corresponds to two solutions of the dyad $BCD$.

# 4 Conclusions

PM of 3-PRRS type is formed by connecting of a moving platform with a fixed base by three passive closing kinematic chains (CKC) of PRRS type with zero DOF. These CKC don't impose geometrical constraints on motion of the moving platform, and its 6 DOF are saved. To describe the geometry and study the kinematics of the PM, two Cartesian coordinate system, fixed at two elements of each kinematic pair, are used. Transformation matrix between these coordinate systems has six parameters, which allows to fully characterize the geometry of the PM. The transformation matrix of the coordinate system fixed at the end of binary link has constant parameters, and it is called the matrix of binary link. The transformation matrix between the coordinate system fixed at the two elements of the kinematic pair is called the kinematic pair matrix, which contains variable parameters characterizing the relative positions of the kinematic pair's elements. The constant and variable parameters of the 3-PRRS type PM are defined. Inverse kinematics of the PM is solved on the basis of solution of the loop-closure matrix equations of the legs. The results of analysis show that the variable parameter of the input prismatic pair is free.

# References

1. Clearly, C., Uebel, M.: Jacobian formulation for a novel 6-DOF parallel manipulator. In: IEEE International Conference on Robotics and Automation, vol. 3, pp. 2377–2382 (1994)
2. Tsai, L.-W., Tahmasebi, F.: Synthesis and analysis of a new class of six-degree-of-freedom parallel manipulators. J. Robot. Syst. **10**(5), 561–580 (1993)
3. Alizade, R.I., Tagiyev, N.R., Duffy, J.: A forward and reverse displacement analysis of a 6-DOF in-parallel manipulator. Mech. Mach. Theory **29**(1), 115–124 (1994)
4. Collins, C.L., Long, G.L.: The singularity analysis of an in-parallel hand controller for force-reflected teleoperation. IEEE Trans. Robot. Autom. **11**, 661–669 (1995)
5. Mimuza, N., Funabashi, Y.: A new analytical system applying 6 DOF parallel link manipulator for evaluating motion sensation. In: IEEE International Conference on Robotics and Automation, pp. 227–333 (1995)

6. Ebert, U., Gosselin, C.M.: Kinematic study of a new type of spatial parallel platform mechanism. In: ASME Design Engineering Technical Conference, Atlanta, 13–16 September 1998
7. Byun, Y.K., Cho, H.S.: Analysis of a novel 6-DOF 3-PPSP parallel manipulator. Int. J. Robot. Res. **16**(6), 859–872 (1997)
8. Behi, F.: Kinematic analysis for a six-degree-of-freedom 3-PRPS parallel mechanism. IEEE J. Robot. Autom. **4**(5), 561–565 (1988)
9. Khali, D., Lee, S.-H., Tsai, K.-Y., Sandor, G.N.: Manipulator configurations based on rotary-liner (R-L) actuators and their direct and inverse kinematics. J. Mech. Transmissions Autom. Des. **110**, 397–404 (1988)
10. Baigunchekov, Z., Izmambetov, M.: Inverse kinematics of six - DOF three – limbed parallel manipulator. In: Advances in Robot Design and Intelligent Control. Proceedings of the 25th Conference on Robotics in Aple-Adria-Danube Region, RAAD 2016, pp. 171–178. Springer, Cham (2016)
11. Baigunchekov, Z., Izmambetov, M.: Geometry and direct kinematics of six-DOF three-limbed parallel manipulators. In: ROMANSY 21-Robot Design, Dynamics and Control. Proceedings of the 21st CISM – IFToMM Symposium, pp. 39–46. Springer, Cham (2016)
12. Baigunchekov, Z., et al.: New Parallel manipulator with 6 degrees-of-freedom. In: Proceedings of the Twelfth IFToMM World Congress in Mechanism and Machine Science, Besanson, France, 17–21 June 2007, vol. 5, pp. 641–646 (2007)
13. Denavit, J., Hartenberg, R.S.: A kinematic notation for lower-pair mechanisms based on matrices. ASME Trans. J. Appl. Mech. **77**, 215–221 (1955). Vol. 22

# Linear Motion Mechanisms for Fine Position Adjustment of Heavy Weight Platforms

Štefan Havlík[1(✉)], Jaroslav Hricko[1], Erik Prada[2], and Jaromír Jezný[2]

[1] Institute of Informatics, Slovak Academy of Sciences, Banská Bystrica, Slovakia
{havlik, hricko}@savbb.sk
[2] ZTS VVU, a.s., Kosice, Slovakia
erik.prada@yahoo.com, jaromir.jezny@ztsvvu.eu

**Abstract.** This paper describes the design, development and experimentation of the linear motion mechanisms for very fine position adjustment of heavy weight platforms. It is proposed the concept that includes the precise electric motor/actuator, the ball-screw motion transducer and the no-backlash motion reducer based on compliant/elastic mechanism. The design, experimental equipment and results from measurements are presented. Such configuration enables to use available ball-screw mechanisms and to reduce the cost of the whole positioning system.

**Keywords:** Compliant mechanisms · Force reducer/amplifier · Fine motion adjustment

## 1 Introduction

Robotic technology gets application practically in all fields of industrial activities. In general, the principal requirements on the parameters and performance of robots are fully satisfied for the majority of nowadays industrial application. Industrial robots are fast, precise and entirely replace human workers in production. They are massively used in various manipulations or other technology works and are able to perform very precise tasks with the repeatable positioning accuracy of $10^{-2}$ mm, or less. Beside yet well technologically proved application there are some special tasks in construction, heavy industry, military or research, where large and heavy weight platforms with variable load should be very precisely adjusted into an exact position/orientation. The cumulative errors done in construction, errors in production of parts and errors due to thermal or flexural distortion are usually greater then the desired accuracy of adjustment. Examples such specific destinations are antennas/mirrors for cosmic or physical research, military homing systems, etc., where regular or occasional adjustment is need for right functions of on-board devices. For this reason the platforms with on board devices (variable load) should make fine correctional motions in order to eliminate all these errors and inaccuracies. This practically means that the platform is the end part of a parallel positioning mechanisms able to perform fine adjustment position/orientation in corresponding degrees of freedom. A typical scheme of such parallel positioning system is shown in Fig. 1.


K. Berns and D. Görges (Eds.): RAAD 2019, AISC 980, pp. 19–25, 2020.
https://doi.org/10.1007/978-3-030-19648-6_3

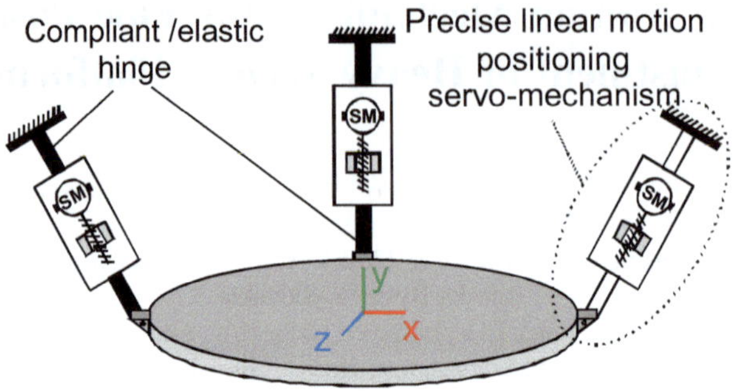


**Fig. 1.** Scheme of the parallel fixation platform

Standard configurations of linear motion positioning systems include the motor with the positional encoder, the no-backlash - high ratio R-R (Rotation to Rotation) reduction gear and the ball screw R-L (Rotation - to Linear) transmission [4]. Specifically, for the high load, the ball and roller screw mechanisms have been developed and are commercially available [1]. As to actuators, the stepping motors are most frequently used in machine tools constructions and other linear positioning systems. The control electronics and algorithms dedicated to these applications are well elaborated and commercially available [2, 3].

The concept, design and development of the positioning mechanisms with the linear motion reducer are described below.

## 2 The Concept

Large majority of available linear positioning systems are made for the range of linear motions that correspond to several diameters of used screws. According to preliminary analysis the fine correctional motions can be assumed within the range of motion 2/3 mm with positional accuracy within $10^{-2}$ mm–$10^{-4}$ mm (better). The axial force/payload capacity is expected about $10^4$ N.

For the majority of commercially available ball-screw positioning mechanisms; the small range of linear motion corresponds to the small rotation angle of the screw. This fact naturally decreases the achievable accuracy (greater step of motion) and results in problems of control; as well. For this reason a further mechanical unit which will reduce the range of linear motion is need. Considering constructional possibilities, evaluation of desired performance characteristics and parameters of available ball-screw positioning mechanisms the concept with the additional L-L (Linear to Linear) motion reducer was adopted as schematically shown in Fig. 2. Beside the purpose of the L-L transmission unit it results in increasing the accuracy of the system, as whole. It is obvious, that no-backlash transmission is principally required.

The task of increasing or reducing linear motion with a no-backlash transmission is the frequent problem well elaborated in micro-mechanics. Mechanisms for this purpose are especially based on elastically compliant structures usually driven by piezoelectric actuators. A broad class of these mechanisms represent "motion amplifiers" exclusively

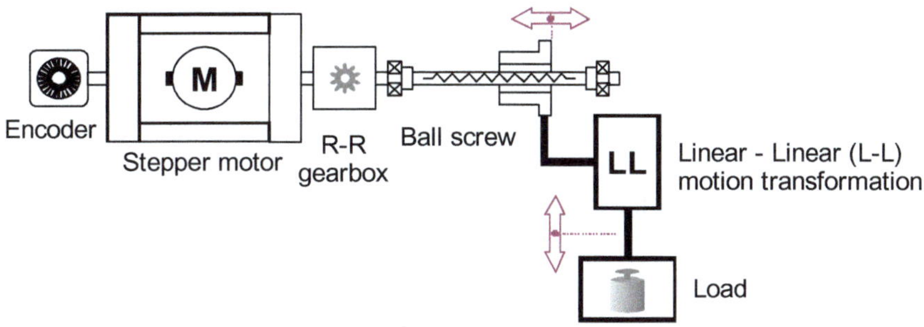


**Fig. 2.** Principle of the servo mechanism with L-L motion reduction

made from a piece of elastic material [5–8, 10, 11]. The main idea of proposed concept is using such a structure in the opposite way; i.e. "as motion reducer/force amplifier". The proposed principle of reducer represents the structure in Fig. 3(a), where revolute joints are replaced by elastic connections. Similar compliant mechanisms are usually made from one piece of elastic material using drilling and milling technologies for making elastic joints in form of double side notches.

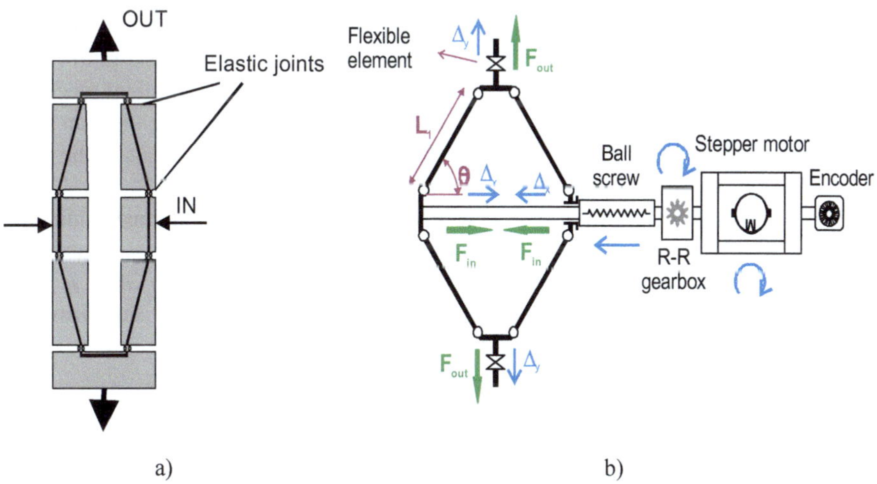


**Fig. 3.** Principle of motion reducer (a) Compliant mechanics, (b) Basic configuration

Solving such idealized compliant structure the displacement ratio $r_d$ will be [5, 10, 12]

$$r_d = \frac{\Delta y}{\Delta x} = \frac{2.\Delta y}{L.\sin\theta - l\sqrt{1 - \left[\frac{1}{L}\left(l.\cos\theta + \frac{\Delta x}{2}\right)\right]^2}} \text{ actual} \quad (1)$$

When displacements are relatively small one can use the next simple relation (2) [7–9]

$$r_d = \frac{\Delta y}{\Delta x} = \cot g\theta \tag{2}$$

then for transformation of forces it is

$$F_x = F_y \cdot r_d + (-)S_x \cdot \Delta x \tag{3}$$

The important parameters for displacement and force ratio are: $l$ - the length of side arms and $\theta$ - the inclination angle. The angle $\theta$ gives the neutral position of the compliant structure. $\Delta x$, $\Delta y$, are actual displacements from the neutral position. The sign (±) depends on direction of lateral motion. $S_x$ is the total stiffness of the flexure in the place and direction of acting force $F_x$.

The first task in design is to find an optimal composition of the geometry and stiffness parameters of the flexure for given transmissions and load.

## 3 Design of Motion Reducer

Primary features of the reducer are predetermined by maximal values of input displacement and acting force. Desired output displacement gives reduction ratio (1), (2) and force transmission (3). The crucial problem in designing any compliant mechanics with flexural joints is always the maximal value of stress/strain in joints that can not be crossed. So building the model and the careful flexural analysis of joints and the whole structure including simulations is need.

The following input conditions were considered in our calculations/simulations: max. input displacement $\Delta x$ = 30 mm, output displacement $\Delta y$ = 3 mm, (i.e. ±1.5 mm), max. load 5 kN, material: Steel 34CrNiNo6, $E$ = 210 GPa, $\mu$ = 0.3, $\rho$ = 7850 kg/m$^3$, $\sigma_{max}$ = 900 MPa. In the first design the length of side arms was chosen $l$ = 100 mm. For this case the desired displacement ratio will be $r_d$ = 0.1; what corresponds to the central inclination angle $\theta$ =84.3 deg.

Several simulation experiments with models represented by parameters $l,\theta$ and various joints using FEM analysis in Comsol Multiphysics have been made.

As showed first calculations the mechanical stress is mainly concentrated in joints and crosses the allowable values. Then, desired displacements under acting of prescribed force can not be achieved if parameters $l$, $\theta$ have to be within constructional limits.

Based on this experience the alternative solution of the compliant mechanisms was proposed, as shown in Fig. 4(a). Here, for the same displacements the stress/strain is distributed along the whole length of arms. Simulation with values of surface stress is shown in Fig. 4(b).

Final configuration of the whole positioning mechanisms fixed on the test-stand for experimental measurements is shown in Fig. 5

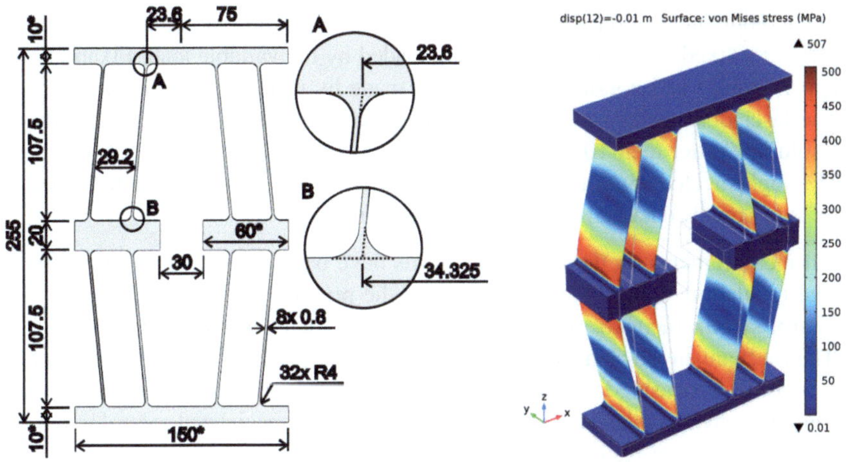


a) b)

**Fig. 4.** Compliant structure of the reducer, (a) Geometry, (b) Stress calculation

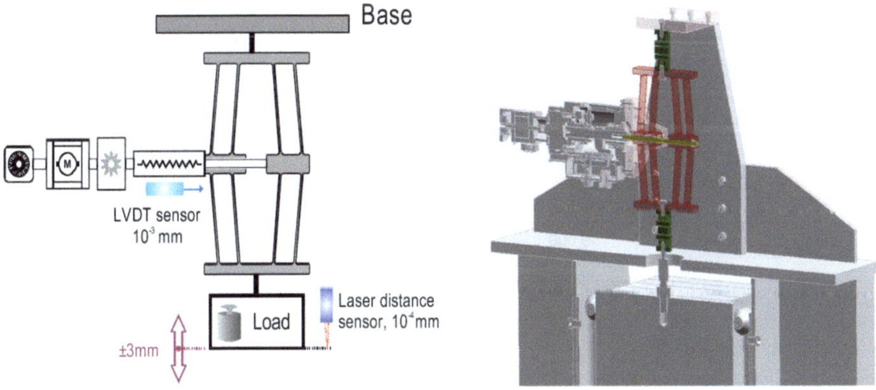


**Fig. 5.** The test-bed configuration of whole positioning mechanism with motion reducer

## 4 Experiments

In order to verify expected performance of the mechanisms several tests have been made. For this purpose the dedicated test bed was designed and constructed. The test equipment includes following parts:

- The actuation unit which consists of mutually integrated devices: the stepper motor, the encoder, the no-backlash R-R high ratio reducer and the ball-screw R-L mechanism with LVDT sensor. The theoretically calculated linear displacement which corresponds of the motor step is $0.266 \times 10^{-3}$ mm.

- The compliant motion reducer/force amplifier according to above design. The reducer is fixed to the robust frame and loaded by the variable load. Fixation on the upper (frame) and bottom (load) sides is made by using 2 d.o.f. flexural hinges.
- The laser distance sensing device for measuring positions of the load body with resolution better then $10^{-4}$ mm.
- The sophisticated control system for stepper motor and the SW tools for processing sensory signals/data and evaluation of measurements.

All measurements on the real mechanisms have been carried out in the room with thermal regulation. Before the measurements the primary positioning mechanism and its control system (actuator with ball-screw mechanisms was calibrated).

The mechanism was loaded by several mass of 0, 300, 600 kg and position of the mass body was measured. Then, for particular experiments the parameters of output motions were evaluated. The similar experiments need specialized metrological laboratory and equipment results from our experiments can be considered as preliminary. An example of the output data from measurement with the load 600 kg is shown in Fig. 6, where the errors in $10^{-4}$ mm from linearity (absolute and delta) are depicted.

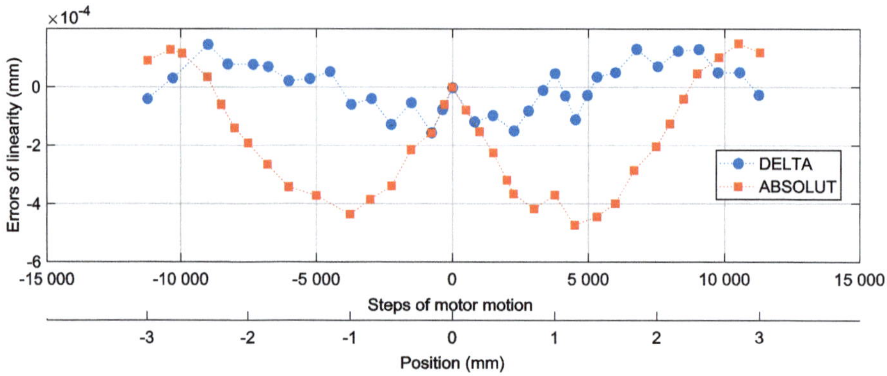


**Fig. 6.** The linearity errors of output motion

## 5 Conclusions

Paper presents the concept of the linear motion mechanisms for precise adjustments of heavy loads/platforms. It is based on insertion of the additional linear no - backlash compliant mechanisms as motion reducer [13]. Proposed configuration enables to reduce input linear displacements and by this way it is possible to use standard ball-screw actuation mechanisms for making very precise correctional displacements within a small range of motions.

**Acknowledgments.** This work is part of the project APVV-14-0076 – "MEMS structures based load cell" supported by the Slovak Grant Agency and the project "**Research and development of intelligent mobile robotic platform** and positioning systems with high accuracy for use in research, development and industry" with the STIMULI program support (ZTS VVU) from the Slovak Ministry of Education.

## References

1. Münch, T.: High Load Ball Screws – The Motion Control Solution with Higher Load Density (2018). https://www.thomsonlinear.com/de/support/25151
2. Erkorkmaz, K., Kamalzadeh, A.: High Bandwidth Control of Ball Screw Drives, Precision Controls Laboratory, Dept. of Mechanical Engineering, University of Waterloo, Waterloo, Canada, Annals of the CIRP, vol. 551112006
3. Kamalzadeh, A.: Precision control of high speed ball screw drives, Thesis Dr of Philosophy, Waterloo, Ontario, Canada (2008)
4. Cloutier, J.: Simulation and control of a ball screw system actuated by a stepper motor with feedback, Thesis The University of Guelph, Guelph, Ontario, Canada, December 2014
5. Ouyang, P.R., Zhang, W.J., Gupta, M.M.: Design of a new compliant mechanical amplifier. In: Proceedings of IDETC/CIE 2005 ASME 2005 International Design Engineering Technical Conferences & Computers and Information in Engineering Conference, 24–28 September 2005, Long Beach, California, USA (2005)
6. Ling, M., Cao, J., Jiang, Z., Lin, J.: Theoretical modeling of attenuated displacement amplification for multistage compliant mechanism and its application. Sens. Actuators A **249**, 15–22 (2016)
7. Ling, M., Cao, J., Zeng, M., Lin, J., Inman, D.J.: Enhanced mathematical modeling of the displacement amplification ratio for piezoelectric compliant mechanisms. Smart Mater. Struct. **25**, 11 pp (2016). 075022
8. Ma, H.-W., Yao, S.-M., Wang, L.-Q., Zhong, Z.: Analysis of the displacement amplification ratio of bridge-type flexure hinge. Sens. Actuators A **132**, 730–736 (2006)
9. Qi, K.-q., Xianga, Y., Fang, C., Zhangb, Y., Chang-song, Yu.: Analysis of the displacement amplification ratio of bridge-type mechanism. Mech. Mach. Theory **87**, 45–56 (2015)
10. Labontiu, N., Garcia, E.: Analytical model of displacement amplification and stiffness optimization for a class of flexure based compliant mechanisms. Comput. Sruct. **81**, 2797–2810 (2003)
11. Chen, J., Zhang, C., Xu, M., et al.: Rhombic micro-displacement amplifier for piezoelectric actuator and its linear and hybrid model. Mech. Syst. Signal Process. **50**, 580–593 (2015)
12. Ling, M., et al.: Enhanced mathematical modeling of the displacement amplification ratio for piezoelectric compliant mechanisms. Smart Mater. Struct. **25** (2016). 075022. ISSN 1361-665X
13. Trebuňa, F., Bocko, J., Pástor, M., Lengvarský, P., Prada, E.: The finite element analysis of high precision positioning system. J. Mech. Eng. Strojnícky časopis **68**(4), 41–48 (2018)

# Compliant Mechanisms for Motion/Force Amplifiers for Robotics

Jaroslav Hricko(✉) and Štefan Havlík

Institute of Informatics, Slovak Academy of Sciences, Banská Bystrica, Slovakia
{hricko,havlik}@savbb.sk

**Abstract.** The force and motion amplifiers are essential mechanical elements used in building small and (micro) robotic devices. This paper brings a short overview of some mechanical structures and their performance characteristics. Three concepts of force amplifiers are analyzed and results from simulations are discussed. The specific design of the linear motion force amplifier/motion reduction for a high-accuracy positioning device with large payload capacity is discussed.

**Keywords:** Mechanical amplifier · Compliant mechanisms · (Micro) robotic devices · Force transducer · Precise positioning

## 1 Introduction

Compact compliant mechanisms take applications in building robotic devices and a precise positioning system. They exhibit advantages, as no friction, lightweight, no noise, no need of lubrication, zero backlashes [1], etc., need in the field of small and micro-robotics. The specific class of these mechanisms represents force or motion amplifiers. In the case of micro-robotic devices, the majority of suitable linear actuators can produce adequate forces, unfortunately with very small output displacement. For instance: the stroke of the piezoelectric actuator is about 0.1%–0.2% of its length [2]. For this reason, motion amplifiers are really needed [2–4].

In principle, such mechanisms can be used as motion or force amplifiers/reducers. Suppose a specific construction of a high accuracy linear positioning device with the high payload capacity. The range of motions should be relatively small. For such case, using lower power linear actuator with force amplifier seems to be much more effective. In general, the connection of an adequate driving unit with a compliant force or motion reducing mechanisms can lead to acceptable results. Such an approach puts some new requirements to design of these devices. According to the reduction ratio, the amplifiers enable to reduce power, mass, and dimension of the driving unit and the cost, as well. In the case of motion reducers, the significant improvement of accuracy can be reached.


K. Berns and D. Görges (Eds.): RAAD 2019, AISC 980, pp. 26–33, 2020.
https://doi.org/10.1007/978-3-030-19648-6_4

## 2 Some Structures of Mechanical Amplifiers

The mechanical amplifiers basically use principal physical elements: differential mechanical amplifier, lever principle, triangular mechanical amplifier [5], etc. Rem.: For the purpose of this paper, there are mentioned mainly compliant mechanisms made from one piece of material, i.e. without any rotary elements like gearwheel, rack-pinion mechanism, rack-pinion paired with piston-like motion [6].

Well, known structures of displacement amplifier in microscale applications are a bridge-type mechanism. There are known two main bridge-type topologies shown in Fig. 1(a), where rigid arms are connected using flexural hinges, or Rhombus-type where slim flexible arms are used (Fig. 1(b)). Both mechanisms exhibit compact symmetrical structures. As a drawback of these structures, one can mention the limited amplification ratio [11]. The enhanced mathematical models of both mechanisms, based on the energy conservation law and the elastic beam theory is exhaustively explained in [7]. The compound bridge type displacement amplifier has a larger lateral stiffness and it is more suited for isolation and protection of an actuator than the ordinary bridge-type amplifier [8]. The design of the high stiffness amplifier inspired by materials of the cellular structure is described in [12]. This approach leads to the design of specific motion amplifiers/reducers of high-level forces. Contemporary one can help in the research of new next-generation materials with tailored mechanical properties such as modulus, strength, strain, and Poisson's ratios. An example of the half-bridge small-scale out-of-plane actuator which enables tactile interfaces with displacement up to 6.3 μm and blocking the force of 16 mN on the surface of the 2 mm$^2$ gives [3].

The lever mechanism is one of most of the ancient mechanisms used by people. Due to its easy realization, it found its use in the field of micro-robotics. One of the few drawbacks is the relatively large stress concentrated in the flexible joints. An example of a compliant lever mechanism is shown on the Fig. 1(d). Its real application is described in [10] where two-stage lever-type amplifier was proposed (Fig. 1(g)). The advantages of bridge-type and lever-type amplifiers (Fig. 1(e)) are used in new amplifier structure [11]. Such device exhibits symmetrical design, a high amplification ratio (34–48) and high load capacity (30 N). The design of new amplifier based on symmetrical five bar profile with a large amplification ratio (24.4) and with high natural frequency (573 Hz) is described in [9] (see Fig. 1(c)). In terms of the displacement amplification and the natural frequency, it was shown that five bar structure is much better than the four bar (bridge-type) amplifiers. Another mechanical amplifier with the integrated piezoceramic actuator with 20:1 stroke amplification is shown in Fig. 1(j).

A general approach to the design of compliant mechanisms based on topology optimization, where only basic information are required (main dimensions, input and output displacements, forces, a location of input, output, basic frame). Such an approach can lead to a very complex structure that can not be easy produced [13].

The specific group of mechanisms that could be used as mechanical amplifiers is mechanism with straight-line output motions (see Fig. 1(k) and (l)). These devices are inspired by known kinematical structures where the rotation of the joints is large, so they can produce smaller movements [14]. The modified version is the Scot-Russell mechanism used in the compliant micro-gripper [15]. This modification of the

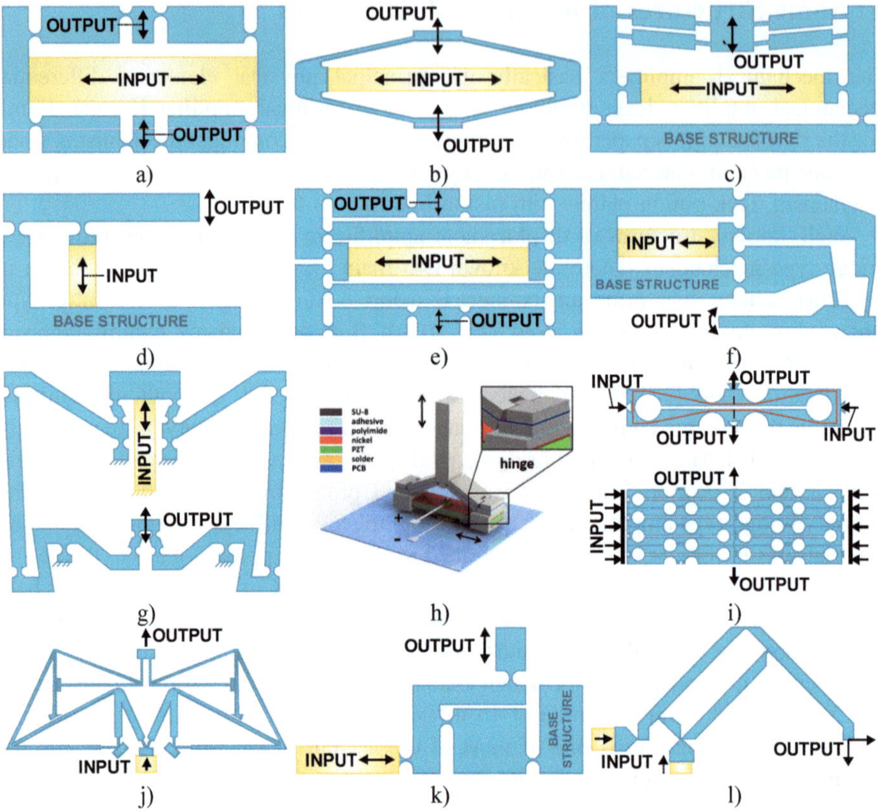


**Fig. 1.** Compliant displacement amplifiers. (a) bridge-type [7], (b) Rhombus type [7], (c) symmetric five bar structure [9], (d) lever mechanism [7], (e) bridge-lever-type amplifier [11], (f) differential amplifier [5], (g) tensural displacement amplifier [10], (h) half-scissor amplifier [3], (i) re-entrant hexagonal honeycomb [12], (j) 20:1 stroke amplification mechanism [13], (k) Scott-Russell mechanism [14], (l) pantograph mechanism [16]

Scot-Russell mechanism provides 3.56 times bigger motion amplification ratio comparing to the conventional structure. The simple topology of the pantograph mechanism for positioning stages is presented [16]. The design concept of the pantograph amplifier with ratio 6 documented by the numerical results, FEM analysis, and experimental results includes [4].

## 3 Design of Motion Reducer/Force Amplifier

The amplification ratio is the most important parameter when starts the design procedure of any amplifier. Mechanical amplifiers presented in Sect. 2, are dominantly designed as amplifiers of linear motions. The mechanical stress/strains in flexural parts depend on actual displacements of mechanisms. In the case of force amplifiers, the effects of external load should be considered too. Naturally, there are several kinematic structures that can satisfy the desired ratio. In order to show some differences three versions of mechanisms designed for the same ratio are evaluated and compared. The desired performance is given as follows:

- the range of input displacement of the linear motion actuator to be 5 mm
- the range of output displacement should be 1.5 mm, what corresponds to amplification ratio $R_{amp} = 0.3$
- the load is given up to 5 kg.

The same material is considered in all versions with characteristics: Polylactic acid (PLA, Young's modulus $E$ is 2852 MPa, density 1237 kg/m$^3$, Poisson's ratio 0.36 and tensile stress at yield is 38 MPa). Other alternative materials for building physical model are Polypropylene (PP) or Polyamide (PA). For these materials and production technology (3D printing) the minimal thickness of the flexure segments should be 0.5 mm, optionally 0.8 mm.

For the purpose of this study, the mathematical models of the mechanisms were built and the two-stage simulation procedure was adopted. In the first stage, the mechanisms are without load and motions with corresponding stress/strains are calculated. The second stage simulations include acting loads (5 kg). Such an approach shows the influence of the load and enables to change of principal dimensions and/or material. Subsequently, physical models of these mechanisms made by 3D print technology have been experimented.

### 3.1 The Two Levers Amplifier

This mechanism connects a linear motion actuator with output parts of positioning mechanisms. It adapts the ranges of motions/forces between both parts of the positioning system. The proposed kinematics is in Fig. 2(a). The similar concept and static model is described in [10]. When considering an "ideal" model the dependence between input displacement $y_{IN}$ and output displacement $y_{OUT}$ is expressed by a simple relation

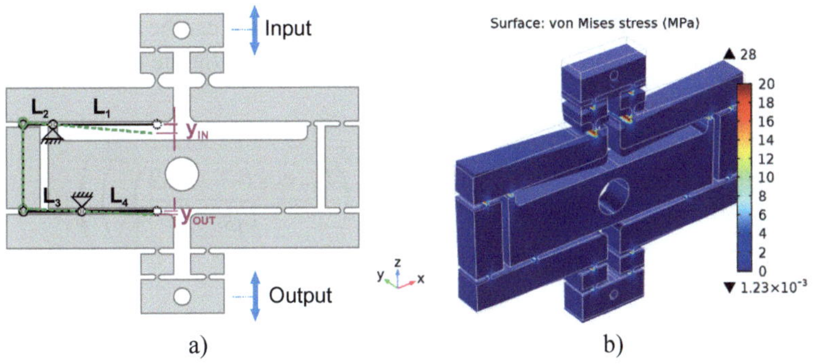


**Fig. 2.** (a) Kinematic scheme of two lever mechanism, (b) stress distribution (von Mises) during motion without the load

$$y_{OUT} = \frac{L_2 \cdot L_4}{L_1 \cdot L_3} y_{IN} \quad \Rightarrow R_{amp} = \frac{y_{OUT}}{y_{IN}} = \frac{L_2 \cdot L_4}{L_1 \cdot L_3} \tag{1}$$

### 3.2 Bridge-Type and Rhombus Mechanical Amplifier

Considering these type of mechanisms the amplification ratio strongly depends on the angle $\theta$ (see Fig. 3.). For instance: the motion amplifiers for piezoelectric actuator have the angle $\theta$ around 20° [17]. For our case of force amplifier, this angle should be more than 45, or, closer to 90°. The change of this angle and the corresponding amplification is related to the range of input motion.

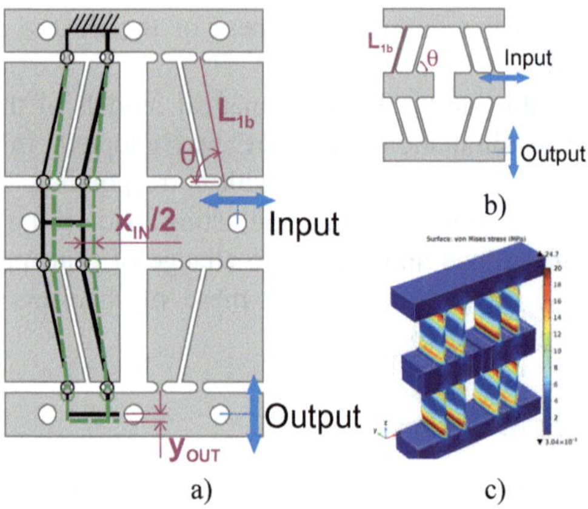


**Fig. 3.** (a) Geometry of the bridge-type mechanical amplifier with kinematic chain, (b) Geometry of the Rhombus-type mechanical amplifier, (c) Distribution of arisen stress (von Mises) in the Rhombus-type mechanism

Ling et al. in [7] expressed the analytical model for the displacement amplification ratio based on the elastic beam method. As they show the idealized geometric model e.g. [17] does not provide adequately accurate results, but it can be used for comparing these different mechanisms. The output displacement based on the idealized geometric model is [17]

$$y_{OUT} = 2\left(L_{1b}\sin\theta - L_{1b}\sqrt{\left(1 - \left(\frac{L_{1b}\cos\theta + x_{IN}/2}{L_{1b}}\right)^2\right)}\right) \tag{2}$$

### 3.3 Results

The design of the mechanisms in the first stage has been focused on carrying out the motion amplification with desired input and output displacement and minimization of the external dimensions with respect to the arisen stress. The second stage was focused to optimize dimensions with respect to the acting load. Comparisons of dimensions of all proposed mechanisms are shown in Table 1. Figure 4 shows the dependence of output displacement and motion amplification ratio with respect to the input displacement.

**Table 1.** Comparison of main parameters of the proposed mechanical amplifiers

| Type | Dimensions (*mm*) | | Scale | Amp. ratio | | Max. stress (*MPa*) | |
|---|---|---|---|---|---|---|---|
| | DS1 | DS2 | | DS1 | DS2 | DS1 | DS2 |
| Double lever[a] | 135 × 90.3 × 10 | 180 × 99.1 × 10 | 1.46 | 0.317 | 0.141 | 36.57 | 37.97 |
| Bridge type[b] | 50 × 100 × 10 | 70 × 170 × 10 | 2.38 | 0.212 | 0.181 | 37.68 | 36.62 |
| Rhombus type[c] | 40 × 57 × 5 × 10 | 40 × 70 × 10 | 1.22 | 0.262 | 0.228 | 37.61 | 31.772 |

**DS1** – design stage 1 (without acting load), **DS2** – design stage 2 (with acting load), **Scale** is calculated as DS2 surface divide DS1 surface.

[a]**DS1** *(mm)*: $L_1$*=48.35;* $L_2$*=9.15;* $L_3$*=21.48;* $L_4$*=36.02*; **DS2** *(mm)*: $L_1$*=67.5;* $L_2$*=10;* $L_3$*=39.6;* $L_4$*=37.7*

[b]**DS1**: $L_1$*=31.22* mm; $\theta$*= 79.2* deg; **DS2**: $L_1$*=60.09* mm; $\theta$*= 80.32* deg

[c]**DS1**: $L_1$*=21.125* mm; $\theta$*= 77.076* deg; **DS2:** *L1=27.39* mm; $\theta$*= 78.51* deg

When compare characteristics of proposed mechanisms can be seen that the smallest influence of the acting load on the stress exhibits the Rhombus type, which provides the smallest dimensions of the design. On the other side, the amplification ratio exhibits the largest range from all of the analyzed structures (from 0.3 to 0.62). Another conclusion is that the dimensions of analyzed mechanisms are relative big despite using flexible materials.

As shown in Table 1 an “ideal” amplification ratio is diametrically different with respect to the desired value ($R_{amp}$ = *0.3*) and FEM simulations too. Comparing to the mechanical amplifiers, as discussed above, the amplification ratio is not constant. The amplification ratio of DS2 mechanisms indicates infinite values as shown in Fig. 4(b). This is due to the flexibility of the material and the topology of the mechanism. The suitable example is the double lever mechanism, where the dependence of the amplification ratio between both design stages is remarkable. The mechanism is also deformed (reshaped) due to the acting load without any input motion.

As a consequence of non-linear output movements, for precise position control, it is necessary to measure/calibrate real displacement characteristics. Within suitable approaches is using strain gauge measurements of loads/displacements, or, it is possible to use wireless (contactless) displacements sensing [18]. Here the resonant circuit is directly integrated into the compliant structure and the actual position is related to the changes of the electromagnetic field.

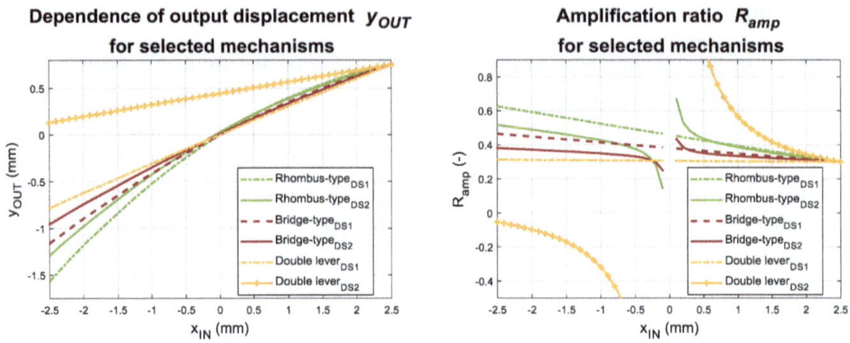


a) b)

**Fig. 4.** (a) Dependence of output displacement $y_{OUT}$ for selected mechanisms (b) Dependence of the motion amplification ratio

## 4 Conclusion

This paper presents some compliant structures of mechanical amplifiers most frequently applied in small positioning devices. The main reason for their use is the adaptation of the displacement/force of actuators to desired motion/force parameters on the side driven mechanisms. There are analyzed three types of mechanisms can be used as force amplifiers/motion reducers for linear motion positioning mechanisms with specific requirements on motion parameters, amplification ratio, dimensions, and load.

As showed calculations for all compared mechanisms their characteristics are very different with respect to specifications. The Rhombus type amplifier enables relatively large input displacements at smallest external dimensions. Except for lever mechanism, the input-output relations for displacement/force are not perfectly linear. So, for the case of high accuracy positioning this fact should be taken into account in control. It is shown that the design process should include three phases: careful simulation/optimization of geometry, building the physical model and measurements of characteristics.

Presented study and comparisons are made mainly from the static point of view. The study of dynamical performances will be a subsequent step in future work.

**Acknowledgment.** This work was supported by the Slovak Research and Development Agency under the contract No.: APVV-14-0076 – "MEMS structures based on load cell", contract No.: APVV-18-0117 and by the national scientific grant agency VEGA under project No.: 2/0155/19 – "Processing sensory data via Artificial Intelligence methods".

## References

1. Howell, L.L.: Compliant Mechanisms. Wiley-IEEE, 459 p. (2001). ISBN 047138478X
2. Dong, W., et al.: Development and analysis of a bridge-lever-type displacement amplifier based on hybrid flexure hinges. Precis. Eng. **54**, 171–181 (2018). ISSN 0141-6359

3. Xie, X., Bigdeli Karimi, M., Liu, S., Myanganbayar, B., Livermore, C.: Micro motion amplifiers for compact out-of-plane actuation. Micromachines **9**(7), 365 (2018)
4. Patil, V.S., Anerao, P.R., Chinchanikar, S.S.: Design and analysis of compliant mechanical amplifier. Mater. Today Proc. **5**(5), 12409–12418 (2018). Part 2. ISSN 2214-7853
5. Zhou, C., Li, J., Duan, J.-A., Deng, G.: Direct-acting piezoelectric jet dispenser with rhombic mechanical amplifier. IEEE Trans. Compon. Packag. Manuf. Technol. **8**(5), 910–913 (2018)
6. Shahosseini, I., Najafi, K.: Mechanical amplifier for translational kinetic energy harvesters. J. Phys. Conf. Ser. **557**, 012135 (2014)
7. Ling, M., et al.: Enhanced mathematical modeling of the displacement amplification ratio for piezoelectric compliant mechanisms. Smart Mater. Struct. **25**, 075022 (2016). ISSN 1361-665X
8. Xu, Q., Li, Y.: Analytical modeling, optimization and testing of a compound bridge-type compliant displacement amplifier. Mech. Mach. Theory **46**(2), 183–200 (2011). ISSN 0094-114X
9. Ouyang, P.R., Zhang, W.J., Gupta, M.M.: Design of a new compliant mechanical amplifier. In: ASME. International Design Engineering Technical Conferences and Computers and Information in Engineering Conference, Volume 7: 29th Mechanisms and Robotics Conference, Parts A and B, pp. 15–24 (2005)
10. Chen, G., Ma, Y., Li, J.: A tensural displacement amplifier employing elliptic-arc flexure hinges. Sens. Actuators A Phys. **247**, 307–315 (2016). ISSN 0924-4247
11. Dong, W., Chen, F., Gao, F., Yang, M., Sun, L., Du, Z., Tang, J., Zhang, D.: Development and analysis of a bridge-lever-type displacement amplifier based on hybrid flexure hinges. Precis. Eng. **54**, 171–181 (2018). ISSN 0141-6359
12. Kim, K., Lee, J., Ju, J., Kim, D.-M.: Compliant cellular materials with compliant porous structures: a mechanism based materials design. Int. J. Solids Struct. **51**(23–24), 3889–3903 (2014). ISSN 0020-7683
13. Kota, S., Hetrick, J., Li, Z., Saggere, L.: Tailoring unconventional actuators using compliant transmissions: design methods and applications. IEEE/ASME Trans. Mechatron. **4**(4), 396–408 (1999)
14. Hricko, J.: Straight-line mechanisms as one building element of small precise robotic devices. Appl. Mech. Mater. **613**, 96–101 (2014). ISBN 978-303835202-0. ISSN 1660-9336
15. Liu, Y., Xu, Q.: Mechanical design, analysis and testing of a large-range compliant microgripper. Mech. Sci. **7**, 119–126 (2016). https://doi.org/10.5194/ms-7-119-2016
16. Ishii, Y., Thümmel, T., Horie, M.: Dynamic characteristic of miniature molding pantograph mechanisms for surface mount systems. Microsyst. Technol. **11**(8–10), 991–996 (2005)
17. Chen, J., Zhang, Ch., Xu, M., Zi, Y., Zhang, X.: Rhombic micro-displacement amplifier for piezoelectric actuator and its linear and hybrid model. Mech. Syst. Sig. Process. **50–51**, 580–593 (2015). ISSN 0888-3270
18. Harťanský, R., Halgoš, J.: The problem of RF radiator with force detector. In: Measurement 2017: 11th International Conference on Measurement, Smolenice, Slovakia, 29–31 May 2017, pp. 139–142. Slovak Academy of Sciences, Bratislava (2017). ISBN 978-80-972629-0-7

# Design and Control of a Flapping Wing System Test Bench

José Zárate(✉) and Hartmut Witte

Technische Universität Ilmenau, 98693 Ilmenau, Germany
{jose-luis.zarate-moya,hartmut.witte}@tu-ilmenau.de
https://www.tu-ilmenau.de/de/biomechatronik/

**Abstract.** The design of an adaptable system to analyze the properties of flight dynamics in flapping wings systems helps to identify their characteristics and allow the use of different control techniques to improve their performance. These devices present difficulties in modeling; therefore, the software to identify the system parameters divided into several stages gives the feasibility to manage these complexities. This work presents the design of a prototype test bench based on mathematical modeling of a balancing beam with a flapping mechanism with wings. This system consists of a bar that has a rotational degree of freedom contained in the vertical and horizontal plane each, in which one of the ends of the flapping mechanism is located. This movement is produced by the thrust force due to wings flapping coupled to a gear set and two motors for each wing. The tunable speed of rotation in the direct current motor allows controlling the force of thrust, managing the bar movement towards a determined angular position measured by an Inertial Measurement Unit (IMU) and two encoders in both axes. The model was obtained experimentally.

**Keywords:** Flapping wings MAV · Biomimetics · Robotics · Test bench · Control

## 1 Introduction

A test bench is a device or system to apply tool tests or study and improve new technologies. This test bench provides all the necessary tools, such as sensors, to analyze the operating parameters of the equipment, testing, find errors and strategies to improve these parameters, if available. In addition, these systems allow you to divide a compound mechanism into different subsystems and test them all together. Complex systems such as flapping-wing robots involving several aerodynamic concepts and flow characteristics such as the Reynolds number are suitable to be developed in a test bench system.

Miniaturization of electronic and electromechanical components, such as motors, power cells, and MEMS accelerometers have led to the development of different flapping wing systems. Some of their advantages are maneuverability and efficiency in low Reynolds number flow. These characteristics have


K. Berns and D. Görges (Eds.): RAAD 2019, AISC 980, pp. 34–42, 2020.
https://doi.org/10.1007/978-3-030-19648-6_5

an influence on the flapping wing robot performance in terms of current and power consumption, maneuverability and performance, for example, landing or hovering [1].

### 1.1 Problem Background

Flapping wing systems with proportions, comparable to their biological prototypes, such as birds and insects, can outperform traditional fixed-wing unmanned aerial vehicles (UAVs) in steady flight [2]. However, the nature of flapping flight creates additional difficulties with the design of devices of this type:

- Disturbances caused by moving parts, such as wings and cranks, affect accelerometer readings, making filters mandatory to use.
- Modeling the flight using a high number of Reynolds is a complex task; therefore, experimental data of the actual flapping system are required.

To solve these problems, a working control system has to be developed, ensuring steady flight in difficult conditions, such as strong wind, which is important for a lightweight flapping system.

### 1.2 Related Works

This task can be solved by using a specialized laboratory complex as mentioned above [3]. There are commercial test benches analogous to 3-DOF helicopters developed by the company Quanser® [4], with one degree of freedom and their QNET 2.0 VTOL Board to study the fundamentals of flight in vertical Take-Off, Landing and control (VTOL). These devices are based on flight systems with propellers, nearly linear behavior, unlike the flapping systems that present disturbances related to limit cycle and wing-beat frequency.

Related to this type of test bench, a control system and its response were experimentally measured by [5]. In this work, the development of a graphical user interface (GUI) for the control of Flapping Wing Vehicles (FWVs) to achieve superior maneuverability is emphasized, and the proposed control interface serves as an effective solution towards interfacing aerial robots and sophisticated software modules. Knüppel et al. [3] studied in their project Tdx-Copter a PID control module and showed that the developed algorithm rather than using an off-the-shelf controller achieved a very stable system and superior flight characteristics, due to direct control of the motors building a quadcopter from scratch.

### 1.3 Considerations

This paper presents the development of a flapping wing system test bench using an inertial measurement unit as an angular position sensor and base of control techniques. Several considerations have been taken into account. The bar that holds the flapping wing mechanism is considered as a despicable mass with a counterweight at the opposite end. The system moves in the vertical plane with one degree of freedom rotating in one point of the bar where the sensor is located. Finally, an Arduino board is used as an interface between the system and the SIMULINK® toolbox.

## 2 Requirements

After studying the related works and the considerations described above, a list of requirements for the development of the test bench was made:

- Support for different mechanisms with varying wingspan and On-board electronics included.
- The ability for 360° rotational work area, to study the mechanism's behavior in different positions.
- The capability to track the angular position in order to create a closed loop regulation system.
- The support of additional devices, such as voltage and current sensors.

Therefore, the device is thought to be a PC-connected laboratory stand with the capability to change flapping frequency and control models via MATLAB SIMULINK®.

The analysis of the requirements with respect to these parameters has led to a set of requirements for the design of the test bench shown in Table 1:

**Table 1.** Flapping wing test bench characteristics

| Parameters | Characteristics |
|---|---|
| Lever length | 6.30 cm (adjustable from the fulcrum to the maximum length) |
| Counterweight mass | Up to 20 g |
| Flapping wing robot mass | 13 g (electronics included) |
| Wingspan | 24 cm |
| Wing area | 96 $cm^2$ |
| Wings covering material | Polyester clear film Mylar 15 Micron |

## 3 System Model

In this section, we will detail the development of the mechanical system model, the equations of motion that represent the rotational drive of the system due to the lifting and pushing forces of the flapping mechanism and the complete system results that define inputs and outputs for the description of the desired behavior.

### 3.1 Mechanical Model of the System

The diagram shown in Fig. 1a represents the structure of the test bench. As mentioned above, the system is modeled as a balancing beam with a flapping mechanism at one end and a counterweight at the opposite end.

The angular position $\theta$ obtained by the IMU represents the elevation generated by force $F$ due to the flapping movement.

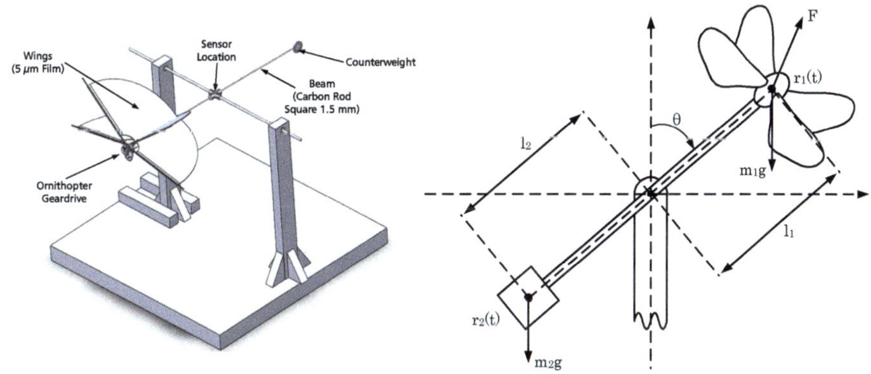


(a) Balancing beam test bench model. (b) Model description.

**Fig. 1.** Flapping wing test bench model.

### 3.2 Equations of Movement and Dynamics

Based on the description in Fig. 1a, the calculation of the system dynamics is presented in the following equations (Fig. 1b):

$$\boldsymbol{r_1}(t) = \begin{bmatrix} l_1(t)\sin(\theta(t)) \\ l_1(t)\cos(\theta(t)) \end{bmatrix}, \tag{1}$$

$$\boldsymbol{r_2}(t) = \begin{bmatrix} -l_2(t)\sin(\theta(t)) \\ -l_2(t)\cos(\theta(t)) \end{bmatrix}, \tag{2}$$

$$\bar{g} = \begin{bmatrix} 0 \\ g \end{bmatrix}. \tag{3}$$

$$L = K - U \tag{4}$$

$$K(\dot{r}(t)) = \frac{1}{2}m_1\dot{r}_1^\top(t)\dot{r}_1(t) + \frac{1}{2}m_2\dot{r}_2^\top(t)\dot{r}_2(t) = \frac{1}{2}\left(m_1 l_1^2 + m_2 l_2^2\right)\dot{\theta}^2(t) \tag{4a}$$

$$U(r(t)) = m_1\bar{g}r_1(t) + m_2\bar{g}r_2(t) = (m_1 l_1 - m_2 l_2)\, g\cos(\theta(t)) \tag{4b}$$

The Lagrangian formulation is defined in Eq. 5, where $q = [\theta(t), \dot{\theta(t)}]^\top$ are generalized system coordinates:

$$\frac{\mathrm{d}}{\mathrm{d}t}\left(\frac{\partial L}{\partial \dot{q}}\right) - \frac{\partial L}{\partial q} = Q. \tag{5}$$

$$\frac{\theta}{\mathrm{F}} = \left( \frac{\left(\frac{l_1}{m_1 l_1^2 + m_2 l_2^2}\right)}{S^2 + \left(\frac{\beta l_1}{m_1 l_1^2 + m_2 l_2^2}\right) S + \left(\frac{g\left(m_2 l_2^2 - m_1 l_1^2\right)}{m_1 l_1^2 + m_2 l_2^2}\right)} \right). \tag{6}$$

## 4 Test Bench Design

With the defined set of components, a 3D model of the swing wing system was built using SolidWorks® software. The initial design is presented in Fig. 2a.

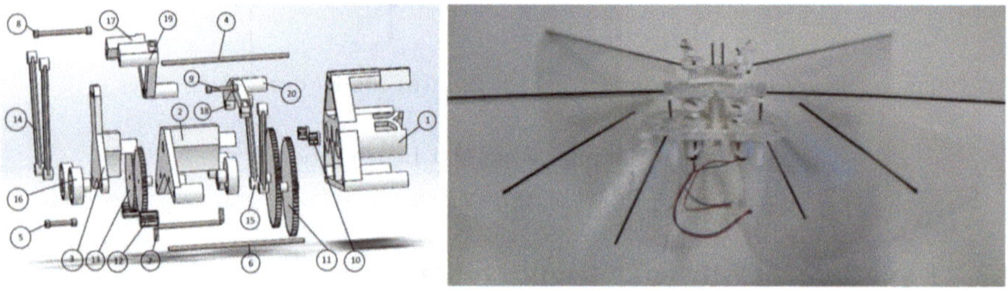

(a) (b)

**Fig. 2.** Flapping wing robot. (a) SolidWorks® model of the flapping wings mechanism with the following components: 1 motor holders, 2 gears and cranks holders, 3 gears and crank lockers, 4, 5, 6, 7, 8, 9 steel shafts, 10, 11, 12, 13 nylon gears, 14, 15 cranks, 16, 17, 18, 19, 20 connecting rods. (b) Assembled flapping wing mechanism with motors.

All the inputs and outputs are soldered to the board, used as a hub for a simple connection of electronic components. Arduino® and current sensors are also placed on the same shield, as shown in Fig. 2b.

Based on the 3D model, the first prototype has been created. It was decided to use a plastic polymer PVC as a manufacturing material for the support structure due to the simplicity of processing to make the first prototype of the test bench.

The physical part of the first prototype of the test bench contains the following components:

- PVC base whose design was made to adjust to the test bench and other elements, such as the shield with electronic components and the control potentiometer.

- Support pillar whose dimension is determined by the length of the lever.
- Cross mounting rod with two versions of the transverse mounting. The first was connected to two balance bars in the similar plane, but after testing the entire system, this version of the cross-section proved to be unstable due to the strong vibrations that occurred in the balance bars with this connection.
- The carbon fiber balance bars that can be added to the construction to move the center of mass away from the main bar axis.
- The main rod that crosses the transverse assembly. One end is used to fix the flapping mechanism, while the other serves as a lever with adjustable counterweight position.
- The axis of rotation is made by a steel bearing and is used to provide a degree of freedom for the system.
- Two plastic bearings, lubricated to ensure minimum friction between the bearings and the axis of rotation. Initial tests have shown that in some cases it is still not enough to provide freedom to the mechanism, therefore, other options should be studied.
- A set of steel rings used as a counterweight.
- Six support angles to ensure the rigidity of the connection between the base and the pillars.
- Two rails for supporting the mobile pillar.

The assembled test bench is shown on Fig. 4.

## 5 Test Bench Control System

To illustrate the control system described in [6], a simulation, presented in Fig. 3a was performed. The system has been developed using the Arduino Toolbox for MATLAB® tool. The signals are managed for the software interacting with the inputs and outputs from the IMU sensor, and the motor driver respectively. All the values are managed by the USB port of the Arduino® board to the PC. To obtain the angular position, the Euler formulas from the acceleration data and a complementary filter based on the gyro data are used in combination to increase the accuracy of the angle at any time as shown in Eq. 7.

$$y = \alpha \times y + (1 - \alpha) \times x, \tag{7}$$

where $\alpha$ is the filter parameter, which is usually chosen with an approximate value of 0.98, applying a high and low pass filter on the gyro signals $y$ and $x$ respectively. Finally, the block filter was considered to reduce the noise in the IMU signal.

## 6 Test and Results

The results obtained in the test bench are shown in Fig. 3b. The system is forced to rise an angular setpoint of 90°, holding the beam in a horizontal position, controlled by the PID block presented in Fig. 3a. The output presents a quasi-stable response trying to follow the input signal. At the time of 150 s a disturbance is applied to analyze the system stability, achieved around 50 s.

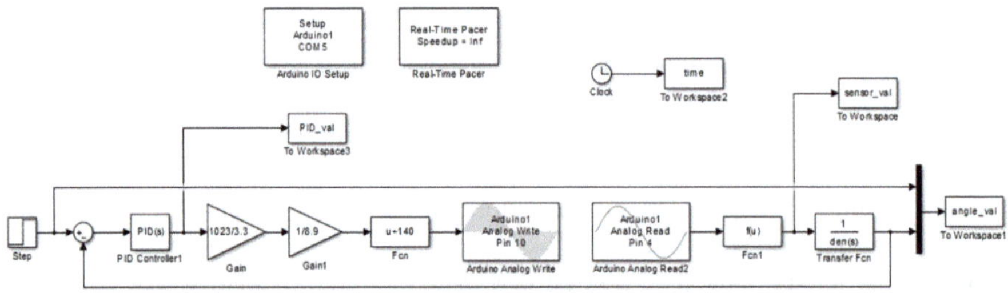


(a) SIMULINK®block diagram.

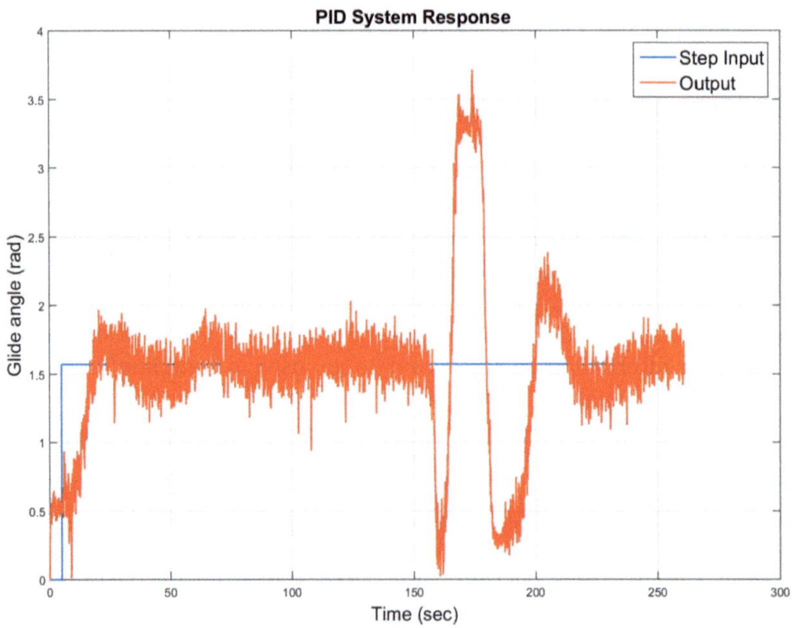


(b) Output response of the system.

**Fig. 3.** Flapping wing test bench control system.

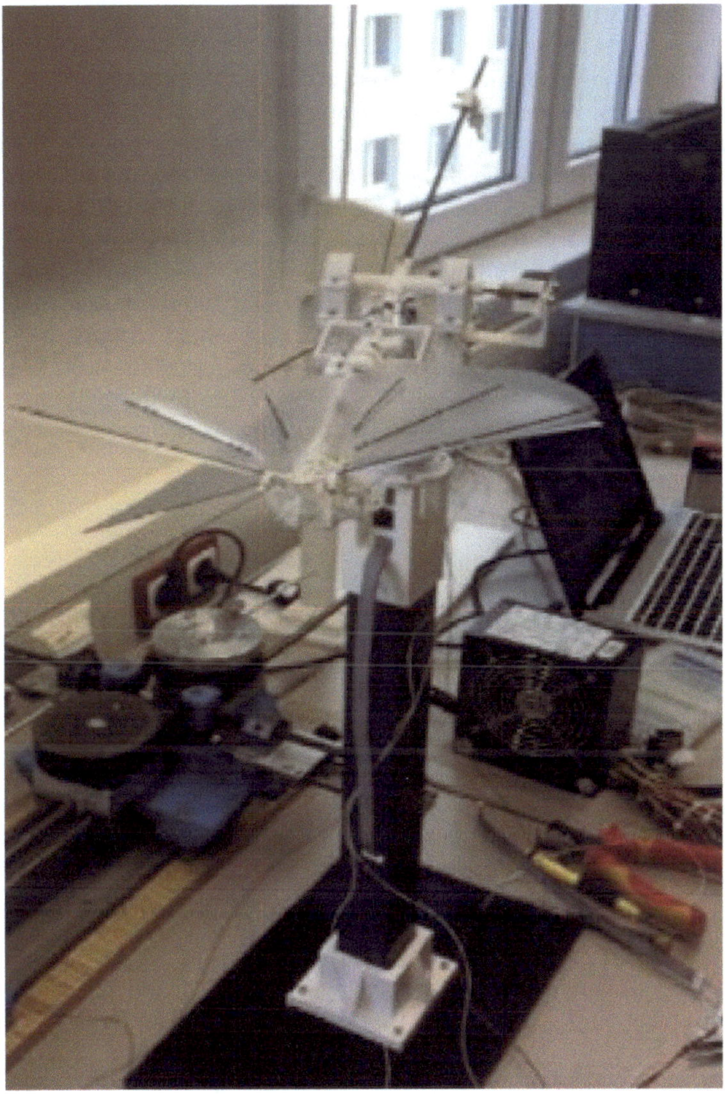

**Fig. 4.** Assembled test bench with power source.

## 7 Discussion

In Fig. 3b, the results show a system that is affected by noise due to the wing flapping frequency with the maximum between 18 to 20 Hz, observing an effect of the limit cycle due to this frequency obtained by the IMU. Due to this effect, the system is sensitive and can be affected by high disturbances and its stability is slow (50 s for recovery) reflected in the performance of the system.

## 8 Conclusion

Based on the dynamics obtained in this test bench, the rotation system has been modeled. Its inputs represent the forces resulting from the lift and thrust, generated by the flapping wings. The angular position was controlled using the PID control strategy. It shows the behavior of the system affected by disturbances related to the oscillation caused by the limit cycle. With respect to external disturbances, the friction force in the pivot is considered small and the trajectory modification shows a slow stabilization. The results have shown an acceptable performance, especially when the external forces try to modify the angular position of the system adjusting its displacement. Subsequent research will focus on testing other control strategies to improve the response of the system.

**Acknowledgment.** The authors would like to gratefully acknowledge support from Biomechatronics Group of Technische Universität Ilmenau.

## References

1. Floreano, D., Zufferey, J.C., Srinivasan, M.V., Ellington, C. (eds.): Flying Insects and Robots. Springer, Heidelberg (2009)
2. Shyy, W., Aono, H., Kang, C.K., Liu, H.: An Introduction to Flapping Wing Aerodynamics, vol. 37. Cambridge University Press, Cambridge (2013)
3. Knüppel, P., Winter, B.: TdxCopter. http://www.beniwinter.com/p/tdxcopter.html
4. Quanser: 3 DOF Helicopter (2017). https://www.quanser.com/products/3-dof-helicopter/
5. Sankarasrinivasan, S., Balasubramanian, E., Yang, L., Hsiao, F.: Autonomous control of flapping wing vehicles using graphical user interface. In: Advances in Computing, Communications and Informatics (ICACCI), pp. 2217–2220. IEEE (2015)
6. Corke, P.: Robotics, Vision and Control: Fundamental Algorithms in MATLAB®, vol. 73. Springer, New York (2011)

# Approximation of Inverse Kinematic Solution of a Metamorphic 3R Manipulator with Multilayer Perceptron

Markos Tzivaridis[1], Vassilis C. Moulianitis[1,2](✉), and Nikos A. Aspragathos[1]

[1] Mechanical Engineering and Aeronautics Department, University of Patras, 26500 Patras, Greece
mark.gvrds@gmail.com, asprag@mech.upatras.gr
[2] Department of Product and Systems Design Engineering, University of the Aegean, 84100 Syros, Greece
moulianitis@syros.aegean.gr

**Abstract.** In this paper, a feedforward neural network is trained towards the generalisation of the inverse kinematics of a 3R metamorhic manipulator with one pseudojoint. A two hidden-layered network is trained with data produced by several anatomies of the metamorphic manipulator and tested to an unforeseen anatomy. The data are separated to aspects per anatomy and combined to one training set. Various configurations of the network is trained using the Levenberg-Marquardt backpropagation method and the best of them is derived. As it will be shown the derived network can achieve very low generalisation errors.

**Keywords:** Metamorphic manipulator · Inverse kinematics · MLP

## 1 Introduction

A family of reconfigurable manipulators are the modular metamorphic serial manipulators that alters their anatomy without dismantling their structure. The modular metamorphic serial manipulators show higher performance for a given task than the known industrial fixed manipulators or even than the reconfigurable manipulators [1, 2]. For each structure of a metamorphic manipulator, a finite number of anatomies could be derived by changing off-line the angle of the passive joints. The forward kinematics of the metamorphic manipulator are determined using the Denavit-Hartenberg (D-H) parameters, that depends on the metamorphic parameters [3]. However, the inverse kinematics problem (IKP) depends on the anatomy of the manipulator, producing different number of solutions based on the metamorphic D-H parameters. Since the possible anatomies belong to a family of manipulator types, a unified approach for the inverse kinematics should be developed. If the passive joints are treated as the active joints the IKP becomes quite complicated. Thus, the kinematic characteristics of the metamorphic structure and the derived anatomies should be investigated to facilitate the choice of the suitable approach.


K. Berns and D. Görges (Eds.): RAAD 2019, AISC 980, pp. 43–50, 2020.
https://doi.org/10.1007/978-3-030-19648-6_6

The IKP has multiple solutions and common industrial robot design's with 6 degrees of freedom (DoF) have at maximum eight (8) analytical solutions. It was proved that the maximum number of inverse kinematics solutions for a general 6 DoF robot is sixteen (16). No-generic geometries divide the space of manipulator geometries into disjoint sets containing the solutions of the IKP [4].

The use of universal approximators based on artificial neural networks (ANN), neuro-fuzzy and genetic algorithms among others have been presented. The configuration space approximated by uniform meshes or by data produced by the Halton sequence where clustered into aspects that took into account the Jacobian matrix. Training data from the derived aspects where fed to ANNs to approximate the inverse kinematics solutions [5–7]. The optimum model parameters used to train ANN for the inverse kinematics of a 3 DoF planar robot was presented [8]. A comparison between Multilayer Perceptron (MLP) and the Radial Basis Function (RBF) for prediction of incremental joint angles with the assistance of forward kinematic relations was proposed [9]. An MLP with backpropagation algorithm was used for the approximation of one set of solutions in a 6 DoF manipulator [10, 11].

In this paper, a feedforward neural network is trained towards the generalisation of the inverse kinematics of the metamorhic manipulator. A two hidden-layered MLP is trained with data produced by several anatomies of the metamorphic manipulator and tested to an unforeseen anatomy. Various configurations of the network is trained and the best of them is derived. As it will be shown the derived network can achieve very low generalisation errors.

The paper is organised as follows: first the proposed method, the ann architecture, the case study metamorphci manipulator and the clustering and training data are presented shortly. Then an application of the case study, discussion of results and concluding remarks are closing this paper.

## 2 The Proposed Method

In this paper, a neural network is trained towards the generalisation of the inverse kinematics of a 3R metamorhic manipulator with one pseudojoint. The flowchart of the method is shown in Fig. 1. The method begins with the analysis of the metamorhic manipulator structure. The D-H parameters of each anatomy are determined with the method presented in [3]. The kinematic analysis determines the joint angles and the metamorphic variables dependance of the Jacobian determinant for every anatomy. A uniform grid of points is formed in the range of these joints and the data are clustered into the aspects for every anatomy. The trained data are produced for a part of an aspect for several anatomies of the metamorphic manipulators using forward kinematics. The neural network have as inputs the given position of the end-effector of the manipulator and the pseudojoint variables. The outputs of the neural network are the corresponding values of the second and third active joint variables. The training data of every anatomy are combined into one dataset and then are used to train the ANN.

In the following, the ANN architecture and the important parameters of it are explained. The case study manipulator kinematics as well as the clustering of data using the Jacobian determinant are presented.

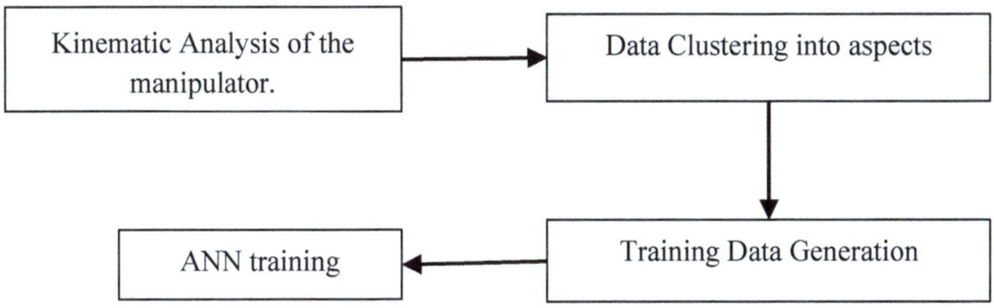


**Fig. 1.** The proposed method

## 2.1 The Metamorphic Manipulator with One Pseudo-Joint

The case study metamorphic manipulator has with three revolute active joints and one revolute pseudo-joint. The manipulator with the geometry parameters and the local coordinate systems are shown in Fig. 2. Let $\theta_1, \theta_2, \theta_3$ and $\theta_{ps}$ be the active joint and the pseudo-joint variables respectively. The D-H parameters of the metamorphic manipulator is shown in Table 1 and are determined with the usual way for all kind of joints of the metamorphic manipulator. However, only the active joints of the metamorphic robot move during the execution of a given task, while the pseudojoints are locked in a prespecified $\theta_{ps}$ forming the metamorphic links between the active joints.

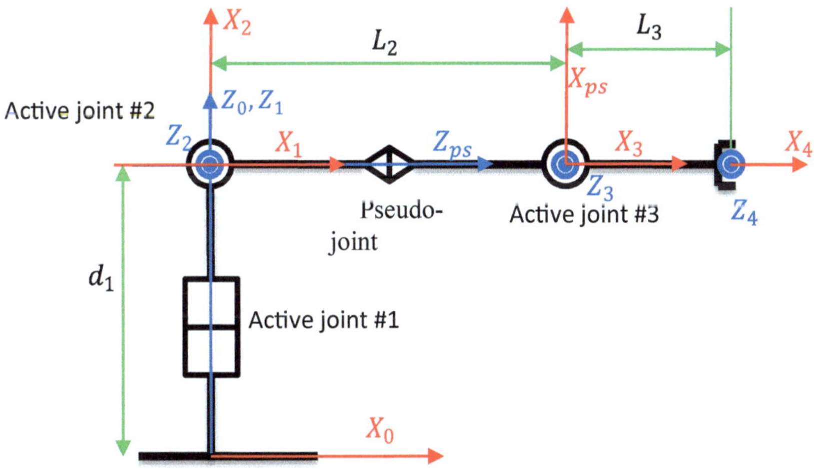


**Fig. 2.** The 3 DoF case study metamorphic manipulator.

**Table 1.** The D-H parameters of the metamorphic structure shown in Fig. 2.

| # | $\alpha_{i-1}$ | $L_{i-1}$ | $d_i$ i | $\theta_i$ | |
|---|---|---|---|---|---|
| 1 | 0 | 0 | $d_1$ | $\theta_1$ | Active joint |
| 2 | $\pi/2$ | 0 | 0 | $\theta_2$ | Active joint |
| 3 | $\pi/2$ | 0 | $L_2$ | $\theta_{ps}$ | Pseudo joint |
| 4 | $-\pi/2$ | 0 | 0 | $\theta_3$ | Active joint |
| 5 | 0 | $L_3$ | 0 | 0 | |

In order to determine the D-H parameters of the metamorphic manipulator the method presented in [3] is utilized. The metamorphic D-H parameters are determined by the dual vectors defined by the k component of the rotational matrix and the position part of the i active joint relative to the 0 frame. Based on this method, it is found that only the $\alpha_2$ parameter if the manipulator depends on the metamorphic variable $\theta_{ps}$ as it shown in Table 2. The $\alpha_2$ parameter changes the orientation of the 3rd joint Z axis relative to the 2nd one. The end-effector is located to different positions with different orientations for the same values of joint variables varied by the change of the $\theta_{ps}$.

**Table 2.** The metamorphic D-H parameters of the manipulator shown in Fig. 2.

| # | $\alpha_{i-1}$ | $L_{i-1}$ | $d_i$ i | $\theta_i$ |
|---|---|---|---|---|
| 1 | 0 | 0 | $d_1$ | $\theta_1$ |
| 2 | π/2 | 0 | 0 | $\theta_2$ |
| 3 | $\theta_{ps}$ | $L_2$ | 0 | $\theta_3$ |
| 4 | 0 | $L_3$ | 0 | 0 |

According to [4] no-generic geometries divide the space of 3R geometries into disjoint sets which have homogeneous topological properties. Four disjoint sets can be produced from a 3R geometry containing the four solutions of the IKP. It has been shown that the determinant of a 3R manipulator does not depend on the first joint variable [15] and the configuration variables that affect the singularity locus are $\theta_2$ and $\theta_3$. Let J be the Jacobian matrix of the manipulator and $|J|$ it's determinant. The geometric Jacobian determinant of the metamorphic manipulator is:

$$|J| = \sin\theta_3\left(L_2L_3^2\cos\theta_{ps}\sin\theta_2\sin\theta_3 - L_3L_2^2\cos\theta_2 - L_2L_3^2\cos\theta_2\cos\theta_3\right) \quad (1)$$

The determination of the aspects and their correspondence to the inverse kinematics solutions is critical for the approximation of all of them and depends on $|J|$. The inverse kinematics solution exists only when the determinant of the Jacobian matrix is not singular. The aspects depend on the metamorphic variable $\theta_{ps}$, and the division of the configuration space is changed. The division of the configuration space in aspects for five anatomies $\theta_{ps} \in \left\{0, 2\frac{\pi}{24}, \ldots, 8\frac{\pi}{24}\right\}$ is shown in Fig. 3. The variance of the aspects is smooth for small change of the $\theta_{ps}$ leading to a smooth change of the workspace.

In order to cluster the data into aspects a uniform grid for the joint variables $[\theta_2 \quad \theta_3]$ per anatomy is produced. The four aspects are determined by the sign of each factor of $|J|$ (see Table 3).

For k non-cuspidal anatomies the combinations of training data per aspect per anatomy is $4^k$. In this paper, a part of the aspect #3 per anatomy is used to train the networks. The investigation of all anatomies is left for future work.

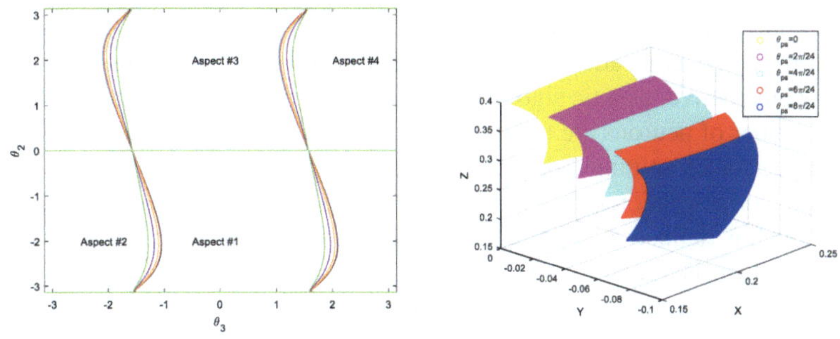


(a) (b)

**Fig. 3.** (a) Aspects of five anatomies of the metamorphic robot ($L_2 = 0.2\,\text{m}$, $L_3 = 0.1\,\text{m}$) (b) Workspace variation according to $\theta_{ps}$, for $\theta_1 = 0$ rad, $\theta_2 \in [-0.7, 0]$ rad and $\theta_3 \in [-1.7, 2]$ rad

**Table 3.** Determination of aspects using (1).

| | $sgn(L_2L_3^2\cos\theta_{ps}\sin\theta_2\sin\theta_3 - L_3L_2^2\cos\theta_2 - L_2L_3^2\cos\theta_2\cos\theta_3)$ | | | |
|---|---|---|---|---|
| | | $-1$ | 0 | 1 |
| | $-1$ | Aspect #1 | Singularity | Aspect #2 |
| $\text{sgn}(\sin\theta_3)$ | 0 | Singularity | Singularity | Singularity |
| | 1 | Aspect #3 | Singularity | Aspect #4 |

The training data for every anatomy are derived using the produced parameters shown in Table 2, using the forward kinematic map to the workspace expressed by the sets of tuples $[x \quad y \quad z \quad \theta_{ps}]$. The tuple contains the Tool Center Point (TCP) position $[x \quad y \quad z]$ and the values of the pseudojoint $\theta_{ps}$. The values of the joint variables $[\theta_2 \quad \theta_3]$ and the corresponding values of the tuples $[x \quad y \quad z \quad \theta_{ps}]$ per anatomy, are stored and used as training data for the ANN.

## 2.2 ANN Architecture

In this paper, a multi-layer perceptron (MLP) with two hidden layers is selected, since according to [12], they are enough for function approximation. A 4-$n_1$- $n_2$-2 architecture is used, with an input layer of four parameters $[x \quad y \quad z \quad \theta_{ps}]$, $n_1$, $n_2$ neurons in the first and second hidden layer respectively and an output layer of two outputs $[\theta_2 \quad \theta_3]$. The purpose of the second hidden layer is to drastically reduce the total required number of nodes [13]. According to [14], a feedforward network with m outputs, $n_1 = \sqrt{(m+2)N} + 2\sqrt{N/(m+2)}$ and $n_2 = m\sqrt{N/(m+2)}$ number of neurons in the first and second hidden layer respectively is able to overfit N training samples with any arbitrarily small error. According to [13], in order to produce an ANN with good generalization the number the structure of the ANN should have fewer nodes than those suggested by [14]. Therefore, the two formulas suggested by [14] determines the upper limit of neurons in every layer. However, the best number of neurons towards generalization should be determined using other techniques (e.g. trial and error, constructive, genetic algorithms etc.). In this paper, several architectures with different number of neurons are trained and tested and the best is found by minimizing the absolute error of untrained data towards generalization.

The activation function employed in the first hidden layer is a logarithmic sigmoid function which is defined over a range of [0, 1]. A linear activation function is used in the second hidden layer.

A large set of training data using forward kinematics of several metamorphic anatomies is obtained for training the ANN. The data are divided in three sets, training (90%), validation (5%) and test (5%). The first set is used to train the ANN using the Levenberg-Marquardt backpropagation method. Fine tuning of the parameters of the network is done using the validation data set ensuring that the network has not overfit the training data. Finally, the test data is used to estimate the predictive power of the ANN by providing an objective measure of performance on a dataset that was not used during the training process for a small part of the dataset.

Finally, test data for an anatomy of a metamorphic robot that is not previously used to train the network is used to test the generalization capability of the network.

## 3 Application and Discussion of Results

The D-H parameters of the case study metamorphic manipulator shown in Fig. 2, are $d_1 = 0.3m$, $L_2 = 0.2m$ and $L_3 = 0.1m$. The anatomies used to train the network are determined by $\theta_{ps} \in \left\{0, \frac{\pi}{24}, 2\frac{\pi}{24}, \ldots, 8\frac{\pi}{24}\right\}$ excluding $\theta_{ps} = \frac{4\pi}{24}$ which is used to verify the generalization of the network.

A part of the aspect #3 is used to produce the training data, where $\theta_2 \in [-0.7, 0]$ rad and $\theta_3 \in [-1.7, 2]$ rad. The uniform grid is created with 0.0087 rad step size, and $\theta_1 = 0$ rad. Based on the number of the training samples N = 47239 (90% of total set), the upper bound of the number of each hidden layer is found to be $n_1 = 651$ and $n_2 = 217$ respectively.

In order to find the best number of neurons, several architectures are trained with fewer neurons. More specific, 121 architectures with $n_1, n_2 \in \{10, 15, \ldots, 60\}$ were trained for 1000 epochs using the sum squared error performance function to be less than $10^{-9}$ rad.

In Table 4, the sum squared error of the training, validation and test error is shown for five ANN architectures. In addition, the maximum absolute generalization error for both joint variables are shown in the last column. The best trained ANN architecture is 4-55-30-2 presenting maximum absolute joint error 0.0001038 rad which is very low showing that the trained network has good generalization performance. According to the performance measures provided during training the best architecture is the 4-55-60-2, but it produces higher maximum absolute error in the untrained data than the 4-55-30-2 one.

**Table 4.** Performance of trained networks

| $n_1 - n_2$ | SSE training ($10^{-3}$ rad) | SSE validation ($10^{-3}$ rad) | SSE test ($10^{-3}$ rad) | Max absolute error (X$10^{-3}$ rad) |
|---|---|---|---|---|
| 55–30 | 0.0125 | 0.0027 | 0.0027 | 0.1038 |
| 55–15 | 0.0265 | 0.0061 | 0.0056 | 0.1110 |
| 50–20 | 0.0531 | 0.0116 | 0.0114 | 0.1206 |
| 55–60 | 0.0065 | 0.0015 | 0.0014 | 0.1284 |
| 60–40 | 0.0116 | 0.0025 | 0.0025 | 0.1416 |

It is shown that the increase of layers and a better tuning of the ANN can produce very low errors compared to previous studies [5] where the best MSE error was in the level of $10^{-1}$ rad.

Although the absolute error is very low an investigation for lower number of neurons than 55 and 30 in the first and second hidden layer should be made in order to find the optimal architecture.

## 4 Conclusions

In this paper, it is shown that the ANN are able to generalize and approximate the inverse kinematics of a metamorphic manipulator. The absolute error of untrained data is very low showing that the method can generalize the inverse kinematics for anatomies that do not have closed solution.

Compared to previous studies the presented method presents very low generalization error.

For future work, an investigation to find the best combination of the data per anatomy should be made. Second, a study on the increase of the metamorphic variables towards the generalization. The employment of fine tuning techniques such as constructive or genetic algorithms to determine the optimal anatomy of the ANN should be made. Finally, the generalization of the ANNs should be investigated in all aspects as well as in wider training data in every aspect.

**Acknowledgement.** Part of this research has been financially supported by General Secretariat for Research and Technology (GSRT) and the Hellenic Foundation for Research and Innovation (HFRI) (Code: 1184).

## References

1. Chen, I.M.: Modular robots. In: Handbook of Manufacturing Engineering and Technology, pp. 2129–2168. Springer, London (2015)
2. Valsamos, C., Moulianitis, V.C., Aspragathos, N.: Index based optimal anatomy of a metamorphic manipulator for a given task. Rob. Comput. Integr. Manuf. **28**(4), 517–529 (2012)

3. Moulianitis, V.C., Aspragathos, N.A., Valsamos, C.: Suboptimal anatomy of metamorphic manipulators based on the high rotational dexterity. In: Advances in Reconfigurable Mechanisms and Robots, pp. 509–519. Springer, Cham (2016)
4. Burdick, J.W.: A classification of 3R regional manipulator singularities and geometries. Mech. Mach. Theory **30**(1), 71–89 (1995)
5. Frisyras, E.K., Moulianitis, V.C., Aspragathos, N.A.: ANNs to approximate all the inverse kinematic solutions of non-cuspidal manipulators. IFAC-PapersOnLine **51**(22), 418–423 (2018)
6. Soylak, M., Oktay, T., Turkmen, İ.: A simulation-based method using artificial neural networks for solving the inverse kinematic problem of articulated robots. Proc. Inst. Mech. Eng. Part E: J. Process Mech. Eng. **231**(3), 470–479 (2017)
7. Hargis, B.E., Demirjian, W.A., Powelson, M.W., Canfield, S.L.: Investigation of neural-network-based inverse kinematics for a 6-DOF serial manipulator with non-spherical wrist. In: ASME 2018 International Design Engineering Technical Conferences and Computers and Information in Engineering Conference pp. V05BT07A048–V05BT07A048. American Society of Mechanical Engineers, August 2018
8. Karkalos, N.E., Markopoulos, A.P., Dossis, M.F.: Optimal model parameters of inverse kinematics solution of a 3R robotic manipulator using ANN models. Int. J. Manuf. Mater. Mech. Eng. (IJMMME) **7**(3), 20–40 (2017)
9. Chiddarwar, S.S., Babu, N.R.: Comparison of RBF and MLP neural networks to solve inverse kinematic problem for 6R serial robot by a fusion approach. Eng. Appl. Artif. Intell. **23**(7), 1083–1092 (2010)
10. Hasan, A.T., Al-Assadi, H.M.A.A.: Performance prediction network for serial manipulators inverse kinematics solution passing through singular configurations. Int. J. Adv. Rob. Syst. **7**(4), 11–24 (2011)
11. Hargis, B.E., Demirjian, W.A., Powelson, M.W., Canfield, S.L.: Investigation of neural-network-based inverse kinematics for a 6-DOF serial manipulator with non-spherical wrist. In: ASME 2018 International Design Engineering Technical Conferences and Computers and Information in Engineering Conference, pp. V05BT07A048–V05BT07A048. American Society of Mechanical Engineers, August 2018
12. Hagan, M.T., Demuth, H.B., Beale, M.H., De Jesús, O.: Neural Network Design, vol. 20. Pws Pub., Boston (1996)
13. Stathakis, D.: How many hidden layers and nodes? Int. J. Remote Sens. **30**(8), 2133–2147 (2009)
14. Huang, G.-B.: Learning capability and storage capacity of two-hidden-layer feedforward networks. IEEE Trans. Neural Networks **14**(2), 274–281 (2003)
15. Staffetti, E., Bruyninckx, H., De Schutter, J.: On the invariance of manipulability indices. In: Advances in Robot Kinematics, pp. 57–66. Springer (2002)

# Mechanical Design, Modeling and Simulation of Human-Size Cable-Driven Over-Actuated Robotic Arm

Aleksandar Rodić[1(✉)], Shunsuke Hioki[2], Marija Radmilović[1], and Miloš Jovanović[2]

[1] Institute Mihajlo Pupin, University of Belgrade, Belgrade, Serbia
aleksandar.rodic@pupin.rs
[2] Faculty of Mechanical Engineering, Yokohama National University, Yokohama, Japan

**Abstract.** This paper presents new mechanical design of a human-sized, lightweight, redundant, over-activated and cable-driven robotic arm dedicated to development of collaborative industrial humanoid. Mechanical design of the robotic arm includes some original technical solutions which reduce total mass of the mechanism, includes passive compliance in the structure and implements redundant number of servo-motors and corresponding mechanical degrees of freedom. Robot motion is achieved by synergy of operation of numerous driving motors that move rob joints by pulling/releasing non-tensile cables. By implementing of the redundant number of driving motors, system achieves enhanced mobility and manipulative capabilities, optimize power consumption and increases robustness of the system against the failures of servo-drives. The set of technical requirements imposed in the paper corresponds to the biological model of the human-size arm. Selected references in this field of research are listed in the paper, too. At the end of paper, some characteristic simulation results are presented, analyzed and discussed with aim to validate the obtained design and theoretical results.

**Keywords:** Lightweight robot · Redundant · Over-actuated · Cable-driven robotic arm · Industrial humanoid

## 1 Introduction

Average weight of the adult human arm is in the range 4–5 kg. At the same time, one arm can carry on and kept the payload of 3–4 kg in a fully abducted position on average no more than 1 min. After that short exercise, the human muscles enter into a state of fatigue or saturation and must be relaxed in order to recover their physical abilities. In robotics, the ability of robot to pick up some load that is lighter or heavier than its own mass is expressed by the special index called 'payload-mass ratio'. This indicator represents the ratio of the robot mass and the payload that robot can carry on


K. Berns and D. Görges (Eds.): RAAD 2019, AISC 980, pp. 51–58, 2020.
https://doi.org/10.1007/978-3-030-19648-6_7

without overload and damage of its' driving motors and mechanical structure. With humans, this index is approximately equal to one. The human arm can normally pick up a load that is even heavier of its own weight. Industrial robots have a significantly higher value of this index than unity. The reason for that is the strict request to make robot structure very stiff which guarantees high precision. The elasticity of structures for industrial robots is not desirable feature.

When talking about the precision of human arm it is only 1–1.5 mm. Human hand has a duty to ensure high accuracy (e.g. less than 0.5 mm) while the task of arms is to perform so-called transfer motion and to bring a robot hand in a convenient location [1]. The aim of this study was to carry out few tests. The goal was to identify the accuracy of human arm with respect to the wrist joint. In the first test, an arbitrary curved graph (contour) was drown on a paper sheet. The paper is placed in front of the subject whose hand was immobilized (fixed) so that hand could hold a pencil just like a stick while the fingers were not used at all to support hand writing. Examinee was asked to repeatedly imitate (copy) by pen the given curved contour. The same experiment was repeated with more subjects, young people of both genders, aged 25–30 years. The drawings are then scanned, digitized in a suitable form, so that they can be analyzed using the appropriate graphical tools (e.g. CorelDraw). By analyzing a large number of digitized test-drawings it was discovered that the average error of copying (repetitions) of the immobilized hand was within the limits of 1–1.5 mm. Another test was performed using a tailor's needle and thread. An ordinary needle for hand sewing was taken, with a small hole at the top for inserting the end of thread. The needle hole was 1 mm and the thread was thinner than the hole. Without using fingers, none of the respondents' hands have failed to put the thread through the narrow eye of a needle. This allowed us to prove the hypothesis that the accuracy of the human hand in the wrist is not better than 1 mm. High accuracy of manual action can be achieved exclusively by capabilities of human hand. Precision of the hand can be improved by training manipulative skills.

## 2 Progress Beyond the State-of-the-Art

In the last 10 years there was lot of successful attempts to build different lightweight robotic arms [2–8]. Many of them have implemented the cable-driven approach to move robot joints. Main difficulty with these systems concerns with structural elasticities and consequently with precision. Another problem regards to the mechanical robustness of the lightweight structure. If we intend to make a lightweight robot arm for industrial purposes we ultimately need to build a reliable and really robust system that can ensure a long life cycle. One of the interesting design in recent time is AMBIDEX robot arm [2] resulting from collaborative R&D on human-robot coexistence. The arm uses innovative cable-driven mechanisms that make any interaction with humans safe. At just 2.6 kg, it weighs less than the average arm of a male adult. AMBIDEX can be operated at a maximum speed of 5 m/s and is capable of carrying up to 3 kg. Because AMBIDEX can be controlled to the same extent as an industrial robot, it has a wide range of applications, from simple carrying to performing complex tasks that require precise manipulation and collaboration. AMBIDEX supports high-speed, wireless, real time control from remote locations using the low latency and high throughput of 5G networks. Our solution to be presented in this paper gives some innovative solutions that simplify mechanical design, increase robustness and reliability of the system but keeping accuracy at the level of the biological model.

## 3 Mechanical Design of Robotic Arm

The robot arm is designed to fulfill the given technical requirements which are discussed in Sect. 1. The actuators of the robotic arm are displaced from joints to the base of robot to reduce the overall weight of the mechanism. Commonly, mass of servo-motors, corresponding gears and housing (mechanical frame) are the heaviest elements of robot mechanism. If robotic arm represents an integral part of a bimanual robotic system, then servo-motors have to be installed in the base of the system. In this specific case, the robot base is the torso of a bimanual system. Moving of the robot arm, compared to the conventional industrial manipulators where the servo-motors are placed directly in the joints of the mechanism, is carried out in a synchronized way by pulling particular cables made of non-tensile material, e.g. Kevlar. Driving cables are switched by one their end over the motorized pulley mounted on the output shaft of the gear reducer while the second end is attached to the corresponding point on the robot link as shown in Fig. 1a. In the biological world, the tendons as muscle extensors are connected to the appropriate places on the bones. The links of the robot arm, in order to be as light as possible and in the same time to have suitable mechanical strength, are made of pair of the aluminum plates with additional frame (Fig. 1a). The housing Al-plates of the robotic arm are connected and fixed to each-other by mean of the internal frame. The Interior of the robot link is used for passage of the pulling cables, power supply cables and sensorial conductors. From the internal side of the robotic arm (Fig. 1a, details 11) the corresponding conveyor plates are fixed (Fig. 1, detail 9). These conductive plates (called "gates") have the small holes deployed by circumference arranged in two parallel columns/rows at distance of 20 mm between the holes (Fig. 1b). Small holes in the cable-conductive plates allow broaching smooth silicone tubes through which pass pulling cables. The poly articular joints of the robot arm are designed as spherical and two double-cylindrical joints. The joints are realized by means of universal joints. The robotic arm has a redundant kinematical structure with 7 DOFs. The axes of joints 22 and 23 (Fig. 1a) do not match each-other and they are mutually shifted by 20 mm. The axes of joints 22 and 24 are collinear (Fig. 1a).

Driving servo-motors of robot mechanism are placed in the robot base as presented in Fig. 2. DC-motors with planetary gear reducers are arranged in a circle and assembled within the cylindrical housing in order to take a minimal space. The motorized pulley is mounted to the output shaft of servo-motor (Fig. 2). One end of the

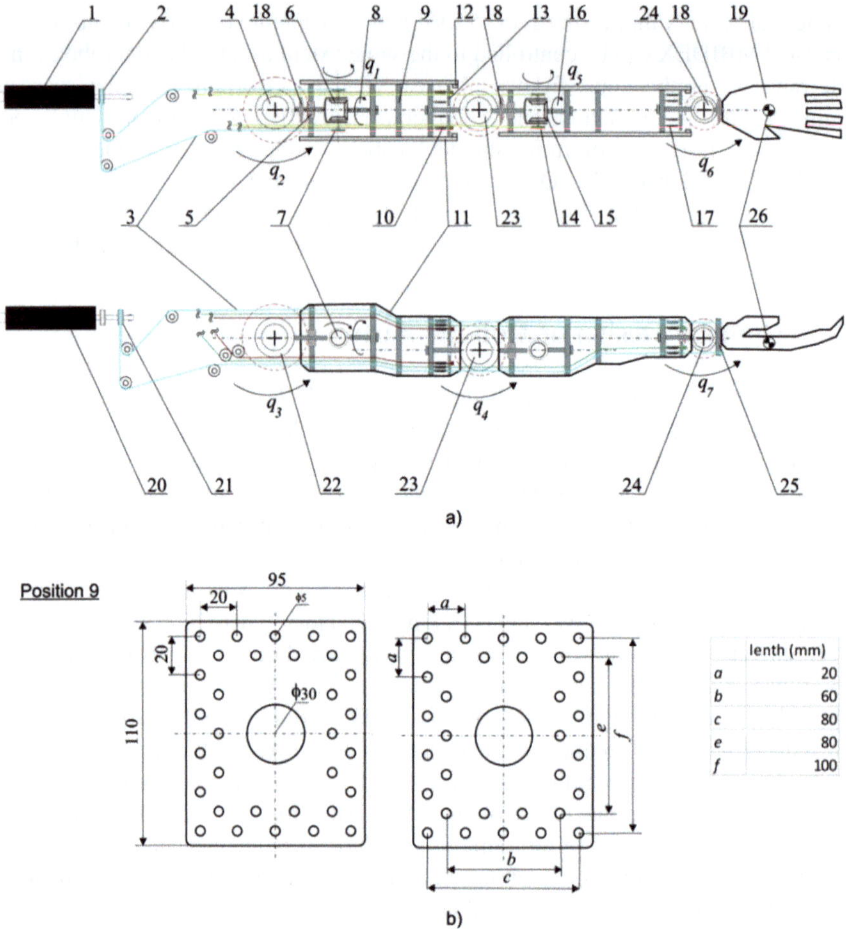


**Fig. 1.** High-level description of the cable-driven, over-actuated robotic arm: (a) mechanism structure and principle of actuation; (b) design of conductive plates for guidance of pulling cables. The meaning of the particular positions in the graphical presentation are: 1-DC-motor; 2-motorized pulley; 3-pulling cables; 4-spherical shoulder joint; 5-ball bearing; 6-bevel gear; 7-drum; 8-upper arm pronation/supination; 9-conductive plate; 10-springs; 11-Al-housing; 12-elastic module limiter; 13-terminal plate of the 1st link; 14-drum; 15-bevel gear; 16-forearm pronation/supination; 17-springs; 18-joint torque sensor (shoulder, elbow, wrist), 19-hand; 20-DC-motor; 21-motorized pulley; 22-shoulder joint; 23-elbow joint; 24-wrist joint; 25-hand flange; 26-hand mass center.

non-tensile cable is wound to the pulley. Another end of the cable is attached to the specific point on the particular robot link. In this way, revolute movement of servo-motors is transmitted to the linear movements of the cable.

## 4 Modeling and Simulation

Mathematical model of robotic arm includes kinematic and dynamic model of the mechanism. Modeling of mechanism was accomplished in the Robotics toolbox [9] for Matlab/Simulink. Parameters of robotic arm were taken from the 3D model of robot mechanism developed in SolidWorks [10]. The lengths of upper arm, forearm and hand

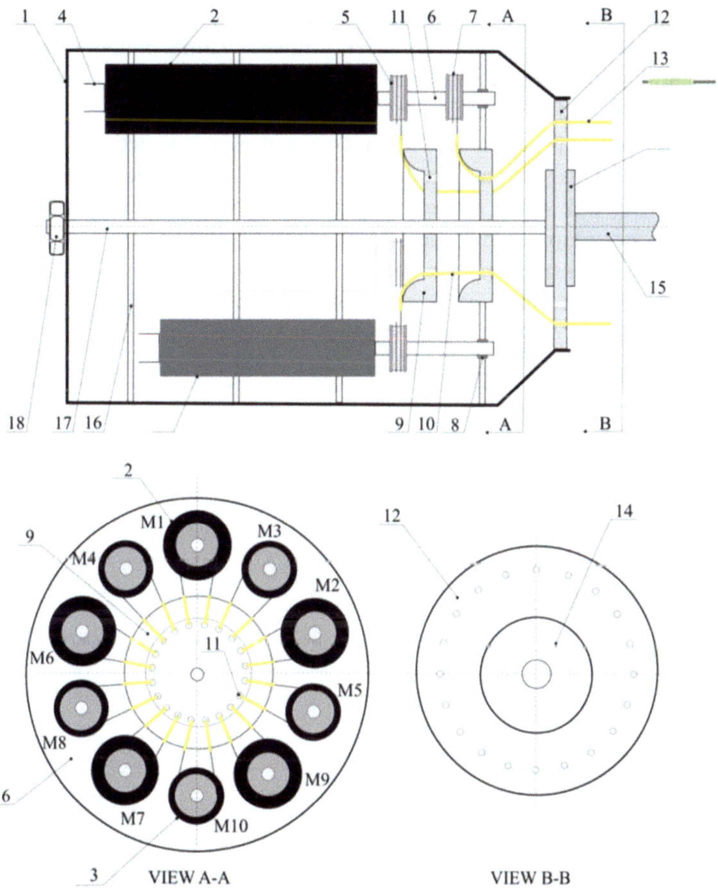


**Fig. 2.** Configuration of the robot power-driving module. Principle of power transmission from servo-motors to robot joints: M1, M2 – upper arm flexion/extension; M3, M4 – forearm flexion/extension; M5 – robot hand flexion/extension; M6, M7 – upper arm abduction/adduction, M8 – robot hand abduction/adduction; M9 – upper arm pronation/supination and M10 – forearm pronation/supination. The meaning of the particular positions in the graphical presentation are: 1-Al-housing, 2-servo-motor of type A; 3-servo-motor of type B; 4-connector terminals; 5-motorized pulley; 6-output shaft (DC-motor and planetary gearhead); 7-non-tensile cables; 8-plain bearing; 9-cable collector and conductor; 10-silicone tube; 11-hole; 12-output conductive plate; 13-cable; 14-flange of the shoulder joint; 15-joint extension; 16-mounting frame; 17-mounting screw; 18-nut.

are: 340, 320 and 210 mm successively. The corresponding masses of robot links, estimated by use of the SolidWorks are: 1847.50, 1172.88 and 838.31 g. Moments of inertia are determined in the SolidWorks, too.

Maximal composite speed of robotic arm 5 m/s and payload of 5 kg was assumed as technical limits and maximal joint torques are calculated for these values. From simulation of robot dynamic model it was established that maximal joint payloads can be overcome by appropriately chosen 10 servo-motors and in the same time to keep minimal space for configuration of the motor group (Fig. 2). The following Faulhaber DC-motors are chosen to fit imposed technical requirements [11]: Series 2657 CR 024 (45 W, 51 mNm), 3272 CR 024 (85 W, 120 mNm). Corresponding planetary gear-heads are chosen from the same producer to satisfy demands on rpm and nominal torque: Series 26A (reduction ratio 64:1) and Series 32A (169:1).

To verify technical performances of the designed over-actuated robotic arm, several simulation experiments were accomplished. Different robot trajectories, in vertical or horizontal plane of task space, different payloads in robot hand and different speeds are imposed to validate robot performances. A characteristic trajectory, presented in Fig. 3, is used to obtain corresponding simulation results shown in Fig. 4 that are used for analysis in the paper. The crucial question with over-actuated systems is how to manage adequate activation of the redundant number of servo-motors (two or more motors drive single degree of freedom).

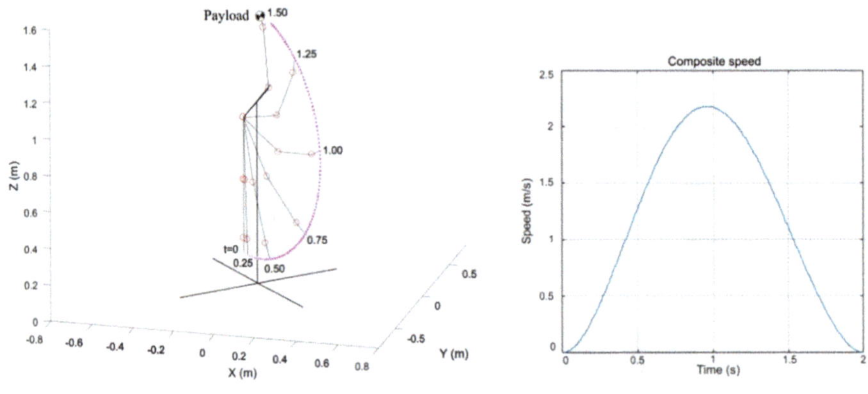


**Fig. 3.** Robot trajectory in the vertical plane with several successive snapshots captured (left side plot); Robot tip speed for robot moving along the trajectory imposed (right side plot).

One possible solution is to activate motors in a selective way where one motor is chosen to be the main drive, while another to be auxiliary. Main motor is subjected to the maximal power-effort while the auxiliary just assists it to complement the missing power or to replace the main actuator in the case of failure. Another solution regards to the uniform distribution of power effort amongst the motors. In this case servo-motors are subjected to the uniform payload. This is more favorable case because it extends the life-cycle of actuators and save energy. The comparison of simulation results for an over-actuated robotic arm, in both considered cases, is given in Fig. 4a and b.

## 5 Conclusions and Future Work

Over actuated robotic system has significant advantages over conventionally actuated (single joint - single motor) robotic systems. The advantages are as follow.

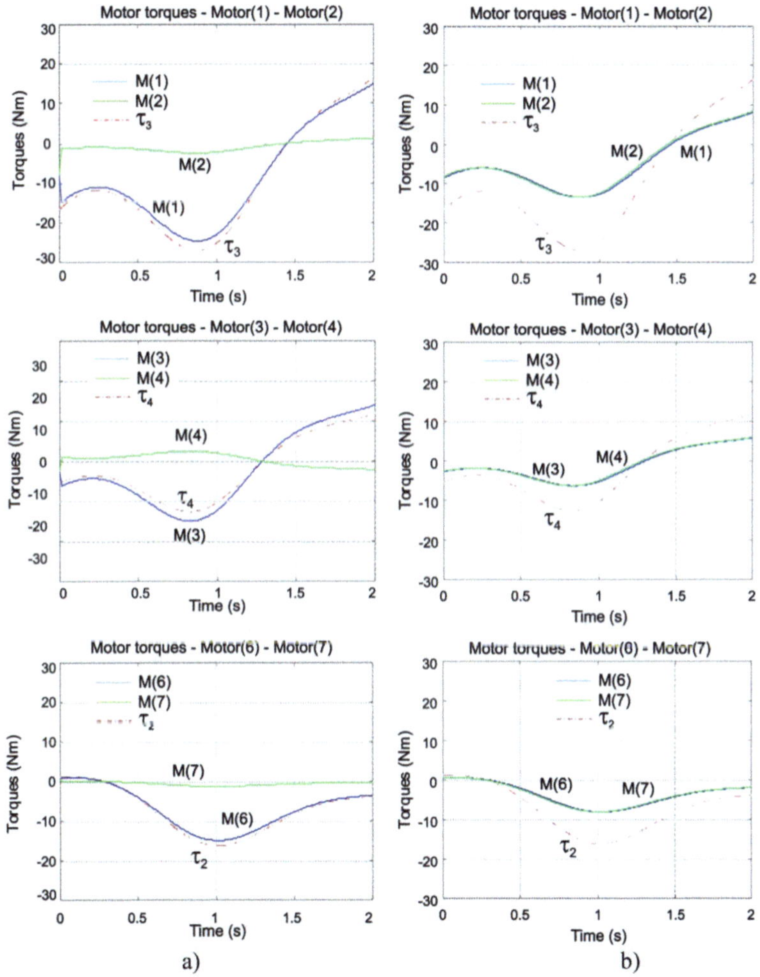


**Fig. 4.** Moments of particular servo-motors that generate corresponding joint torques in the robotic arm during robot motion along the imposed trajectory presented in Fig. 5. Motors M1 and M2 generate joint torque $\tau_3$ for upper arm flexion/extension. Motors M3 and M4 generate joint torque $\tau_4$ for forearm flexion/extension. Motors M6 and M7 generate joint torque $\tau_2$ for upper arm abduction/adduction. Two experimental cases are considered: (a) selective activation of servo motors, and (b) uniform activation of servo-motors.

When an over-actuated system has some failure of particular servo-motor, the remaining drive(s) take over their tasks and system continues to operate even with a reduced technical capacity. The second, various optimization control (intelligent) algorithms, regulating servo-motors load distribution can be applied with over-actuated systems. These intelligent algorithms minimize power consumption, run motors in a nominal regime or just generate uniform torques distribution at the servo-motor output shafts. The third advantage refers to the compact design and savings of the configuration space as presented in Fig. 2. Use of a redundant number of smaller, less powerful servo-motors reduce the necessary space for motor configuration compared to the case of use large, powerful driving units.

**Acknowledgements.** This work was carried out within R&D project "Development and experimental verification of performance of bi-manual robot for collaboration with humans" from the strategic program with PR China under no. 2017–2019. 401-00-00589/2018-09.

## References

1. Rodić, A., Miloradović, B., Popić, S., Urukalo, D.J.: On developing lightweight robot arm of anthropomorphic characteristics. In: Bleuler, H., Pisla, D., Rodic, A., Bouri, M., Mondada, F. (eds.) New Trends in Medical and Service Robots. Mechanisms and Machine Science, vol. 3. Springer, Cham (2015). ISBN: 978-3-319-23831-9, Book ID: 332595 _1_En
2. AMBIDEX Cable-Driven Robot Arm. https://www.naverlabs.com/en/storyDetail/12. Accessed 02 June 2019
3. Yang, G., Lin, W., Kurbanhusen, M.S, Pham, C.B., Yeo, S.H.:. Kinematic design of a 7-DOF cable-driven humanoid arm: a solution-in-nature approach. In: Journal IEEE/ASME International Conference on Advanced Intelligent Mechatronics (AIM) Monterey, CA, pp. 24–28 (2005)
4. Lum, G.Z., Mustafa, S.K., Lim, H.R., Lim, W.B., Yang. G., Yeo, S.H.: Design and motion control of a cable-driven dexterous robotic arm. In: Proceedings of the 2010 IEEE Conference on Sustainable Utilization and Development in Engineering and Technology. Petaling Jaya, Malaysia (2010). https://doi.org/10.1109/student.2010.5686997
5. Lens, T., Kirchhoff, J., von Stryk, O.: Dynamic modeling of elastic tendon actuators with tendon slackening. In: 12th IEEE-RAS International Conference on Humanoid Robots (Humanoids 2012), Osaka, Japan (2012). https://doi.org/10.1109/humanoids.2012.6651608
6. Palli, G., Borghesan, G., Melchiorri, C.: Modeling, identification, and control of tendon-based actuation systems. IEEE Trans. Rob. **28**(2), 277–290 (2012)
7. Rost, A., Verl, A.: The QuadHelix-Drive - an improved rope actuator for robotic applications. In: 2010 IEEE International Conference on Robotics and Automation, 3–7 May 2010, Anchorage, USA (2010). ISSN: 1050-4729, https://doi.org/10.1109/robot.2010.5509764
8. Potkonjak, V., Svetozarevic, B., Jovanović, K., Holland, O.: The puller-follower control of compliant and noncompliant antagonistic tendon drives in robotic systems. Int. J. Adv. Rob. Syst., 143–155 (2011). https://doi.org/10.5772/10690
9. Corke, P.I.: Robotics, Vision & Control. Springer (2017). ISBN 978-3-319-54413-7. Accessed 02 June 2019
10. SolidWorks 3D model. https://www.solidworks.com/category/3d-cad. Accessed 02 June 2019
11. Faulhaber. https://www.faulhaber.com/en/search/products/?tx_solr%5Bpage%5D=2. Accessed 02 June 2019

# Robot Control

# Torque-Based Velocity Control for Safe Human-Humanoid Interaction

Dmitriy Shingarey, Lukas Kaul, and Tamim Asfour(✉)

Karlsruhe Institute of Technology (KIT),
Institute for Anthropomatics and Robotics (IAR),
High Performance Humanoid Technologies (H2T),
Adenauerring 2, 76131 Karlsruhe, Germany
asfour@kit.edu

**Abstract.** Torque-controlled robots are essential for safe human-robot interaction (HRI). In this paper we address the question of how joint level torque control can be implemented and seamlessly integrated into tasks that require precise velocity control. We present a control scheme that takes a desired velocity and torque as input to generate control output driving the joints at this velocity, and simultaneously realizing compliant behavior for safe HRI. We propose a novel method to integrate torque and velocity control into a single controller. The controller consists of an inner torque control loop embedded in an outer velocity control loop, and is hence called *Torque-Based Velocity Control* (TBVC). Experiments demonstrating the performance of the proposed control scheme are carried out on the humanoid ARMAR-6.

**Keywords:** Humanoid robotics · Torque control

## 1 Introduction

Robotic manipulators with serial kinematics, such as the arms of a humanoid robot, typically consist of a chain of rigid elements and individually controllable joints [1]. To achieve a desired behavior or motion of the end-effector, the joint actuators must be controlled precisely and synchronously. Common control modes are position, velocity and torque control [2]. There also exist approaches to combine different control modes, such as torque, position and impedance in a unifying framework (e. g. [3]). However, it is generally not straight-forward to simultaneously control both joint torques and joint velocities in a seamless manner. This is mainly due to implementations of torque and velocity control as two separate control loops that do not take each other's effects into account.

In this paper we focus on the merits of velocity and torque control and propose a novel method to combine them into one integrated joint level control mode for electric actuators. We call this control mode *Torque-Based Velocity Control* (TBVC).

For a wide range of tasks that robot manipulators typically do, such as reaching, grasping and manipulating by following predefined trajectories in a


K. Berns and D. Görges (Eds.): RAAD 2019, AISC 980, pp. 61–68, 2020.
https://doi.org/10.1007/978-3-030-19648-6_8

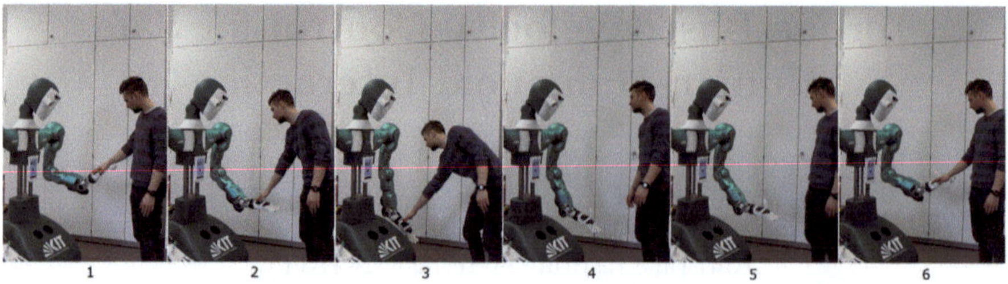


**Fig. 1.** Snapshots of the experiments used for collecting the data presented in Sect. 3. The human initially pushes the elbow into the extended position (1, 2, 3). Upon release, it bends with the desired velocity (4, 5). The human stops the motion through physical contact and the robot applies the desired torque (6).

controlled environment free of unforeseen obstacles, high-gain velocity control is the method of choice for joint-level control. For many of the more complex tasks that modern robots are faced with, such as interacting with soft objects or even humans, velocity control without torque or force control is not suitable. The ability to directly command desired torques to the actuators of a robot is therefore of advantage [4,5]. Joint-level torque control allows for the implementation of mechanical compliance not only at the end-effector but for the whole body [6]. This feature is fundamental when physically interacting with humans, where compliance of the robot facilitates intuitive physical interaction and, most importantly, can limit interaction forces and thereby dramatically increases safety [7].

However torque-control is not straight-forward to use for *trajectory tracking*. Tracking a desired trajectory in torque control mode can be achieved by using inverse dynamics. These methods rely on accurate dynamic models and require perfect torque tracking [8,9]. Another way to achieve trajectory tracking in torque control mode is to use torque-based position tracking, i. e. (active) impedance control schemes [10]. Achieving good position/velocity tracking performance with this method requires proper selection of the impedance parameters. This is generally difficult, partially due to the configuration-dependent dynamics of the robot, and partially due to the effects of the environment's dynamics when in contact [1].

In this paper we propose a torque-based velocity controller (TBVC), a control method that provides the ease-of-use of joint-level velocity control for trajectory tracking combined with the safe interaction characteristics of compliant torque control when in contact with the environment.

## 2 Control Method

The presented control method relies on accurate and reliable torque-control on the joint level. For the demonstrated experiments it furthermore requires model-based feed-forward gravity compensation to achieve feedback linearization,

i. e. the computation of gravity induced torques. This section is therefore split into two main parts: We will first very briefly describe the key components of the actuator and the Low-Level Torque Control (LLTC) with a brief introduction of gravity compensation and active damping for torque controlled robot arms. Secondly we will describe how active damping can be expanded to TBVC.

### 2.1 Low-Level Torque Control (LLTC)

**Hardware and Modeling.** Achieving accurate torque-control on the joint level highly depends on the hardware that is used, since this hardware will affect the modeling and ultimately the controller synthesis. The hardware setup in our case consists of a brushless DC motor, a strain-wave gear with a reduction ratio of 1:160 and a custom, strain gauge based output torque sensor. Current control of the motor is handled by a commercial motor controller. A detailed description of the actuation units is presented in [11].

**Controller.** With standard methods of robust control theory for plant analysis and control synthesis we find that the system can be stabilized with a PI-controller that converts a desired torque to an input current $i_{set}$ and achieves good torque tracking over a wide range of torques and actuator speeds.

**Gravity Compensation.** Using a model of the robot arm, the joint torques that are induced by gravity can be computed by summing over the cross product of the gravitational forces and vectors from the rotational axis of the joint in question to the center of masses of all following links. Applying the so computed torques to the robot joints using joint-level torque control results in a linearization of the torque feedback with respect to the joint position.

**Active Damping.** Using torque control in a kinematic chain where all the joints along the chain influence each other can lead to oscillations even if each joint-level torque controller on its own is stable. To mitigate this problem we add a velocity dependent term to influence the controller set point, which we call active damping. The actual shaft velocity $\omega_{act}$ is multiplied by a positive constant value $k_d$ and the result gets subtracted from the torque set point. Using suitable values for $k_d$ on the joint level suppresses torque-oscillations in the actuators along the kinematic chain but preserves the good torque tracking capabilities of the torque controllers.

### 2.2 Torque-Based Velocity Controller

The idea of torque based velocity control directly arises from the concept of active damping introduced in Sect. 2.1. Active damping introduces a velocity feedback loop around the inner torque-control loop, which allows influencing the speed of the actuator, while the underlying control mode is still torque control

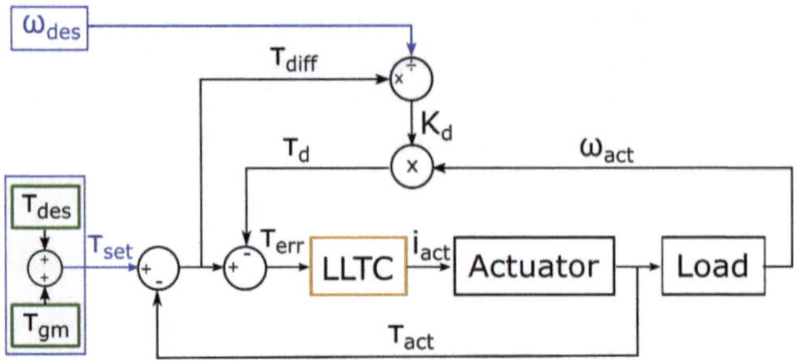


**Fig. 2.** Block diagram of the joint-level torque-based velocity control scheme.

with all its benefits of safe interaction and backdrivability. To achieve a desired velocity, the active damping parameter $k_d$ needs to be set accordingly, effectively implementing a velocity control loop. This turns the single input control system (desired torque) into a dual input system (desired torque $\tau_{des}$, desired velocity $\omega_{des}$).

The underlying idea can be summarized as follows: If the system is at rest and not moving, for example when it is restricted due to physical contact, the desired torques should be acting in the joints. If the system is freely moving with the desired velocity, only the gravity compensating torques should be acting in the joints, resulting in zero acceleration and joint motions with the desired velocities. We propose a mapping between the two states in which the system is (i) not moving at all and therefore applying the desired torques and (ii) moving with the desired velocity and therefore not applying any torque other than what is necessary to compensate for influences of gravity.

**Naive Approach.** From the basic idea, an appropriate $k_d$ can be computed from the two inputs $\tau_{des}$ and $\omega_{des}$ from the following condition

$$\omega_{act} k_d = \tau_{des} \text{ for } \omega_{act} = \omega_{des} \tag{1}$$

as follows:

$$k_d = \frac{\tau_{des}}{\omega_{des}} \tag{2}$$

This naive approach leads to a control scheme where velocity feedback is used to adjust the input of the underlying torque controller. Here, $k_d$ is constant for constant $\tau_{des}$ and $\omega_{des}$. While this control scheme does not interfere with the torque controller, it does not lead to satisfactory results with respect to velocity tracking as unmodelled effects can not be compensated.

**Proposed Approach.** To overcome the velocity control insufficiencies of the naive scheme, we preserve the basic idea but modify the implementation slightly to make the controller more robust against model inaccuracies and unmodeled dynamic effects. For this purpose we replace $\tau_{des}$ in the computation of $k_d$ as shown in Eq. 2 by the difference $\tau_{diff}$ of the set torque (the sum of the model-based gravity compensating torque $\tau_{gm}$ and desired torque $\tau_{des}$) and the measured joint torque $\tau_{act}$. This modified control scheme is depicted in Fig. 2 where $k_d$ is computed as

$$k_d = \frac{\tau_{diff}}{\omega_{des}}. \tag{3}$$

Note that the active damping term $k_d$ now depends on the torque measurement and is therefore constantly changing. Including the torque feedback into the velocity feedback loop in this manner results in the desired robustness.

**Comparison of the Two Approaches.** To illustrate why the proposed approach leads to better results in terms of robustness against modeling errors it is useful to examine the torque error $\tau_{err}$ that will be processed by LLTC in detail. The definition of $\tau_{err}$ is as follows:

$$\tau_{err} = \tau_{set} - (\tau_{act} + k_d\omega_{act}) \tag{4}$$

By substituting Eq. 3 in Eq. 4 and replacing $\tau_{diff}$ with $\tau_{set} - \tau_{act}$, $\tau_{err}$ takes the following form:

$$\tau_{err} = \tau_{set}(1 - \frac{\omega_{act}}{\omega_{des}}) - \tau_{act}(1 - \frac{\omega_{act}}{\omega_{des}}) \tag{5}$$

It can be seen from Eq. 5 that if there is no motion (i. e. $\omega_{act} = 0$), the equation simplifies to $\tau_{err} = \tau_{set} - \tau_{act}$ and only the desired torque will be applied (state I). If $\omega_{act} = \omega_{des}$ the torque error $\tau_{err}$ will be zero (state II).

By definition, the set torque $\tau_{set}$ is the sum of the model based gravity torque $\tau_{gm}$ and desired torque $\tau_{des}$. $\tau_{act}$ can be presented as the sum of the real gravity torque $\tau_{grav}$ and other dynamical torques $\tau_{dyn}$. Combining all of this leads to the error term

$$\tau_{err} = \tau_{des}(1 - \frac{\omega_{act}}{\omega_{des}}) + (\tau_{gm} - \tau_{grav})(1 - \frac{\omega_{act}}{\omega_{des}}) - \tau_{dyn}(1 - \frac{\omega_{act}}{\omega_{des}}) \tag{6}$$

Equation 6 shows that both the effects of the dynamic torques $\tau_{dyn}$ as well as the difference of model-based and actual gravity torques ($\tau_{gm} - \tau_{grav}$) vanish as $\omega_{act}$ reaches $\omega_{des}$.

In contrast, the error term for the naive approach does not exhibit this property as there is no favorable scaling term for the effects of dynamics and model errors:

$$\tau_{err} = \tau_{des}(1 - \frac{\omega_{act}}{\omega_{des}}) + \tau_{gm} - \tau_{grav} - \tau_{dyn} \tag{7}$$

As can be seen from Eq. 7, only the desired torque $\tau_{des}$ will be compensated as $\omega_{act}$ reaches $\omega_{des}$. Dynamical torques as well as the difference between modeled and real gravity will affect the torque error $\tau_{err}$, ultimately resulting in a deviation of actual velocity $\omega_{act}$ from the desired velocity $\omega_{des}$.

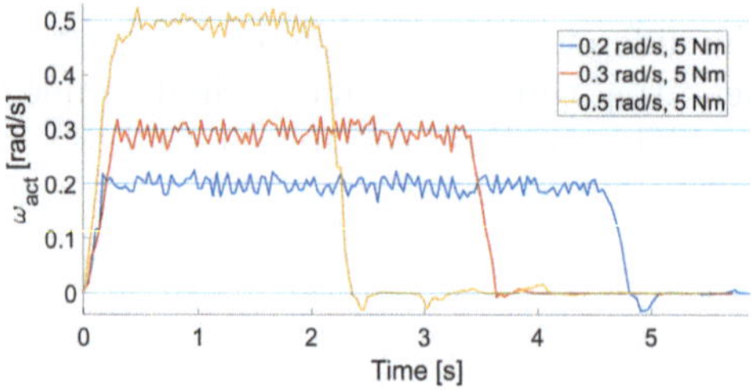


**Fig. 3.** Velocity plots for different target joint velocities at $\tau_{des} = 5\,\mathrm{Nm}$.

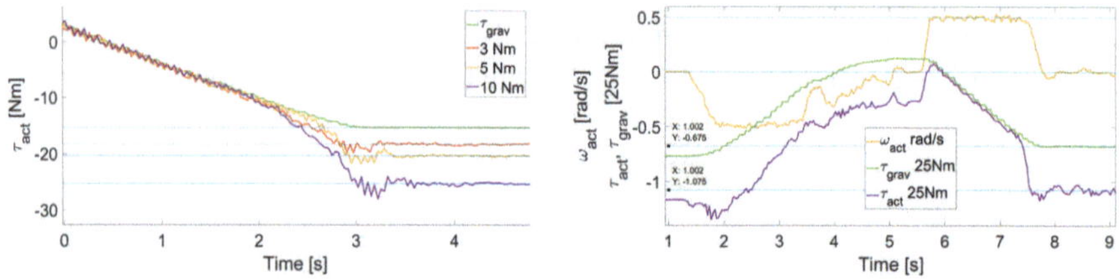


(a) Measured joint torque during three robot experiments with identical desired velocities (0.3 rad/s) and different desired torque values (3, 5 and 10 Nm).

(b) Measured torque (purple), model based computed gravity torque (green) and rotational joint velocity (yellow) over one validation experiment.

**Fig. 4.** Torque data for three validation experiments.

# 3 Experimental Validation

Experimental validation of the proposed control method was carried out on the humanoid robot ARMAR-6 [12] as follows: With the elbow joint in TBVC mode with desired torque $\tau_{des}$ and desired velocity $\omega_{des}$, the left arm was manually extended so that the forearm pointed downwards (1, 2, 3 in Fig. 1). In this extended initial position, the arm was released. The elbow joint then started to bend with the desired angular velocity. Before the joint reached the 90° bent position it was stopped by a human, blocking the robot's forearm with his hand (4, 5, 6 in Fig. 1). During these experiments the joint torque, joint position and joint velocity were recorded.

**Velocity Tracking.** Figure 3 shows the velocity curves for three of these experiments. The plots start when the joint is released in the extended position. The controller then adjusts the joint torque so that the desired velocity (horizontal reference lines) is reached and held throughout. Contact with the human is made and the joint rapidly slows down to zero velocity, exerting the desired torque (together with the gravity compensating torque).

**Torque Tracking.** Figure 4a shows torque curves for three experiments with identical desired velocities but different desired torque. Similar to the plots shown in Fig. 3, the torque curves start at $t = 0$ when the joint is released. During the initial free movement of the arm, control is governed by the velocity control, and the joint torque very closely matches the gravity torque. After impact at about $t = 3$ s and at zero velocity, control is governed by the torque component, and the measured torques are the sum of gravity compensation and the additionally desired torques. The light blue horizontal lines show the desired steady-state torque values with magnitudes of $|\tau_{grav}| + 3\,\text{Nm}$, $|\tau_{grav}| + 5\,\text{Nm}$ and $|\tau_{grav}| + 10\,\text{Nm}$.

**Backdrivability.** The proposed torque control method allows for safe physical human-robot interaction in the sense that the robot stops its motion and only applies a pre-specified torque when contact with a human or any other physical object (see Fig. 4a) occurs. Further, the robot is compliantly backdrivable against its intended direction of motion. The robot is in fact backdriven in the beginning of all of the experiments presented in the previous sections. Figure 4b shows joint torque and velocity over an entire experiment, where the moment of release in the extended position (where the other plots start) is at about $t = 5.8$ s. Once the joint is stopped at about $t = 7.6$ s, the velocity becomes 0 and the magnitude of the actual joint torque becomes $0.4 \times 25\,\text{Nm} = 10\,\text{Nm}$ higher than the torque needed for gravity compensation. 10 Nm was the desired torque during this experiment.

**Practical Considerations.** A few considerations when applying the described control method in practice need to be taken into account to achieve reliable and predictable behavior as described above. In the case of backdriving the joint, $k_d$ might become negative, leading to undesirable robot behavior. We therefore use the absolute values of $k_d$, which in turn necessitates manipulating the sign of $\tau_{des}$ to enable bi-directional joint movement. Since $k_d$ is a ratio, it is not defined for $\omega_{des} = 0$ and becomes very large for $\omega_{des} \approx 0$. In practice, an effective countermeasure is to cap $k_d$ and to adapt the PI-controller gains to changes of the $k_d$, i. e. using a $k_d$-adaptive LLTC.

## 4 Conclusion

We presented a joint-level control scheme for simultaneous joint torque and joint velocity control. The control scheme is derived from the desire to have the joint rotating with a desired speed when freely moving, and applying a specified torque when restricted in its motion by physical contact with its environment, e. g. a human. Since the inner control loop of this control scheme is a torque control loop that is enclosed by a closed velocity control loop, we call this scheme Torque-Based Velocity Control (TBVC). This controller can replace common stiff velocity control modes and enhance safety in human-robot interaction situations.

Evaluation experiments with the elbow joint of the full-sized humanoid robot ARMAR-6 were conducted in which a human safely interacted with the robot. We showed that the TBVC behaves as expected both when freely moving (applying gravity compensating torque and maintaining the desired velocity) and when in contact (applying gravity compensating torque plus desired torque).

## References

1. Siciliano, B., Khatib, O.: Springer Handbook of Robotics. Springer, Heidelberg (2008)
2. Spong, M.W., Hutchinson, S., Vidyasagar, M., et al.: Robot Modeling and Control, vol. 3. Wiley, New York (2006)
3. Albu-Schäffer, A., Ott, C., Hirzinger, G.: A unified passivity-based control framework for position, torque and impedance control of flexible joint robots. Int. J. Rob. Res. **26**(1), 23–39 (2007)
4. Asfour, T., Schill, J., Peters, H., Klas, C., Bucker, J., Sander, C., Schulz, S., Kargov, A., Werner, T., Bartenbach, V.: ARMAR-4: a 63 DOF torque controlled humanoid robot. In: IEEE-RAS International Conference on Humanoid Robots (Humanoids), pp. 390–396 (2013)
5. Englsberger, J., Werner, A., Ott, C., Henze, B., Roa, M.A., Garofalo, G., Burger, R., Beyer, A., Eiberger, O., Schmid, K., et al.: Overview of the torque-controlled humanoid robot TORO. In: 2014 14th IEEE-RAS International Conference on Humanoid Robots (Humanoids), pp. 916–923. IEEE (2014)
6. Hyon, S.H., Hale, J.G., Cheng, G.: Full-body compliant human-humanoid interaction: balancing in the presence of unknown external forces. IEEE Trans. Rob. **23**(5), 884–898 (2007)
7. Albu-Schäffer, A., Haddadin, S., Ott, C., Stemmer, A., Wimböck, T., Hirzinger, G.: The DLR lightweight robot: design and control concepts for robots in human environments. Ind. Rob. Int. J. **34**(5), 376–385 (2007)
8. Sciavicco, L., Siciliano, B.: Modelling and Control of Robot Manipulators. Springer, London (2012)
9. Hutter, M., Remy, C.D., Hoepflinger, M.A., Siegwart, R.: High compliant series elastic actuation for the robotic leg ScarlETH. In: Field Robotics, pp. 507–514. World Scientific (2012)
10. Mosadeghzad, M., Medrano-Cerda, G.A., Saglia, J.A., Tsagarakis, N.G., Caldwell, D.G.: Comparison of various active impedance control approaches, modeling, implementation, passivity, stability and trade-offs. In: 2012 IEEE/ASME International Conference on Advanced Intelligent Mechatronics (AIM), pp. 342–348. IEEE (2012)
11. Rader, S., Kaul, L., Weiner, P., Asfour, T.: Highly integrated sensor-actuator-controller units for modular robot design. In: IEEE International Conference on Advanced Intelligent Mechatronics (AIM), pp. 1160–1166 (2017)
12. Asfour, T., Kaul, L., Wächter, M., Ottenhaus, S., Weiner, P., Rader, S., Grimm, R., Zhou, Y., Grotz, M., Paus, F., Shingarey, D., Haubert, H.: ARMAR-6: A collaborative humanoid robot for industrial environments. In: IEEE/RAS International Conference on Humanoid Robots (Humanoids), pp. 447–454 (2018)

# On an Analytic Generation of Null Space Spanners in Robotics

Ignacy Duleba(✉) and Iwona Karcz-Duleba

Faculty of Electronics, Wroclaw University of Science and Technology, Janiszewski St. 11/17, 50-372 Wroclaw, Poland
{ignacy.duleba,iwona.duleba}@pwr.edu.pl

**Abstract.** In this paper three algorithms are presented to generate automatically bases of null spaces (kernels) for symbolic matrices resulting from some tasks of robotics. Orthogonal (orthonormal) bases of kernels can be derived if required. Additionally, a switching procedure is described to continuously transfer from a currently used basis, which becomes ill-conditioned, to another, well-conditioned basis. The algorithms are illustrated on computing the Jacobian matrix kernel of an exemplary 4DoF planar pendulum manipulator to point out their advantages and disadvantages.

**Keywords:** Null space · Basis · Orthogonality · Analytic form · Robotics

## 1 Introduction

A basis, spanning a null space (kernel) of a given matrix is required in various tasks of robotics. When the matrix contains real numbers only, a standard Gaussian elimination method allows one to generate the basis [5]. The Gram-Schmidt orthogonalization procedure is applicable [2] when an orthogonal (orthonormal) basis is needed. However, a vast majority of matrices encountered in robotics have got elements being configuration-dependent functions. Those matrices appear in kinematics and dynamics, for manipulators, wheeled or free-floating mobile robots, for elastic or solid structures.

For manipulators, the basis of a Jacobian matrix kernel is required in repeatable inverse kinematics [4,7,8], while for mobile wheeled robots or free floating robots the kernel is required to define a driftless control system resulting from nonholonomic constraints [3]. A naive strategy to generate the kernel (i.e. adding some vector-rows composed of undetermined coefficients, formulate and solve equations resulting from orthogonality of added vectors with those from the given matrix and also independent of each other) is not particularly good as the resulting (polynomial) equations are difficult to solve. More reliable procedures are iterative and will be presented later on. The paper is organized as follows. In Sect. 2 the theory and algorithms to generate bases of a kernel for


K. Berns and D. Görges (Eds.): RAAD 2019, AISC 980, pp. 69–76, 2020.
https://doi.org/10.1007/978-3-030-19648-6_9

a given matrix in a symbolic form are presented. In Sect. 3 the algorithms are illustrated on the 4-DoF planar pendulum manipulator. Section 4 concludes the paper.

## 2 Preliminaries and Algorithms Generating Kernels of a Symbolic Matrix

Although all further derivations are quite general, to fix attention on a single task, we will concentrate on generating kernel bases of a Jacobian matrix derived from forward kinematics in robot manipulators. Forward kinematics $\boldsymbol{k}$ maps a configuration $\boldsymbol{q}$ into a point $\boldsymbol{x}$ in a task-space $\mathbb{X}$

$$\mathbb{Q} \ni \boldsymbol{q} \rightarrow \mathbb{X} \ni \boldsymbol{x} = \boldsymbol{k}(\boldsymbol{q}), \tag{1}$$

where $\dim \mathbb{Q} = n \geq m = \dim \mathbb{X}$. Based on kinematics (1), the $(m \times n)$ Jacobian matrix is defined

$$\boldsymbol{J}(\boldsymbol{q}) = \frac{\partial \boldsymbol{k}(\boldsymbol{q})}{\partial \boldsymbol{q}}. \tag{2}$$

It is assumed that Jacobian kernel bases are generated at non-singular configurations, i.e. $\operatorname{rank} \boldsymbol{J}(\boldsymbol{q}) = m$, and only redundant manipulators are considered with a redundancy degree equal to $r = n - m > 0$ as only in this case (and outside singular configurations) the kernel is not trivial. Notice that non-singularity of Jacobian $\boldsymbol{J}$ does not mean that there exists a fixed set of $m$ its columns independent of each other at each configuration. It may happen that the set changes from one configuration to another. Task 1 is to generate analytically any basis of the kernel of the Jacobian matrix $\boldsymbol{J}(\boldsymbol{q})$, i.e. an $(r \times n)$ matrix $\boldsymbol{J}^{\perp}(\boldsymbol{q})$ satisfying

$$\forall \boldsymbol{q} \in \mathbb{Q}: \quad \boldsymbol{J}(\boldsymbol{q})(\boldsymbol{J}^{\perp}(\boldsymbol{q}))^T = \boldsymbol{0}_{m \times r}, \tag{3}$$

where the superscript $'T'$ denotes a transpose matrix. Task 2, being a special instance of Task 1, is to generate an orthogonal (orthonormal) basis valid at each configuration excluding singular ones. Having generated an orthogonal basis $\boldsymbol{J}^{\perp}(\boldsymbol{q})$, it is enough to normalize its rows to get the orthonormal basis.

The first try to solve Task 1 is to split the matrix $\boldsymbol{J}$ into two sub-matrices

$$\boldsymbol{J} = \left[\boldsymbol{J}_m \ \boldsymbol{J}_r\right], \tag{4}$$

where $\boldsymbol{J}_m$ is the $(m \times m)$ matrix, $\operatorname{rank} \boldsymbol{J}_m = m$ and $\boldsymbol{J}_r$ is the $(m \times (r = n - m))$ matrix. To simplify notations, in Eq. (4) it is assumed that the first $m$ columns of $\boldsymbol{J}$ form $\boldsymbol{J}_m$ (by rearranging components of the configuration vector $\boldsymbol{q}$ columns of $\boldsymbol{J}$ can be permuted). In this case, using Eq. (3)

$$\boldsymbol{J}^{\perp} = \left[\boldsymbol{Z}_m \ \boldsymbol{Z}_r\right] \Rightarrow \boldsymbol{J}_m \boldsymbol{Z}_m^T + \boldsymbol{J}^r \boldsymbol{Z}_r^T = \boldsymbol{0} \Rightarrow \boldsymbol{Z}_m^T = -\boldsymbol{J}_m^{-1} \boldsymbol{J}_r \boldsymbol{Z}_r^T, \tag{5}$$

with matrices $\boldsymbol{Z}_m, \boldsymbol{Z}_r$ of appropriate sizes. Assuming $\boldsymbol{Z}_r = \boldsymbol{I}_r$ (the $(r \times r)$ identity matrix), the matrix spanning the null space is as follows

$$\boldsymbol{J}^{\perp} = \left[-\boldsymbol{J}_r^T (\boldsymbol{J}_m^{-1})^T \ \boldsymbol{I}_r\right]. \tag{6}$$

This approach has been presented in [3,6]. Unfortunately, it may happen that although the matrix $\boldsymbol{J}$ is of the full-rank, still a single sub-matrix $\boldsymbol{J}_m$ can not be selected globally, as for the exemplary matrix $\boldsymbol{J} = \begin{bmatrix} 0 & 0 & 1 \\ \cos(\xi) & \sin(\xi) & 0 \end{bmatrix}$. Nevertheless, the idea behind this approach, will be used in Algorithm 1.

The second straightforward method to solve Task 1 is to use an explicit expression for the null space of the Jacobian matrix [9]

$$\boldsymbol{I}_n - \boldsymbol{J}^{\#}(\boldsymbol{q})\boldsymbol{J}(\boldsymbol{q}), \quad \text{where} \quad \boldsymbol{J}^{\#} = \boldsymbol{J}^T(\boldsymbol{J}\boldsymbol{J}^T)^{-1} \tag{7}$$

denotes the pseudo-inverse of $\boldsymbol{J}$. Unfortunately, the method has got its own disadvantages as there are $n$ columns in the matrix (7), but only $r < n$ columns are required and computations (and final expressions) are very complicated due to inverting the manipulability matrix $\boldsymbol{J}\boldsymbol{J}^T$. Therefore other approaches to get the kernel will be searched for.

Later on, to simplify notations, the configuration dependence of matrices will be frequently omitted. An important preliminary step to compute the (orthogonal) basis of $\boldsymbol{J}$ kernel is to transform the Jacobian into the simplest possible form to make it easier to process further. Useful transformations follow:

1. a linear change of coordinates,
2. pre-multiplying Jacobian by a $(m \times m)$ non-singular matrix $\boldsymbol{L}$

$$\tilde{\boldsymbol{J}} = \boldsymbol{L}\boldsymbol{J} \tag{8}$$

   which allows to manipulate with rows (addition, subtraction, scaling). It can be easily verified that $\boldsymbol{J}^{\perp} = \tilde{\boldsymbol{J}}^{\perp}$ as

$$\boldsymbol{L}\boldsymbol{J}(\tilde{\boldsymbol{J}}^{\perp})^T = \boldsymbol{0} \quad \Rightarrow \quad \boldsymbol{J}(\tilde{\boldsymbol{J}}^{\perp})^T = \boldsymbol{0} \quad \Rightarrow \quad \boldsymbol{J}^{\perp} = \tilde{\boldsymbol{J}}^{\perp}, \tag{9}$$

3. post-multiplying Jacobian by a $(n \times n)$ non-singular matrix $\boldsymbol{R}$

$$\tilde{\boldsymbol{J}} = \boldsymbol{J}\boldsymbol{R} \tag{10}$$

   to modify its columns. This time $\boldsymbol{J}^{\perp}(\boldsymbol{q})$ is retrieved from $\tilde{\boldsymbol{J}}^{\perp}(\boldsymbol{q})$ using

$$\boldsymbol{J}\boldsymbol{R}(\tilde{\boldsymbol{J}}^{\perp})^T = \boldsymbol{0} \quad \Rightarrow \quad \tilde{\boldsymbol{J}}^{\perp} = \boldsymbol{J}^{\perp}(\boldsymbol{R}^{-1})^T \quad \Rightarrow \quad \boldsymbol{J}^{\perp} = \tilde{\boldsymbol{J}}^{\perp}\boldsymbol{R}^T. \tag{11}$$

All aforementioned transformations can be applied repeatedly, when the goal is to get a basis of the kernel of $\boldsymbol{J}$. However, when the basis should be orthogonal, only the first two transformations are allowed or some special forms of $\boldsymbol{R}$ (eg. a permutation of columns).

Now, an auxiliary algorithm, Algorithm 1, will be presented to determine a kernel of a given block matrix. This algorithm is a building block (subroutine) for constructing other algorithms an its steps are presented in Listing 1.

Let us notice that the square matrix $\begin{bmatrix} \boldsymbol{A} & \boldsymbol{A}_1 \\ \boldsymbol{B} & \boldsymbol{B}_1 \end{bmatrix}$ is non-singular exactly at those configurations where its sub-matrix $\boldsymbol{A}$ (assuming that $\boldsymbol{A}$ is a configuration-dependent). Consequently, no other singularities besides those of $\boldsymbol{A}$ are introduced by the procedure.

**Algorithm 1.** (basic and general)

**Task:** For a given block matrix $[\boldsymbol{A}\ \boldsymbol{A}_1]$ with $(u \times u)$, $(u \times w)$ sub-matrices, respectively, and non-singular $\boldsymbol{A}$, its kernel is to be determined, i.e. a block matrix $[\boldsymbol{B}\ \boldsymbol{B}_1]$ with $(w \times u)$, $(w \times w)$, sub-matrices, respectively, satisfying

$$[\boldsymbol{A}\ \boldsymbol{A}_1] \begin{bmatrix} \boldsymbol{B}^T \\ \boldsymbol{B}_1^T \end{bmatrix} = \boldsymbol{0}. \tag{12}$$

**Step 1.** Transform (12) into

$$\boldsymbol{B}^T = -\boldsymbol{A}^{-1}\boldsymbol{A}_1\boldsymbol{B}_1^T, \tag{13}$$

**Step 2.** Take $\boldsymbol{B}_1 = \boldsymbol{B}_1^T = \det(\boldsymbol{A})\boldsymbol{I}_w$, where $\boldsymbol{I}_w$ denotes $w$-dimensional identity matrix, to get

$$\boldsymbol{B}^T = -\text{adj}(\boldsymbol{A}) \cdot \boldsymbol{A}_1, \quad \text{where } \text{adj}(\boldsymbol{A}) = \det(\boldsymbol{A}) \cdot \boldsymbol{A}^{-1}. \tag{14}$$

A special form of $\boldsymbol{B}_1$ couples simplicity with avoiding denominators in resulting expressions (when there are no denominators in sub-matrices $\boldsymbol{A}, \boldsymbol{A}_1$, which is a typical case in robotics).

Now, a construction will be presented to prepare the (transformed) Jacobian for processing with Algorithm 1. Let $\mathbb{S}$ denote a set of all $m$ element sub-sets taken from $\mathbb{P} = \{1, \ldots, n\}$. $\mathbb{S}_i$ denotes the $i$-th element of $\mathbb{S}$ while $\overline{\mathbb{S}}_i = \mathbb{P} - \mathbb{S}_i$, its complement. $\boldsymbol{J}_{\mathbb{S}_i}/\boldsymbol{J}_{\overline{\mathbb{S}}_i}$ is the $(m \times m)/(m \times r)$ matrix composed of columns of $\boldsymbol{J}$ with indices from $\mathbb{S}_i/\overline{\mathbb{S}}_i$. A dominating set $\mathbb{S}_i^\star(\boldsymbol{q})$ is selected from the condition

$$|\det(\boldsymbol{J}_{\mathbb{S}_i^\star}(\boldsymbol{q}))| = \max_{\mathbb{S}_i \in \mathbb{S}} |\det(\boldsymbol{J}_{\mathbb{S}_i}(\boldsymbol{q}))|. \tag{15}$$

When two or more elements of $\mathbb{S}$ satisfy (15), any of them is taken. From the $\boldsymbol{J}(\boldsymbol{q})$ full-rank assumption, one has $\forall \boldsymbol{q}\, |\det(\boldsymbol{J}_{\mathbb{S}_i^\star})(\boldsymbol{q})| > 0$. In practice many sets (possibly different at different configurations $\boldsymbol{q}$) $\mathbb{S}_i$ satisfy $|\det(\boldsymbol{J}_{\mathbb{S}_i})(\boldsymbol{q})| > 0$. The purpose of the optimal selection process is to get a set of columns of $\boldsymbol{J}$ well-conditioned as long as possible (as $\boldsymbol{J}(\boldsymbol{q})$ varies continuously with $\boldsymbol{q}$).

Using Algorithm 1, it is easy to calculate a basis of the Jacobian matrix kernel by splitting the Jacobian into $[\boldsymbol{A}, \boldsymbol{A}_1] = [\boldsymbol{J}_{\mathbb{S}_i}, \boldsymbol{J}_{\overline{\mathbb{S}}_i}]$ with $u = m$ and $w = n - m = r$. A little bit harder is to get the orthogonal basis. In this case, iterative Algorithm 2 is applied with steps given in Listing 2.

**Algorithm 2.** (orthogonal kernel of the $(m \times n)$ matrix $\boldsymbol{J}$)

**Step 1.** Currently processed matrix $\boldsymbol{J}_{ext}$ attains its initial value $\boldsymbol{J}_{ext} = \boldsymbol{J}$
**Step 2.** Repeat Steps 3–5 for $i = 1, \ldots, r = n - m$.
**Step 3.** Select $(m + i - 1)$ independent columns of $\boldsymbol{J}_{ext}$; their indices are collected in the set $\mathbb{S}$. Define $\overline{\mathbb{S}} = \{1, \ldots, n\} - \mathbb{S}$.
**Step 4.** Apply Algorithm 1, for a sub-matrix $\boldsymbol{A}$ composed of $\boldsymbol{J}_{ext}$ columns selected based on $\mathbb{S}$, and sub-matrix $\boldsymbol{A}_1$ using selector $\overline{\mathbb{S}}$, to get the matrix $[\boldsymbol{B}\ \boldsymbol{B}_1]$ as an output.
**Step 5.** Select a single row of the latter matrix and append it to the matrix $\boldsymbol{J}_{ext}$.
**Step 6.** Rows appended in Step 5, span the searched orthogonal basis.

Remarks concerning Algorithm 2:

- the basis constructed is orthogonal as in each iteration of Algorithm 2 a vector perpendicular to rows of $\boldsymbol{J}_{ext}$ is generated (not only to rows of the Jacobian $\boldsymbol{J}$), and it contributes to a sub-basis of the kernel,
- In Step 3 there is a freedom in selecting independent columns. It is advised to take columns either described by as simple expressions as possible or to process in parallel many possible sets of columns and to select that one which is the most suitable (similarly to the construction used to determine $\mathbb{S}_i^\star(\boldsymbol{q})$).

Algorithm 3, being a modification of Algorithm 2, is aimed at minimizing computational complexity, as computations are carried out on matrices with smaller sizes than in Algorithm 2. In Algorithm 3 only Steps 3–5, which are different then in Algorithm 2, are presented, cf. Listing 3.

**Algorithm 3.** (orthogonal kernel) faster version of Algorithm 2

**Step 3.** Select $(m+i-1)$ independent columns of $\boldsymbol{J}_{ext}$. Their indices define a set $\mathbb{S}$. From the remaining columns, $\overline{\mathbb{S}} = \{1,\ldots,n\} - \mathbb{S}$, select exactly one denoted as $i^\star$.
**Step 4.** Apply Algorithm 1, for a sub-matrix $\boldsymbol{A}$ composed of columns of $\boldsymbol{J}_{ext}$ indexed with $\mathbb{S}$, and the single $i^\star$-th column sub-matrix $\boldsymbol{A}_1$ to get a single row matrix $[\boldsymbol{B}\ \boldsymbol{B}_1]$ as an output.
**Step 5.** Supplement this row with zeroes at elements with indices $\overline{\mathbb{S}} - i^\star$ and append the resulting row to the matrix $\boldsymbol{J}_{ext}$.

The kernel basis generated with presented algorithms is heavily based on an initial selection of columns of the Jacobian matrix. The selection is aimed at keeping the matrix $\boldsymbol{A}$ (cf. Algorithm 1) nonsingular and well-conditioned, but usually a configuration $\boldsymbol{q}$ varies (as a trajectory in a configuration space is generated). It may happen that at a configuration $\boldsymbol{q}_{\text{switch}}$, a basis $\boldsymbol{C}$, corresponding to the selected non-singular matrix $\boldsymbol{A}$, becomes ill-conditioned and should be switched to another well-conditioned kernel basis matrix $\boldsymbol{D}$. A clear indicator that the switching is necessary is that

$$|\det(\boldsymbol{J}_{\mathbb{S}_i^\star}(\boldsymbol{q}_{\text{switch}}))| < \epsilon \tag{16}$$

where $\epsilon$ is a small parameter. Notice that switching should not be applied when another set of columns becomes better conditioned than $\mathbb{S}_i^\star$ to avoid possible undesirable chattering (frequent changes of the kernel bases). Steps of the switching procedure are presented in Algorithm 4.

## 3 Simulations

Computer calculations are performed for kinematics of the $n$-DoF planar pendula with a two dimensional, positional task-space $(x,y)^T$ given by the expression

$$\begin{bmatrix} x \\ y \end{bmatrix} = \begin{bmatrix} \sum_{i=1}^n l_i c_{1,i} \\ \sum_{i=1}^n l_i s_{1,i} \end{bmatrix}, \tag{18}$$

**Algorithm 4.** To continuously switch from the Jacobian kernel basis $\boldsymbol{C}$ to $\boldsymbol{D}$

**Step 1.** Let two, real-value full-rank orthonormal $(r \times n)$ matrices $\boldsymbol{C}$, $\boldsymbol{D}$ be given (actually bases of a kernel).
**Step 2.** Select a set $\mathbb{S}_r$ of $r$ independent columns of the matrix $\boldsymbol{C}$ to form a square matrix $\boldsymbol{C}(\mathbb{S}_r)$ with a non-zero determinant
**Step 3.** Compute the real-valued rotation matrix $\boldsymbol{R}$ (up to orientation)

$$\boldsymbol{R}\boldsymbol{C}(\mathbb{S}_r) = \boldsymbol{D}(\mathbb{S}_r) \quad \Rightarrow \quad \boldsymbol{R} = \boldsymbol{D}(\mathbb{S}_r)\boldsymbol{C}(\mathbb{S}_r)^{-1}, \tag{17}$$

where $\boldsymbol{D}(\mathbb{S}_r)$ is a square matrix composed of columns $\mathbb{S}_r$ of $\boldsymbol{D}$. In fact $\det(\boldsymbol{R}) = \pm 1$.
**Step 4.** Process later (as $\boldsymbol{q}$ varies) with the Jacobian kernel basis computed as $\boldsymbol{R}^{-1}\boldsymbol{D}(\boldsymbol{q})$ where $\boldsymbol{R}^{-1} = \boldsymbol{C}(\mathbb{S}_r)\boldsymbol{D}(\mathbb{S}_r)^{-1}$.

where $c_{k,j} = \cos(\sum_{i=k}^{j} q_i)$, $s_{k,j} = \sin(\sum_{i=k}^{j} q_i)$, and $l_i$ are link lengths ($c_{k,k} = c_k$, $s_{k,k} = s_k$). The Jacobian matrix is given as

$$\boldsymbol{J}(\boldsymbol{q}) = \begin{bmatrix} -\sum_{i=1}^{n} l_i s_{1,i} & -\sum_{i=2}^{n} l_i s_{1,i} & \dots & -l_n s_{1,n} \\ \sum_{i=1}^{n} l_i c_{1,i} & \sum_{i=2}^{n} l_i c_{1,i} & \dots & l_n c_{1,n} \end{bmatrix}. \tag{19}$$

A compact expression to describe all singular configurations of the $n$-DoF planar pendula

$$\sum_{i=2}^{n} \cos(2(\sum_{j=2}^{i} q_j)) = n, \tag{20}$$

is satisfied for

$$q_i = p_i\,\pi, \quad \text{where } p_i \in \mathbb{N}, \quad i = 2, \dots, n, \quad \text{and } q_1 \text{ coordinate attains any value.} \tag{21}$$

After rotating the Jacobian (19) with the non-singular matrix $\begin{bmatrix} c_1 & -s_1 \\ s_1 & c_1 \end{bmatrix}$ and multiplying the first row by $-1$, the coordinate $q_1$ is excluded from the transformed Jacobian matrix

$$\tilde{\boldsymbol{J}}(\boldsymbol{q}) = \begin{bmatrix} \sum_{i=2}^{n} l_i s_{2,i} & \sum_{i=2}^{n} l_i s_{2,i} & \dots & l_n s_{2,n} \\ l_1 + \sum_{i=2}^{n} l_i c_{2,i} & \sum_{i=2}^{n} l_i c_{2,i} & \dots & l_n c_{2,n} \end{bmatrix}. \tag{22}$$

Later on, the dimension $n = 4$ is selected to keep computations not too extensive but still non-trivial. The first task is to get any basis of the Jacobian kernel. Selecting last two columns of $\tilde{\boldsymbol{J}}(\boldsymbol{q})$, one gets $det(\mathbb{S}_{\{3,4\}}) = l_3 l_4 s_4$. Consequently, when $|s_4| > \epsilon$ with $\epsilon$ a small parameter, the kernel is described by the matrix

$$\boldsymbol{J}^{\perp}(\boldsymbol{q}) = \begin{bmatrix} -l_3 l_4 s_4 & 0 & l_4(l_3 s_4 + l_2 s_{3,4} + l_1 s_{2,4}) & -l_2(l_3 s_3 + l_4 s_{3,4}) - l_1(l_3 s_{2,3} + l_4 s_{2,4}) \\ 0 & -l_3 l_4 s_4 & l_4(l_3 s_4 + l_2 s_{3,4}) & -l_2(l_3 s_3 + l_4 s_{3,4}) \end{bmatrix} \tag{23}$$

A few interesting observations can be made:

- generally, at a random configuration, the rows of $\boldsymbol{J}^{\perp}$ are not perpendicular to each other,
- for $q_4 \in \{0, \pi\}$ the rows become dependent,
- for $q_3, q_4 \in \{0, \pi\}$, the last row vanishes,
- at singular configurations, cf. (21), the rows vanish simultaneously,
- when two first columns of $\tilde{\boldsymbol{J}}(\boldsymbol{q})$ in (22) are selected, $\det(\mathbb{S}_{\{1,2\}}) = -l_1(l_2 s_2 + l_3 s_{2,3} + l_4 s_{2,4})$ and, generally, a set of configurations where kernel bases corresponding to selection $\{1, 2\}$ and $\{3, 4\}$ are degenerated simultaneously is relatively small. Therefore switching between kernel bases, corresponding to different selection sets, is possible and useful.

The second task is to generate an orthogonal basis of the null-space of $\tilde{\boldsymbol{J}}(\boldsymbol{q})$ given in (22) when columns $\{3, 4\}$ are selected at the beginning of Algorithm 3. The second row of matrix (23) is the first of the orthogonal $\boldsymbol{J}^{\perp}$. The second row is too extensive to be presented, therefore its 4-th element only is given below

$$\begin{aligned} &l_4\{l_2 s_2(-l_3(-l_1 + l_3 c_{2,3} + l_4 c_{2,4}) s_4 + l_1 l_2 s_{3,4}) \\ &+ (l_3(2l_1 + l_2 c_2) s_4 + l_1 l_2 s_{3,4})(l_3 s_{2,3} + l_4 s_{2,4})\}. \end{aligned} \tag{24}$$

To illustrate the switching procedure, the configuration $\boldsymbol{q} = (45°, 45°, 0°, 0.1°)^T$ is selected and all link lengths of the pendulum are set to $l_i = 1$. At this configuration, the kernel matrix $\boldsymbol{C}$ corresponding to the selection set $(3, 4)$ becomes ill-conditioned (the subscript $n$ denotes row-normalized versions of the matrices)

$$\boldsymbol{C} = \begin{bmatrix} 0 & -0.0017 & 0.0035 & -0.0017 \\ 0.000018 & -0.0050 & 0.0025 & 0.010 \end{bmatrix}, \quad \boldsymbol{C}_n = \begin{bmatrix} 0 & -0.41 & 0.82 & -0.41 \\ 0.0016 & -0.44 & 0.22 & 0.87 \end{bmatrix}, \tag{25}$$

and it has to be replaced by the selection set $(1, 2)$ generating the kernel matrix $\boldsymbol{B}$

$$\boldsymbol{D} = \begin{bmatrix} -0.0017 & 1.41 & -2.12 & 0 \\ 0.010 & -1.51 & -1.00 & 6.51 \end{bmatrix}, \quad \boldsymbol{D}_n = \begin{bmatrix} -0.00068 & 0.56 & -0.83 & 0 \\ 0.0015 & -0.22 & -0.15 & 0.96 \end{bmatrix}. \tag{26}$$

The rotation matrix $\boldsymbol{R}$, computed according to Eq. (17), is equal to

$$\boldsymbol{R} = \begin{bmatrix} -0.91 & -0.42 \\ -0.42 & 0.91 \end{bmatrix}, \quad \boldsymbol{C}_n = \boldsymbol{R}^{-1} \boldsymbol{D}_n. \tag{27}$$

and its determinant is equal to $\det(\boldsymbol{R}) = -1$.

It should be eventually mentioned that a computational algebra software (like Mathematica function `NullSpace[]` [1]) has got build-in procedures to compute a kernel analytically. However, the presented algorithms generate more readable expressions for kernel bases and give some flexibility is selecting various parameters.

## 4 Conclusion

In this paper a few algorithms have been presented to generate analytic bases of matrices of robotics outside singular configurations. The construction of kernel bases strongly depends on selecting some columns of the matrices. As the columns are configuration-dependent, some extra singular configurations can be introduced by the selection process. In order to retrieve well-conditioning of the basis, switching between bases corresponding to different sets of columns is advised. Simulations, carried out on the model of the 4-DoF planar pendulum, illustrated usefulness of the proposed algorithms.

## References

1. Abell, M.L., Braselton, J.P.: Mathematica by Example, 5th edn. Academic Press, Amsterdam (2017)
2. Björck, A.: Numerics of Gram-Schmidt orthogonalization. Linear Algebra Appl. **197–198**, 297–316 (1994)
3. Duleba, I.: Algorithms of Motion Planning for Nonholonomic Robots. Wroclaw Univ. of Science and Technology Publ., House, Wrocław (1998)
4. Duleba, I., Karcz-Duleba, I.: A comparison of methods solving repeatable inverse kinematics for robot manipulators. Arch. Control Sci. **28**, 5–18 (2018)
5. Kincaid, D., Cheney, W.: Numerical Analysis: Mathematics of Scientific Computing, 2nd edn. Brooks/Cole Publishing Co., Pacific Grove (1996)
6. Ott, C.: Cartesian Impedance Control of Redundant and Flexible-Joint Robots. Springer Tracts in Advanced Robotics. Springer, Heidelberg (2008)
7. Ratajczak, J.E., Tchon, K., Janiak, M.: Approximation of Jacobian inverse kinematics algorithms: differential geometric vs. variational approach. J. Intell. Rob. Syst. **68**(3/4), 211–224 (2012)
8. Roberts, R., Maciejewski, A.A.: Repeatable generalized inverse control strategies for kinematically redundant manipulators. IEEE Trans. Autom. Control **38**, 689–699 (1993)
9. Spong, M., Vidyasagar, M.: Robot Dynamics and Control. Wiley, New York (1989)

# Velocity Motion Path Control of Redundant Robot Arms

Evgeniy Krastev(✉)

Faculty of Mathematics and Informatics,
St. Kliment Ohridski University of Sofia,
5 James Bourchier Blvd., 1164 Sofia, Bulgaria
eck@fmi.uni-sofia.bg

**Abstract.** This paper proposes a novel, integrated method for kinematic motion control of a redundant robot arm subject to a prescribed velocity profile of the end effector in task space. Unlike most of the existing solutions this method assumes that the velocity profile of motion is defined in task space and not in joint space. It can be seamlessly extended to support other optimization criteria for the redundant robot arm motion. The capability to optimize the manipulability measure of the robot motion in addition to following a prescribed velocity profile in task space is another important novelty of the proposed solution. The mathematical model provides efficient software implementation. It is tested in different scenarios for kinematic path control of well-known in the literature reference kinematic models of industrial robots. The obtained numerical results allow to compare and evaluate the quality of the obtained solutions with respect to the existing ones. The obtained results can be extended to computing the joint motion of a non-redundant robot arm for a given velocity profile of the end-effector motion.

**Keywords:** Redundant robot arm · Velocity profile · Kinematics · Robotic toolbox · Motion path control · Manipulability measure

## 1 Introduction

Robot arms are classified by their mechanical structure as articulated robots because they have at least three rotational joints. Industrial robot arms usually have a greater number $n$ of joints allowing the end-effector to perform tasks with flexibility typical for a human hand. Maintaining a greater value of the manipulability measure [1] during a task execution enables greater flexibility in avoiding collision with obstacles and maximizing the distance from kinematic singularities [2, 3]. The manipulability measure can be increased by adding redundant degrees of freedom to the kinematic scheme of the robot arm with respect to the description of the desired end-effector movement in task space. This allows to satisfy basic requirements for flexibility in the execution of a wide range of applications of robots in industry such as welding, painting or operating manipulators on a mobile platform. The robot arm has redundant degrees of freedom for a given task when the number $n$ of joints is greater that the number $m, m \leq 6$ of the parameters selected to define the linear and angular positions of the end-effector in task space.


K. Berns and D. Görges (Eds.): RAAD 2019, AISC 980, pp. 77–85, 2020.
https://doi.org/10.1007/978-3-030-19648-6_10

From a human's perspective it is more convenient to represent a work operation of the robot by means of the parameters describing the task space. Repetitive operations are defined by the geometric path of the operation and the velocity profile of the end-effector motion along this path. For example, the seam of a welding operation can be modeled as a curve in a 4- dimensional task space, where the movement of the welding torch along the direction of the seam is specified by three linear position parameters and one angular position parameter. The ability of a redundant robot arm to reconfigure like a human hand is essential for achieving the necessary quality of welding [4, 5]. In applications like this one it is very important to follow a smooth velocity profile during the end-effector motion.

Unlike the human's view of a work operation existing literature sources most often consider robot motion control in joint space, where mathematical methods for trajectory planning [6, 7], iterative learning control algorithms [8, 9] or optimization of redundant robot arm motion can be directly implemented [10, 11]. A drawback of this approach is that the computed motion in the joints of redundant robot arm [12] not necessarily results in the desired velocity profile of the tool in the end-effector. Additionally, it makes it difficult to optimize the redundant robot arm motion in terms of flexibility or other criteria subject to a given velocity profile in task space.

This paper presents a numerically efficient solution to the problem for kinematic control of a redundant robot arm satisfying a velocity profile along a preplanned path in task space. The solution is obtained in the framework of a mathematical model for kinematic path control presented in related research papers [13]. It employs intuitive vector space methods [14] to build a uniform geometrical representation for motion control both in task space and in an extended space of joint configurations. It allows to introduce kinematic motion control in both spaces independently from the fixed geometrical objects over which the motion takes place. The kinematic motion control serves to optimize the redundant robot arm motion. Here we obtain motion control that enables a redundant robot motion to satisfy a given velocity profile of the tool in the end-effector along a given path in task space. The kinematic control can also adjust the manipulability measure, while preserving the given velocity profile.

This paper is organized as follows. In the following section we shortly introduce the mathematical model for kinematic path control and formulate the problem statement. The new results in applying the proposed solution are presented in Sects. 3 and 4. The kinematic path control that enables a redundant robot arm to follow a prescribed geometric path with a given velocity profile is obtained in Sect. 3. Numerical experiments with the Robotics Toolbox [12] are discussed in Sect. 4.

## 2 Problem Statement

Denote by $n$ the degrees of freedom of a robot arm, where $Int\, Q_M = \{q = (q_1, \ldots, q_n) : \; q_i \in [a_i, b_i] \quad \text{for } i = 1, 2, 3, \ldots, n\}$ represents the interior of the set of the robot arm configurations. Let's assume the robot arm must execute a task described in terms of a path $\gamma$ represented parametrically in a $m(m \leq 6)$ dimensional

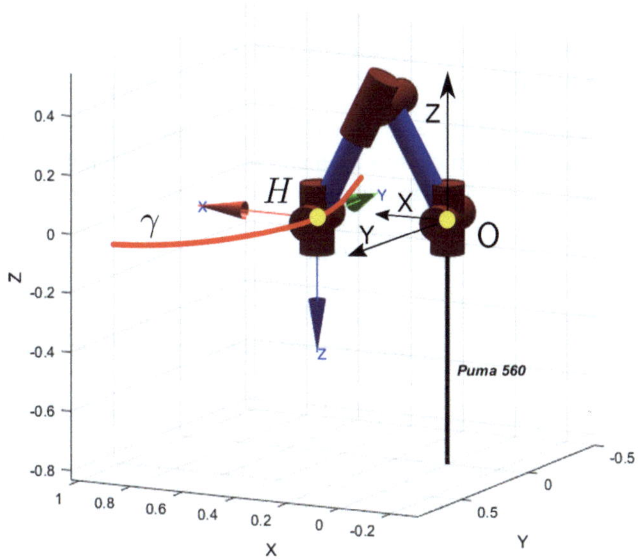


**Fig. 1.** Representation of a work operation in task space.

task space, where the robot arm is redundant ($n > m$) with respect to the assigned operation (Fig. 1).[1]

$$\gamma : \Lambda \to R^m, \quad \Lambda = [\lambda_1,\ \lambda_2]. \tag{1}$$

The coordinates in task space represent the position of a characteristic point $H$ of the end-effector and the orientation of a coordinate system $Hxyz$ (roll-pitch-yaw representation) fixed to the end-effector with respect to the base frame Oxyz. It is natural to assume that $\gamma(\Lambda) \subset W \subset R^m$, where $W$ is the workspace of the robot arm determined by the forward kinematics $F : Int\, Q_M \to W$. This guarantees that the end-effector can track the points on the path γ during the task execution and in this case we say that the path γ is feasible for the robot arm.

Apparently, the motion of the end-effector over a given geometric path γ can be accomplished in many ways. Note that the geometric path and the motion over γ represent two different types of concerns and they are treated separately from each other.

Let the motion required for the selected work operation is defined by the time dependency $\lambda = \lambda(t), \mathrm{t} \in [t_0, t_1]$, where $\gamma = \gamma(\lambda)$ is a parametric representation of $\gamma$ for $\lambda \in [\lambda_1, \lambda_2]$. Since the path $\gamma$ is feasible then there exists a robot arm motion

$$q(t) = (q_1(t), \ldots, q_n(t)) : \mathrm{t} \in [t_0, t_1] \tag{2}$$

such that the effector moves along the path γ subject to $\lambda = \lambda(t), \mathrm{t} \in [t_0, t_1]$.

[1] The figures in this paper make use of the Robotics Toolbox for MATLAB [12].

Both the motion $q(t)$ in the joints and the motion $\lambda(t)$ of the end effector concern the robot arm motion as whole. We address this area of interest by considering $q(t)$ and $\lambda(t)$ in a common vector space.

**Definition.** The set $\mathrm{Q} = Int\ Q_M \times R$ we call extended space of configurations.

Thus, the robot arm motion obtains a general form of representation in Q by means of $q^*(t) = (q(t), q_{n+1}(t)), t \in [t_0, t_1]$, where for uniformity we denote

$$q_{n+1}(t) = \lambda(t) \tag{3}$$

Apparently, all the robot arm motions $q^*(t)$ for which the end-effector tracks the path $\gamma$ belong to the $n - m + 1$ dimensional smooth manifold $\mathcal{B}$ defined as follows:

$$\mathcal{B} = \{q^* = (q, q_{n+1}) : F(q) - \gamma(q_{n+1}) = 0, \quad q^* \in \mathrm{Q}\} \tag{4}$$

Hence the robot arm motion $q^*(t)$ executes over the geometric object $\mathcal{B}$ in a similar way the end-effector motion executes over the geometric path $\gamma$. Then by controlling the vector field $\hat{\boldsymbol{U}}_t = (\boldsymbol{u}^T, u_{n+1})^T : \mathcal{B} \to R^{n+1}$ over $\mathcal{B}$ we can control the quality of the robot motion. Assume that Jacobian matrix $J(q) = D(F)/D(q)$ of the forward kinematics $F(q)$ has full rank and denote by $P = E - J^+J$ the projector $J(q)$ into the Null space of $J(q)$, $E$ is the unity $nxn$ matrix and $J^+$ is the Moore-Penrose pseudoinverse matrix of the Jacobian $J(q)$. Then the motions of a redundant robot arm over $\mathcal{B}$ in (4) satisfy the following differential equation [13]:

$$\dot{\boldsymbol{q}}^* = K\hat{\boldsymbol{U}}_t, t \in [t_o, t_1] \tag{5}$$

where $K$ is a $(n+1)x(n+1)$ block matrix in the form

$$K = \begin{bmatrix} P & \xi \\ 0 & 1 \end{bmatrix} \tag{6}$$

and the $n x 1$ vector $\xi$ in (5) is defined as follows:

$$\xi = J^+ \frac{\partial \gamma}{\partial(\boldsymbol{q}_{n+1})} \tag{7}$$

Note that the component $\xi u_{n+1}$ is responsible for moving the end-effector of the robot arm from one point to another point on path $\gamma$, while the component $\boldsymbol{Pu}$ determines the degree of reconfiguration of the robot arm. In the following section we use these properties of $\xi u_{n+1}$ and $\boldsymbol{Pu}$ to obtain a robot arm motion tracking a desired velocity profile over the geometric path $\gamma$ and optimize the manipulability measure.

## 3 Velocity Profile in Task Space

The velocity profile is an important motion characteristic for a wide range of industrial applications like welding and painting. It is used to determine the size of the linear and angular speed of the end-effector motion at each moment of time. Most often the velocity profiles are trapezoidal, where we can identify a sequence of acceleration, steady motion and deceleration areas. In order to avoid jerks the edges of transition between these areas are usually approximated by smooth curves [5]. Without loss of generality we assume that $\frac{\partial\gamma}{\partial(q_{n+1})}$ in (7) is a unit vector tangent to $\gamma$ at $q_{n+1}$. Then the size of this vector will be determined by the value of the function $u_{n+1} = u_{n+1}(t)$. A sample velocity profile is displayed on Fig. 2.

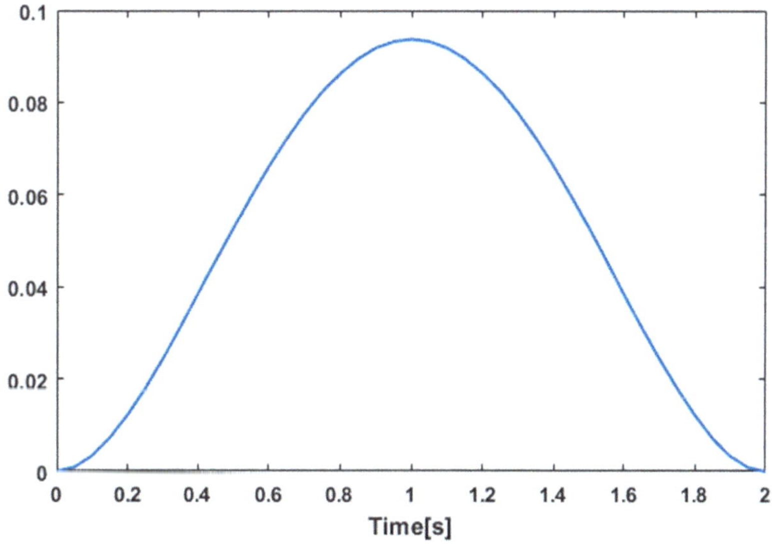


**Fig. 2.** Sample velocity profile.

This velocity profile makes use of a fifth order polynomial to approximate the edges of transition in the trapezoidal profile [12]. Taking in consideration (3) and (7) we can interpret $u_{n+1}$ as $u_{n+1} = \dot{\lambda}(t), t \in [t_o, t_1]$. This way by denoting the zero vector as $\boldsymbol{o}^T$ in $\hat{\boldsymbol{U}}_t = \left(\boldsymbol{o}^T, \dot{\lambda}(t)\right)^T$ we can compute from (5) the robot motion producing a given velocity profile of motion in task space and the respective $\lambda = \lambda(t)$. In the general case the first $n$ components of $\hat{\boldsymbol{U}}_t$ can be employed to optimize the manipulability measure, while satisfying the given velocity profile motion. We include this case in consideration in the following section.

## 4 Numerical Experiments of the Proposed Solution

The proposed method for velocity motion control has been tested to solve several tasks of varying difficulty with the Puma 560 kinematic model making use of the Robotics Toolbox for MATLAB. This robot has $n = 6$ rotational joints with Denavit-Hartenberg parameters displayed in Table 1. The initial configuration [3.2631, 2.0791, 0.5992, 0.0000, 0.4633, 0.1215] of the robot arm is displayed on Fig. 1, where the coordinates of point $H$ are [0.4, 0.2, 0]. Here we consider a work task where the point $H$ must track a segment of a straight line with endpoints [0.4, 0.2, 0] and [0.5, −0.3, 0] for $t \in [0, 2]$, where the $X$ axis of $Hxyz$ rotates by $\frac{\pi}{2}[rad]$ during that motion. The required velocity profile of end-effector motion along the points of this line is displayed on Fig. 2. The task parameters in this problem setting comprise the coordinates of $H$ and the angle of rotation around the axis $X$ of $Hxyz$, $(m = 4)$. Therefore, the robot arm will execute this task with two redundant degrees of freedom.

**Table 1.** Denavit-Hartenberg parameters for Puma 560.

| Joint [rad] | $d_i$[m] | $a_i$[m] | $\alpha_i$[rad] | $\theta_i$ [rad] |
|---|---|---|---|---|
| $q_1$ | 0 | 0 | 1.5708 | 0 |
| $q_2$ | 0 | 0.4318 | 0 | 0 |
| $q_3$ | 0.15005 | 0.0203 | −1.5708 | 0 |
| $q_4$ | 0.4318 | 0 | 1.5708 | 0 |
| $q_5$ | 0 | 0 | −1.5708 | 0 |
| $q_6$ | 0 | 0 | 0 | 0 |

The solution of (5) for $\hat{\boldsymbol{U}}_t = \left(\boldsymbol{o}^T, \dot{\lambda}(t)\right)^T$ yields the motion in joint space (Fig. 3) corresponding to motion with the given velocity profile in task space (Fig. 4). Note that the $x$ coordinate on Fig. 4 increases from 0.4 to 0.5 and the $y$ coordinate decreases from 0.2 to −0.3. Both coordinates change smoothly in accordance with the selected velocity profile. It is also interesting to investigate how the manipulability measure changes during the robot motion subject to different values of the first $n$ components of the kinematic control $\hat{\boldsymbol{U}}_t$. The results displayed on Fig. 5 show that for one and the same velocity profile the manipulability of a redundant robot arm can be optimized by properly selecting the first $n$ components of the kinematic control $\hat{\boldsymbol{U}}_t$.

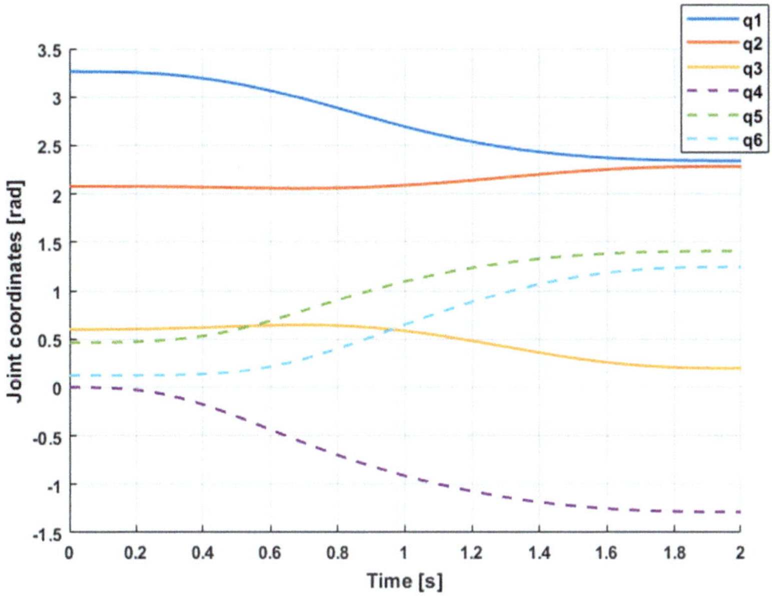


**Fig. 3.** Motion in joint space.

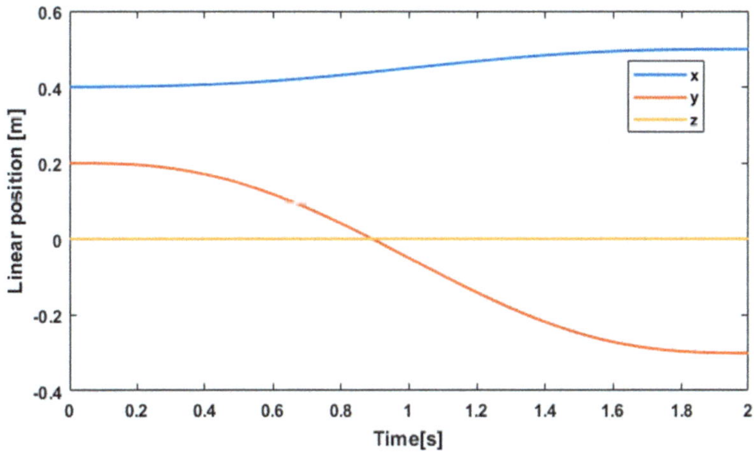


**Fig. 4.** Motion in task space.

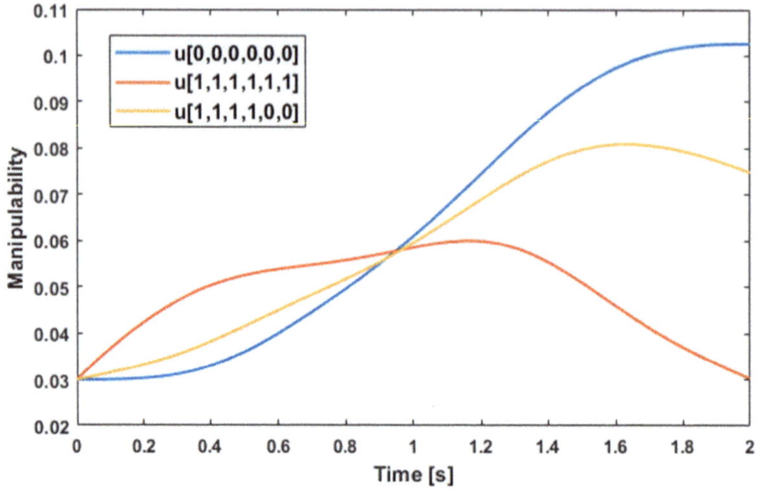


**Fig. 5.** Optimization of the manipulability measure.

## 5 Conclusion

This paper proposes a novel, integrated solution for kinematic motion control of a redundant robot arm subject to a prescribed velocity profile of the end effector in task space. Unlike most of the existing solutions this method assumes that the velocity profile of motion is defined in task space and not in joint space. It can be seamlessly extended to support other optimization criteria for the redundant robot arm motion.

The capability to optimize the manipulability measure of the robot motion in addition to following a prescribed velocity profile in task space is another important novelty of the proposed solution. The mathematical model provides efficient software implementation. It is tested in different scenarios for kinematic path control of well-known in the literature reference kinematic models of industrial robots. The obtained numerical results allow to compare and evaluate the quality of the obtained solutions with respect to the existing ones. The obtained results can be extended to computing the joint motion of a non-redundant robot arm for a given velocity profile of the end-effector motion. The computation of the initial configuration with the greatest manipulability measure and computing the kinematic control that maximizes the manipulability measure in the velocity profile problem are subject of separate research.

**Acknowledgement.** This work is supported by the Fund for Scientific Research at Sofia University "St. Kliment Ohridski".

## References


1. Yoshikawa, T.: Manipulability of robotic mechanism. Int. J. Robot. Res. **4**(2), 3–9 (1985)
2. De Luca, A.: Robots with kinematic redundancy. Robotics II Department of Computer Control and Management Engineering (2013)
3. Siciliano, B., Khatib, O.: Springer Handbook of Robotics, 2nd edn. Springer, Cham (2016)
4. International Federation of Robotics: Korte Friesland takes the next step with its new welding robot system, 23 May 2017. https://ifr.org/case-studies/industrial-robots/korte-friesland-takes-the-next-step-with-its-new-welding-robot-system. Accessed 22 Jan 2019
5. Zhao, R.: Trajectory planning and control for robot manipulations. Doctoral dissertation, Université Paul Sabatier-Toulouse III (2015)
6. Spong, M.W., Vidyasagar, M.: Robot Dynamics and Control, 2nd edn. Wiley India Pvt. Limited, New York (2004)
7. Zlajpah, L., Petri, T.: Virtual guides for redundant robots using admittance control for path tracking tasks. In: Advances in Service and Industrial Robotics, vol. 67, pp. 13–23. Springer (2019)
8. Yovchev, K.: Iterative learning control for precise trajectory tracking within a constrained workspace. In: Advances in Service and Industrial Robotics, vol. 67, pp. 483–492. Springer (2019)
9. Yovchev, K.: Finding the optimal parameters for robotic manipulator applications of the bounded error algorithm for iterative learning control. J. Theoret. Appl. Mech. **47**(4), 3–11 (2017)
10. Flacco, F., De Luca, A., Khatib, O.: Control of redundant robots under hard joint constraints: saturation in the null space. IEEE Trans. Rob. **31**(3), 637–654 (2015)
11. Valsamos, C., Wolniakowski, A., Miatliuk, K., Moulianitis, V.C.: Minimization of joint velocities during the execution of a robotic task by a 6 D.o.F. articulated manipulator. In: Advances in Service and Industrial Robotics, vol. 67, pp. 368–375. Springer (2019)
12. Corke, P.: Robotics, Vision and Control. Fundamental Algorithms in MATLAB, 2nd edn. Springer, Cham (2017)
13. Krastev, E.: Mathematical model for motion control of redundant robot arms. WSEAS Trans. Syst. **16**, 36–42 (2017)
14. Calafiore, G., Ghaoui, L.E.: Optimization Models. Cambridge University Press, Cambridge (2014)

# Reduction of Trajectory Encoding Data Using a Deep Autoencoder Network: Robotic Throwing

Zvezdan Lončarević(✉), Rok Pahič, Mihael Simonič, Aleš Ude, and Andrej Gams

Jožef Stefan Institute, Jamova cesta 39, 1000 Ljubljana, Slovenia
zvezdan.loncarevic@ijs.si

**Abstract.** Autonomous learning and adaptation of robotic trajectories by complex robots in unstructured environments, for example with the use of reinforcement learning, very quickly encounters problems where the dimensionality of the search space is beyond the range of practical use. Different methods of reducing the dimensionality have been proposed in the literature. In this paper we explore the use of deep autoencoders, where the dimensionality of autoencoder latent space is low. However, a database of actions is required to train a deep autoencoder network. The paper presents a study on the number of required database samples in order to achieve dimensionality reduction without much loss of information.

**Keywords:** Deep autoencoder · Reinforcement learning · Robotic throwing

## 1 Introduction and Related Work

Learning and adaptation of robotic skills has been researched in depth over the last decade. However, as [11] reports, the search space of learning of complete actions/skills from scratch is too large, making learning from scratch unfeasible.

Several methods have been proposed to reduce the size of the learning space. For example, the search space was reduced with accurate initial demonstrations [5], or by confining it to the space between an over- and under- attempt [9]. Policy learning has been developed for continuous actions [4]. Information has been reduced by only keeping the principal components of data [1] and specific methods of trajectory encoding, such as locally weighted projection regression [14], leading to latent spaces [1].

Z. Lončarević—Holder of Ad futura Scholarship for Postgraduate Studies of Nationals of Western Balkan States for Study in the Republic of Slovenia (226. Public Call).


K. Berns and D. Görges (Eds.): RAAD 2019, AISC 980, pp. 86–94, 2020.
https://doi.org/10.1007/978-3-030-19648-6_11

Latent spaces of deep autoencoders has also been proposed as means of dimensionality reduction of robotic actions and skills [2], specifically in combination with policy representations in the form of dynamic movement primitives [6]. In our previous paper on this topic we have shown that skill learning (using reinforcement learning) in latent space of deep autoencoders is faster than learning in the space of the trajectory [10].

However, the problem of generating the database for training the autoencoder network remains. While in [10] we used simulated results to train the autoencoder network, in this paper we explore how much data is actually needed for faithful representation of an action. To do this, we used the humanoid robot Talos to perform throwing actions with one of its arms, where the trajectory of throwing was encoded in the task space. We applied reinforcement learning (RL [8]) to learn throwing at the desired spot. We recorded all the roll-outs and used them to make our database. By learning to throw at different locations, we quickly acquired a large database, and then used parts of it to see how accurately we can perform throws at untrained locations. Testing of different database sizes and deep autoencoder latent space sizes can provide an insight whether training a small database in the real world makes sense, or whether it is better to train a large database in simulation and then apply RL in the latent space of an autoencoder trained with this database. This paper reports on the first steps of researching in this direction, with the research on the required simulated database.

The rest of the paper is organized as follows. In the next Section we first show how the task-space trajectories of throwing are encoded and how reinforcement learning using PoWER is applied to refine a given task. Section 3 gives details on autoencoders. The experiments are described in Sect. 4, followed by results in Sect. 5 and a Conclusion.

## 2 Trajectory and Learning

Robot motion trajectories can be written in either joint space or in Cartesian space. Cartesian space trajectories are typically easier to interpret for humans, specially in high degree-of-freedom robots. In this paper the trajectories of motion were encoded in task space of the end-effector, in the form of Cartesian space dynamic movement primitives (CDMPs). The original derivation of CDMPs was proposed in [13]. For clarity we provide a short recap of CDMPs below.

### 2.1 Cartesian Space Dynamic Movement Primitives

CDMPs are composed of two parts: position and orientation part. The position part of the trajectory is the same as in standard DMPs. On the other hand, the orientation part of the CDMP is represented by unit quaternions. Unit quaternions require special treatment, for both the nonlinear dynamic equations and the integration of these equations.

The following parameters compose a CDMP: weights $\boldsymbol{w}_k^p$, $\boldsymbol{w}_k^o \in \mathbb{R}^3$, $k = 1, \ldots, N$, which represent the position and orientation parts of the trajectory, respectively; trajectory duration $\tau$ and the final desired, goal position $\boldsymbol{g}^p$ and orientation $\boldsymbol{g}^o$ of the robot. Variable $N$ sets the number of radial basis functions that are used to encode the trajectory. The orientation is in CDMP represented by a unit quaternion $\boldsymbol{q} = v + \boldsymbol{u} \in \mathrm{S}^3$, where $\mathrm{S}^3$ is a unit sphere in $\mathbb{R}^4$. $v \in \mathbb{R}$ is its scalar and $\boldsymbol{u} \in \mathbb{R}^3$ its vector part. To encode position ($\boldsymbol{p}$) and orientation ($\boldsymbol{q}$) trajectories we use the following differential equations:

$$\tau\dot{\mathbf{z}} = \alpha_z(\beta_z(\mathbf{g}^p - \mathbf{p}) - \mathbf{z}) + \mathbf{f}_p(x), \tag{1}$$

$$\tau\dot{\mathbf{p}} = \mathbf{z}, \tag{2}$$

$$\tau\dot{\boldsymbol{\eta}} = \alpha_z\left(\beta_z 2\log\left(\boldsymbol{g}^o * \overline{\boldsymbol{q}}\right) - \boldsymbol{\eta}\right) + \mathbf{f}_o(x), \tag{3}$$

$$\tau\dot{\boldsymbol{q}} = \frac{1}{2}\boldsymbol{\eta} * \boldsymbol{q}, \tag{4}$$

$$\tau\dot{x} = -\alpha x. \tag{5}$$

Parameters $\boldsymbol{z}$, $\boldsymbol{\eta}$ denote the scaled linear and angular velocity ($\boldsymbol{z} = \tau\dot{\boldsymbol{p}}$, $\boldsymbol{\eta} = \tau\boldsymbol{\omega}$). For details on quaternion product $*$, conjugation $\overline{\boldsymbol{q}}$, and the quaternion logarithm $\log(\boldsymbol{q})$, see [13]. The nonlinear parts, termed also forcing terms, $\mathbf{f}_p$ and $\mathbf{f}_o$ are defined as

$$\mathbf{f}_p(x) = \boldsymbol{D}_p\frac{\sum_{k=1}^N \mathbf{w}_k^p\Psi_k(x)}{\sum_{k=1}^N \Psi_k(x)}x, \tag{6}$$

$$\mathbf{f}_o(x) = \boldsymbol{D}_o\frac{\sum_{k=1}^N \boldsymbol{w}_k^o\Psi_k(x)}{\sum_{k=1}^N \Psi_k(x)}x. \tag{7}$$

Forcing terms contain parameters $\boldsymbol{w}_k^p$, $\boldsymbol{w}_k^o \in \mathbb{R}^3$. They have to be learned, for example directly from an input Cartesian trajectory $\{\boldsymbol{p}_j, \boldsymbol{q}_j, \dot{\boldsymbol{p}}_j, \boldsymbol{\omega}_j, \ddot{\boldsymbol{p}}_j, \dot{\boldsymbol{\omega}}_j, t_j\}_{j=1}^T$. The scaling matrices $\boldsymbol{D}_p$, $\boldsymbol{D}_o \in \mathbb{R}^{3\times 3}$ can be set to $\boldsymbol{D}_p = \boldsymbol{D}_o = \boldsymbol{I}$. Other possibilities are described in [13]. The nonlinear forcing terms are defined as a linear combination of radial basis functions $\Psi_k$

$$\Psi_k(x) = \exp\left(-h_k\left(x - c_k\right)^2\right). \tag{8}$$

Here $c_k$ are the centers and $h_k$ the widths of the radial basis functions. The distribution of weights can be, as in [12], $c_k = \exp\left(-\alpha_x\frac{k-1}{N-1}\right)$, $h_k = \dfrac{1}{(c_{k+1} - c_k)^2}$, $h_N = h_{N-1}$, $k = 1, \ldots, N$. The time constant $\tau$ is set to the desired duration of the trajectory, i. e. $\tau = t_T - t_1$. The goal position and orientation are usually set to the final position and orientation on the desired trajectory, i. e. $\boldsymbol{g}^p = \boldsymbol{p}_{t_T}$ and $\boldsymbol{g}^o = \boldsymbol{q}_{t_T}$. Detailed CDMP description and auxiliary math are explained in [13].

## 2.2 PoWER

The goal of the paper is to show that autonomous learning, in our case reinforcement learning, can be faster in the latent space. We use Policy Learning by Weighting Exploration with the Returns (PoWER) [7] RL algorithm. It is an Expectation-Maximization (EM) based RL algorithm that can be combined with importance sampling to better exploit previous experience. It tries to maximize the expected return of trials using a parametrized policy, such as the aforementioned CDMPs. We use PoWER because it is robust with respect to reward functions [7]. Furthermore, it also uses only the terminal reward, and no intermediate rewards.

## 3 Deep Autoencoder Network

An antoencoder is a neural network used for unsupervised encoding of data. This neural network is comprised of two parts: an encoder and a decoder network. Using the encoder network part, the data is encoded. Because the number of neurons in the hidden layers is lower than in the input layer, this forces the data through a bottleneck or latent space. In this space the most relevant features are extracted. Latent space is therefore often also called feature space. One of typical applications of autoencoders is for dimensionality reduction [2].

The decoder network part reconstructs the feature representation so that the output data $\boldsymbol{\theta}'$ matches the input data $\boldsymbol{\theta}$. Training of the parameters of the autoencoder ($\boldsymbol{\theta}^\star$) is described by

$$\boldsymbol{\theta}^\star = \arg\min_{\boldsymbol{\theta}^\star} \frac{1}{n} \sum_{i=1}^{n} L(\boldsymbol{\theta}^{(i)}, \boldsymbol{\theta}'^{(i)}) \tag{9}$$

$$= \arg\min_{\boldsymbol{\theta}^\star} \frac{1}{n} \sum_{i=1}^{n} L(\boldsymbol{\theta}^{(i)}, h(g(\boldsymbol{\theta}^{(i)}))), \tag{10}$$

where $L$ is Euclidian distance between the input and output vectors and $n$ is the number of samples. Figure 1 shows an illustration of an arbitrary autoencoder architecture.

## 4 Experimental Evaluation

We experimentally tested how many database entries for training the autoencoder network are needed to faithfully represent the input data, with the goal to show faster learning in the latent space.

The learning experiments were performed on a simulated humanoid robot Talos, using seven degrees of the robot's left arm for the throwing. A passive elastic spoon was used to extend the range of the throw of the ball. An image sequence of a successful throw with the simulated Talos is shown in Fig. 2.

We first implemented learning in the space of the CDMP parameters. Starting from an initial task-space trajectory demonstration, the RL using PoWER modifies the policy so that the throw results in the ball landing in the basket. The distance and the angle of the throw have to be learned, while the learning takes place in all 6 DOF of the CDMP trajectory encoding. In the CDMP

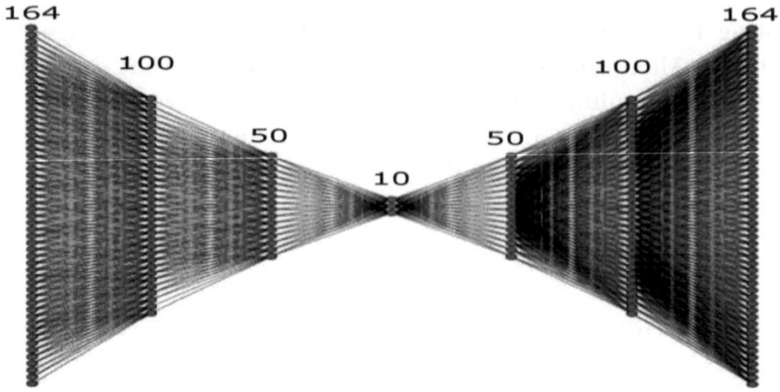


**Fig. 1.** Illustration of the autoencoder structure with five hidden layers. Note that the number of neurons per layer in the used autoencoder is too high for an effective illustration, that way we plotted in each layer three times less neurons. The depicted number of neurons per layer does not match the number we used (164 for input and output layers, 100, 50, 10, 50, 100 for hidden layers).

**Fig. 2.** Still images of the simulated throwing experiment using the Talos humanoid robot in Gazebo dynamical simulation. A sequence of a successful throw is shown.

parameter space we learn parameters $\boldsymbol{\theta}$ for the next, $n+1$ experiment, $\boldsymbol{\theta}_{n+1}$, where

$$\boldsymbol{\theta} = \left[\boldsymbol{\theta}_x^T, \boldsymbol{\theta}_y^T, \boldsymbol{\theta}_z^T, \boldsymbol{\theta}_{u_1}^T, \boldsymbol{\theta}_{u_2}^T, \boldsymbol{\theta}_{u_3}^T\right]^T, \tag{11}$$
$$\boldsymbol{\theta}_{(.)} = \left[\mathbf{w}^T, y_{0,}, g\right]^T, \ (.) = x, y, z, u_1, u_2, u_3.$$

We used $N = 25$ basis functions per dimension. After taking into the consideration start and goal pose, the size of learning space sums up to 164 parameters.

To create the database for autoencoder training, we recorded all the roll-outs of the learning and the corresponding landing spots of the ball.

We used (11) as the input data for the autoencoder network. This sets the input and output layer sizes to 164. The autoencoder was comprised of 5 hidden layers with 100, 50, $L$, 50, and 100 neurons. We varied latent layer size $L$ between 2 and 10 for different experiments. Activation function for each hidden layer is $\mathbf{y} = \tanh(\mathbf{W}\boldsymbol{\theta}^{\#} + \mathbf{b})$, where $\boldsymbol{\theta}^{\#}$ is the input into a neuron network layer and $\boldsymbol{\theta}^{\star} = \{\mathbf{W},\ \mathbf{b}\}$ are the autoencoder parameters. Note that the input is different for each layer, because it is the output of the previous layer. The activation function of the output layer was linear. After the training we split the autoencoder in the encoder and the decoder parts. The encoder maps input into the latent space $\boldsymbol{\theta}^l = g(\boldsymbol{\theta})$ and the decoder maps from latent space to the output $\boldsymbol{\theta}' = h(\boldsymbol{\theta}^l)$, i. e., again into CDMP parameters that describe the robot cartesian trajectories.

In latent space we also use PoWER [7] to learn $\boldsymbol{\theta}^l_{n+1}$. However, the learning space in this case is only L-dimensional. The values of parameters in latent space $\boldsymbol{\theta}_l$ define the DMP parameters, and therefore the shape of the trajectory on the robot, through the decoder network

$$\boldsymbol{\theta}'_{n+1} = h(\boldsymbol{\theta}^l_{n+1}). \tag{12}$$

## 5 Results

From the graphs of mean square error for trajectory positions and quarterion orientations (Fig. 3) we can see the expected trend that bigger database and bigger latent space reduced the error. Similar was reporten in [3]. As we can see in our case, the trend settles for the databases bigger than 200 examples and latent space size bigger than 4. These are the most efficient training parameters in our case. However, we can see certain improvements for much bigger latent space size and database size (e.g. 10 dimensions and 800 examples), but this leads to much bigger costs of generating input data and would slow down the learning process.

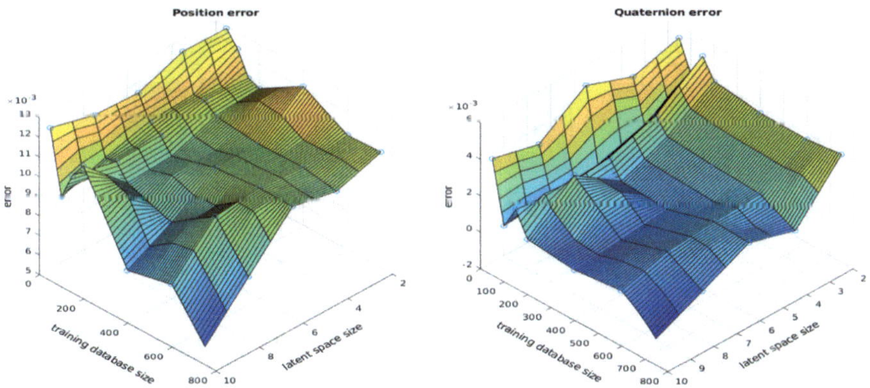


**Fig. 3.** Mean square error of the trajectory position and quaternion orientation in respect to database size and latent space size

As a proof of concept we tested our RL algorithm on three different reward systems (exact, unsigned, signed [10]) and two different search spaces (task space and latent space). Learning was done 25 times for all the cases. For the case of latent space RL we have chosen the neural network with the latent space size 4 that was trained on 200 shots.

Figure 4 shows the error convergence through the iterations of RL. Error is given as the absolute distance in meters between the basket and the landing spot

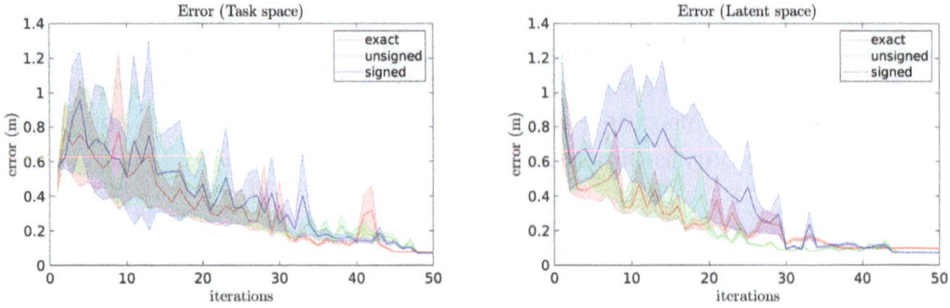


**Fig. 4.** Average error of throwing through the iterations of learning. RL in configuration space is shown in the left plot, and in latent space in the right plot. In both plots, the exact reward is shown with the red line, the unsigned reward with the green line and the signed with the blue line. Shaded areas show the corresponding distributions for all the reward systems.

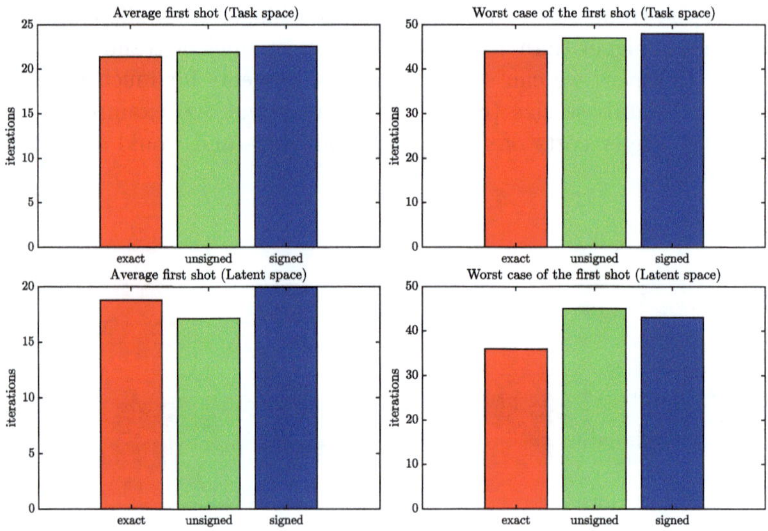


**Fig. 5.** Average number of throws until the first hit (left) and maximal number of throws until first hit for different reward systems.

of the ball. Graphs show that all the reward systems in both task and latent space converged successfully. It shows that reduced reward systems (unsigned and signed) were able to converge to the target almost equally fast as the exact reward system. Learning the parameters in latent space outperformed learning in task space in all the cases, no matter the reward.

The average convergence rates for different reward systems and cases are shown in Fig. 5. The top left graph shows the average iteration of the successful shot in the case of RL in task space and bottom left graph shows the average iteration of the successful shot for RL in the latent space. On the right side maximal number of iterations needed for the successful accomplishment of the task is shown for the both task space (top graph) and latent space (bottom graph).

## 6 Conclusion

Apart from confirming the results shown in [10], that the simplified reward systems can work equally good as exact reward if only terminal reward is available, we have also shown that the search space size for the RL can be successfully reduced using the neural networks even for complex systems such as a high degree-of-freedom robot arm. All the experiments were conducted on a simulated robot, that behaves slightly differently than the real system would, but it is faster to generate the required training database. In the future we plan to test this approach on the real robot as well.

## References


1. Bitzer, S., Vijayakumar, S.: Latent spaces for dynamic movement primitives. In: 2009 9th IEEE-RAS International Conference on Humanoid Robots, pp. 574–581, December 2009
2. Chen, N., Bayer, J., Urban, S., van der Smagt, P.: Efficient movement representation by embedding dynamic movement primitives in deep autoencoders. In: 2015 IEEE-RAS 15th International Conference on Humanoid Robots (Humanoids), pp. 434–440, November 2015
3. Chen, N., Karl, M., Smagt, P.V.D.: Dynamic movement primitives in latent space of time-dependent variational autoencoders. 2016 IEEE-RAS 16th International Conference on Humanoid Robots (Humanoids), vol. 2, no. 3, pp. 629–636 (2016)
4. Deisenroth, M.P., Neumann, G., Peters, J.: A survey on policy search for robotics, pp. 388–403 (2013)
5. Gams, A., Petrič, T., Do, M., Nemec, B., Morimoto, J., Asfour, T., Ude, A.: Adaptation and coaching of periodic motion primitives through physical and visual interaction. Robot. Auto. Syst. **75**, 340–351 (2016). http://www.sciencedirect.com/science/article/pii/S0921889015001992
6. Ijspeert, A., Nakanishi, J., Pastor, P., Hoffmann, H., Schaal, S.: Dynamical movement primitives: Learning attractor models for motor behaviors. Neural Comput. **25**(2), 328–373 (2013)
7. Kober, J., Peters, J.: Policy search for motor primitives in robotics. Mach. Learn. **84**(1–2), 171–203 (2011)
8. Kober, J., Bagnell, J.A., Peters, J.: Reinforcement learning in robotics (2013)
9. Nemec, B., Vuga, R., Ude, A.: Efficient sensorimotor learning from multiple demonstrations. Adv. Robot. **27**(13), 1023–1031 (2013)
10. Pahič, R., Lončarević, Z., Ude, A., Nemec, B., Gams, A.: User feedback in latent space robotic skill learning. In: 2018 IEEE-RAS 18th International Conference on Humanoid Robots (Humanoids), pp. 270–276, November 2018
11. Schaal, S.: Is imitation learning the route to humanoid robots? Trends Cogn. Sci. **3**(6), 233–242 (1999)

12. Ude, A., Gams, A., Asfour, T., Morimoto, J.: Task-specific generalization of discrete and periodic dynamic movement primitives. IEEE Trans. Robot. **26**(5), 800–815 (2010)
13. Ude, A., Nemec, B., Petrič, T., Morimoto, J.: Orientation in cartesian space dynamic movement primitives. In: 2014 IEEE International Conference on Robotics and Automation (ICRA), pp. 2997–3004, May 2014
14. Vijayakumar, S., D'souza, A., Schaal, S.: Incremental online learning in high dimensions. Neural Comput. **17**(12), 2602–2634 (2005). https://doi.org/10.1162/089976605774320557

# Shared Impedance Control Based on Reinforcement Learning in a Human-Robot Collaboration Task

Min Wu(✉), Yanhao He, and Steven Liu

Technische Universität Kaiserslautern, 67663 Kaiserslautern, Germany
{mwu,yanhao,sliu}@eit.uni-kl.de

**Abstract.** In this work a shared impedance control scheme for a hybrid human-robot team is designed for transporting a rigid workpiece to a desired position. Within the scope of proposed control structure, both human and robot are regarded as mechanical impedance and their parameters are adapted continuously in real-time. Reinforcement learning is used to find an impedance parameter set for the whole team to optimize a task-orient cost function. Then the learned parameters are further adjusted by taking human's disagreement into consideration. The proposed method is aimed to reduce human's control effort during collaboration and be flexible to variation of the task or environment. Experimental results are presented to illustrate the performance.

**Keywords:** Physical human-robot collaboration · Impedance control · Reinforcement learning

## 1 Introduction

Human robot collaboration has recently received considerable attention in both academic community and technology companies. Its applications are found not only productive in the industrial field, but also profitable and inspiring in the area of daily personal service. In many cases robot and human contact with each other and form a tightly coupled dynamical system to accomplish a task. This is specially characterized as physical human robot interaction (pHRI) [1]. Safety and cognition are the two main research fields in terms of pHRI. The safety issue has been intensively researched in the last years and successfully handled by designed pre- and post reactions of collision [3]. However, cognition is one skill that many robots are still not able to provide jet.

One feature of the cognitive robot is that it can understand human's intension and make proactive contribution in a joint task with human. This concept leads to a shared control strategy [10], where both human and robot "share" a common control task and should contribute control effort together at the same time. How to allocate and adapt the role of each participant to exploit strength of both human and robot is becoming an interesting research topic [5].


K. Berns and D. Görges (Eds.): RAAD 2019, AISC 980, pp. 95–103, 2020.
https://doi.org/10.1007/978-3-030-19648-6_12

Within the structure of pHRI, contact force is an important interaction modality for dynamic role adaption between human and robot. Hence, it should be particularly considered and handled in robot control. Impedance control [4] provides an effective solution for that since it is able to deal with both motion and force control problem at the same time. Moreover, previous research [12] shows that the human motor control could also be described by an impedance control scheme. Based on this assumption, a pHRI scenario could be physically interpreted as a shared impedance control, in which both human and robot are modeled as mechanical impedance and generate force/torque together to influence the dynamic of the payload. Under this concept, the role allocation between participants could be achieved by varying their impedance parameters.

Many previous researches aim to find a proper adaptive law according to different features. In [2] the damping factor is tuned based on the end-effector's velocity. In [7] the control law of robot is adjusted by taking feedback of contact force. In [13] the authors first study the human motor control scheme and then implement in robot to achieve a human-like adaption on the impedance parameters. From our point of view the adaption of the impedance parameters should take both task requirement and human's intention into consideration. Hence, we attend to combine different task-relied and human-relied features together in a cost function and try to solve it as an optimization problem.

Conventional optimal control is usually model-based and not-adaptive. It is not sufficient for pHRI since there exists high uncertainty in human control goals and control mechanisms. Machine learning techniques, especially reinforcement learning, provides promising solution. It is model-free so that control policy is evaluated and updated on-line by taking feedback from enviroment. In [9] the authors proposed a reinforcement learning based adaptive control in a pHRI task. An Actor-Critic structure with two neural networks is used for estimating human's control policy and updating the robot control. In our approach, instead of analyzing behavior of human and robot separately, we consider human-robot as a team, define a task-orient cost function, and obtain the optimal control strategy using reinforcement learning. After that the robot further adjusts its control parameter regarding to the disagreement of human, which can be observed from the contact force.

In summary, the contributions of this paper are: (1) we integrate reinforcement learning in the conventional impedance control loop to solve an optimization problem, which is defined based on study on the performance indexes of human-robot collaboration. (2) we further modify the learned control parameters by estimating the degree of human's acceptance on the robot's control action. The proposed method is validated in real experiment.

The rest of this paper is organized as follows: In Sect. 2 the model and structure of the proposed shared impedance control method are overviewed. In Sect. 3 the approach is discussed in detail and the experimental validation results are shown in Sect. 4. At the end the conclusions are drawn in Sect. 5.

# 2 Control Structure Overview

## 2.1 Problem Formulation

In this work we consider human-robot collaboration task of jointly manipulating a rigid object, whose dynamic is described by:

$$\boldsymbol{M}_o\ddot{\boldsymbol{x}} + \boldsymbol{C}_o\left(\dot{\boldsymbol{x}}, \boldsymbol{x}\right) = \boldsymbol{G}\boldsymbol{u}, \tag{1}$$

where $\boldsymbol{x}$ describes the object's configuration. $\boldsymbol{M}_o$ is the inertia Matrix, $\boldsymbol{C}_o\left(\dot{\boldsymbol{x}}, \boldsymbol{x}\right)$ represents the centrifugal and gravity term, $\boldsymbol{G} = \left(\boldsymbol{G}_h, \boldsymbol{G}_r\right)$ represents the grasping matrix, $\boldsymbol{u} = \left(\boldsymbol{u}_h, \boldsymbol{u}_r\right)^T$ are the generated wrench of both human and robot.

The control task for the human-robot team is formulated as: generating proper control effort $\boldsymbol{u}$ to achieve an accurate, efficient and smooth collaboration. It aims to (1) maintain the desired object motion, (2) keep the internal force acts on the workpiece small, (3) adjust their motions collaboratively if the constraints of environment change.

## 2.2 Overview of the Shared Control Structure

The whole system is modeled as a shared impedance control scheme (see Fig. 1), in which human and robot are both regarded as mechanical impedance (contains spring and damper) and jointly generate required wrench to influence the object's dynamic. Furthermore, we describe the interactive behavior between human and robot as cooperatively solving an optimal control problem that contains different objective/cost functions. Analytically the solution of an optimal control problem could be determined by solving Hamilton-Jacobian-Bellman (HJB) equations. However, it requires complete knowledge of the system dynamic and is difficult to solve on-line. Moreover, both system dynamic and performance objectives

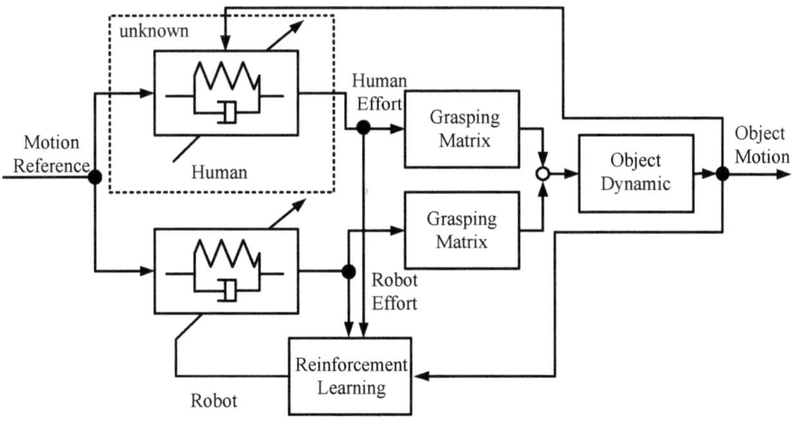


**Fig. 1.** Overview of the shared control system

may change over time. With regard to these difficulties, we use reinforcement learning, which continuously evaluates and updates the control policy, or in our case, the impedance parameter set. Note that a model of the system is not required in reinforcement learning.

Since there exists high uncertainty in human control goals and control mechanisms, it can not be guaranteed that the learned parameters completely match to human's expectation. In several previous researches, people attend to identify human's impedance parameter [13], cost function [6] and control effort [9] in order to have more knowledge on the human control behavior. In our approach, instead of estimating human's control strategy, we try to adjust the robot's impedance by taking feedback of human's response. In pHRI, it is commonly agreed that the interaction force plays an important role in human intention recognition. Hence, we further take the norm of interaction force as a measurement of disagreement, and use this information to modify the robot control parameter.

## 3 Approach

In this section first the reinforcement learning algorithm is introduced. Then we integrate it into the shared control structure and discuss its physical properties. We further modify the control parameter by estimating the degree of disagreement of the human operator. Finally, the resulting algorithm is summarized.

### 3.1 Reinforcement Learning

Firstly a task orient quadratic cost function is defined as follows:

$$r(\tilde{\boldsymbol{x}}, \dot{\boldsymbol{x}}, \boldsymbol{u}) = \boldsymbol{X}^T \boldsymbol{Q} \boldsymbol{X} + \boldsymbol{u}^T \boldsymbol{R} \boldsymbol{u}, \tag{2}$$

where $\boldsymbol{X} = (\tilde{\boldsymbol{x}}, \dot{\boldsymbol{x}})^T$, $\tilde{\boldsymbol{x}} = \boldsymbol{x} - \boldsymbol{x}_d$ represents motion error of the object, $\boldsymbol{u} = (\boldsymbol{u}_h, \boldsymbol{u}_r)^T$ represents the control input of both human and robot. $\boldsymbol{Q}$ and $\boldsymbol{R}$ are semi-positive definite weighting matrices.

In practice it is interesting to ask how the weighting matrices should be chosen property to build a meaningful cost function. According to the study on performance measurements for physical human-robot collaborations in [11], acceptable forces are crucial at first glance. Afterwards motion characteristics, e.g. velocity, acceleration, would be considered. It is more acceptable for human that the motion parameters (as well as their derivatives) do not change rapidly. Then as counterpart of the motion profile, accuracy parameters such as motion errors would be included. At last the effort of robot would be considered.

Correspondingly, the value function, which represents the conditional expected value of future cost when starting in state $\boldsymbol{X}$ at time $k$ by applying control input $\boldsymbol{u}$, is written as:

$$V(\boldsymbol{X}_k) = r(\boldsymbol{X}_k, \boldsymbol{u}_k) + \gamma \sum_{n=0}^{\infty} \gamma^n r(\boldsymbol{X}_{k+n+1}, \boldsymbol{u}_{k+n+1}) \tag{3}$$

where $\gamma \in (0, 1]$ is a discount factor.

Then the collaborative task of the human-robot team is defined as an optimal control problem, i.e. finding an optimal control input $\boldsymbol{u}^\star$ so that:

$$\boldsymbol{u}^\star = \arg\min_{\boldsymbol{u}} V(\boldsymbol{X}). \tag{4}$$

In this work, we use policy iteration with value function approximation [8] to achieve the optimal control input. In practice, one of the commonly used methods is to represent the value function approximately by a linear combination of features. Here we choose motion error $\tilde{\boldsymbol{x}}$ and velocity $\dot{\boldsymbol{x}}$ as features so that the approximated value function is defined as:

$$\hat{V}(\boldsymbol{X}, \boldsymbol{W}) = \tilde{\boldsymbol{x}}^T \boldsymbol{W}_1 \tilde{\boldsymbol{x}} + \dot{\boldsymbol{x}}^T \boldsymbol{W}_2 \dot{\boldsymbol{x}} = \boldsymbol{W}^T \varphi(\boldsymbol{X}), \tag{5}$$

where $\boldsymbol{W}_1, \boldsymbol{W}_2$ are diagonal weighting matrices, $\boldsymbol{W}$ contains all their diagonal elements and $\varphi(\boldsymbol{X}) = \boldsymbol{X} \otimes \boldsymbol{X}$.

The control input $\boldsymbol{u}$ is adapted along gradient of the approximated value function, namely:

$$\boldsymbol{u} = -\frac{\gamma}{2} \boldsymbol{R}^{-1} \boldsymbol{G}^T \nabla \varphi^T (\boldsymbol{X}) \boldsymbol{W} \tag{6}$$

and will converge toward the optimal solution.

### 3.2 Adaptive Impedance Control

In this subsection we first interpret the proposed control scheme in a physical way. By substituting $\nabla\varphi^T$ into (6), the control input $\boldsymbol{u}$ is rewritten as:

$$\boldsymbol{u} = -\underbrace{\gamma \boldsymbol{R}^{-1} \boldsymbol{G}^T \boldsymbol{W}_1}_{\boldsymbol{K}} \tilde{\boldsymbol{x}} - \underbrace{\gamma \boldsymbol{R}^{-1} \boldsymbol{G}^T \boldsymbol{W}_2}_{\boldsymbol{D}} \dot{\boldsymbol{x}}, \tag{7}$$

which has the same form as a mechanical spring-damper system. Hence the reinforcement learning phase could be concluded as: to learn an optimal virtual impedance parameter set $\{\boldsymbol{K}, \boldsymbol{D}\}$, which generates an elastic force to cancel the motion error and achieve sufficient damping.

However, in reality the learned impedance parameter might not satisfy the human's expectation. As mentioned at the beginning of this section, the optimization is performed based on several performance indexes and their weightings, which depend on some general studies on pHRI and could vary much from one operator to another. Hence, we design an additional adaption phase, in which the impedance parameter of robot is adjusted by the response of human on current distribution of the total impedance.

According to previous study [7], interaction force contributes significantly to the sensory communication in pHRI and could be used as measurement of "disagreement". If user keeps applying large force, we assume that he/she indents to have more control. In this situation the robot should decrease its impedance to behave compliantly. Depending on these assumptions, we further define a discount factor $\alpha$ as follows:

$$\alpha = \begin{cases} 1 - \frac{\|\boldsymbol{f}\|}{f_t}, & \text{if } \|\boldsymbol{f}\| < f_t \\ 0. & \text{else} \end{cases} \tag{8}$$

where $\boldsymbol{f}$ represents the interaction force and threshold $f_t > 0$ is a design parameter. Then the desired robot impedance parameters are calculated as:

$$\{\hat{\boldsymbol{K}}, \hat{\boldsymbol{D}}\} = \{(\boldsymbol{K})^{\alpha}, (\boldsymbol{D})^{\alpha}\}. \tag{9}$$

At last, depending on the robot dynamic equation, the joint control input is determined by:

$$\boldsymbol{\tau} = \boldsymbol{J}^T(\boldsymbol{q})(-\hat{\boldsymbol{K}}\tilde{\boldsymbol{x}} - \hat{\boldsymbol{D}}\dot{\boldsymbol{x}}) + \boldsymbol{M}(\boldsymbol{q})\ddot{q} + \boldsymbol{C}(\dot{q}, \boldsymbol{q})\dot{q} + \boldsymbol{g}(\boldsymbol{q}), \tag{10}$$

where $\boldsymbol{\tau}$ is the vector of input joint torques, $\boldsymbol{q}$ is the vector of joint angles, $\boldsymbol{M}$ is the inertia matrix, $\boldsymbol{C}$ is the matrix describing centrifugal and Coriolis terms, $\boldsymbol{g}$ is the vector of gravitational moments, $\boldsymbol{J}$ is the Jacobian matrix that describes the velocity kinematic.

As in all the proposed control algorithm for robot is summarized in Algorithm 1:

---

**Algorithm 1.** Proposed adaptive impedance control algorithm

---

**Input:** object state $\boldsymbol{X}_k$, interaction force $\boldsymbol{f}_k$ at current time $k$
**Output:** robot torque control input $\boldsymbol{\tau}_{k+1}$ for the next step $k+1$
1: Calculate $\boldsymbol{r}_k$ and $\boldsymbol{\Phi}_k$
2: **while** $\|\hat{\boldsymbol{W}}_k^{n+1} - \hat{\boldsymbol{W}}_k^{n}\| \geq \varepsilon$ and $n \leq n_{max}$ **do**
3: $\quad \boldsymbol{\Phi}_k = \varphi(\boldsymbol{X}_k) - \gamma\varphi(\boldsymbol{X}_{k+1})$
4: $\quad \boldsymbol{W}_k^{n+1} = \boldsymbol{W}_k^n - l\boldsymbol{\Phi}_k\left((\boldsymbol{W}_k^n)^T\boldsymbol{\Phi}_k - r\left(\boldsymbol{X}_k, \boldsymbol{u}_k\right)\right)$ (Value function approximation)
5: **end while**
6: Calculate robot impedance parameters using (7)
7: Adjust the impedance parameters using (9)
8: Calculate the control input $\boldsymbol{\tau}_{k+1}$ in (10)

---

## 4 Experimental Validation

In this section, a practical experiment is performed on a Franka Emika Panda robot with 7-DOFs and a two-finger gripper. Moreover, the robot is able to estimate the contact force acts on its end-effector based on measured joint torque and its dynamic model. For simplification the experiment only contains 2D translation motion. As shown in Fig. 2, the human operator and the robot collaboratively transfer an object along a prescribed path between two points A and B repeatedly. If the task varies, the human will make a corrective motion to bring the object to point C, which is unknown to the robot (red dashed line in Fig. 2).

After that the robot should regenerate its motion to return the original path (blue dashed line in Fig. 2). We expect that the robot impedance would be low when human tries to take the object to point C since he attends to increase the arm impedance to have more control. On the contrary, if human has no intension on motion correction, the robot should keep a high impedance in order to reduce the tracking error.

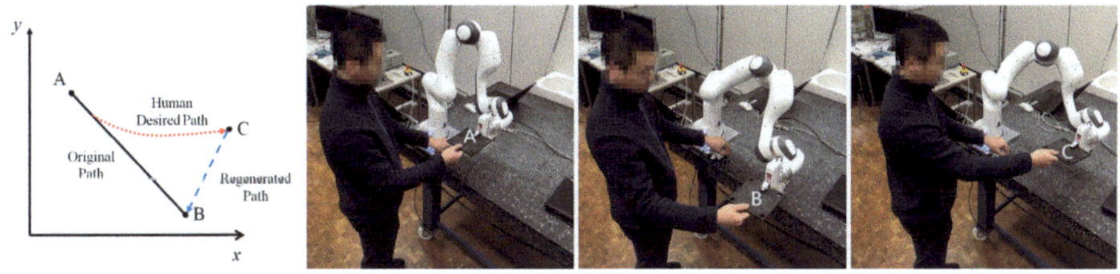


**Fig. 2.** Experiment setup, from left to right: (1) sketch of the task, (2) starting position A, (3) end position B, (4) human desired new position C

The robot end-effector's trajectory is generated by a fifth-order polynomial (quintic polynomial) as a function of time by taking A and B as initial- and end position. Moreover, if robot is too far away from the original path due to human's intervention, a new point-to-point motion, i.e. from its current position to end position B, is generated in order to help the robot back to the original task.

Figure 3 shows the experiment results. Note that we only plot for y-dimension since in x-dimension the results are similar. The gray zone describes the phase that human moves the object to position C. At first (20–36 s) the robot maintains a high stiffness in order to reduce the tracking error. Meanwhile the human operator dose not generate large force, which means he is generally agree with the current control strategy.

In the gray zone (15–20 s), human tries to dominate the task by increasing the force. At the same time robot decreases its stiffness to reduce the interaction force so that the human can easily move the robot with a small force (≈5N). Note that one can observe light oscillation in both force and stiffness signal in the human-dominate phase. The main reason is that during this phase, from the robot's point of view the motion error is large, hence it tries to increase its stiffness inquiringly to bring the object back to the prescribed trajectory. As long as human disagree with that it decreases the stiffness again. This tentative motion changes fast according to the short running period of learning algorithm (10 ms). Hence, the signals appear to be noisy.

In the last phase (20–36 s), human reduces the force sine the corrective motion is finished. One can observe that the robot increases its stiffness again and drives the object back to the original trajectory. Based on above results, the proposed method is able to automatically adapt the control parameter to different situations.

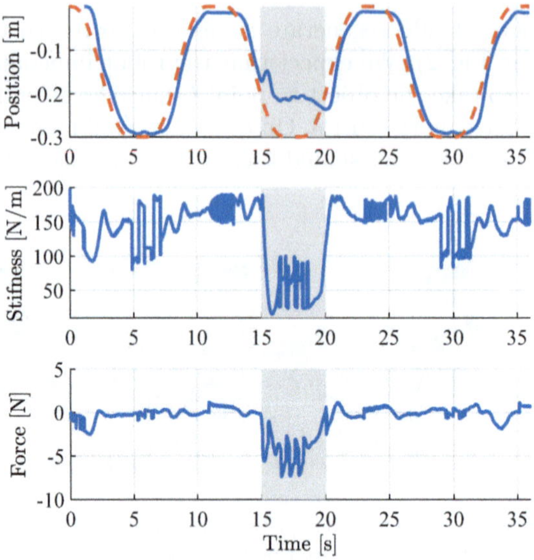


**Fig. 3.** Experiment results in y-dimension: (from top to bottom) (1) reference (dash line) and actual position (solid line) of the object, (2) adaption of robot stiffness, (3) estimated interaction force.

## 5 Conclusion

This paper proposes a shared impedance control strategy for human-robot collaboration. Under our concept both human and robot are modeled as mechanical impedance and coupled by the object's dynamic. The control algorithms contains two phases: (1) reinforcement learning is used to find an optimal impedance parameter set according to prescribed task-orient cost function. (2) The learned robot impedance parameters are further adjusted based on the estimated contact force, which is treated as a measurement of disagreement from human. The proposed method is validated by experiment. The results shows that this method enables reduction of human's effort and is flexible to variation in the task.

Future work includes: (1) a further study on the performance indexes of a human-robot collaboration task. What is exactly a "good" collaboration is still an open question. Furthermore we believe that the performance measurement should be adapted to task variation and human's exception as well. (2) integration of human motion prediction and intension recognition in the control scheme and used for real-time task adaption and trajectory planning.

**Acknowledgement.** We would like to thank Dr. Daniel Görges for his helpful advice and comments.

## References


1. Ajoudani, A., Zanchettin, A.M., Ivaldi, S., Albu-Schäffer, A., Kosuge, K., Khatib, O.: Progress and prospects of the human-robot collaboration. Auton. Robots **42**(5), 957–975 (2018)
2. Ficuciello, F., Villani, L., Siciliano, B.: Variable impedance control of redundant manipulators for intuitive human-robot physical interaction. IEEE Trans. Robotics **31**(4), 850–863 (2015)
3. Haddadin, S.: Towards Safe Robots: Approaching Asimov's 1st Law. PhD thesis, Rheinisch-Westfälischen Technischen Hochschule Aachen (2011)
4. Hogan, N.: Impedance control: an approach to manipulation: Part II–implementation. J. Dyn. Syst. Meas. Control **107**(1), 8 (1985)
5. Jarrassé, N., Sanguineti, V., Burdet, E.: Slaves no longer: review on role assignment for human-robot joint motor action. Adapt. Behav. **22**(1), 70–82 (2014)
6. Köpf, F., Inga, J., Rothfuß, S., Flad, M., Hohmann, S.: Inverse reinforcement learning for identification in linear-quadratic dynamic games. IFAC-PapersOnLine **50**(1), 14902–14908 (2017). 20th IFAC World Congress
7. Kucukyilmaz, A., Sezgin, T.M., Basdogan, C.: Conveying intentions through haptics in human-computer collaboration. In: 2011 IEEE World Haptics Conference, pp. 421–426, June 2011
8. Lewis, F., Vrabie, D., Syrmos, V.: Optimal Control. EngineeringPro collection. Wiley, New York (2012)
9. Li, Y., Tee, K.P., Yan, R., Chan, W.L., Wu, Y.: A framework of human-robot coordination based on game theory and policy iteration. IEEE Trans. Robot. **32**(6), 1408–1418 (2016)
10. Mörtl, A., Lawitzky, M., Kucukyilmaz, A., Sezgin, M., Basdogan, C., Hirche, S.: The role of roles: physical cooperation between humans and robots. Int. J. Robot. Res. **31**(13), 1656–1674 (2012)
11. Schmidtler, J., Bengler, K.: Fast or accurate? -performance measurements for physical human-robot collaborations. Procedia Manufacturing **3**, 1387–1394 (2015). ) 6th International Conference on Applied Human Factors and Ergonomics (AHFE 2015) and the Affiliated Conferences, AHFE 2015
12. Svinin, M., Goncharenko, I., Lee, H., Yamamoto, M.: Modeling of human-like reaching movements in the manipulation of parallel flexible objects. In: 2016 IEEE/RSJ International Conference on Intelligent Robots and Systems (IROS), pp. 4830–4835, October 2016
13. Yang, C., Ganesh, G., Haddadin, S., Parusel, S., Albu-Schaeffer, A., Burdet, E.: Human-like adaptation of force and impedance in stable and unstable interactions. IEEE Trans. Robot. **27**(5), 918–930 (2011)

# Gain Scheduled PID Force Control of a Robotic Arm for Sewing Fabrics

Ioannis H. Misios, Panagiotis N. Koustoumpardis(✉), and Nikos A. Aspragathos

Robotics Group, Mechanical and Aeronautics Engineering Department, University of Patras, Patras, Greece
koust@upatras.gr

**Abstract.** In this paper, a gain scheduled PID force feedback controller of a manipulator is designed and implemented for the robotized sewing of fabrics, using a commercial sewing machine. The proposed manipulator controller should keep a constant tension of the fabric, to achieve high quality of cloths seams, as the length of the fabric is shortened along the sewing process. For the controller design, a non-linear model of the fabric is considered, varying with the actual length between the grasping and sewing points. The model is based on a simplified Kelvin-Voigt model with non-linear spring and damper coefficients, which also depend on the type and the length of the fabric and are estimated experimentally. The Model-based PID tuning process from Simulink is used, for tuning and scheduling the gains of the PID controller corresponding to fabric's different actual lengths. The determined sets of gains are used for controlling on-line an Adept Cobra s800 robot to manipulate a woven piece of fabric during the sewing process, where the proposed approach is tested. The proposed force control approach maintains a stable tensional force on the fabric and therefore the quality of the seam and hence of the cloth could be enhanced. Finally, it is compared with a PID controller with constant gains.

**Keywords:** Robot force control · Sewing fabrics · PID control · Flexible objects

## 1 Introduction

Nowadays robotized automation has been adopted and established in most of the production processes with an exception of the apparel and cloth making industries, where manual work dominates. One of the reasons for this delay in automation is the fabrics behavior: the non-linearity and the variation of their mechanical properties and their very low bending rigidity that cause the infiniteness of states that fabrics could take in their manipulation. In the sewing process, the problems are even more accentuated, as the disturbances caused by the feed-dog's mechanism, the varying velocities and the different types of seams should be taken into account. One of the main factors for producing qualitative cloths, is the preservation of a constant tensional force, that should be applied on the fabric throughout the robotized sewing, which is a very complicated and challenging task for the robotic community.


K. Berns and D. Görges (Eds.): RAAD 2019, AISC 980, pp. 104–114, 2020.
https://doi.org/10.1007/978-3-030-19648-6_13

There is a limited number of research attempts on robotized sewing of fabrics. Gershon [1] presented the FIGARO project, which was the first approach for automated sewing. In that work, the goal of constant tension force was accomplished correlating the angular velocity of the feed-dog mechanism with the linear velocity of an end-effector of a robot guiding the fabric. The gains of a PI robot controller were adjusted by trial and error, which should be modified for a new type of fabric. Kudo et al. [2] used two cooperative robots for handling the fabric, which enables the rotation of the fabric around the needle. Both robots were equipped with force sensors mounted on their wrists for controlling the tension force between robots but not along the sewing line. In addition, vision feedback is used for the manipulation of the sewed fabric. In recent years, Koustoumpardis and Aspragathos [3, 4] used a SCARA robot for the automatic sewing of two-ply fabrics. A neural controller was designed to control the applied tensional force during the robotized sewing. The desired tensional force for each type of fabric was decided by a combination of neural and fuzzy systems fed by the extensibility of the fabric. In another work [5], the control of the tensional force has been based on a fuzzy model reference adaptive controller. In a most recent attempt, Schrimpf [6] presented the automation of the sewing process using two or three robots for manipulating the fabric. A PID controller with constant gains was used to control and synchronize the velocities of the robots with that of sewing machine according to the force sensors measurements, while the fabric properties are not involved in the whole procedure. Several attempts appeared to automate the sewing process by modifying the mechanism of the sewing machine. Book et al. [7] presented the most important design, where the original feed-dog mechanism of the sewing machine is replaced by a new servomechanism.

In order to model the fabrics behavior, it is required to experimentally measure their mechanical properties and especially their extensibility, which is the most important factor for the sewing task. For this reason, several measuring devices have been developed, with more important ones the KESF [8], FAST-3 [9] and FAMOUS [10]. The difference between them is the time required to measure the extensibility of the fabric, but they have low flexibility and need human intervention. Koustoumpardis et al. [4] recently introduced an automated method, which is embedded on the robotized sewing cell, to estimate the extensibility of both single and double ply fabrics. A lot of research works, on modelling fabrics and especially fabric extensibility, have been published. The most of them, approached the fabric viscoelastic behavior using springs, dampers and a combination of them. Liu [11] modelled the fabric behavior using generalized Kelvin-Voigt and Maxwell models. These approaches, for modelling fabrics, are very useful for designing PID controllers with constant gains. But, due to the nature of fabrics (non-linearities etc.), the adaptation of the controller parameters is needed. PID controller gain scheduling is a process, where the controller gains change in order to adapt to the variations of the controlled plant. An extend bibliography review and analysis on gain scheduling can be found in [12]. Stilwell [13] presented a method of non-linear controller gains adaption in order to stabilize both linear and non-linear changing systems.

The present paper introduces an approach for the design and implementation of a gain scheduled PID controller, for a robotic manipulator, for keeping constant the tensile force of a fabric during the sewing process. Experiments are conducted to model the extensibility and the damping of the fabric using a Kelvin-Voigt model with non-linear spring and damper. The robot motion is controlled by the gain scheduled PID controller with force feedback by a force sensor mounted at robot wrist. The adaptation of the PID controller gains is implemented according to the fabric stiffness and damping, which are changing during the sewing process.

## 2 The Proposed Method and Modelling of the Fabric

In the sewing process, shown in Fig. 1, the fabric is clamped from the one side by the sewing machine's feed dog and from the other side by the pneumatic gripper, which is mounted at the robot wrist and after the force sensor. Once the sewing machine is started the manipulator should guide the fabric to keep a desired constant tensional force at the fabric. The aim of this research is to improve the sewing quality by controlling the fabric tension force, during the robotized sewing process, with an adaptation of the gains of a manipulator PID controller (outer loop controller). The principal difficulty for achieving this, is the fabric's modelling due to its non-linear behavior and its properties variation, while the distance $L$ (referred also as actual length of the fabric) shortens when it is sewed, as shown in Fig. 1. Furthermore, the sewing machine's feed-dog mechanism, causes a periodical external disturbance which is transferred to the controller via the force sensor and affects the attainment of a constant tension force. This disturbance has pulse form with variant width and frequency depending on the stitch (type and length) and the variable velocity of the feed-dog mechanism. A force sensor mounted on the robot wrist feeds back the force to the PID controller, which controls the robot end-effector displacement and hence the fabric tensional force.

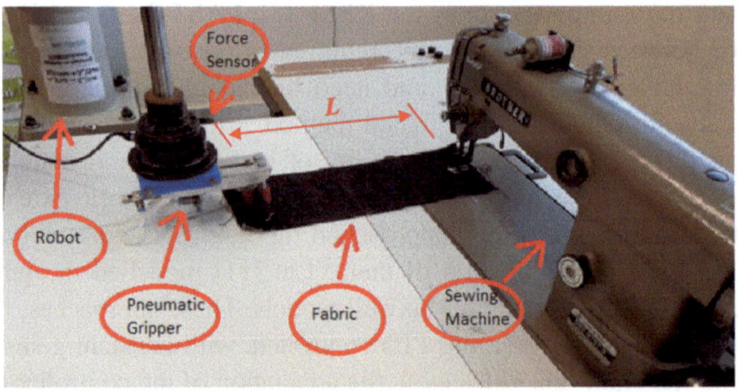


**Fig. 1.** The sewing process setup and the tensile test. Distance $L$ (between the grasping and sewing point) defines the actual length of the fabric in which the tensional force is applied.

For adapting the gains of the PID controller, the estimation of the properties that change with $L$ is performed experimentally and a model of the fabric with nonlinear damping coefficient (C) and spring coefficient (K) is developed. To estimate the fabric properties, two tests were designed and conducted: one for measuring the fabric extensibility and the other to measure its damping. These tests are embedded to the sewing process, as described in [4].

The experimental setup for the tensile loading-unloading test is shown in Fig. 1. The same experimental equipment, which is used for the sewing process, is used for the tensile test too. At the beginning, the tensile loading test was performed two times for two different distances *L* (137 mm and 247 mm) to identify the fabric linear area.

Force elongation diagrams (Fig. 2) show that the gradient is constant until approximately 5N, which defines the fabric linear area. Then tensile loading-unloading tests in this area (until 5N) were conducted for different lengths of the fabric. We assumed that 4 tests is a sufficient number for a 400 mm piece of fabric. For every length (297 mm, 247 mm, 137 mm and 93 mm) the robot extends the fabric until the linear area limit (5N) and then unloading it until the force sensor measures a zero force, as shown in Fig. 3. The diagram shows that there is a hysteresis loop. Therefore, in order to model the fabric behavior two functions are defined: one expressing fabric's loading and the other for its unloading phase, which are calculated using the Matlab Curve Fitting Toolbox as given by Eqs. (1 and 2) and illustrated in Fig. 4a and b respectively. The coefficients $p_{xy}$ are available and could be disposed if needed.

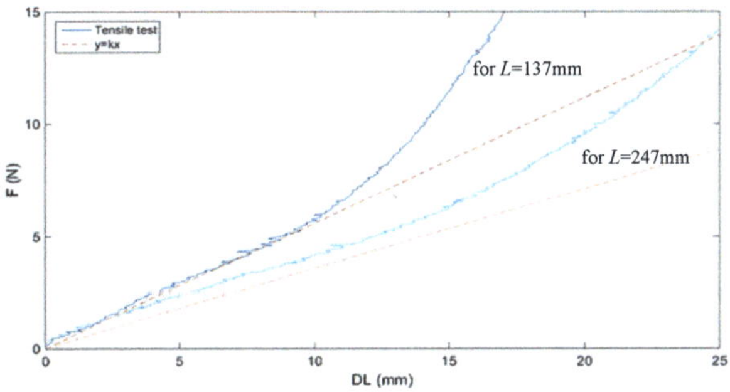


**Fig. 2.** Tensile tests for $L$ = 137 mm and $L$ = 247 mm, where DL (extension of the fabric) is the displacement from neutral potion.

Loading polynomial:

$$
\begin{aligned}
DL(L,F) &= p_{00} + p_{10} * L + p_{01} * F + p_{20} * L^2 + p_{11} * L * F \\
&\quad + p_{02} * F^2 + p_{30} * L^3 + p_{21} * L^2 * F + p_{12} * L * F^2 + p_{03} * F^3
\end{aligned} \tag{1}
$$

Unloading polynomial:

$$
\begin{aligned}
DL(L,F) &= p_{00} + p_{10} * L + p_{01} * F + p_{20} * L^2 + p_{11} * L * F + p_{02} \\
&\quad * F^2 + p_{30} * L^3 + p_{21} * L^2 * F + p_{12} * L * F^2 + p_{03} * F^3 \\
&\quad + p_{40} * L^4 + p_{31} * L^3 * F + p_{22} * L^2 * F^2 + p_{13} * L * F^3 + p_{04} * F^4
\end{aligned} \tag{2}
$$

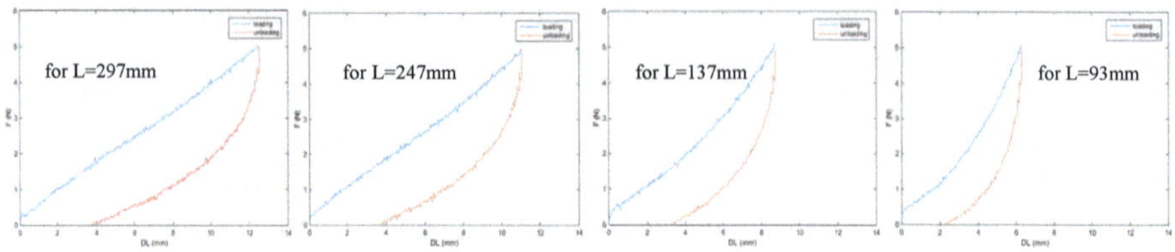


**Fig. 3.** Tensile loading-unloading tests for actual fabric length: L = 297, 247, 137 and 93 mm.

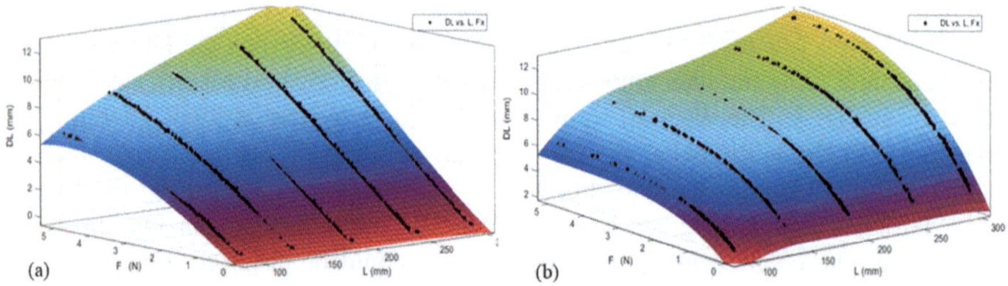


**Fig. 4.** Tensile curves for: (a) loading (b) unloading

The fabric damping is caused by the material of the yarns (fibers) and by the friction between them. In this research, the fabric damping ratio was estimated using the logarithmic decrement method [14] (Fig. 5(c)) by a decreasing oscillations experiment, as shown in Fig. 5a and b. In this experiment we hang the fabric by the robot gripper from the one side and in the other side we hang a small mass (120 g), so that it can oscillate in its linear area. Then the fabric-mass system is pulled down until the linear area limit (5N) of the fabric and then released to oscillate. This test is executed for four different actual lengths L = 335, 237, 180 and 120 mm. The diagram for one of these tests (L = 335 mm) is shown in Fig. 5c.

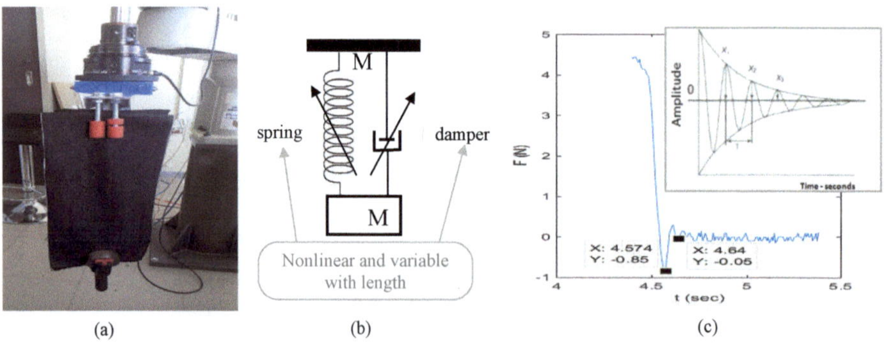


**Fig. 5.** Damping experimental setup (a), Kelvin-Voigt model with nonlinear and time varying elements (b) and damping test for actual length of fabric L = 335 mm (c).

Based on logarithmic decrement method the fabric damping factor (Eq. 3) is estimated using the amplitude of two sequential high or low peaks $x_1$ and $x_2$ (Fig. 5c).

$$\zeta = \frac{1}{2\pi} ln \frac{x_1}{x_2} \tag{3}$$

where $x_1$, $x_2$ are the amplitudes of two adjacent peaks (Table 1).

**Table 1.** Damping factor for different lengths

| Damping factor $\zeta$ (%) | Fabric actual length L (mm) | | | |
|---|---|---|---|---|
| | $L_1 = 335$ | $L_2 = 237$ | $L_3 = 180$ | $L_4 = 120$ |
| $\zeta_1$ | 45.09 | 33.1 | 27.61 | 17.48 |

Using the Matlab Curve Fitting Toolbox we estimate the polynomial (Eq. 4) by fitting the experimental data. The polynomial expressing the fabric damping factor versus its length is presented in Fig. 6.

$$\zeta(L) = 3.66 \cdot 10^{-8} L^3 - 2.58 \cdot 10^{-5} L^2 + 6.93 \cdot 10^{-3} L - 0.348 \tag{4}$$

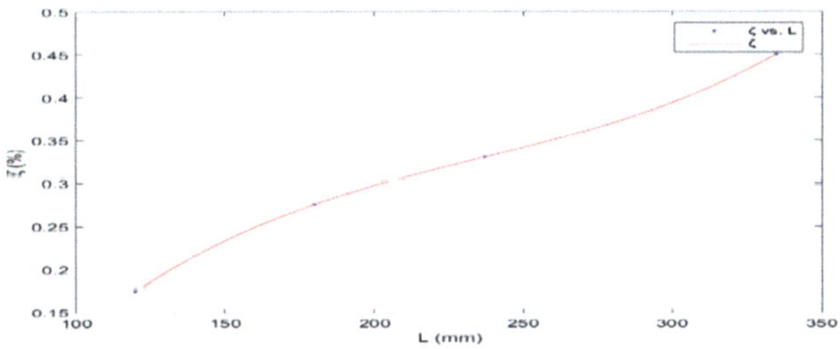


**Fig. 6.** Damping factor ($\zeta$) versus actual length of fabric (L)

For the purpose of modeling the fabric's nonlinear behavior, Kelvin-Voigt model consisted by nonlinear elements is used. The differential equation of this model is:

$$M \cdot \ddot{DL} + C(L, F) \cdot \dot{DL} + K(L, F) \cdot DL = F \tag{5}$$

where *DL* is the extension of the fabric for a specific L, M: fabric mass 20gr, C(*L*, *F*): nonlinear damping coefficient, K(*L*, *F*): nonlinear spring coefficient, *F*: external force measured by the force sensor.

Since the presented method is restricted in an approximately linear area (0–5N) of the fabric then the Hooke's law could be used to calculate K(*L*, *F*), which arises from loading or from unloading polynomial graph (Fig. 4a and b). Coefficient C(*L*, *F*) is calculated by ζ(*L*) and K(*L*, *F*). Because the mass of fabric is insignificant (20 g), we consider the term $M \cdot \ddot{DL} \cong 0$. Therefore, (Eq. 5) is simplified to:

$$C(L,F) \cdot \dot{DL} + K(L,F) \cdot DL = F \tag{6}$$

The differential equation (Eq. 6) varies while K, ζ coefficients change which depend on the fabric length (*L*) and the applied force (*F*). In order to estimate all the possible combinations of the coefficients K and C during sewing process, a block diagram in Matlab/Simulink synthetic environment is elaborated. Coefficients K, C were calculated as follows:

- Get as entry the distance between the robot end-effector and the sewing needle (*L*), and the applied force (F).
- Compare the F with the desired force ($F_{reference}$). If $F \geq F_{reference}$ then the calculate the displacement by (Eq. 2), while, if $F < F_{reference}$ by (Eq. 1).
- Calculate K using Hooke's law.
- Calculate damping ratio (ζ) by (Eq. 4).
- Calculate the damping coefficient (C) by ζ and K.

Thus, considering as requirement that $F = F_{reference}$ we calculate the coefficients of the differential equation K, C for different distances *L*. Using Curve Fitting Toolbox of Matlab the K-L and C-L diagrams are derived, as shown in Fig. 7.

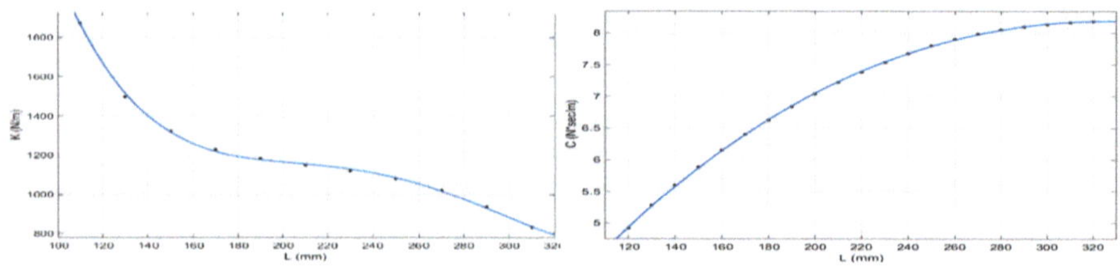


**Fig. 7.** The spring coefficient *K* (for the case of loading polynomial graph) and the damping coefficient *C* variations, relative to distance L.

## 3 Gain Scheduling of the Controller for the Sewing Task

In this work, a PID controller is designed and its tuning is done in Matlab/Simulink program, as shown in Fig. 8. This control scheme is an outer loop to the robot internal controller. The PID controller Transfer Function is shown in Fig. 8:

$$G_{PID}(z) = K_P + K_I \frac{T_s z}{z-1} + K_D \frac{z-1}{T_s z}$$

$T_s$: *sampling period*

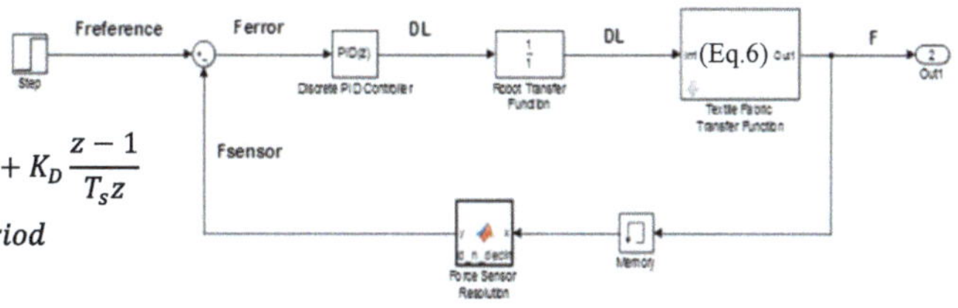


**Fig. 8.** Control system to tune controller gains

The control system receives as feedback the measurements from the force sensor. In this closed loop system, a force sensor resolution block is added to simulate the actual resolution of the real force sensor (resolution 0.05). Furthermore, the robot transfer function is set equal to one. The controller receives the error signal $(F_{reference} - F_{sensor})$ and supplies the system with the corresponding end-effector displacement, which is equivalent to DL. The input $(F_{reference})$ of the closed loop is unit step. According to the Model-based PID tuning process for Matlab/Simulink, the following specifications were defined so that the system could reduce the feedbacked tension force to the desired one during the time of one stitch (0.4 s):

Initial Value: 0, Final Value: 1, Step Time: 1 s, Rise Time: 1.2 s, Settling Time: 1.4 s, Percent Overshoot: 10%, Percent Settling: 2%, Percent Rise: 85%.

In order to achieve fast convergence according to the above specifications we set the mean square error functions (according to the step, settling, rise and end time) for which we are searching its total minimum using *patternsearch* of Matlab. For this total minimum, the gains $K_P, K_I, K_D$ of the PID controller are the optimum for a specific actual length of fabric. For different fabric lengths we develop the functions $K_P(L), K_I(L)$ and $K_D(L)$ using Curve Fitting Toolbox, which are:

$$K_P(L) = 4.336 \cdot 10^{-8} L^3 - 2.66 \cdot 10^{-5} L^2 + 7.598 \cdot 10^{-3} L - 0.3192 \tag{7}$$

$$K_I(L) = 1.592 \cdot 10^{-7} L^3 - 1.137 \cdot 10^{-4} L^2 + 3.192 \cdot 10^{-2} L - 1.368 \tag{8}$$

$$K_D(L) = -4.236 \cdot 10^{-9} L^3 + 3.061 \cdot 10^{-6} L^2 - 7.551 \cdot 10^{-4} L + 0.04821 \tag{9}$$

Because the length of the fabric is shortened, it is appropriate to adapt the PID gains using a 'gain scheduling' method. Getting started with the fabric with the longest actual length (L = 320 mm) the optimum gains are estimated as described above. Afterwards, for the next length (L = 310 mm) we check if the gains from the previous length (L = 320 mm) meet all the standards about system response and if not then we estimate the new optimum gains for the current actual length (L = 310 mm). This process is repeated until the minimum fabric actual length (L = 120 mm). By this way we can find the lengths of fabric, where the gains of the PID must change and which must be the optimum new values as shown in Fig. 9. The simulation of the whole sewing process was conducted in Matlab/Simulink, as presented in Fig. 10a.

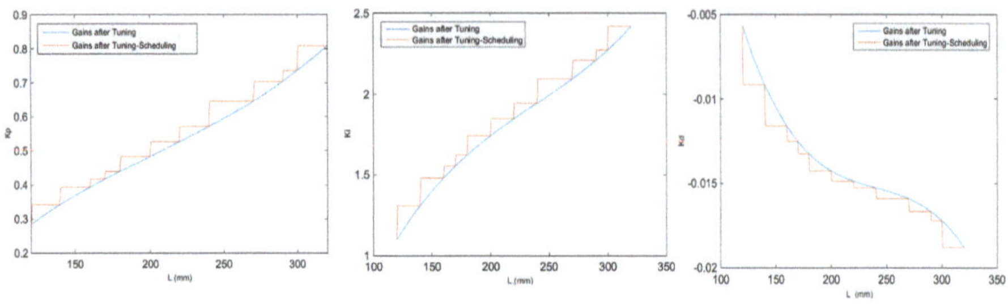


**Fig. 9.** Variation of the $K_P$, $K_I$ and $K_D$ gains during sewing

# 4 Experimental Results of the Robotized Sewing

The experimental cell consists of an Adept Cobra s800 robot equipped with an ATI Industrial Automation force sensor mounted on its wrist. The resolution of the force sensor is 0.05N and the sampling frequency 20 Hz. Also, we assigned the period for every stitch at 0.4 s and the stitch length of the sewing machine to remain constant during the whole sewing process at 2.5 mm. A woven elastic jean (parallelogram shaped 400 × 120 mm) is used for the experiment.

For the needs of the experimental process, the fabric is positioned and fastened manually in the sewing machine. At the beginning, the PID controller is activated in order to pre-stress the fabric at 2N force ($F_{reference}$) which is assumed to be the desired tensional force for the specific type of fabric [4]. Then the sewing machine is activated and remains on, until distance $L$ becomes less than 120 mm. This experiment was repeated several times so as to find the optimal times that the three gains should be changed. At first the experiment was executed for 1 set and then for 2, 4, 8, 16, 32, 64, 128 and 256 sets of gains. Sewing effectiveness was reviewed via calculating the MSE of the force signal, which is plotted in Fig. 10b, which reveals that the most effective sewing was achieved for approximately 32 sets of gains (the lowest value of MSE in Fig. 10b). Hence, the case of changing the gains 32 times during the sewing is used.

Comparing Fig. 11a, b we mark that PID with adaptable gains (Fig. 11a) can control the tension force in a range of ±1N even in small actual lengths of the fabric (at the end of the sewing process) opposed to the PID with constant gains (Fig. 11b),

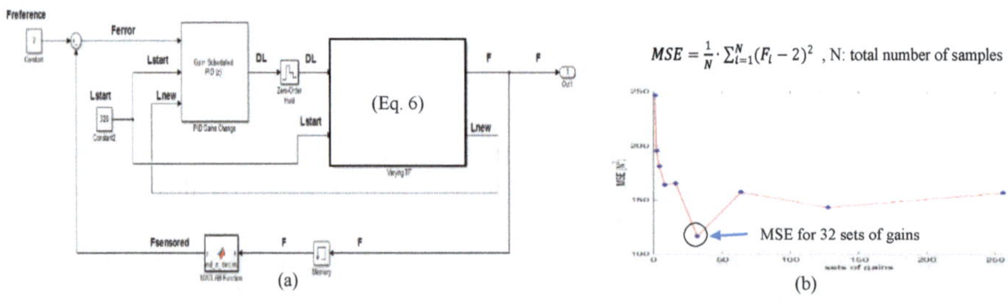


**Fig. 10.** Gain scheduled PID control scheme for the robotized sewing (a) and Mean Square Error of force signal for related sets of gains (b).

where when the length of the fabric is very small (the extensibility of the fabric is very small) then the controller response is worsened (tension force range ≅ ±2N).

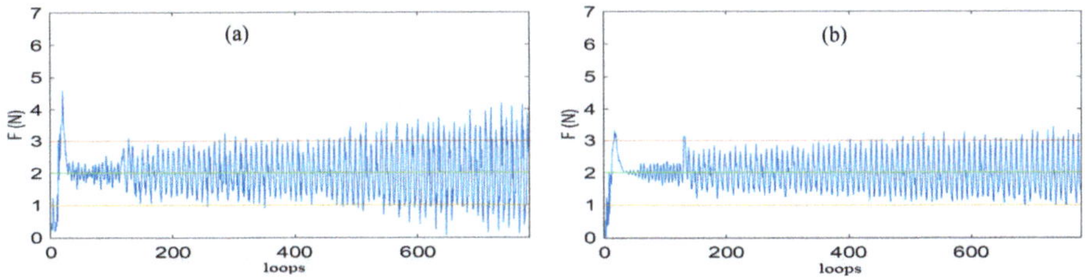


**Fig. 11.** Robotized real time sewing with (a) 1 set of gains (a PID controller with constant gains), (b) 32 sets of gains throughout the sewing process.

As shown in Fig. 11 there is an oscillation in the force sensor signal during the sewing process. To explain it, the Power Spectral Density (PSD) via Welch method was used to find the dominant frequency of the force signal, which is f = 2.344 Hz. This is too close to the operating frequency of the sewing machine ($f_s = 2.5\,\text{Hz}$). So, we conclude that the system cannot cut off the frequencies of the external disturbance. Furthermore, plotting both the velocity of the sewing machine and robot's one versus time (Fig. 12), we notice that robot velocity (blue line) differs from sewing machine's one (red line). There is an overshoot, which the PID controller is trying to smooth it out during the period of one stitch. To improve the controller performance, a moving average filter (four values window size) is implemented, which improves slightly the disturbance rejection, but the controller response is delayed. However, the small variation of the applied tensional force on fabric (±1N) could be assumed acceptable for the seam quality. The presented method could be easily applied on other types/sizes of fabrics, by following the steps of the method, starting with the necessary two experiments for extensibility and damping, as described in Sect. 2.

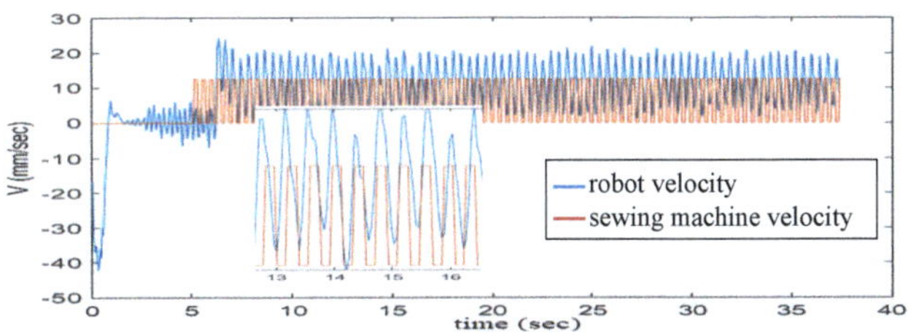


**Fig. 12.** Both robot (blue line) and sewing machine (red line) velocities versus time

## 5 Conclusion and Future Work

A gain scheduled PID controller is introduced for the robotized sewing of a piece of fabric (woven elastic jean) using an industrial sewing machine. The tuning and scheduling of the controller were performed offline using Matlab/Simulink by incorporating experimental tensile and damping tests. From the results, when comparing the gain scheduled PID response with the one where a typical PID with constant gains is used, it can be clearly seen the improvement of the controller response. The achieved reduced oscillations of the fabric tensional force, especially at the end of the sewing process, could improve the final quality of the seam.

As a future work, to improve the robotized sewing and hence the cloths seam quality, a detailed model of the fabric will be considered, that will be based on a complex structure of more than one springer/damping elements. On the other hand, the use of an admittance controller with variable stiffness and considering also the apparent mass of the robot is planned.

## References

1. Gershon, D.: Strategies for robotic handling of flexible sheet material. Mechatronics **3**(5), 611–623 (1993)
2. Kudo, M., Nasu, Y., Mitobe, K., Borovac, B.: Multi-arm robot control system for manipulation of flexible materials in sewing operation. Mechatronics **10**(3), 371–402 (2000)
3. Koustoumpardis, P.N., Aspragathos, N.A.: Robotized sewing of fabrics based on a force neural network controller. In: 4th International Conference on Intelligent Robotics and Applications, Lecture Notes in Computer Science, ICIRA 2011, Germany, vol. 7101, LNAI, pp. 486–495. Springer, Heidelberg (2011)
4. Koustoumpardis, P.N., Aspragathos, N.A.: Intelligent hierarchical robot control for sewing fabrics. Robot. Comput. Integr. Manuf. **30**(1), 34–46 (2014)
5. Triantafyllou, D., Koustoumpardis, P., Aspragathos, N.: Model reference fuzzy learning force control for robotized sewing. In: 19th Mediterranean Conference on Control and Automation, MED 2011, Corfu, Greece, June 20–23, 2011
6. Schrimpf, J., Wetterwald, L.E.: Experiments towards automated sewing with a multi-robot system. In: Proceedings - IEEE International Conference on Robotics and Automation, pp. 5258–5263 (2012)
7. Book, W.J., Killpack, M., Huggins, J., Collins, T., Dickerson, S.: UPL-2620, pp. 1–10 (2010)
8. Kawabata, M., Niwa, S.: Objective measurement of fabric mechanical property and quality. J. Cloth. Sci. Technol. **3**(1), 7–18 (1995)
9. Minazio, P.G., Minazio, P.G.: FAST – fabric assurance by simple testing (1996)
10. Stylios, G.K.: New measurement technologies for textiles and clothing. Int. J. Cloth. Sci. Technol. **17**(3/4), 135–149 (2005)
11. Liu, Y., Hu, H.: Compressive mechanics of warp-knitted spacer fabrics. Text. Res. J. **86**(1), 3–12 (2016)
12. Leith, D.J., Leithead, W.: Survey of Gain-Scheduling Analysis & Design
13. Stilwell, D.J., Rugh, W.J.: Interpolation methods for gain scheduling. In: IEEE Conference on Decision and Control, pp. 3003–3008 (1998)
14. Inman, D.J.: Engineering Vibration, pp. 43–48. Pearson Education Inc., Upper Saddle (2008). ISBN 0-13-228173-2

# Industry Robotics

# Implementation of an Autonomous Tool Trolley in a Production Line

Heiko Engemann[1(✉)], Sriram Badri[2], Marius Wenning[2], and Stephan Kallweit[3]

[1] Tshwane University of Technology, Pretoria 0183, South Africa
engemann@fh-aachen.de
[2] Chair of Production Engineering of E-Mobility Components (PEM), RWTH Aachen University, 52074 Aachen, Germany
[3] Faculty of Mechanical Engineering and Mechatronics, University of Applied Sciences Aachen, 52064 Aachen, Germany

**Abstract.** In recent years, the rising demand for customised products has changed the requirements for industrial production. Nonetheless human workers are still an important part of the production process, especially in the area of final assembly. In the production of small series as well as customer-configured products, tools for different work tasks are used on different workstations at various instances of time. This paper documents the development of an autonomous tool trolley (ATT), which makes the production equipment effortless available directly at the required location. The implemented control system is based on the Robot Operating System (ROS) and provides autonomous navigation functionalities. An experiment compares different sensor concepts used for localisation. The experiments are performed in the unstructured environment of an industrial production line. The best preforming sensor setup is used for the task of autonomous navigation. A final experiment proves that the positioning accuracy of the developed ATT is suitable for the tool delivery task.

**Keywords:** Mobile robotics · Industrial production · Robot Operating System

## 1 Introduction

While western countries dominated the global production most of the 20th century, today emerging countries are putting increasing pressure on high-wage countries. Significantly cheaper labour costs cause a competitive advantage for low-wage countries [1]. Numerous product variations and rapid response to customer requirements through short-cycle products set the spotlight on increasing the production flexibility [2]. Automation allows lowering the price per piece decoupled from labour costs. However, highly automated production facilities are very limited regarding the product variance. The ability to adapt to new external and internal circumstances, while leaving enough room for further developments


K. Berns and D. Görges (Eds.): RAAD 2019, AISC 980, pp. 117–125, 2020.
https://doi.org/10.1007/978-3-030-19648-6_14

describes an efficient production flexibility [3]. Since automation is yet not as flexible as humans are, the seamless integration of humans into production systems can compensate the robot's inflexibility. On the other hand, automation increases workers' productivity and helps optimizing blue-collar workers' satisfaction, i.e. due to less physical stress [4]. Therefore, automation devices enable elder employees to participate less independently from their physical fitness, which is becoming more important considering an aging society [5].

Therefore, autonomous robots are increasingly becoming part of flexible industrial production. In recent years great efforts were made to integrate these robots into the production process. Different approaches in the area of mobile manipulation [6,7] or collaborative human robot interaction [8–10] show the great potential. However, such assistance systems are not widely used in real industrial applications, mainly caused by the high safety requirements and the high overall complexity of such systems [11].

In the production of small series as well as customer-configured products, tools for different work tasks are used on different workstations at various instances of time. The tool trolley is a prime example for an often-used asset in highly flexible manufacturing tasks. The frequent use and the need at different stations proposes an automation of the tool supply. Depending on the organisational form of the shop floor, the autonomous tool trolley (ATT) promises a high reduction of walking distances.

## 2 System Overview

The mechanical and electrical design are based on low cost components and a standard tool trolley chassis. A software framework is implemented based on the middle-ware Robot Operating System (ROS) [12]. The developed system provides autonomous functionalities for a collision free navigation in the highly dynamic production environment.

In order to automate a common tool trolley an under body was developed. The under body contains the motors and gearboxes as well as a corresponding lead battery (s. Fig. 1).

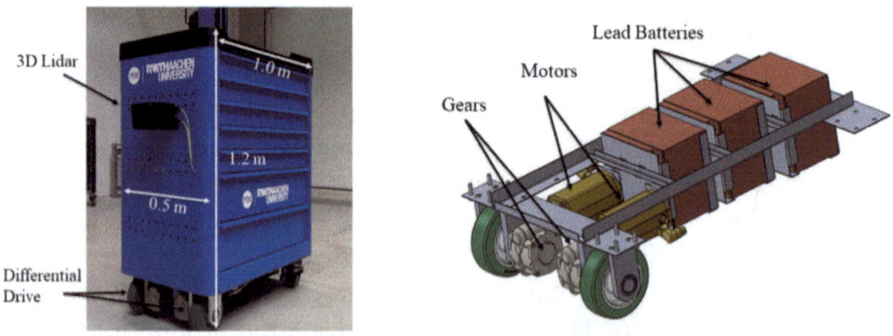


**Fig. 1.** Automated tool trolley (left) and under body (right).

The ATT has a differential kinematics, including a damping system, to reduce the effects of ground unevenness. Encoders, an IMU and a 3D lidar acquire internal and external sensor data during operation.

## 3 Autonomous Navigation

Figure 2 shows a schematic of the software framework of the ATT. The following sections document the implementation of the three core functionalities of autonomous navigation in the software framework.

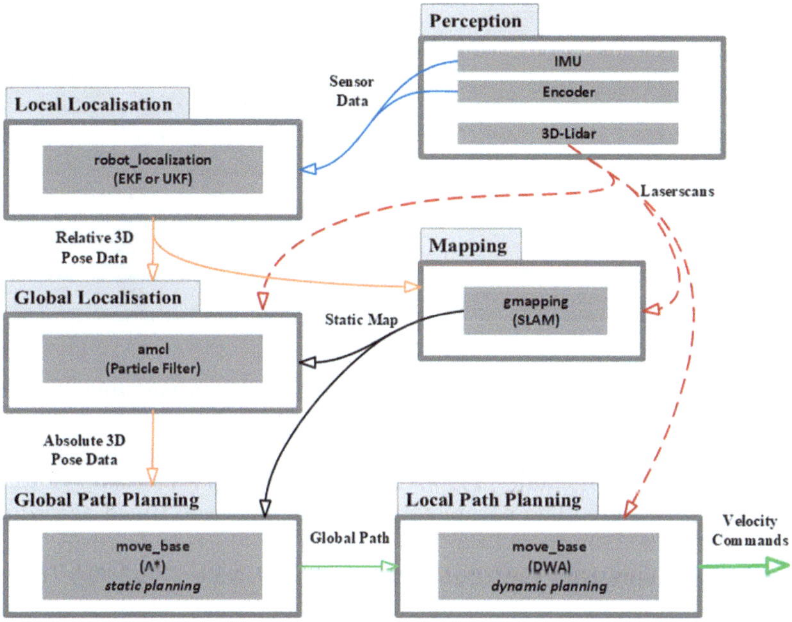


**Fig. 2.** Software framework of the automated tool trolley.

The **local localisation** provides information about the current pose of the ATT in relation to its previous pose. Therefore the ROS package *robot_localization* [13] is used. It provides an implementation of an Extended Kalman Filter (EKF) as well as an Unscented Kalman Filter (UKF) [14]. The EKF represents the probability density function (PDF) as a Gaussian Random Variable (GRV) and propagates it through the first order linearization of the non-linear system. The UKF avoids this linearization process and uses a deterministic sampling approach called Unscented transform. There is no clear evidence as to which filter performs the best and is more dependent on the application [15–18]. Hence, both filter approaches were validated as part of the local localisation. The local pose estimation is used for the mapping procedure. In addition, it represents the initial pose for the global localisation.

The **global localisation** provides information about the current pose of the robot in a fixed reference frame. In case of the ATT represented by a occupancy grid map of the production environment. The global localization is based on the adaptive Monte Carlo localisation described in [19]. For the implementation the ROS package *amcl*[1] is used.

Figure 3 left shows the **occupancy grid map** of the production environment. It was generated by using the ROS package *gmapping*[2], which is based on the work described in [20] and [21].

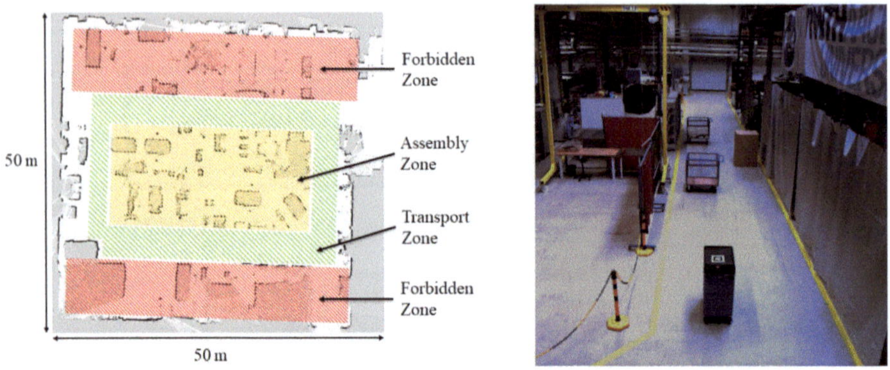


**Fig. 3.** Occupancy grid map (left) and partly occupied transport zone (right).

The SLAM process includes a global localisation step and a loop closure for the mapping procedure. The parameters of gmapping were tuned according to the work in [22] and [23]. In addition, the different production zones are marked. The green area represents the transport zone. This zone is used to transport goods to the workstations of the manual assembly zone, which is marked in yellow. The red marked areas are forbidden zones, as these are test benches and welding workshops.

The task of **path planning** is based on the ROS stack *navigation*[3], which provides an implementation of different modern global and local path planners. The utilised global planner is based on the A* approach described in [24]. The applied local planner is based on the work in [25]. It provides the motion commands in consideration of collision avoidance based on the data of the 3D Lidar. Figure 3 right shows an example situation of the ATT navigating in the transport zone, which is partly occupied by typical dynamic obstacles.

[1] http://wiki.ros.org/amcl.
[2] http://wiki.ros.org/gmapping.
[3] http://wiki.ros.org/navigation.

# 4 Experiments

The experimental setup consists of the high-resolution camera Manta-G609B mounted at a tripod. The camera system is pointing downwards to detect an AR-marker. The marker is attached to the top of the ATT at a fixed pose. Figure 4 shows the used measurement setup. The camera frame is static. The frame, which represents the ATT, moves accordingly to the results of the localisation process. If the AR-marker is inside the field of view (FOV) of the camera system, the 6D pose of the marker is calculated using the ROS package *ar_track_alvar*[4]. The pose of the AR-marker represents the ground-truth pose of the ATT.

**Fig. 4.** Experimental setup.

A preceding experiment ensures the traceability of the measurement system. A ruler is used as the reference standard. It is manufactured following the *DIN 866-B* and provides a tolerance of 100 $\mu m$. The ruler is positioned with an accuracy of ± one pixel along the x- and y-axis in the camera frame. At chosen positions the relative displacement error is measured. Therefore, alongside a 7×7 grid, with a cell size of 100 mm, the position provided by *ar_track_alvar* is compared to the reference position. In addition, the accuracy of the measurement system is determined over 400 measurements at each chosen position. The traceability of the measurement system results in a measurement deviation of 1.5 mm ± 0.5 mm in x-direction and a measurement deviation of 2.1 mm ± 0.5 mm in y-direction for a relative displacement measurement.

## 4.1 Localisation Error

To estimate the local localisation error, the ATT is controlled remotely. The ATT performs one circular round through the navigation zone. The 3D Pose (x, y, theta), provided by the local localisation process, is compared with the

---

[4] http://wiki.ros.org/ar_track_alvar.

displacement measured by the camera setup. The whole procedure is repeated 20 times. The different local localisation approaches are preformed offline, based on the recorded sensor data. In addition, four different types of IMUs are tested: MPU6050, Xsens MT-i1, Fairchild FMT1000, Razor SEN-10736.

Table 1 shows the accuracy of the local localisation based on encoder and IMU data.

**Table 1.** Local localisation error based on encoder data. Mean error (ME) and the standard deviation (SD) over the 20 performed runs.

| Filter approach | $ME_x[m]$ | $ME_y[m]$ | $ME_{theta}[deg]$ | $SD_x[m]$ | $SD_y[m]$ | $SD_{theta}[deg]$ |
|---|---|---|---|---|---|---|
| Encoder | | | | | | |
| EKF | 15.45 | 4.24 | 70.87 | 1.07 | 1.10 | 4.94 |
| UKF | 15.45 | 4.35 | 70.76 | 1.06 | 1.09 | 4.9 |
| **Encoder & IMU (Xsens MT-i1)** | | | | | | |
| EKF | 0.30 | 0.13 | 2.57 | 0.18 | 0.15 | 1.81 |
| UKF | 0.30 | 0.14 | 2.42 | 0.18 | 0.15 | 1.79 |
| Encoder & IMU (Fairchild FMT1000) | | | | | | |
| EKF | 1.44 | 0.41 | 5.8 | 0.56 | 0.25 | 2.95 |
| UKF | 1.44 | 0.42 | 5.8 | 0.56 | 0.25 | 2.96 |
| Encoder & IMU (Razor SEN-10736) | | | | | | |
| EKF | 3.35 | 1.56 | 16.9 | 4.96 | 2.94 | 26.38 |
| UKF | 4.02 | 2.64 | 18.09 | 6.37 | 5.09 | 26.13 |
| Encoder & IMU (MPU6050) | | | | | | |
| EKF | 10.52 | 5.19 | 39.49 | 7.27 | 7.48 | 31.66 |
| UKF | 10.51 | 5.18 | 39.46 | 7.28 | 7.47 | 31.57 |

The filtered data of the encoder and the Xsens IMU provide the best local localisation results.

### 4.2 Positioning Error

The experimental procedure to evaluate the positioning error is performed accordingly to the localisation experiment. As mentioned in Sect. 3 a global localisation method is used to enlarge the local localisation approach. The corner points of the reference path are set as intermediate goal poses for each test run. One set of runs is performed in clockwise direction. An additional set of runs is performed in counter-clockwise direction. Table 2 shows the experimental results.

For this experiment the goal tolerance of the local path planner was set to 0.1 m in x- and y-direction and 2 degree for theta. A lower goal tolerance resulted in a recurring situation, where the ATT was not able to reach the goal

**Table 2.** Positioning error in autonomous navigation. Mean error (ME) and the standard deviation (SD) over the 20 performed runs.

| Direction | $ME_x[m]$ | $ME_y[m]$ | $ME_{theta}[deg]$ | $SD_x[m]$ | $SD_y[m]$ | $SD_{theta}[deg]$ |
|---|---|---|---|---|---|---|
| Clockwise | 0.10 | 0.08 | 2.6 | 0.05 | 0.05 | 1.8 |
| Counter-clockwise | 0.05 | 0.08 | 1.2 | 0.02 | 0.06 | 0.9 |

position, instead it started to oscillate. The oscillating behaviour indicates, that the motor encoder combination isn't precise enough to properly execute the velocity commands given by the local path planner. In addition, the counter-clockwise direction results in a lower mean error, which also indicates that the accuracy depends heavily on the motors itself.

The positioning error of the ATT is smaller than the previous estimated localisation accuracy, which is caused by the global localisation process. In fact, the autonomous navigation was not able to reach the global localisation accuracy throughout the test runs.

## 5 Conclusion

This work documents the development of an autonomous tool trolley, whereby the focus is on the implementation of a software framework for the task of autonomous navigation. An experiment was performed to determine the local localisation accuracy of different sensor setups as well as filtering approaches. The combination providing the best results was used throughout an additional experiment to estimate the positioning error of the autonomous tool trolley.

With a positioning error of $0.05\,\text{m} \pm 0.02\,\text{m}$ in x- and $0.08\,\text{m} \pm 0.06\,\text{m}$ y-direction as well as $1.2$ degree $\pm\,0.9$ degree for theta the functionality of autonomous navigation was implemented successfully. The used differential kinematic is easy to implement and well supported by the framework ROS. It meets the requirements of the ATT navigating through the transport zones (s. Fig. 3) of the production environment.

However, the reachable goal tolerance of the ATT isn't sufficient for highly dynamic and mainly occupied areas, like the assembly zone. By implementing a more precise motor and encoder combination the positioning error can be further reduced towards the global localisation accuracy. In addition, a holonomic kinematic would offer a higher manoeuvrability.

The ATT increases blue-collar worker's productivity while keeping the flexibility on a high level. The commissioning of tools reduces the worker's productive time. Moreover, the manual movement of heavy tool trolleys afflicts especially elder humans.

**Acknowledgments.** The results in this paper were achieved within the project "A4BLUE: Adaptive Automation in Assembly for BLUE collar workers satisfaction in Evolvable context" funded by the European Commission (H2020); Grant Agreement Number 723 828.

## References


1. Tseng, M.: Industry development perspectives: global distribution of work and market. CIRP 53rd General Assembly, Montreal, Canada (2003)
2. Asadi, N., Fundin, A., Jackson, M.: The essential constituents of flexible assembly systems: a case study in the heavy vehicle manufacturing industry. Global J. Flex. Syst. Manage. **16**(3), 235–250 (2015)
3. Toni, A.D., Tonchia, S.: Manufacturing flexibility: a literature review. Int. J. Prod. Res. **36**(6), 1587–1617 (1998)
4. Fletcher, S., Johnson, T., Larreina, J.: Putting people and robots together in manufacturing: are we ready? (2017)
5. Hall, A., Tiemann, M.: Abschlussbericht BIBB/BAuA-Erwerbstätigenbefragung 2012, p. 20–21 (2012)
6. Madsen, O., Bøgh, S., Schou, C., Andersen, R.S., Damgaard, J.S., Pedersen, M.R., et al.: Integration of mobile manipulators in an industrial production. Ind. Robot Int. J. **42**(1), 11–18 (2015)
7. Saenz, J., Penzlin, F., Vogel, C., Fritzsche, M.: VALERI–a collaborative mobile manipulator for aerospace production. In: Advances in Cooperative Robotics, p. 186–195. World Scientific (2017)
8. Cherubini, A., Passama, R., Crosnier, A., Lasnier, A., Fraisse, P.: Collaborative manufacturing with physical human-robot interaction. Robot. Comput.-Integr. Manuf. **40**, 1–13 (2016)
9. Morioka, M., Sakakibara, S.: A new cell production assembly system with human-robot cooperation. CIRP Ann. **59**(1), 9–12 (2010)
10. Tsarouchi, P., Matthaiakis, A.S., Makris, S., Chryssolouris, G.: On a human-robot collaboration in an assembly cell. Robot. Comput. Integr. Manuf. **30**(6), 580–589 (2017)
11. Saenz, J., Elkmann, N., Gibaru, O., Neto, P.: Survey of methods for design of collaborative robotics applications-Why safety is a barrier to more widespread robotics uptake. In: Proceedings of the 2018 4th International Conference on Mechatronics and Robotics Engineering, p. 95–101. ACM (2018)
12. Quigley, M., Conley, K., Gerkey, B.P., Faust, J., Foote, T., Leibs, J., et al.: ROS: an open-source Robot Operating System. In: ICRA Workshop on Open Source Software (2009)
13. Moore, T., Stouch, D.: A generalized extended Kalman filter implementation for the robot operating system. In: Proceedings of the 13th International Conference on Intelligent Autonomous Systems (IAS-13), Springer, Cham (2014)
14. Wan, E.A., Van Der Merwe, R.: The unscented Kalman filter for nonlinear estimation. In: Adaptive Systems for Signal Processing, Communications, and Control Symposium 2000, AS-SPCC, The IEEE 2000, p. 153–158. IEEE (2000)
15. Kumar, S., Prakash, J., Kanagasabapathy, P.: A critical evaluation and experimental verification of Extended Kalman Filter, Unscented Kalman Filter and Neural State Filter for state estimation of three phase induction motor. Appl. Soft Comput. **11**(3), 3199–3208 (2011)
16. Orderud, F.: Comparison of Kalman filter estimation approaches for state space models with nonlinear measurements. In: Proceedings of Scandinavian Conference on Simulation and Modeling, p. 1–8 (2005)
17. LaViola, J.J.: A comparison of unscented and extended Kalman filtering for estimating quaternion motion. In: Proceedings of the 2003 American Control Conference, 2003, vol. 3, p. 2435–2440. IEEE (2003)

18. St-Pierre, M., Gingras, D.: Comparison between the unscented Kalman filter and the extended Kalman filter for the position estimation module of an integrated navigation information system. In: IEEE Intelligent Vehicles Symposium, p. 831–835. Citeseer (2004)
19. Fox, D.: KLD-sampling: adaptive particle filters and mobile robot localization. In: Advances in Neural Information Processing Systems (NIPS) (2001)
20. Grisetti, G.: Improving grid-based slam with rao-blackwellized particle filters by adaptive proposals and selective resampling. In: Proceedings of the IEEE International Conference on Robotics & Automation (ICRA), p. 2443–2448 (2005)
21. Grisetti, G., Stachniss, C., Burgard, W.: Improved techniques for grid mapping with Rao-Blackwellized particle filters. IEEE Trans. Robot. **23**, 34 (2007)
22. Abdelrasoul, Y., Absh, S., Sebastian P.: A quantitative study of tuning ROS gmapping parameters and their effect on performing indoor 2D SLAM. In: 2016 2nd IEEE International Symposium on Robotics and Manufacturing Automation (ROMA), p. 1–6 (2016)
23. Zaman, S., Slany, W., Steinbauer, G.: ROS-based mapping, localization and autonomous navigation using a Pioneer 3-DX robot and their relevant issues. In: Electronics, Communications and Photonics Conference (SIECPC), 2011 Saudi International, p. 1–5. IEEE (2011)
24. Brock, O., Khatib, O.: High-speed navigation using the global dynamic window approach. In: Proceedings of 1999 IEEE International Conference on Robotics and Automation, vol. 1, p. 341–346. IEEE (1999)
25. Fox, D., Burgard, W., Thrun, S.: The dynamic window approach to collision avoidance. IEEE Robot. Autom. Mag. **4**(1), 23–33 (1997)

# Developing a Production Scheduling System for Modular Factory Using Constraint Programming

Hoonseok Park[1], Jumyung Um[1(✉)], Jae-Yoon Jung[1], and Martin Ruskowski[2]

[1] Department of Industrial and Management Systems Engineering, Kyung Hee University, 1732 Deogyeong-daero, Giheung-Gu, Yongin-si, Gyeonggi-do, South Korea
{hoonseok, jayum, jyjung}@khu.ac.kr
[2] IFS-Innovative Factory Systems, German Research Center for Artificial Intelligence (DFKI GmbH), Trippstadter Str. 122, 67663 Kaiserslautern, Germany
Martin.Ruskowski@dfki.de

**Abstract.** This paper presents a production scheduling system that optimizes operations of a modular factory. The proposed system consists of a database, a scheduling optimizer and an interface connecting two other components. A scheduling model for optimal schedules plays a key role in the scheduling system. In this research, we aim to develop the scheduling model using constraint programming, and design a data mart storing the related operational data. In particular, we set the modular factory environment to the hybrid flow shop and apply the constraint programming for scheduling. We provide a case study to test the performance of the proposed scheduling model with the synthesized dataset based on a modular factory environment in *SmartFactory*$^{KL}$.

**Keywords:** Smart factory · Flexible manufacturing · Production scheduling · Hybrid flow shop · Constraint programming

## 1 Introduction

Modularizing factory systems have improved due to the demand of fast time-to-market and customization, and they are recognized as a critical factor of smart factory. The concept of modular factory represents the paradigm of smart factory, and defines the structure and interfaces operations. This concept has been proposed and has evolved by *SmartFactory*$^{KL}$ since 2005 [1].

Satisfying the goal of Industry 4.0, modular factory has to achieve the reduction of reconfiguration time, adapt time of production individualization, flexible line configuration, fast decision making by distributed control, and high utilization of manufacturing resources.

In terms of resource utilization, modular factory harmonizes traditional facilities within the concept of optimized schedule. For this reason, the scheduling of modular


K. Berns and D. Görges (Eds.): RAAD 2019, AISC 980, pp. 126–133, 2020.
https://doi.org/10.1007/978-3-030-19648-6_15

factory needs innovative approaches because of new requirements that are not considered in traditional factory.

Scheduling manufacturing systems is essential to enhance productivity and avoid the delay of delivery. It can be achieved by efficiently utilization of the limited resource, such as time, machine, and human and energy [2]. For example, when a new production order comes, the problem of order sequencing for each machine to reduce the lead time of whole resources should be solved.

From the perspective of the scheduling algorithm, a manufacturing system is defined as various models with resource, layout, task flow and other constraints. The scheduling modeling problems can be categorized for single machine system, parallel machine system, job shop, flow shop, flexible flow shop and open shop. Among them, modular factory belongs to mixed flow production which includes more than one scheduling modeling problem.

In this research, we design a modular factory to the hybrid flow shop because this scheduling problem is suitable to represent the characteristic of modular factories such as high flexibility and scalability.

The remainder of this paper is structured as follows. Section 2 describes the proposed production scheduling system architecture and the scheduling model using the constraint programming(CP). Section 3 provides the case study to evaluate the proposed methodology on the modular factory in *SmartFactory*$^{KL}$ and the experimental results. Finally, the conclusion of this research is described in Sect. 4.

## 2 Methodology

The proposed production scheduling system architecture for modular factory consists of four components as shown in Fig. 1. The modular factory manufactures the product according to order information such as product type, quantity and due time. Each type of product has different cycle time from start to end of the process. These kinds of data called the operation data are collected from the production system and they are stored in the operational databases.

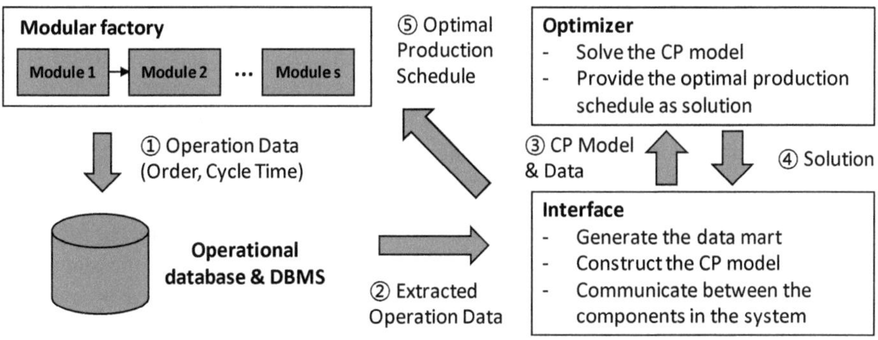


**Fig. 1.** The system architecture of the production scheduling system for the modular factory.

To construct a scheduling model, the interface connects to the operational database, and then generates the data mart based on the extracted operation data. This data mart contains records with appropriate schema to establish the scheduling model. After that, the interface calls an optimizer to solve the model. With the scheduling model, the optimizer interprets the data coming through the interface, and returns the best solution in the configured time. When a scheduling solution is given, the interface sends the schedule to the modular factory. And run this procedure repeatedly even if system is in operation mode.

## 2.1 Production Scheduling for a Modular Factory

There are a number of variant which describes scheduling problems such as single machine, parallel machines, flow shop and job shop. In this research, we design a modular factory to the hybrid flow shop because this scheduling problem is suitable to represent the characteristic of modular factories such as high flexibility and scalability.

The structure of the hybrid flow shop has $s$ stages is shown as Fig. 2. This scheduling problem has the combined features of parallel machines and flow shop. Each stage $t$ has one or more machines in parallel, and at least one of the stages has more than one machine. All of $n$ jobs are processed through the same production flow. In this structure, it is assumed that the machines in all the stages are identical.

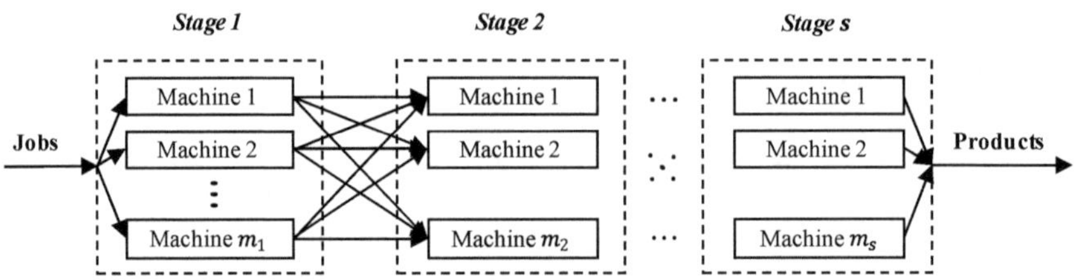


**Fig. 2.** The structure of a hybrid flow shop with $s$ stages

## 2.2 Constraint Programming Modelling

CP is one of promising approaches to solve scheduling problems and it achieved good results in many application areas [3]. CP provides the useful expression that can formulate temporal and capacity constraints [4]. To formulate the hybrid flow shop scheduling problem in CP, we determine the modelling components like interval variables, objective functions and constraints as follows.

- **Interval variables:** In this problem, we set two interval variables, $job_{jt}$ and $machine_{jk_t}$, to avoid the overlapped and redundant job sequence in the stage with parallel machines. $job_{jt}$ represents the time interval of job $j$ processed by a certain machine in a stage $t$. In the similar context, $machine_{jk_t}$ represents the time interval of machine $k$ in stage $t$ to process job $j$. CP optimizer solves the sequence of these interval variables that is unknown in the scheduling problem.

- **Objective functions:** There are many objective functions in scheduling such as maximum completion time (a.k.a. makespan), tardiness, lateness, flow time and execution cost. Given a schedule, $C_j$ is the completion time of job $j$ and *makespan*, denoted $C_{max}$ is the maximum completion time of job $j$ in the last stage of the shop (stage $m$). Let $d_j$ be the due date of job $j$ and the *tardiness*, $T_j = \max\{C_j - d_j, 0\}$. In this research, three objective functions, *makespan* $C_{max}$, *total tardiness* $\sum T$ and the sum of two $C_{max} + \sum T$ are adopted to reflect the real-world situation, which generally consider multiple key performance indicators. They are also performance metrics to measure cases when the solution is applied to factory operations. In the semantic of CP, *makespan* and *total tardiness* can be rewritten as follows.
  - *Makespan*: $C_{max}= max(endOf(job_{js}))$, *where* $j = 1, \ldots, n$
  - *total tardiness*: $\sum T= sum(max\{0, endOf(job_{js}) - d_j\})$ *for all* $j = 1, \ldots, n$
- **Constraints:** CP provides a set of specialized modeling features to describe the constraints for the scheduling problem. Three temporal and sequential constraints need to be described for the proposed hybrid flow shop problem as follows.
  - *endBeforeStart*($job_{jq}$, $job_{jt}$, $z$): The a job $j$ operation in the certain stage $q$ must be done at least $z$ time units before the start of job operation in the consecutive stage $t$. This constraint describes the characteristic of the flowshop that each stage can only start after the previous stage finishes. In this problem, we set $z = 0$ because it is assumed that there is no lead time between two stages in the production line.
  - *alternative*($job_{jt}$, $\{machine_{j1_t}, \ldots, machine_{jm_t}\}$): If interval $job_{jt}$ is present then exactly one of intervals $\{machine_{j1_t}, \ldots, machine_{jm_t}\}$ is chosen to process it and $job_{jt}$ is starts and ends at the same time with this chosen interval $machine_{jk_t}$. This constraint states the stage with parallel machines.
  - *noOverlap*($\{machine_{1k_t}, \ldots, machine_{nk_t}\}$): We assume that a machine in a modular factory can perform only one operation at a time. In this context, the *noOverlap* constraint makes a sequence of interval variables $machine_{jk_t}$ as a chain of non-overlapping intervals regarding machine $k$ at stage $t$.

# 3 Case Study

## 3.1 Modular Factory in *SmartFactory$^{KL}$*

The modular factory in *SmartFactory$^{KL}$* consists of modules which are provided from the multi-vendors and connected through standardized interfaces [5]. This production line is developed to realize the highly modular and scalable production system as a reference for Industry 4.0 [6]. Therefore, the modular factory in *SmartFactory$^{KL}$* is appropriate to evaluate the proposed scheduling method.

This production line produces 8 types of personalized business card cases designed with four components such as bottom plate, retaining spring, lid and inlay. Every type of product has the same manufacturing route through 7 stages as described in Fig. 3. The various colors of the lid enable to provide the customized product, but it causes the difference of processing time among the product types. This time difference makes the bottleneck between the adjacent stages, and reduce the productivity of the whole production line.

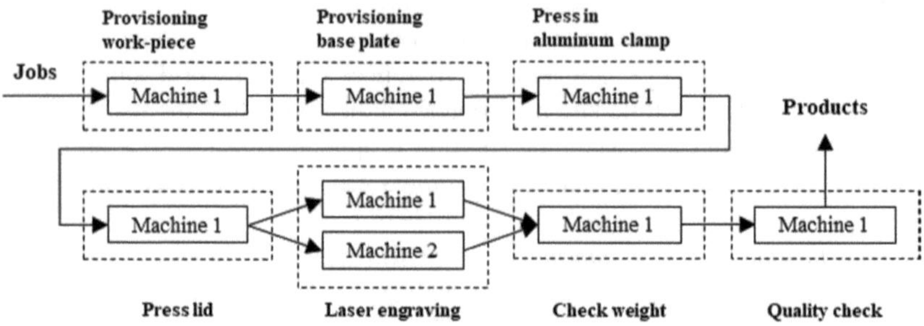


**Fig. 3.** An example of production process that has two machines in the laser engraving stage

Each module in the modular factory has just a machine. Therefore, one of the ways to solve aforementioned problem is to add an additional machine to the bottleneck stages, and then rearrange the job sequence based on the optimized schedule from the algorithmic scheduling model. For example, we can install two machines on the laser engraving stage as shown in Fig. 3. Since the modular factory also aims at high flexibility and scalability compared to the traditional manufacturing systems, we reflect these characteristics to the test scenario for the proposed scheduling system.

Because of the lack of real order data, a synthetic dataset is prepared for this experiment. Although it is important to validate the usefulness of the solution in the real-world situation, there is a limitation to cover all cases. Therefore, we construct 12 scenarios with consideration on several factors such as the number of machines per stage, the number of orders per product type, release time/due time of orders. To reflect the stochastic environment, we set the due time when the release time is zero, as below. These scenarios are described in Table 1.

- $(due\ time) = (release\ time) + uniform(cycleTime_p,\ 1.2 \times cycleTime_p)$

where, given product type $p$, $cycleTime_p$ is the cycle time of the $p$.

**Table 1.** The synthetic production scenarios for testing the CP scheduling model

| Factors | Levels |
|---|---|
| Number of orders per product type | 5, 10, 15, 20 |
| Number of machines per stage | 1, 2, 3 |

## 3.2 Results

The scheduling model for 12 scenarios were solved by the optimizer when the time limit is set on 60 s. The three kinds of objective function are adopted to optimize the schedule with regarding to each scenario. We could evaluate the schedule using makespan and total tardiness as performance metric, as shown in Fig. 4: the makespan is in the left-side and the tardiness is in the right-side.

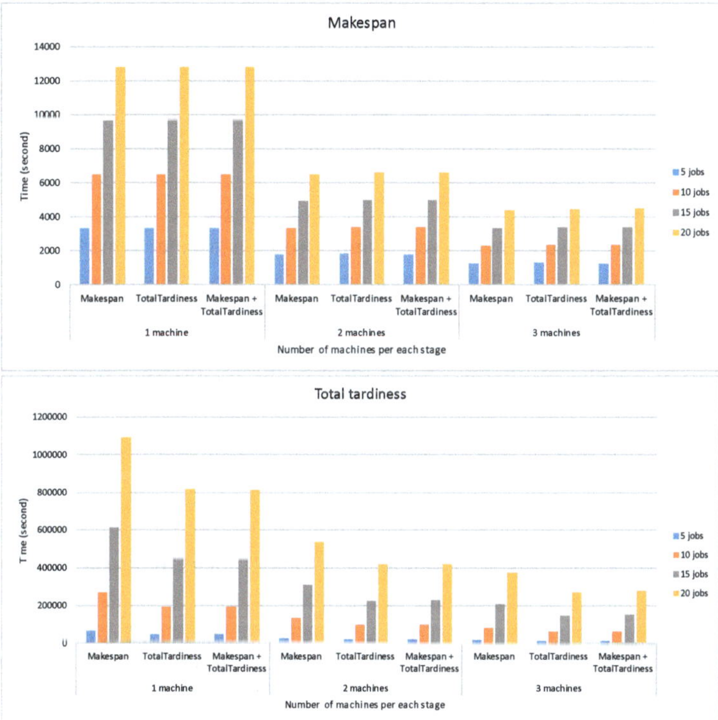


**Fig. 4.** The performance comparison of the three objective functions for 12 scenarios. The two performance metrics are shown as makespan (top) and total tardiness (bottom).

As can be seen, there is monotonic patterns between two factors and two performance metrics. both makespan and total tardiness are decreasing when the number of machines per each stage is increasing. On the contrary, the increasing of amount of job makes growth of the performance metrics.

The objective functions are directly connected to the performance metrics. For example, if the objective function of certain scheduling model is makespan, then the solution from this model has minimum makespan as a performance metric. It should have smaller makespan than a scheduling model which has the objective function as the total tardiness.

Interestingly, with regarding to the makespan, this performance metric is less affected from the difference of objective function. On the other hand, the minimum total tardiness is achieved when the objective function is also total tardiness, significantly. This phenomenon caused by the stochastic setting for the due time while cycle time is deterministic. This setting describes more realistic situation. Therefore, it shows that the appropriate setting of objective function related to interest has more advantage in the real-world with uncertainty.

## 4 Conclusion

In this paper, we presented the system architecture and the scheduling model using CP for the production scheduling system of a modular factory. The proposed system architecture was developed to define the dataflow and functionality of each component for the production scheduling system.

Interval variables, objective functions and constraints were considered to formulate the modular factory based on the scheduling problem as the hybrid flow shop. To evaluate the performance of the scheduling model, the synthetic dataset was used considering the features of the production line such as like high flexibility and scalability.

In the experimental results, we could identify the relationship between stochastic setting and objective functions. If there is uncertainty in the temporal variable like cycle time and due time, the setting of objective function is getting important.

The main contribution of this study is on a dealing of the practical scheduling problem in the systematic perspective for the modular factory. Furthermore, we provided the case study based on the reference modular factory in $SmartFactory^{KL}$.

Although, we try to consider the realistic problem, there is still a number of limitations. For example, we assume no intermediate buffer, no lead time between the modules and no preemption. It makes the gap between the theory and the practice of scheduling. More studies are needed to overcome those limitations. Furthermore, it is also important to consider the real-time scheduling problem for the modular factory. The concept of other approaches such as mathematical programming, heuristics and meta-heuristics might be help to solve this challenge. This is the aim of future work.

**Acknowledgements.** This research was supported by the KIAT (Korea Institution for Advanced of Technology) grant funded by the Korea Government (MOTIE: Ministry of Trade Industry and Energy). (Advanced Training Program for Smart Factory, No. N0002429) and was also funded by European Union Framework Programme for Research and Innovation Horizon 2020 under Grant agreement no 723909.

## References

1. Zuehlke, D.: SmartFactory-Towards a factory-of-things. Ann. Rev. Control **34**(1), 129–138 (2010)
2. Harjunkoski, I., Grossmann, I.E.: Decomposition techniques for multistage scheduling problems using mixed-integer and constraint programming methods. Comput. Chem. Eng. **26** (11), 1533–1552 (2002)

3. Laborie, P., Rogerie, J., Shaw, P., Vilim, P.: IBM ILOG CP optimizer for scheduling 20+ years of scheduling with constraints at IBM/ILOG. Constraints **23**(2), 210–250 (2018)
4. Baptiste, P., Pape, C.L., Nuijten, W.: Constraint-Based Scheduling: Applying Constraint Programming to Scheduling Problems. Springer, New York (2012)
5. Kolberg, D., Hermann, J., Mohr, F., Bertelsmeier, F., Engler, F., Franken, R., Kiradjiev, P., Pfeifer, M., Richter, D., Salleem, M., Schibli, E., Staub-Lang, P.: SmartFactoryKL System Architecture for Industrie 4.0 Production Plants, SmartFactoryKL, Whitepaper SF-1.2:04 (2018)
6. Wayer, S., Schmitt, M., Ohmer, M., Gorecky, D.: Towards Industry 4.0 - Standardization as the crucial challenge for highly modular, multi-vendor production systems. Int. Fed. Autom. Control (IFAC)-Paperonline **48**(3), 579–584 (2015)

# Robotic System Reliability Analysis and RUL Estimation Using an Iterative Approach

Francesco Aggogeri[1], Riccardo Adamini[1], Panagiotis Aivaliotis[2], Alberto Borboni[1], Amit Eytan[3], Angelo Merlo[4], István Németh[5], Claudio Taesi[1], and Nicola Pellegrini[1(✉)]

[1] Department of Mechanical and Industrial Engineering, Università degli Studi di Brescia, Brescia, Italy
nicola.pellegrini@unibs.it
[2] Laboratory for Manufacturing Systems & Automation, University of Patras, Patras, Greece
[3] We Plus S.p.A, Orbassano, Italy
[4] Ce.S.I. Centro Studi Industriali, Cologno Monzese, Italy
[5] Department of Manufacturing Science and Engineering, Budapest University of Technology and Economics, Budapest, Hungary

**Abstract.** This paper presents a novel methodology to evaluate robotic system reliability and Remaining Useful Life (RUL) integrating FMECA (Failure Modes, Effects and Criticality Analysis), life data analysis and data-driven & model-based methods. Starting from the FMECA analysis, the methodology proposes to identify the main critical components of new parts or systems, using life data analysis. A database collects and shares data directly from the field on similar systems and applications. Data are stored and managed via a web-based interface, the user may obtain them in real time as needed, when a modification in the robot or production cells occurs. Information are captured through a set of appropriate sensors, selected and located studying historical life data. From this dataset, RUL of components may be estimated using data-driven methods and model-based approaches. Then, the RUL results are shared with ERP systems to optimize production resources and maintenance activities and with FMECA again, to improve new projects in a closed loop. A preliminary application of the methodology is proposed on an anthropomorphic robot integrated in a production cell. This research is a part of PROGRAMs: PROGnostics based Reliability Analysis for Maintenance Scheduling, H2020-FOF-09-2017-767287.

**Keywords:** Robotics · Reliability analysis · Simulation · Residual Useful Life

## 1 Introduction

Over the past years, the applications of robotics have rapidly expanded in the industrial environment, improving the performance level of processes and the human safety. High levels of Reliability and Maintainability (R&M) are required to satisfy the market needs, in particular when robots are integrated in production cells and lines. Although the state of the art on robotics is considerable, studies on robot reliability are partial [1].


K. Berns and D. Görges (Eds.): RAAD 2019, AISC 980, pp. 134–143, 2020.
https://doi.org/10.1007/978-3-030-19648-6_16

Literature presents a small number of researches on algorithm development [2], statistical life data analysis [3], perfect and imperfect maintenance approaches [4–7] and reliability performance evaluation using neural network or fuzzy method [8].

The main concern related to the reliability assessment is to determine whether equipment is repairable or non-repairable. In this way, a robot needs to be decomposed into a set of subsystems, classified as repairable or no-repairable elements, as illustrated in Fig. 1. The robot consists of five major subsystems: the controller, the robot arms, the end effectors, the drive mechanism and the sensors [9, 10]. The controller is the brain of the robot system. The microprocessor, the encoder as well as the software and the programming in the controller are interconnected to accomplish a specific task [11, 12]. The robot arms aim to support static and dynamic input load for shoulder, elbow and wrist assembly. The end effectors are mechanisms attached to the wrist of the arm of the robot to grasp and handle parts or tools. The drive mechanism has hydraulic, pneumatic and servomechanism items. Recent systems are also equipped with servomotors as a drive mechanism. The performance of a servomotor depending on the payload and the degree of freedom of the robot. Finally, the sensors may be divided into three classes: internal, external and interlock sensors [13].

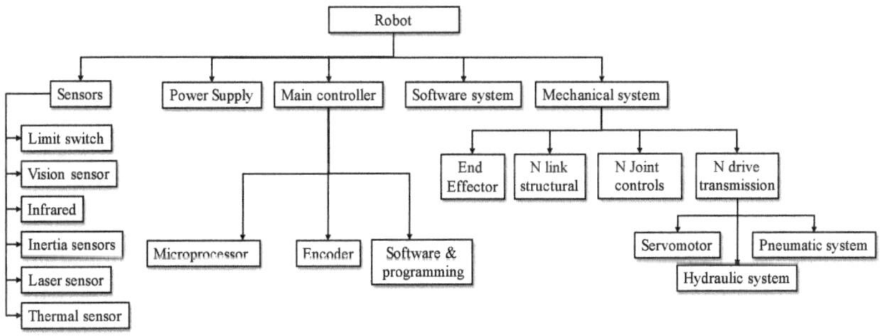


**Fig. 1.** Industrial Robot System breakdown

As a result of repairable and non-repairable components, interconnected in the robotics system, the RUL (Residual Useful Life) analysis is essential to avoid any degradation performance of the production system. In industrial environment, it is decisive to estimate the failure rate, the MTBF (Mean Time Between Failure) and the MTTR (Mean Time To Repair) of the elements using effective techniques. Their estimation permits to optimize the production plan and the maintenance activities, increasing plant efficiency and reducing costs.

The MTBF affirmed by Fanuc robots and ABB robots are in the range 60k–80k hours respectively, based on nonstop tasks. Despite these results, a methodology that overcomes the gap due to missing information from the field is required to:

- support the designers in selecting the most appropriate and reliable elements to increase the performance of new robotic cells and applications;
- plan production activities and resources, avoiding any breakdown or failure;
- optimize the maintenance tasks, reducing time to repair and spare part costs.

The proposed methodology is based on the integration of FMECA (Failure Modes, Effects and Criticality Analysis), Life Data Analysis and Data-driven and Model-based methods in order to determine the RUL of a robotic system. The approach is a loop that collects information from the field, processes data to estimate RUL and, finally, shares the results with ERP systems - to plan production and maintenance activities –and with FMECA to optimize new projects.

## 2 A Closed Loop Methodology for Life Component Prognostics Based on Reliability Analysis

Remaining Useful Life is an evaluation of the residual lifetime of an element, analyzing the occurrence of a failure. RUL estimation is based on four points: (i) the electromechanical systems age is influenced by their usage and environmental conditions; (ii) the damage accumulation is a monotonic progression; (iii) the failure event was anticipated by a detectable signals, (iv) the correlation between degradation signals and reliability performance need to be statistically significant. Methods of degradation analysis include physical examination, monitoring of damaging features (e.g., vibrations, acoustic emissions, oil temperature, energy consumption), and non-destructive inspections. Direct methods are expensive and not always effective due to the low occurrence rate of the failure event. Methods for indirect evaluation are based on models of physics of failure or life data analysis of a component. Figure 2(a) describes the iterative loop of the proposed methodology. For a new system (e.g. robotic cell), the initial phase is a failure investigation using the FMECA, defining the critical breakdowns. The existing FMECA commercial tools are not able to completely use updated data and information (e.g. MTBF and MTTR) in real time, compromising the results. For this reason, FMECA has to share a database, where life data of existing and operating components in the field are upload and processed. LDA permits to assess the current state of robot reliability performance, identifying strengths and weaknesses of subassemblies and components. Appropriate sensors and devices, selected by historical analysis on the occurred failures, collect life data. Then, using data-driven and physical-based degradation models, a hybrid RUL is predicted, supporting both FMECA analysis and production-maintenance decisions. The user needs to identify appropriate prognostics models based on physical insight and operational conditions to increase the accuracy of the failure evolution.

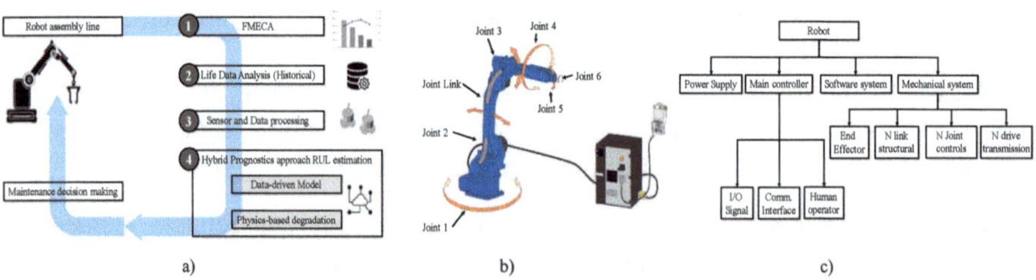


**Fig. 2.** (a) Roadmap of the proposed closed loop methodology, (b) scheme of the physical industrial robot structure Joint 1-Swing, Joint 2-Lower Arm, Joint 3-Upper Arm, (c) and the morphological functional decomposition

## 2.1 FMECA and Risk Analysis Evaluation

An effective estimation of RUL is based on a full awareness of the element degradation mechanisms under different stresses and environmental conditions. Robot components show fatigue failures that may depend on the operational conditions but may also be determined by the human responsibility. To manage sub-system failures, users have to recognize and classify failure modes and their causes. This approach permits to design and develop new parts or systems more reliable to perform the required production tasks. FMECA is a recognized methodology focused on the identification of potential failure modes of a product or process, evaluating potential failure modes and assessing the risk of their impact [14–20]. An effective application of FMECA may help designers to identify and prioritize the corrective actions inhibiting the failure mode. One of the main outcome of FMECA is the Risk Priority Number (RPN) indicator, obtained by the multiplication of Severity, Occurrence and Detection rates. Severity defines the gravity of the effect of a potential part failure. Occurrence estimates the likelihood that a specific failure mode will occur, while Detection is the capacity to detect the failure. Parts and components with a high RPN have to be managed with priority for corrective and improvement actions. In particular, RPN may support the definition of the maintenance strategies of a component, since it includes the evaluation of the impact of the fault component and the time to repair or replace it [21]. In this way, the use of the Criticality Matrix helps to identify graphically the critical components and assess the risk.

## 2.2 Data-Driven Methods and Model-Based Approach for RUL Estimation

Data-driven methods need to collect and study multidimensional dataset that includes different information (e.g. environmental state, direct-indirect measures of degradation) from a population (e.g. robotic cells). Diagnosis aims to define the status of a component while prognosis is the prediction of the component future state and duration. This evaluation depends on a broad range of factors, influencing the accuracy of the RUL calculation. Data-driven methods apply Machine Learning techniques (e.g. ARIMA, Bayes Network, Fuzzy c-Means, k-Means) based on recursive phases of training-learning of the system, collecting and processing data and information continuously. These iterative processes allow to increase the accuracy of RUL estimation, limiting the amount of data-knowledge. They are easy to be implemented with limited costs, nevertheless the physical interpretation may be a complex task and a significant tuning period is often required. The Model-based (M-B) approach requires the development of a mathematical model able to represent the system use and degradation. In this way, the operational data are indispensable to fine-tune model parameters. The aim is to correlate the kinematic-dynamic and material models to the component degradation. This approach permits to extrapolate an accurate RUL estimation, nevertheless the model validation may be expensive and time consuming, requiring a high knowledge of the system and the physical lows, making it no sustainable in industrial application.

## 3 Robot LDA for Reliability Analysis

In this section, a preliminary application of the methodology is proposed. A robotic cell is considered, composed by an anthropomorphic robot and 3 machine tools. In the design phase, it is required to analyse the robot reliability in terms of cell breakdowns and micro-stops, having a negative influence on the system availability and throughput, reducing the production rate. A robot system consists of a number of subsystems that are serial-connected with an interface for communication with the environment, as shown in Fig. 2. The robot arm has different links and joints, usually the articulated robots have 5–6 joints, however more auxiliary axes may be integrated.

For serial connected subsystems, the failure of one component determines the whole system breakdown. Considering a complex structure with N serial linked items, each of them has exponential failure times with rate $\lambda_i$, i: (1, 2, N). The overall failure rate $\lambda_s$ is the sum of the failure rates of each item $\lambda_i$, as follows:

$$\lambda_S = \sum\nolimits_{i=1}^{N} \lambda_i \tag{1}$$

The robot $MTBF_R$ is the sum of the inverse of $MTBF_i$ of the considered items:

$$\frac{1}{MTBF_S} = \sum\nolimits_{i=1}^{n} \frac{1}{MTBF_i} \tag{2}$$

There are different types of potential failures: (i) internal hardware failures (mechanical unit, drive, gear), (ii) internal software failures (control system), (iii) external component failures (equipment, sensors, grippers), (iv) human errors, that require the production cell stop and the operator intervention. The robotic cell availability is calculated as follows:

$$\text{Availability} = \frac{MTBF}{MTBF + MTTR} \tag{3}$$

Figure 3 shows the FMECA analysis focused on the robot. This tool has been developed to share data with a database that collects information directly from the field on similar operating robotic cells. The designer may evaluate life data in real time and define the critical components for the risk analysis.

FMECA toolbox also proposes the Criticality Matrix Table (Fig. 4), that states the main critical components in terms of severity and occurrence. The criticality number is an indicator of the frequency of a failure mode. It defines the component risk considering the failure rate, as follows:

$$C_m = \beta \cdot \alpha \cdot \lambda_p \cdot t \tag{4}$$

where $C_m$ is the Failure mode criticality number, $\beta$ is the conditional probability of the current failure mode's failure effect, $\alpha$ is the failure mode ratio, $\lambda_p$ is the component failure rate and t is the duration of applicable mission phase (expressed in hours or operating cycles).

**FMECA - FAILURE MODE EFFECT CLITICALITY ANALYSIS**

| FMECA Number | Product Name: | Model: | FMECA Facilitator: | | Prepared by: |
|---|---|---|---|---|---|
| X123Y2018 | Robot 6-axis | Alpha | Nicola Pellegrini | | Nicola Pellegrini |
| System, Subsystem or Component | | Technical drawings: | FMECA Start Date | FMECA Revision Date | Issued Date |
| Robot 6-axis | | .dwg | 04/05/2018 | 08/03/2019 | 0 |

Start to FMECA | SEV Report | OCC Report | DET Report

| No | Main Level | Item / Function Part / Operation | Severity | Occurrence | Detection | MTBF | RPN | Potential Failure Mode(s) | Potential Causes of Failure | Potential Effect(s) of Failure | Current Design and Machinery Controls |
|---|---|---|---|---|---|---|---|---|---|---|---|
| 1 | Robot 6-axis | S-axis Motor | 7 | 6 | 7 | 8662 | 294 | Performance Unsuitable | Load conditions not conformed to | Machine failure | Laboratory |
| 2 | Robot 6-axis | U-axis Motor | 6 | 6 | 6 | 19600 | 216 | Burning of primary circuit | Lack of cooling fluid AND failure of | Machine failure | Laboratory |
| 3 | Robot 6-axis | L-axis Motor | 5 | 5 | 7 | 20000 | 175 | Working not properly | Thermal effects excessive (due to | Lost of accuracy | Laboratory |
| 4 | Robot 6-axis | S-axis Reducer | 7 | 5 | 6 | 23400 | 210 | Early wear and vibration | Cutting fluid (cooling lubricant) | Reduction of life, lost of accuracy, | Design review (check-list) |
| 5 | Robot 6-axis | U-axis Reducer | 6 | 5 | 5 | 31200 | 150 | Early wear and vibration | Pre-load inadequate | Reduction of life, lost of accuracy, | Design review (check-list) |
| 6 | Robot 6-axis | L-axis Reducer | 5 | 4 | 4 | 36000 | 80 | Early wear and vibration | Pre-load inadequate | Reduction of life, lost of accuracy, | Design review (check-list) |
| 7 | Robot 6-axis | HW/SW system | 3 | 4 | 5 | 38700 | 60 | Working not properly | Thermal effects excessive (due to | Lost of accuracy | Laboratory |
| 8 | Robot 6-axis | CPU board | 5 | 4 | 4 | 2628000 | 80 | Working not properly | Thermal effects excessive (due to | Lost of accuracy | Laboratory |
| 9 | Robot 6-axis | Power supply | 4 | 3 | 4 | 2728000 | 48 | Working not properly | Thermal effects excessive (due to | Lost of accuracy | Laboratory |
| 10 | Robot 6-axis | I/O board | 5 | 3 | 3 | 2789000 | 45 | Working not properly | Thermal effects excessive (due to | Lost of accuracy | Laboratory |
| 11 | Robot 6-axis | Main board | 6 | 4 | 4 | 2928000 | 96 | Working not properly | Thermal effects excessive (due to | Lost of accuracy | Laboratory |
| 12 | Robot 6-axis | Shock sensor | 5 | 3 | 3 | [illegible] | 45 | Working not properly | Thermal effects excessive (due to | Lost of accuracy | Laboratory |
| 13 | Robot 6-axis | Robot structure | 9 | 1 | 3 | 8628000 | 27 | Working not properly | Thermal effects excessive (due to | Lost of accuracy | Laboratory |

**Fig. 3.** FMECA analysis report

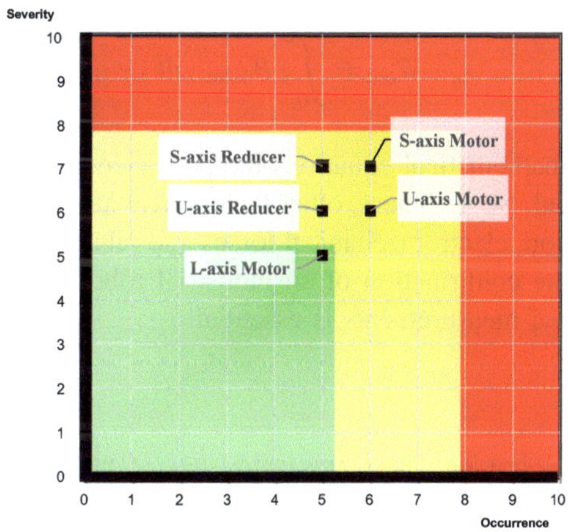


**Fig. 4.** Criticality Matrix representation of an anthropomorphic robot system

The result of FMECA analysis highlights the servo-motor as a critical part, in particular "S – Swing or Swivel" axis or "Joint 1" (Fig. 3), that may compromise the robotic cell performance. In addition, the servo-motor replacement may be expensive. The main identified failure modes are overheating, over current, electrical surges, humidity and vibrations. From the dataset, LDA is performed based on data collected from the maintenance activities of similar systems and applications, as shown in Table 1.

**Table 1.** Robot dataset for LDA analysis

| Robot type | Structure | DOF | Payload [kg] | Repeatability [mm] | Controller |
|---|---|---|---|---|---|
| Anthropomorphic | Vertically articulated | 6 | 3.0 to 165.0 | ±0.2 to ±0.08 | Direct transfer of I-O to the manipulator |

Figures 5 and 6 illustrate the S-axis motor life data considering events occurred in the last 4 years. The results show the instantaneous MTBF and the cumulative MTBF. LDA has been executed using Reliasoft software that estimates parameters at C50 (solid line) and C90 (dotted line) confidence level. Table 2 summarizes the reliability parameters estimated from LDA. This study suggests to equip the robotic cell with sensors able to detect signals (e.g. temperature, vibrations, humidity, torque) in order to develop a RUL model based on data-driven methods of S-axis motor. In the same way, a M-B approach may be implemented as described in [22, 23], the active-power (P) utilized by the robot's system is stated in Eq. 5. This equation shows that the payload and robot speed impact on the power consumption since it is a function of the robot torque (T), the angular speed (ω) and the mechanical-electrical effectiveness ($\eta_{mec}$,$\eta_{el}$) of the drives. As presented in Eq. 6, the energy spent ($W_{act}$) by a robot is the integral of active-power over time [0, $t_f$] [23].

$$\mathrm{P}_{robot} = \sum\nolimits_{i=1}^{n} T_i \cdot \omega_i \cdot \frac{1}{\prod \eta_{mec,i} \cdot \eta_{el,i}} \tag{5}$$

$$W_{act} = \int_0^{t_f} P_{robot} \cdot dt \tag{6}$$

Using the stated mathematical equations the power consumption rate of the robot system can be predicted. Nevertheless, other factors may affect the power-consumption of the robot (e.g. friction, electromechanical losses and vibration) in the real conditions [24]. Consequently, the confirmation of an industrial robot's model using a machine learning study with real measurements is essential.

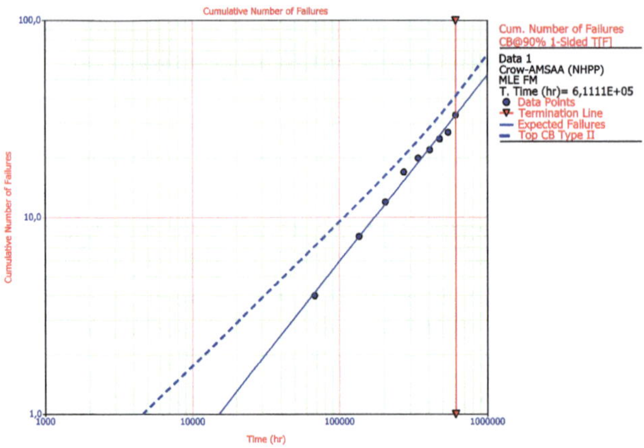


**Fig. 5.** S-axis motor of 6 DoF robot: Cumulative number of Failure – Operative time (hours); Confidence level 50 (solid line), Confidence level 90 (dotted line)

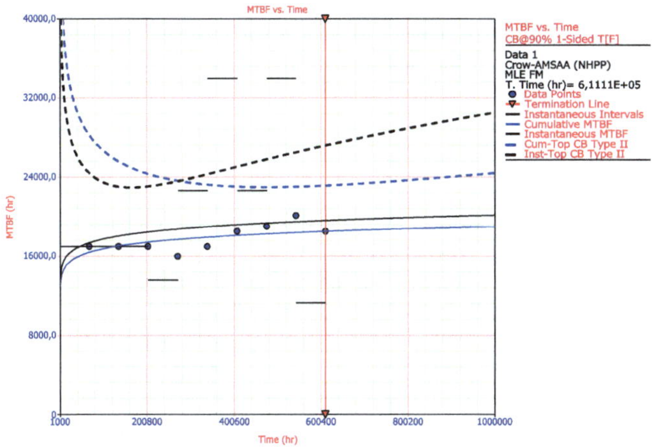


**Fig. 6.** S-axis motor of 6 DoF robot: Cumulative MTBF (blue), Instantaneous MTBF (black); Confidence level 50 (solid line), Confidence level 90 (dotted line)

**Table 2.** S-axis motor reliability estimation based on Life Data Analysis

| Model | Shape parameter | λ | Scale parameter (h) | Cumulative MTBF (h) | Analysis method |
|---|---|---|---|---|---|
| Crow-AMSAA | β: 0.9445 | 1.01E−4 | η: 17.1k | 19.6k | MLE |

## 4 Conclusions

This paper aims to present a closed loop methodology to estimate robot life based on reliability analysis. Starting from the FMECA analysis, the methodology proposes to identify the main critical components of new parts or systems, using life data analysis. A database collects and shares data directly from the field on similar systems and applications. Information are captured through a set of appropriate sensors, selected and located studying historical data. From this dataset, RUL of components may be estimated using data-driven methods and model-based approaches. The RUL results need to be shared with ERP systems to optimize production resources and maintenance activities and with FMECA again, to improve new projects in a closed loop. The paper presents a preliminary application of the methodology. An anthropomorphic robot has been considered to be integrated in a production cell. The FMECA analysis has been developed to identify the main critical parts that affected the reliability performance, by using RPN index and Criticality matrix representation. The S-motor assembly showed the low performance. This analysis has been executed using data collected from the field. In particular, a set of sensors has been identified to detect the critical factors that influence the component behaviour. From these signals, a number of models may be developed in order to estimate RUL to optimize production and maintenance activities and future new projects using FMECA.

**Acknowledgements.** This paper is established on the work that has been in progress in PROGRAMS project (www.programs-project.eu), which has received funding from the European Union's Horizon 2020 Research and Innovation Programme.

## References

1. Dhillon, B.S., Fashandi, A.R.M., Liu, K.L.: Robot systems reliability and safety: a review. J. Qual. Maintenance Eng. **8**(3), 170–212 (2002)
2. Musto, J.C.: A two-level strategy for optimizing the reliability of redundant inverse kinematic solutions. J. Intell. Robot. Syst. **33**, 73–84 (2002)
3. Carlson, J., Murphy, R.R., Nelson, A.: Follow up analysis of mobile robot failures. In: Proceedings of the IEEE Annual Reliability and Maintainability Symposium, vol. 5, pp. 4987–4994 (2004)
4. Hoshino, S., Seki, H., Ota, J.: Optimal maintenance strategy in fault-tolerant multi-robot systems. In: IEEE/RSJ International Conference on Intelligent Robots and Systems, 25–30 September 2011, pp. 2314–2320 (2011)
5. Bererton, C., Khosla, P.: An analysis of cooperative repair capabilities in a team of robots. In: Proceedings IEEE International Conference on Robotics and Automation, pp. 476–486 (2002)
6. Dhillon, B.S., Li, Z.: Stochastic analysis of a system with redundant robots, one built-in safety unit, and common-cause failures. J. Intell. Robot. Syst. **45**, 137–155 (2006)
7. Carreras, C., Walker, I.D., Nieto, O., Cavallaro, J.R.: Robot reliability estimation using interval methods. In: Proceedings of the International Workshop on Applications of Interval Analysis to Systems and Control, pp. 371–385 (1999)
8. Sharma, S.P., Sukavanam, N., Kumar, N., Kumar, A.: Reliability analysis of complex robotic system using petri nets and fuzzy lambda-tau methodology. Eng. Comput. **27**(3), 354–364 (2010)
9. Owen, T.: Robotics: Introduction, Programming and Projects by James L. Fuller. Macmillan Publishing Co, Oxford (1991). Robotica. 11.05: 487–488 (1996)
10. Rehg, J.A.: Introduction to Robotics in CIM Systems, vol. 5. Prentice Hall, New Jersey (1997)
11. Aggogeri, F., Borboni, A., Merlo, A., Pellegrini, N., Ricatto, R.: Real-time performance of mechatronic PZT module using active vibration feedback control. Sensor **16**(10), 1577 (2016)
12. Carlson, J., Murphy, R.R.: Reliability analysis of mobile robots. In: Proceedings of the IEEE Annual Reliability and Maintainability Symposium, vol. 1, pp. 274–281 (2003)
13. Aggogeri, F., Borboni, A., Faglia, R., Merlo, A., Pellegrini, N.: A kinematic model to compensate the structural deformations in machine tools using fiber Bragg grating (FBG) sensors. Appl. Sci. **7**(2), 114 (2017)
14. Spreafico, C., Russo, D., Rizzi, C.: A state-of-the-art review of FMEA/FMECA including patents. Comput. Sci. Rev. **25**, 19–28 (2017)
15. US MIL-STD-1629: Procedures Performing a Failure Mode, Effects & Criticality Analysis
16. ISO/IEC 61508: Functional Safety of Electrical/Electronic/Programmable Electronic Safety-related Systems
17. SAE J1739: Potential Failure Mode and Effects Analysis in Design (Design FMEA) and Effects Analysis for Machinery (Machinery FMEA)
18. IEC 60812: Analysis techniques for system reliability - Procedure for failure mode and effects analysis (FMEA)

19. Aggogeri, F., Borboni, A., Merlo, A., Pellegrini, N., Ricatto, R.: Vibration damping analysis of lightweight structures in machine tools. Materials **10**(3), 297 (2017)
20. Bongiorno, J.: Improving, FMEAs can transform compliance into competitive advantage. Qual. Dig. **20**(10), 37–40 (2000)
21. ANSI/RIA R15.06-2012: American National Standard for Industrial Robots and Robot Systems- Safety Requirements
22. Choset, H., Lynch, K.M., Hutchinson, S., Kantor, G.A., Burgard, W., Kavraki, L.E., Thrun, S.: Principles of Robot Motion: Theory, Algorithms, and Implementation. MIT Press, London (2005)
23. Rassõlkin, A., Hõimoja, H., Teemets, R.: Energy saving possibilities in the industrial robot IRB 1600 control. In: Proceedings of the 7th International Conference-Workshop Compatibility and Power Electronics, pp. 226–229 (2011)
24. Sergaki, E.S., Stavrakakis, G.S., Pouliezos, A.D.: Optimal robot speed trajectory by minimization of the actuator motor electromechanical losses. J. Intell. Robot. Syst. **33**, 187–207 (2002)

# Edge Computing in Smart Production

Jumyung Um[1,2(✉)], Volkan Gezer[1], Achim Wagner[1], and Martin Ruskowski[1]

[1] DFKI, Trippstadter Str. 122, 67663 Kaiserslautern, Germany
{volkan.gezer,achim.wagner,martin.ruskowski}@dfki.de
[2] Kyung Hee University, Deogyeong-daero 1732, Yongin-si 17104, Republic of Korea
jayum@khu.ac.kr

**Abstract.** Flexibility in mass-customized manufacturing can be supported significantly by the introduction of Cyber-Physical Production System and the connection of production modules to AI (artificial intelligence) Cloud services. Even though there exist standardized protocols from device to IT system, there are still challenges for the synchronization between cyber-model and physical object, and the application of decision making in the cyber-model. Although high performance machine learning services make the Cloud a preferred computation node, possible unstable connection with manufacturing resources enforce new service distribution approaches in the network. This paper proposes an Edge Computing architecture which is the mediator between machines, by providing local Cloud services with fast response time and preprocessing resources for a vast amount of data. As an illustrative example the selected Edge service pre-processes data form an augmented reality device in order to communicate with the cyber-model in real time. The Edge platform controls the computing resources and prioritizes all processes of Edge Services for a dynamic update of production lines and human-machine-interaction.

**Keywords:** Smart factory · Edge computing · Human-machine interaction · Cyber-Physical system

## 1 Introduction

Artificial intelligence (AI) is a key technology to fulfil the requirements of mass customization, personalized products, and fast-shifting market demands. Digital and AI services may help to operate a number of local micro factories, which is mayor step towards keeping production closed to the marketplace. One concept is the connection of production systems to the Industrial Internet of Things (IIoT) by defining Cyber-Physical Production System (CPPS). Such systems provide a direct link from the data lake to the Cloud, in which AI services are hosted for supporting manufacturing facilities in a flexible way. However, there are a number of challenges for using AI services in a manufacturing environment.

A CPPS contains a physical part such as a production module and a digital representation (digital twin) of the physical properties and processes, which can bring to operation using specific computing devices such as Edge computers. The Edge


K. Berns and D. Görges (Eds.): RAAD 2019, AISC 980, pp. 144–152, 2020.
https://doi.org/10.1007/978-3-030-19648-6_17

computers are also able to connect the CPPS with the IIoT and Cloud platforms utilizing specific interfaces and protocols. AI services included in the edge devices typically perform a number of additional tasks, such as control and optimization of physical processes, monitoring and decision making for supervising manufacturing processes as well as providing auxiliary services to human operators. Typically, the services have to fulfill the real-time requirements of the production environment. However, the processing of a huge amount of data as required by many AI services is limited by the computing resources of Edge-devices. Thus, one of the main challenges is to find a suited architecture for reducing complexity and for distributing services between Edge and Cloud under real-time conditions.

To overcome these problems, various system architectures have been proposed in the notion of Cyber-Physical systems. Zuehlke [1] suggested the innovative approaches linking with the fourth industrial revolution and the Smart Factory by reducing the complexity and achieving faster planning and setup. Tao et al. [2] emphasized the role of digital twins enhancing training performance of machine learning of a manufacturing resource. A lever to increase productivity and competitiveness is performed towards a reduction of "time to production" and "time to optimization" [3]. Weyer et al. [4] proposed a digital twin structure connected with a series of factory simulators so that all decision-making stages are updated online. Gorecky et al. [5] proposed technological assistance of human operators by human-machine-interaction through a Cyber-Physical system. Finally, Volkan et al. [6] proposed an edge computing system as a solution of implementing Cyber-Physical system. Um et al. [7] also shows the usage of auxiliary devices for manual operators supported by object detection service running on the edge device. These studies will be the technological basis to create an CPPS that can utilize AI services on the shop-floor.

This paper discusses the two technologies: (1) how human interaction systems and automatic mission systems are connected and (2) how to implement synchronized computing services. Section 2 introduces this concept, while Sect. 3 presents case studies. Section 4 discuss the results and the advantages of the solution compared to the actual plant environment.

## 2 Systematic Architecture of Edge Computing for CPPS

Edge computing is believed to solve problems with respect of Quality of Service (QoS) and latency, that the current Cloud computing systems have [6]. Edge computing keeps computing power close to the field, allowing continuous service provision even when the network connection to the Cloud is lost.

### 2.1 Requirements

Edge computing devices are the bridge between the cloud services and the auxiliaries of human operators who are responsible for each process on the shop-floor. In contrast, factory operators make high level decisions for controlling the overall factory systems and for updating the programs of machine controllers. Edge servers have to provide **intelligent** functions and high-level information close to the human decision making process. Further, they have to secure the seamless communication with all controllers by improving the **interoperability** in order to reduce the manual works of re-programming machines.

The complexity of factory information processing is resolved by providing extensibility to the factory network. Additional data processing capability e.g. for filtering raw data is provided in order to abstract the data for subsequent transmission. An edge server transmits, in parallel, a huge number of signals generated by machine controllers and auxiliary sensors. Each service of the edge server **filters** raw data and **abstracts** them into high-level event messages for sending them to the cloud. When a new machine is introduced in the production line, the edge server also accepts it without re-programming the pipeline of the data processing by providing function for **extensibility** of the factory network.

Real-timeliness is divided into time-sensitiveness and zero-disconnection. **Time-sensitiveness** is important for satisfactory service results, regardless of the real-timeliness of a task. For example, the emergency-stop, which is a mandatory function for safety, must react within 10 ms even in case of high network traffic. A further requirement related to edge servers is **zero-disconnection** for handling issues, which may occur in the cloud due to connection or infrastructure problems. Restarting IT service causes the stop of production line leading to high cost. In the reliability context, maturity, fault tolerance, and recoverability are considered [8].

## 2.2 Components

To build an edge server requires several components to be defined. Some of these components are already available in operating systems such as real time Linux. Some of the others must be adjusted or implemented for a seamless execution. Figure 1 shows the components that work together in edge computing. The API will be the initial contact point with general purpose inputs and outputs (IOs) or external end devices and sensors. If the request is genuine, authenticated and valid, it will be passed to the message router for initial processing.

Depending on the available resources, that are monitored by the resource manager, the edge server will either execute the request in the edge server or offload it to another available server. The chosen server could be another edge server, if exists in the network, or the cloud. Decisions about the system specification of edge computing are made taking latencies, transmission times, CPU usage, running tasks, deadlines, and the priorities of the requests into account. If the server decides that the request should be handled here depending on the priority and deadline, the orchestrator adjusts the resources to execute it on-time. It can create or/and use virtual processors for the new or previous tasks to isolate them from each other, limit the execution capacity or pause them. These operations should only be performed by the server, without external modification.

The edge server architecture shown in Fig. 1 can be applied to one single server or to a network of multiple servers. In the scope of this paper, the edge network consists of a single server; hence the message router will only be used to execute the task in the

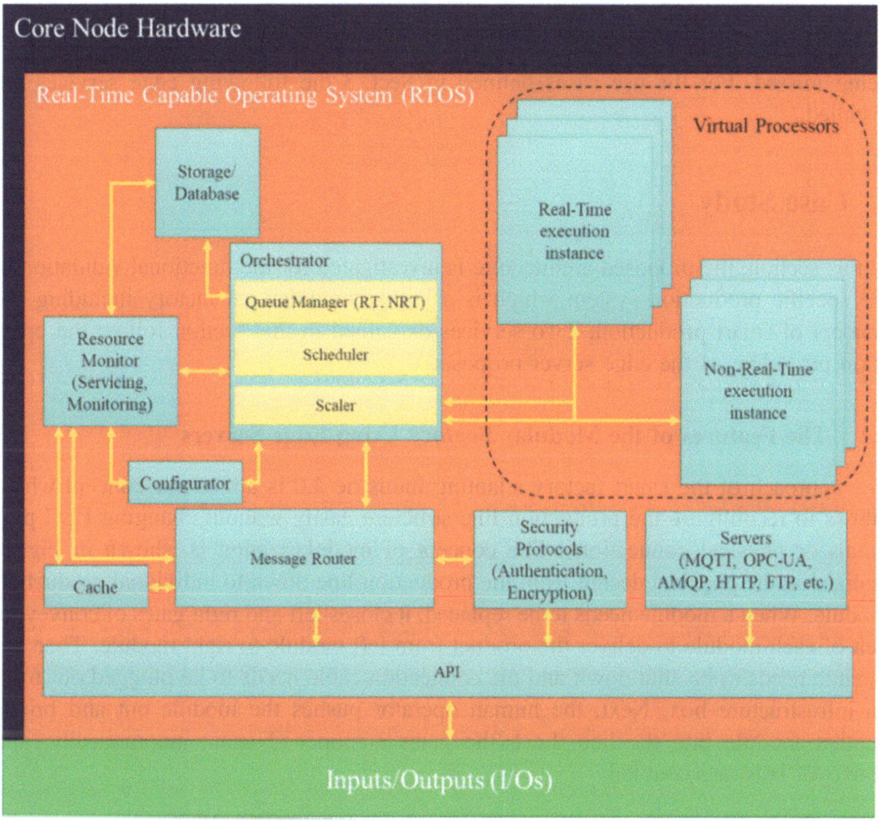


**Fig. 1.** Proposed server architecture for edge computing

edge server or the cloud. The resource monitor component, however, will only track the resources of the current server. Depending on the priorities of the tasks, resources will be allocated for on-time execution.

## 2.3 Edge Services

Software available in the edge server are called edge services. Edge services can be implemented in any programming language. Once they are installed or deployed in the edge server, they need to be registered in the device. Execution of the unregistered services directly from outside is not allowed in order to keep orchestration and maintaining resources on the edge server itself. Once they are registered with specified parameters, they can be called or requested via the available API. Services can be requested from edge servers multiple times. Each requested instance of services is called a task. Tasks in the edge server can have the following states: running, waiting, blocked, paused, completed, killed. The status can be changed via the operating system or the edge server components for optimum performance. The task can also be killed by the request from the caller. The edge services are also prioritized to decide the worst-case behavior in case the server runs out of resources or if the deadline is in risk of being missed. For the use cases defined in Sect. 3 the following edge services are installed.

# 3 Case Study

In this section, the proposed architecture is investigated for the functional validation in the specific production system which is a highly modularized factory including the concept of smart production. Two services explained in this section follow the operation procedure of the edge server proposed.

## 3.1 The Features of the Modular Factory Using Edge Servers

The approach of the smart factory adapting Industrie 4.0 is a modular concept which allows to reconfigure the production line structure easily without changing PLC programs or physical connections. The concept of modularization is, shown in Fig. 2, realized in the physical design from the production line down to individual production module. When a module needs to be replaced, it closes left and right gates of conveyors belt of each module to deliver the product from left module to right module. Then the module needs to be shut down and the connection cable needs to be plugged out from an infrastructure box. Next, the human operator pushes the module out and brings another module into the line. Last, the gates are opened again automatically, and conveyor belts are coupled.

**Fig. 2.** Modular production line of Smart Factory-KL

## 3.2 Implementation of Edge Services

### 3.2.1 3D Visualizer Service

The modular factory is equipped by a 3D visualizer [9] and a plant simulation to support quick information flow to a worker, who can decide on appropriate actions rapidly. The mechanism of the edge service enables the reconfiguration of the control software in real-timely without additional effort along with the physical modules reconfiguration, i.e. the edge service of the 3D visualizer supports the plug-n-play functionality. The sequence of the deployment of a service and docking animations are shown in Fig. 3. The module manager registers the adaptor of a new module in order to communicate with 3D visualizer client and activates it automatically. The adaptor follows the series of status of the edge service as the response to the request of 3D visualizer client. This case study illustrates the sequence of docking and undocking. Products on conveyors and gates of modules are reacted to the message of each module adaptor reading the status of physical module. Figure 3(b), (c) and (d) shows the changes of modules in the 3D visualizer.

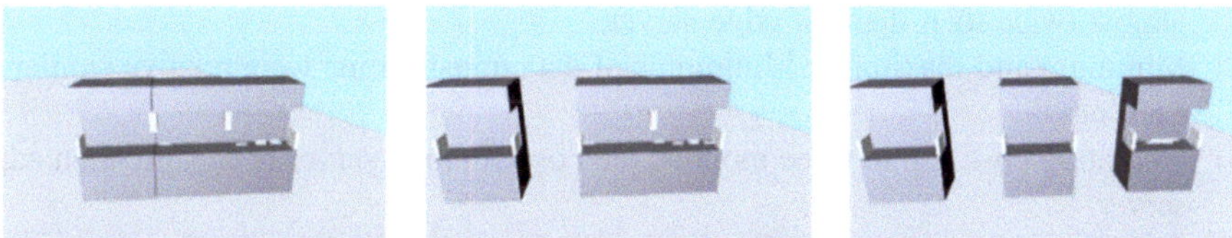

**Fig. 3.** Screen shots of module animations

### 3.2.2 Object Detection Service

The object detection service implemented here includes a high-performance data processing depicted in Fig. 4 – This edge service is the example of showing the advantage in the running status. A deep learning-based edge service was implemented in which a set of algorithms convert camera streaming in order to recognize individual parts presented in front of smart glasses. The principle advantages of the introduction of proposed edge architecture were (1) to ensure the robust performance of the wearable device without battery shortage and (2) to extend to new wearable device programmed in different languages by only developing an app to transmit the camera streaming into the edge device. The addition of this edge service assists human operators during manual assembly of new products and repair of machines after unexpected events with little training time.

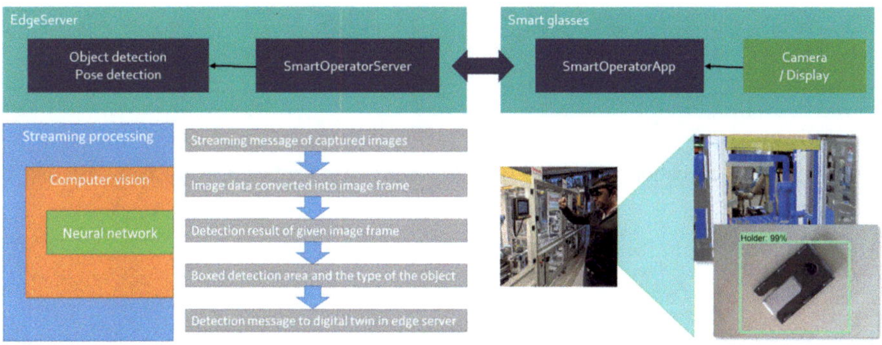


**Fig. 4.** System architecture of object detection service

### 3.3 Performance Comparison with Cloud Service

This section compares the performance of two services in the Cloud and the Edge. The 3D Visualizer integrates the components OPC-UA client, MQTT broker, module manager, and 3D Web client. These four modules start sequentially in an edge server while object detection algorithm is running. The overall cycle time is measured for the estimation of process execution time within the edge server. Smart glasses transmit the camera streaming with 1920 × 1080-pixel resolution for 30 s, in which the size of data is 121 MB. Both edge server and cloud server tested have quad cores and 16 GB RAM. The Table 1 shows the performances and conditions.

The results of the performance comparison are presented in Table 1.

(1) Average and deviation: Average time and standard deviation of cloud server were almost twice than them of edge server.
(2) Minimum and maximum: Minimums of data transfer time were relative similar to each other.
(3) Operating cost: Up to three months, the cost of cloud server is cheaper than edge server.
(4) Distributed control: the utilization of distributed task allocation let the overall performance of object detection increase almost three times. Worst case and best case are shown in Fig. 5.

**Table 1.** Performances and conditions of data transfer of edge server and cloud server

| Edge server | Edge server | Cloud server |
|---|---|---|
| Average time to transfer video (sec) | 12.93 | 33.55 |
| Standard deviation of transfer speed | 0.98 | 19.24 |
| Min/Max of data transfer time (sec) | 11.44/14.89 | 14.3/60.42 |
| Cost for hardware and service charge (€) | Less than 500 once | ~ 146 per month |
| Location of smart glasses and server | Same building | Same country |

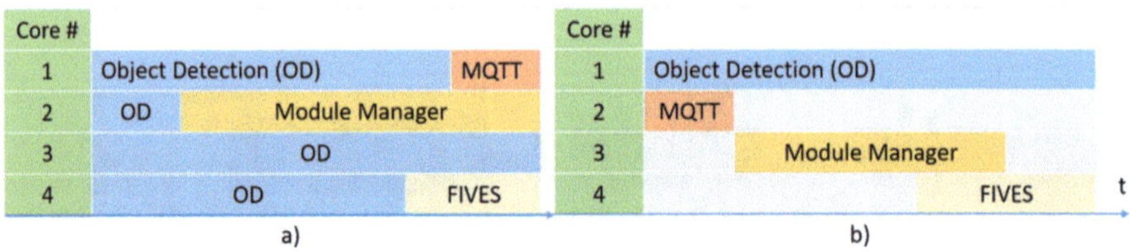


**Fig. 5.** Example execution of edge server in a quad-core hardware in (a) best case and in (b) worst cases

## 4 Discussion and Conclude Remarks

Edge computing is believed to solve latency and reliability problems of the current cloud solutions, enhancing the Quality of Service (QoS). Additionally, it keeps the connection persistent in case of network problems and still assists completing the request.

This paper uses only one Edge server in the network and the Cloud. Therefore, the tasks that cannot be executed due to the available resources are offloaded to the Cloud. Results of the use case explained in the paper give the following conclusions: The first use case shows the series of software modules to represent digital twins with actual behavior of production system, which is displayed with a 3D visualizer. Whenever a production module is changed, the digital twin can follow these amendments. Thanks to the orchestrator of the Edge server, all software modules keep their own data consistent, even if high computing power in demanded on object detection services. The second use case shows the user friendliness in terms of the reduced time to find product related data and access to this data. The unified interface using smart glasses, Hololens enables faster and easier access to the operator. In terms of AI service, thanks to the Edge server, the computation can be performed in Edge Tier for faster response and for a lower cost. In the future, this case study shall be enhanced by integrating further digital twin and object detection services together and to use the Edge Computing in context-based resource allocation.

**Acknowledgements.** This research was funded in part by the H2020 program of European Union, project numbers 723094 (project FAR-EDGE) and 723909 (project AUTOWARE). The responsibility for this publication lies with the authors. The project details can be found under project websites at: http://www.far-edge.eu and http://www.autoware-eu.org.

## References

1. Zuehlke, D.: Smart factory—towards a factory-of-things. Annu. Rev. Control **34**(1), 129–138 (2010)
2. Tao, F., Cheng, J., Qi, Q., Zhang, M., Zhang, H., Sui, F.: Digital twin-driven product design, manufacturing and service with big data. Int. J. Adv. Manuf. Technol. **94**(9–12), 3563–3576 (2018)
3. Mourtzis, D., Papakostas, N., Mavrikios, D., Makris, S., Alexopoulos, K.: The role of simulation in digital manufacturing: applications and outlook. Int. J. Comput. Integr. Manuf. **28**(1), 3–24 (2015)
4. Weyer, S., Meyer, T., Jumyung, U., Ohmer, G.: Open semantic meta-model as a cornerstone for the design, engineering and management of CPS-based Factories. In: 4th Automation ML User Conference (2016)
5. Gorecky, D., Schmitt, M., Loskyll, M., Zühlke, D.: Human-machine-interaction in the industry 4.0 era. In: 2014 12th IEEE International Conference on Industrial Informatics (INDIN), pp. 289–294. IEEE (2014)
6. Gezer, V., Um, J., Ruskowski, M.: An introduction to edge computing and a real-time capable server architecture. Int. J. Adv. Intell. Syst. (IARIA) **11**(7), 105–114 (2018)

7. Um, J., Popper, J., Ruskowski, M.: Modular augmented reality platform for smart operator in production environment. In: 2018 IEEE Industrial Cyber-Physical Systems (ICPS), pp. 720–725. IEEE (2018)
8. ISO/IEC 25010:2011: International Organization for Standardization, March 2011. Accessed Aug 2018
9. Um, J., Klaus, F., Spieldener, T., Kolberg, D.: Development a modular factory with modular software components. Procedia Manuf. **11**, 922–930 (2017)

# Neural Adaptive Control of a Robot Joint Using Secondary Encoders

Jonas Weigand(✉), Magnus Volkmann, and Martin Ruskowski

Technische Universität Kaiserslautern,
Lehrstuhl für Werkzeugmaschinen und Steuerungen, Kaiserslautern, Germany
jonas.weigand@mv.uni-kl.de
https://www.mv.uni-kl.de/en/wskl/

**Abstract.** This paper aims to reduce gearbox errors on industrial robots with a feed forward, neural adaptive control. The algorithm combines two networks, one for control and one for system identification. In order to achieve a high precision and generality on untrained data, a Runge-Kutta Neural Network is used for black-box identification of a nonlinear robot joint. Secondary encoders as additional angle sensors measure the gearbox error and are used for supervised learning. The presented algorithm is capable of online application and reduces gearbox errors in a nonlinear simulation.

**Keywords:** Robot control · Secondary encoders · Robot accuracy · Neural adaptive control · Runge Kutta neural network

## 1 Introduction

Using industrial robots in machining processes offers great potential regarding process flexibility while remaining low cost [1–3]. The main drawback of robots compared to machining tools is the lack of Tool-Center-Point (TCP) accuracy. One way of improving this accuracy is using secondary encoders (SE) on the main axes of the robot in order to reduce gearbox errors [4,5]. [6] has shown, that SE are able to improve robot accuracy in a cascade control using stiff control parameters and optimizing SE for disturbance compensation only. In order to overcome the drawbacks of a stiff cascade control configuration, the guiding behaviour of the axes needs to be improved by a feed forward control. However, a model based control needs good model knowledge, which is difficult to obtain. [2] shows that especially the friction model of the gearbox is complicated, e.g. because the gearbox temperature changes more than 80 °C during machining. Furthermore, a good control needs additional sensor information, which is usually not available on standard robots and imply a costly update. Therefore, this paper aims to overcome the drawbacks of bad model knowledge by using black-box system identification and adjusting the control accordingly. Due to nonlinear physical effects of the robot, e.g. gearbox friction and backlash, the algorithm


K. Berns and D. Görges (Eds.): RAAD 2019, AISC 980, pp. 153–161, 2020.
https://doi.org/10.1007/978-3-030-19648-6_18

has to be capable of nonlinear system identification. In addition, the algorithm has to be ready for online usage.

This paper is organized as follows. First, a concept - neural adaptive control - is introduced and explained. Next, system identification using Runge-Kutta Neural Networks (RKNN) for nonlinear modelling is presented. Afterwards, a Multi-Layer-Perceptron (MLP) as feed forward controller is shown. Finally, a nonlinear simulation model of the robot axis is presented and simulation results using the proposed algorithm are demonstrated.

## 2 Concept

An overview of neural adaptive control is given in [7,8]. The core idea used in this paper is to separate the tasks of system identification and controller design. The signal concept is shown in Fig. 1. A controller network $c(\cdot)$ is used to modify the feed forward reference trajectory $r$, which is in our case the arm reference angle $\varphi_R$. Depending on the reference, the controller creates a modified reference arm angle $u \mathrel{\widehat{=}} \varphi_U$. Before the controller is trained, $\varphi_R$ is equal to $\varphi_U$. Using an unmodified reference as input is possible because a closed loop system is utilized. For training the controller in a simulation environment, a model of the measured system is needed. In a comparative study [9] of multi-axis robot system identification, RKNN achieve a high prediction accuracy while remaining a good generalization. RKNN are explained in [10] and combine neural networks with the Runge-Kutta differential equation solver. The RKNN is trained to approximate the measured arm angle $y \mathrel{\widehat{=}} \varphi_A$ with the simulated arm angle $\hat{y} \mathrel{\widehat{=}} \hat{\varphi}_A$ as good as possible.

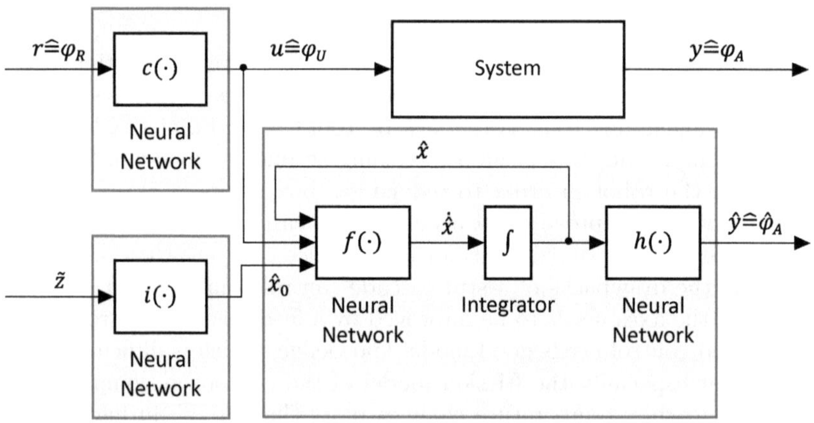


**Fig. 1.** Complete control structure. A RKNN is used for online black-box system identification. Using the RKNN, a feed forward controller network is trained and used online.

## 3 Runge-Kutta Neural Networks

For the state network $f(\cdot)$, the output network $h(\cdot)$ and the initial network $i(\cdot)$, MLPs with a single hidden layer are applied [11]. Using a MLP, we obtain the network equations

$$\dot{\hat{x}}(t) = f(\cdot) = W_F^o \sigma(W_{FX}^h \hat{x}(t) + W_{FU}^h u(t) + b_F^h) + b_F^o \tag{1}$$

$$\hat{y}(t) = h(\cdot) = W_H^o \sigma(W_{HX}^h \hat{x}(t) + b_H^h) + b_H^o \tag{2}$$

$$\hat{x}_0 = i(\cdot) = W_I^o \sigma(W_{IZ}^h \tilde{z}_k + b_I^h) + b_I^o \tag{3}$$

with the continuous state vector $x(t)$, the estimated state vector $\hat{x}(t)$, input $u(t)$ and estimated output $\hat{y}(t)$. In the following sections, the measured and estimated variables are transferred to discrete-time variables with time index $k$. The activation function $\sigma(\cdot)$ is a hyperbolic tangent. Weight matrices and bias terms use the following index convention. Low-indices $F$, $H$ and $I$ are used for $f(\cdot)$, $h(\cdot)$ and $i(\cdot)$ network respectively. Low-indices $X$, $U$ and $Z$ represent state $\hat{x}$ signal, input $u$ signal and regression vector $\tilde{z}_k$ respectively. The regression vector consists of a measured input and output sequence $\tilde{z}_k = [u_{k-N_1}, u_{k-N_1+1}, \dots, u_{k-1}, y_{k-N_2}, y_{k-N_2+1}, \dots, y_{k-1}]$ with $N_1 \in \mathbb{N}^+$ and $N_2 \in \mathbb{N}^+$. High-indices $o$ and $h$ denote output and hidden layer respectively. For a compact notation, all model network weights and bias terms are summarized in $\Theta_M = [W_I^o, W_{IZ}^h, b_I^h, b_I^o, W_F^o, W_{FX}^h, W_{FU}^h, b_F^h, b_F^o, W_H^o, W_{HX}^h, b_H^h, b_H^o]$. Calculating (1) requires a solution of the differential equation. This is done by using the 4-stage Runge-Kutta algorithm (RK) [10]. Incorporating RK leads to

$$\begin{aligned}
v_1 &= W_F^o \sigma\left(W_{FX}^h \hat{x}_k + W_{FU}^h u_k + b_F^h\right) + b_F^o \\
v_2 &= W_F^o \sigma\left(W_{FX}^h \left(\hat{x}_k + \frac{h}{2} v_1\right) + W_{FU}^h u_k + b_F^h\right) + b_F^o \\
v_3 &= W_F^o \sigma\left(W_{FX}^h \left(\hat{x}_k + \frac{h}{2} v_2\right) + W_{FU}^h u_k + b_F^h\right) + b_F^o \\
v_4 &= W_F^o \sigma\left(W_{FX}^h \left(\hat{x}_k + h v_3\right) + W_{FU}^h u_k + b_F^h\right) + b_F^o \\
\hat{x}_{k+1} &= \hat{x}_k + \frac{h}{6} v_1 + \frac{h}{3} v_2 + \frac{h}{3} v_3 + \frac{h}{6} v_4 \\
\hat{y}_{k+1} &= W_H^o \sigma\left(W_{HX}^h \hat{x}_{k+1} + b_H^h\right) + b_H^o
\end{aligned} \tag{4}$$

with the time step length $h$. The input $u_k$ is assumed to be constant during one time interval. The RKNN is trained using *fmincon* from the *MATLAB optimization toolbox* with Sequential Quadratic Programming (SQP). The cost function for training is

$$\arg\min_{\Theta_M} \left( \left( \sum_{j=k-N_3+\max(N_1,N_2)}^{k} \left(y_j - \hat{y}_j\left(\Theta_M\right)\right)^2 \right) + \lambda_1 \cdot \Theta_M^2 \right) \tag{5}$$

using (3) and (4) for calculating $\hat{y}_k$. The regularisation parameter is $\lambda_1$. The number of time steps used for training the model is $N_3 \in \mathbb{N}^+$ with $N_3 \gg N_1$ and $N_3 \gg N_2$.

## 4 Controller

In [12], a chaotic Lur'e Problem is solved using a similar concept. Based on that work, a MLP is chosen feed forward control network $c(\cdot)$ and is defined by

$$u_k = c(\cdot) = W_C^o \sigma \left( W_{CR}^h \tilde{r}_k + b_C^h \right) + b_C^o \tag{6}$$

with the partial reference trajectory $\tilde{r}_k = [r_k, r_{k+1}, \ldots, r_{k+N_4}]$ with an arbitrary number of future time steps $N_4 \in \mathbb{N}^+$. Given $\tilde{r}_k$, the MLP has access to the information about the future reference trajectory. Next, the weights of network $c(\cdot)$ are trained with reinforcement learning to modify the trajectory. Note that the network has no information about the actual state of the system. For a compact notation, all weights and bias terms of network $c(\cdot)$ are summarized in $\Theta_C = [W_C^o, W_{CR}^h, b_C^h, b_C^o]$. The optimization is also done by the *MATLAB optimization toolbox*, with the cost function

$$\arg\min_{\Theta_C} \left( \left( \sum_{j=k}^{k+N_5-N_4} \left( r_j - \hat{y}_j \left( \Theta_C \right) \right)^2 \right) + \lambda_2 \cdot \Theta_C^2 + \left( \lambda_3 \cdot \sum_{j=k}^{k+N_5-N_4-1} \left( u_{j+1} - u_j \right)^2 \right) \right) \tag{7}$$

and the weighting factors $\lambda_2$ and $\lambda_3$. The simulated output $\hat{y}_k$ is calculated using (2), (3) and (4). The first term $(r_j - \hat{y}_j(\Theta_C))^2$ is the core of (7), training the input $u_k$ from (6) to optimise the simulated output to the unmodified reference trajectory. The second term in (7) is used for regularisation and the third term minimizes the input change rate. The number of time steps used for training the controller is $N_5 \in \mathbb{N}^+$ with $N_5 \gg N_4$.

## 5 Simulation Model

The presented algorithm is tested with a closed loop nonlinear model of a single robot joint. Due to the longest lever, axis 2 of a *KUKA Quantec Ultra SE* is chosen. The complete model is presented in [6] and this paper will cover a summary only. The open loop robot joint model is a two-mass-oscillator including nonlinear stiffness and nonlinear friction. Figure 2 shows the model with the physical relations described in Table 1. The real robot is pictured in Fig. 3.

The motor torque serves as actuating variable and generates an acceleration of the motor inertia which rotates the gearbox. The gearbox, acting as a nonlinear spring, encounters the motor torque, overcomes the gearbox friction and accelerates the arm inertia. Using the model in Fig. 2, the nonlinear model can be obtained by calculating the sum of torques around motor and arm inertia with the friction torque $\tau_F$ and the elastic spring torque $\tau_E$.

$$J_M \cdot \ddot{\varphi}_M = \tau_M - \frac{\tau_E}{u_G} \tag{8}$$

$$J_A \cdot \ddot{\varphi}_A = -\tau_L - \tau_F + \tau_E \tag{9}$$

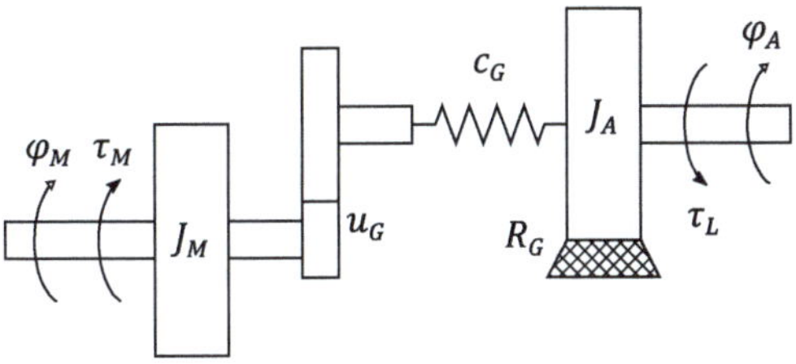


**Fig. 2.** Model of the robot joint as two-mass-oscillator

**Fig. 3.** KUKA Quantec Ultra SE

**Table 1.** Robot model variables and parameters

| Description | Type | Symbol | Value | Unit |
|---|---|---|---|---|
| Motor torque | State variable | $\tau_M$ | | Nm |
| Motor angle | State variable | $\varphi_M$ | | rad |
| Motor angle speed | State and output variable | $\dot{\varphi}_M$ | | rad/s |
| Arm angle | State and output variable | $\varphi_A$ | | rad |
| Arm angle speed | State variable | $\dot{\varphi}_A$ | | rad/s |
| Load torque | Distortion variable | $\tau_L$ | | Nm |
| Modified reference arm angle | Input variable | $\varphi_U$ | | rad |
| Reference arm angle | Guiding variable | $\varphi_R$ | | rad |
| Motor inertia | Parameter | $J_M$ | 0.0145 | $\text{kgm}^2$ |
| Arm inertia | Parameter | $J_A$ | 1092 | $\text{kgm}^2$ |
| Gearbox ratio | Parameter | $u_G$ | 203.52 | – |
| Backlash angle | Parameter | $\phi_B$ | $0.15 \cdot 10^{-3}$ | rad |
| Lost-motion angle | Parameter | $\phi_{LM}$ | $0.29 \cdot 10^{-3}$ | rad |
| Offset torque | Parameter | $\tau_{E,0}$ | 210 | Nm |
| Lost-motion stiffness | Parameter | $c_{LM}$ | $1.44 \cdot 10^6$ | Nm/rad |
| Linear stiffness | Parameter | $c_G$ | $8.94 \cdot 10^6$ | Nm/rad |
| Coulomb friction | Parameter | $\tau_{C,0}$ | 500 | Nm |
| Viscous friction | Parameter | $R_G$ | 764 | Nms/rad |
| Proportional speed gain | Parameter | $P_1$ | 2 | Nms/rad |
| Integral speed gain | Parameter | $I_1$ | 16 | $\text{Nms}^2/\text{rad}$ |
| Proportional position gain | Parameter | $P_2$ | 8192 | 1/s |

Motor friction as well as all electro-mechanical effects are neglected. Since the *KUKA Quantec* has a hydraulic counterweight acting on axis 2, both counterweight and gravity effects are neglected in this work. Temperature effects that have an influence on the friction are neglected, too, and require further research.

The stiffness is modelled with backlash, lost-motion, and linear elasticity. Lost-motion describes an effect in between backlash and linear elasticity, where not all tooth flanks are yet in full contact. During backlash $\tau_E$ equals 0. In the lost-motion range, the stiffness is modelled linear with a lower slope and an offset. In the full-contact range, the stiffness is modelled as a linear function with an offset. The stiffness coefficients are defined as $c_{LM}$ and $c_G$, for lost-motion and full-contact respectively. We further define the angle ranges $\phi_B$ and $\phi_{LM}$, for backlash and lost-motion respectively. With the torsion angle

$$\Delta\varphi = \varphi_M / u_G - \varphi_A \tag{10}$$

the following condition is defined.

$$\tau_E = \begin{cases} 0 & \text{for } |\Delta\varphi| \leq \phi_B \\ c_{LM}\,\Delta\varphi - c_{LM}\,\frac{\phi_B}{2}\,\text{sign}(\Delta\varphi) & \text{for } \phi_B < |\Delta\varphi| \leq \phi_{LM} \\ c_G\,\Delta\varphi + \tau_{E,0}\,\text{sign}(\Delta\varphi) & \text{for } \phi_{LM} < |\Delta\varphi| \end{cases} \tag{11}$$

The gearbox friction is modelled as coulomb and viscous friction. According to [2,13,14] the Stribeck effect can be neglected. With the linear friction constant $R_G$ and the coulomb friction torque $\tau_{C,0}$ the friction conditions for the model are defined.

$$\tau_F = \begin{cases} 0 & \text{for } \quad \dot{\varphi}_A = 0 \\ R_G\,\dot{\varphi}_A + \tau_{C,0}\,\text{sign}(\dot{\varphi}_A) & \text{otherwise} \end{cases} \tag{12}$$

In order to augment the open loop model of the axis to a closed loop model, a feed back controller is introduced. This cascade controller is configured as P-PI-controller

$$\tau_M = \left(P_1 + \frac{I_1}{s}\right) \cdot \left(P_2 \cdot (-\varphi_A + \varphi_U) - \dot{\varphi}_M\right) \tag{13}$$

with the Laplace transform variable $s$ and the control parameters $P_1$, $I_1$ and $P_2$. The controller uses both, the measured arm angle $\varphi_A$ obtained by the SE and the motor speed $\dot{\varphi}_M$. The modified arm reference angle $\varphi_U$ is the input whereas the measured arm angle $\varphi_A$ is the output of the closed loop system.

## 6 Simulations Results

The main result of this paper is presented in Fig. 4. The simulation is divided into three subsections. First, from 0 to 1 s, the input-output relation of the system is only measured and the unmodified reference is used as input. Then, from 1 to 1.5 s, a RKNN is trained for 500 iterations. Using the RKNN, a controller network is trained for 100 iterations. Finally, from 1.5 to 2.5 s, the controller network is used online.

The training result of the RKNN is shown in Fig. 5. With a Maximum-Error of 0.016°, the RKNN achieves 96% accuracy with respect to the angle amplitude of 0.52°. Unlike other training algorithms, the data set is not split into a training and validation set. This is for simplicity only. A second data set for validation would require the effort of measuring a second, physical trajectory. However, a validation data set implies the advantages of a better evaluation of the trained network. Since the training data set of the RKNN needs to cover the complete system dynamics, a rather long data set is chosen. To shorten calculation time, the number of time steps in the training data set are reduced by an interpolation. Therefore, the step length for training can be unequal to the step length of data logging. The training of the control network is done on both, random created trajectories and the unmodified reference trajectory. Using different trajectories improves the generality of the controller network.

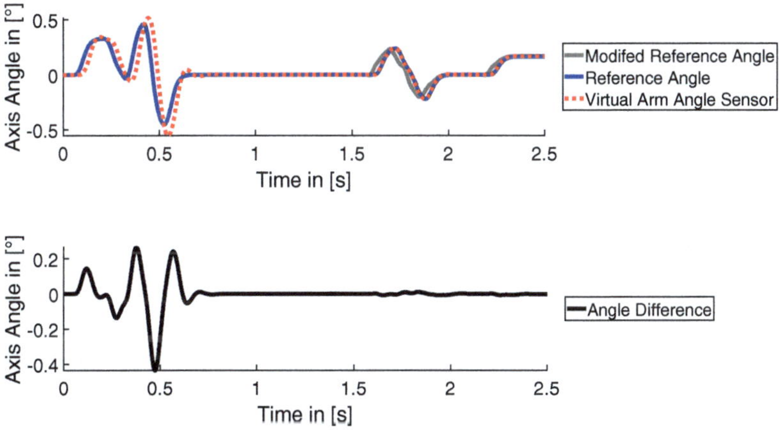


**Fig. 4.** Training and test of proposed algorithm

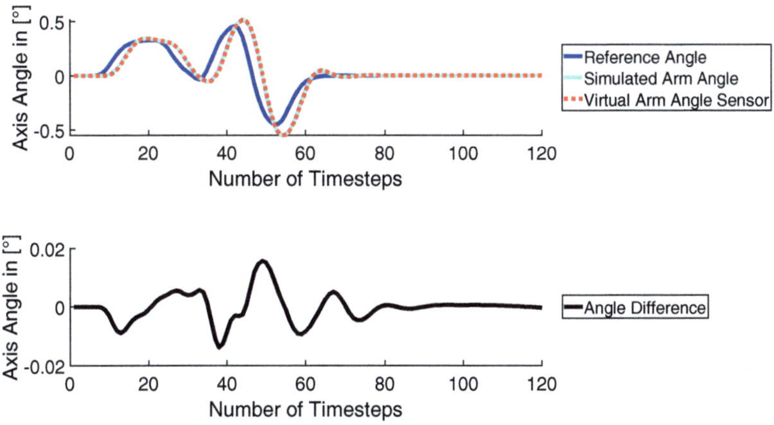


**Fig. 5.** Reference and system identification

## 7 Conclusion

All in all, the proposed algorithm achieves a Root-Mean-Square-Error between the reference and actual arm angle of 0.0026° and a Maximum-Error of 0.012°. With respect to the Maximum-Error without this algorithm of 0.121°, this is an improvement of 90%. The simulation results prove that it is possible to define the robot joint as a black-box and to optimize it with the help of RKKN without physical model knowledge. The algorithm achieves to overcome the challenges of backlash, lost-motion and nonlinear stiffness in simulation.

Further research is required how stability constrains can be incorporated. Only with this advancement, the network controller could be designed as feed back controller and an open loop model could be applied. As a starting point, the NLq Theory [15] presents a unified framework for neural adaptive control with global asymptotic stability criteria. Furthermore, the continuous online update of both, the model and the control network, need further research. Finally, the algorithm needs to be implemented and tested on the real robot.

## References

1. Abele, E. (ed.): Spanende Bearbeitung mit Industrierobotern: Forschungsprojekt ADVOCUT - Entwicklungen und Industrietransfer. Abschlußbericht BMBF-Verbundprojekt. Meisenbach, Bamberg (2007)
2. Thümmel, M.: Modellbasierte Regelung mit nichtlinearen inversen Systemen. Dissertation, Technische Universität München (2006)
3. Rösch, O.: Steigerung der Arbeitsgenauigkeit bei der Fräsbearbeitung metallischer Werkstoffe mit Industrierobotern. Dissertation, Technische Universität München (2015)
4. Uhlmann, E., Reinkober, S., Hollerbach, T.: Erhöhung der genauigkeit beim fräsen mit industrierobotern. ZWF Zeitschrift für wirtschaftlichen Fabrikbetrieb **113**(9), 576–579 (2018)
5. Devlieg, R.: High-accuracy robotic drilling/milling of 737 inboard flaps. SAE Int. J. Aerosp. **4**(2), 1373–1379 (2011)
6. Belz, S.: Nichtlineare Modellierung, flachheitsbasierte Vorsteuerung und abtriebsseitige Regelung eines Robotergelenkes. Technische Universität Kaiserslautern, Studienarbeit (2018)
7. Hagan, M., Demuth, H.: Neural networks for control. In: Proceedings of the American Control Conference (1999)
8. Polycarpou, M.M.: Stable adaptive neural control scheme for nonlinear systems. IEEE Trans. Autom. Control **41**(3), 447–451 (1996)
9. Efe, M., Kaynak, O.: A comparative study of neural network structures in identification of nonlinear systems. Mechatronics (1998)
10. Wang, Y.J., Lin, C.T.: Runge-kutta neural network for identification of dynamical systems in high accuracy. IEEE Trans. Neural Networks **9**(2), 294–307 (1998)
11. Deflorian, M.: Versuchsplanung und Methoden zur Identifikation zeitkontinuierlicher Zustandsraummodelle am Beispiel des Verbrennungsmotors. Dissertation, Technische Universität München (2011)
12. Suykens, J., Vandewalle, J.: Absolute stability criterion for a lur'e problem with multilayer perceptron nonlinearity. In: International Workshop on Nonlinear Dynamics of Electronic Systems (1996)

13. Reiner, M.: Modellierung und Steuerung von strukturelastischen Robotern. Dissertation, Technische Universität München (2010)
14. Kurze, M.: Modellbasierte Regelung von Robotern mit elastischen Gelenken ohne abtriebsseitige Sensorik. Dissertation, Technische Universität München (2008)
15. Suykens, J., de Moor, B.L.R., Vandewalle, J.: NLq theory: a neural control framework with global asymptotic stability criteria. J. Neural Netw. **10**(4), 615–637 (1997)

# Task-Dependent Energetic Analysis of a 3 d.o.f. Industrial Manipulator

Paolo Boscariol[1(✉)], Lorenzo Scalera[2,3], and Alessandro Gasparetto[3]

[1] DTG, Università degli Studi di Padova, Vicenza, Italy
paolo.boscariol@unipd.it
[2] Faculty of Science and Technology, Free University of Bozen-Bolzano, Bolzano, Italy
lorenzo.scalera@unibz.it
[3] DPIA, Università degli Studi di Udine, Udine, Italy
gasparetto@uniud.it

**Abstract.** In this paper a preliminary analysis of the energetic performance of an industrial manipulator is presented. In particular, the paper investigates the effects of the trajectory planning on the overall energy consumption of the manipulator in a pick & place task, focusing also on the location of the path within the workspace. An electro-mechanical model of the actuators and the inverse dynamic model of the robot have been developed and used to estimate the robot energy consumption when executing a basic motion task. Results are then collected into energy consumption maps, showing how the location of the task within the robot workspace affects the energetic performance of the robot.

**Keywords:** Energy efficiency · Trajectory planning · Robotic manipulator · SCARA robot · Workspace

## 1 Introduction

One of the most relevant challenge in modern robotics is the reduction of energy consumption, especially when high speed operations and high volumes of production are needed. In recent years, the increasing of energy costs and the growing of environment awareness have driven engineers and researchers to find new solutions for energy saving. In this context, a classification and analysis of different methods and techniques for enhancing the energy performances of industrial manipulators can be found in [8]. Possible approaches are the design of lightweight robots [1,3,20], the use of regenerative power storage systems [12], the exploitation of the manipulator natural dynamics [10,13], the robot model selection [9], and the planning of energy efficient trajectories.

Examples of energy-efficient motion planning approaches can be found in [14], where both time-optimal and energy-optimal trajectories are investigated for pick-and-place motion with a 6 degree-of-freedom (d.o.f.) manipulator, and in [11], where an energy saving method for industrial machines using simple motion


K. Berns and D. Görges (Eds.): RAAD 2019, AISC 980, pp. 162–169, 2020.
https://doi.org/10.1007/978-3-030-19648-6_19

trajectories is presented. Furthermore, in [7], point-to-point trajectories based on standard primitives have been employed for the overall energy reduction of a typical 1 d.o.f. mechatronic system, whereas a trajectory planning approach for the energy saving in a redundant robotic cell has been adopted in [4]. Alternatively, energy efficiency can be improved by reducing the actuator effort thorough motion design, as performed for industrial robots in [17,19].

In literature several performance measures have been investigated to estimate the behavior of industrial manipulators, either by using global or local measures. Referring to the latter category, in [18], the application and graphical visualization of different indexes are applied to a SCARA manipulator, whereas in [6] a task-dependent performance index has been introduced and adopted to optimize the location of a pick-and-place task for an industrial parallel robot. This studies have been motivated by the fact that local performance indexes can be useful for highlighting the relationship between the robot and the task definition, and in particular they often provide a guidelines in the choice of the areas of the workspace in which 'better' robot performance can be achieved [15]. Robot performance can be measured by performance indexes focusing on some measurement of dexterity, force or speed exertion capability, or manipulability, but currently there are, to the best of Authors' knowledge, no performance index that are focused strictly on energy consumption. With this final goal in mind, the aim of this paper is to provide a preliminary analysis of the energetic performance of a 3 d.o.f. SCARA manipulator. The effects of the path and trajectory planning on the overall energy consumption are evaluated for a specific task: a common pick-and-place operation, as performed in [5]. On the basis of an electro-mechanical model of the manipulator, the influence of the location of the task within the robot workspace, as well as of the choice of motion law, are evaluated. Simulation results provide, for each considered trajectory, energy consumption maps for the performance evaluation and the optimal robot positioning.

## 2 Dynamic and Electro-Mechanical Model

The industrial manipulator under consideration in this work is a 3 d.o.f. SCARA robot, designed according to the common RRP architecture. The dynamic model of the robot can be written by attributing to each link its specific inertial properties. The dynamic equations of motion, which describe the joint torques $\boldsymbol{\tau}_j$ as a function of the joint variables $\boldsymbol{q} = [q_1, q_2, q_3]^T$, can be derived by using the Lagrangian formalism, leading to the usual formulation:

$$\boldsymbol{\tau}_j = \boldsymbol{M}(\boldsymbol{q})\ddot{\boldsymbol{q}} + \boldsymbol{C}(\boldsymbol{q}, \dot{\boldsymbol{q}}) + \boldsymbol{f}_v\,\dot{\boldsymbol{q}} + \boldsymbol{F}_c\,\mathrm{sign}(\dot{\boldsymbol{q}}) \tag{1}$$

in which $\boldsymbol{M}(\boldsymbol{q})$ is the mass matrix of the manipulator, matrix $\boldsymbol{C}(\boldsymbol{q}, \dot{\boldsymbol{q}})$ accounts for the centrifugal and Coriolis effects, $\boldsymbol{f}_v$ is the diagonal matrix of viscous friction coefficients, and $\boldsymbol{F}_c$ is the diagonal matrix of Coulomb friction forces. In order to derive the electro-mechanical model of the actuators of the robot, assuming that brushless motors are used as in [4,16], the motor torques vector $\boldsymbol{\tau}_m$ is related to the motor armature current vector $\boldsymbol{I}(t)$ as:

$$\boldsymbol{\tau}_m(t) = \boldsymbol{k}_t\, \boldsymbol{I}(t) \tag{2}$$

where $\boldsymbol{k}_t$ is the diagonal matrix of the individual motor torque constants. The armature model can be then introduced to describe the voltage drop $\boldsymbol{V}(t)$ across the motors of all the joints of the manipulator:

$$\boldsymbol{V}(t) = \boldsymbol{R}\boldsymbol{I}(t) + \boldsymbol{k}_b\, \dot{\boldsymbol{q}}_m(t) \tag{3}$$

where matrices $\boldsymbol{R}$ and $\boldsymbol{k}_b$ include each individual motor winding resistance and the back-emf constants, respectively, while $\dot{\boldsymbol{q}}_m$ is the motor velocities vector. Finally, the motors input energy $E_{robot}$ required to perform a generic operation over the time interval $t \in [t_a, t_b]$ can be found by computing the time integral of the instantaneous electric power drawn by the robot, expressed by the voltage-current product:

$$E_{robot} = \int_{t_a}^{t_b} \boldsymbol{V}^T(t)\, \boldsymbol{I}(t)\, dt \tag{4}$$

Equation 4 represents the net energetic balance of the robot within the time frame $[t_a, t_b]$, so that is the energy is evaluated without making any distinction between positive and negative values of the voltage-current product. This feature is however not very realistic, since the vast majority of industrial robots are equipped with non-regenerative motor drives and negative electric power is simply dissipated on a braking resistor. Accordingly, to avoid subtracting the energy dissipated on the braking resistor from the energy balance, the computation of the integral of Eq. 4 is limited to positive values of the current-voltage product.

## 3 Task-Dependent Energetic Performance Evaluation

The dynamic model introduced in Sect. 2 is here used to analyze the impact of the choice of the task on the energetic performance of the robot. The analysis is limited to a basic but commonly performed task, i.e. the translation of the robot end-effector along a straight line, as performed in pick & place operation. Given a fixed execution time and a fixed total displacement, such a task can be performed in infinite ways, provided that the initial and final point of the trajectory are left free. Additionally, the design of the task gains an additional degree of freedom if the motion law is left free as well. Under this assumptions the path of the end-effector of the robot can be parametrized by two values: the real positive value $d$, which measures the distance, defined on the horizontal plane $\{X, Y\}$ between the base of the robot and the mid-point of the trajectory, and $\alpha$, which measures the angular distance between the line described by the path and the line along which $d$ is measured: a graphical reference is provided in Fig. 1.

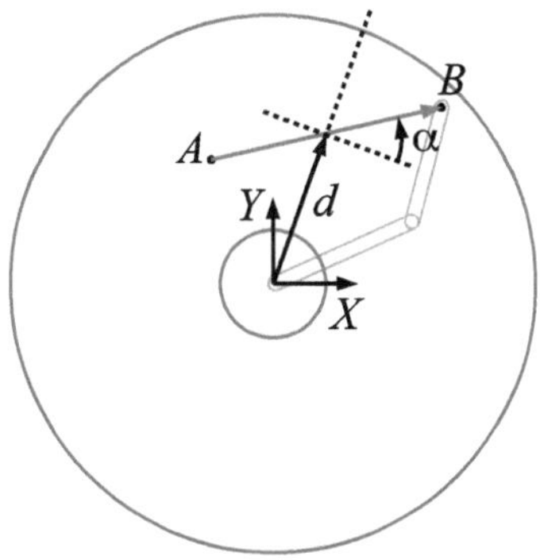


**Fig. 1.** Parametrization of the path within the workspace of the robot

The base of the robot is set on the origin of the $\{X, Y\}$ reference frame, and the workspace of the robot is delimited by the two gray circles shown in Fig. 1. The task is represented by a vector that connects the starting point $A$ with the final point $B$. Given the radial symmetry of the robotic configuration, all possible tasks are defined by values of $d$ ranging from 0.1 m to 0.8 m, and for values of $\alpha$ between 0 and $\pi$ radians. The kinematic and dynamic properties of the manipulator used for the computation of the energy costs are shown Table 1. The distance between point $A$ and $B$ is set, for all experiment, to be equal to 0.3 m and the total execution time is set to 0.5 s. The motion of the third axis of the robot is not included in the analysis, given that it is not influenced by the choice of the path. Referring to the motion profile, countless choices are available. Here four well-known options are taken into consideration: the trapezoidal speed profile, the harmonic speed profile, the third and fifth-degree polynomial profiles [2]. In all cases the motion design is performed in the operational space to ensure that the robot end-effector is moved along a straight line. For each of the four

**Table 1.** Mechanical and electric parameters of the SCARA robot

| Parameter | Joint 1 | Joint 2 | Joint 3 |
|---|---|---|---|
| Link length | 0.45 m | 0.35 m | - |
| Link mass | 14 kg | 18 kg | 2 kg |
| Gear ratio | 1/80 | 1/50 | 1/30 |
| Motor inertia | $3 \cdot 10^{-4}$ kgm$^2$ | $2 \cdot 10^{-4}$ kgm$^2$ | $1 \cdot 10^{-4}$ kgm$^2$ |
| Viscous friction coefficient | 0.001 Nms/rad | 0.001 Nms/rad | 0.001 Nms/rad |
| Coulomb friction force | $2 \cdot 10^{-2}$ Nm | $2 \cdot 10^{-2}$ Nm | $2 \cdot 10^{-2}$ Nm |
| Motor winding resistance | 3 Ω | 3 Ω | 3.5 Ω |
| Motor back-emf constant | 0.6 Vs/rad | 0.6 Vs/rad | 0.6 Vs/rad |
| Motor torque constant | 0.6 Nm/A | 0.6 Nm/A | 0.6 Nm/A |

choices of the motion profile, a sequence of trajectories are simulated and the resulting energy consumption data are collected and plotted in the energy maps shown in Figs. 2 and 3. Figure 2(a) shows a contour plot of the total energy delivered to the robot when moving according to a trapezoidal speed profile: each point corresponds to a specific choice of $d$ and $\alpha$ and the the energy consumption is measured in Joules. Darker shades of purple represent the tasks associated with higher energy consumption, while the white areas represent tasks that are either unfeasible, due to workspace limitations, or that imply unreasonable energetic requirements. The analysis is then repeated by using the harmonic profile for computing the contour plot in Fig. 2(b), and then by using the third and fifth degree polynomial profiles for the energy maps in Fig. 3(a) and (b), respectively. All maps also include a small black circle, which identifies the minimum energy task: in all four cases the energetic optimality is found for values of $d$ close to 0.6 m and for values of $\alpha$ in the neighborhood of 0.5 rad. According to the task parametrization, as it can be seen in Fig. 1, $\alpha = 0$ corresponds to 'rightwards' motion along the tangential direction, while $\alpha = \pi/2$ refers to an 'outward' motion along the radial direction. The analysis of the energy maps in Fig. 2 and 3 suggest that, as expected, the different motion profiles result in different energy consumptions, and that the best choice, among the one analyzed here, is the third degree polynomial function. Despite the relative differences between the absolute values of the energy consumption, all maps show a similar 'shapes', meaning that the higher energetic efficiency is found for similar values of $d$ and $\alpha$ in all cases. It can be therefore inferred that the total energy consumption associated with a single task is affected by an higher degree by the task collocation within the workspace, rather than by the choice of the specific motion profile. In other words, in the case under consideration, as far as the energy consumption in concerned, the critical choice is the definition of the task position within the workspace.

All four maps show that the best choices for the definition of the task are found within and area whose shape resembles an 'hysteresis loop', and that gen-

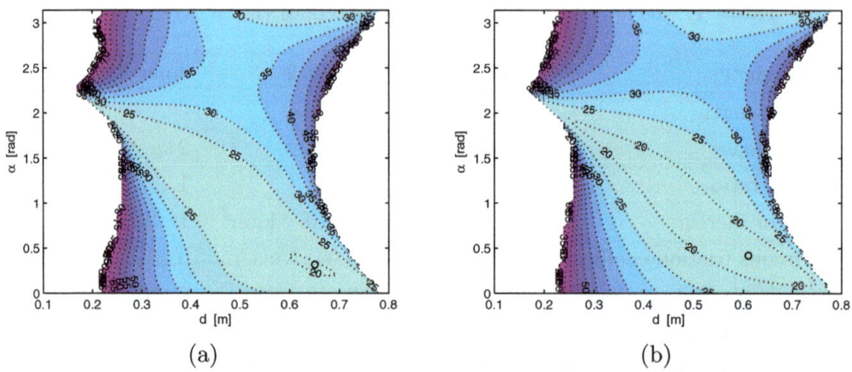


**Fig. 2.** Energy consumption maps: trapezoidal (a) and harmonic (b) motion profiles

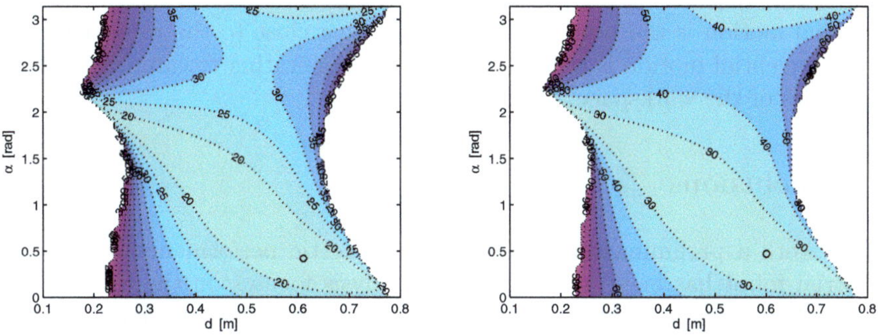


(a) (b)

**Fig. 3.** Energy consumption maps: third order (a) and fifth order (b) polynomial motion profiles

erally higher efficiency is found when operating the robot towards the external areas of the workspace. The influence of the motion direction is more clearly understood by looking at Fig. 4: the red arrows represent the optimal displacements for 12 uniformly distributed distances from the base of the robot, and the blue arrow shows the overall optimal solution, which corresponds to the parameters $d = 0.61$ m and $\alpha = 0.4189$ rad. Each arrow connects point A and B, i.e. the initial and final point of the trajectory, according to the parametrization shown in Fig. 1. All of them refer to the third degree motion profile, i.e. the less energetically expensive among the ones tested here. Figure 4 suggests that

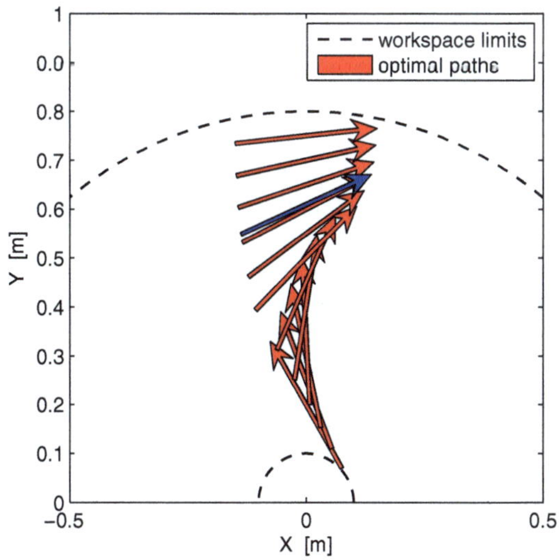


**Fig. 4.** Optimal paths with varying distance from the base: optimal solution shown in red color

the radial motion is the best choice when working close to the robot base, and that a tangential motion is the best choice when operating towards the external boundaries of the workspace.

## 4 Conclusions

In this work a preliminary analysis of the energetic performance of a 3 d.o.f. industrial robot has been presented. The analysis focuses on the estimation of the energy consumption of the robot when executing a basic motion task, highlighting the influence of the choice of the motion profile and of the task positioning within the workspace. The numerical results indicate that the latter has a major impact on the energy consumption, and therefore the energy maps provided here can be used to define the optimal robot-workpiece relative positioning for improving the energy efficiency.

## References

1. Belotti, R., Caneva, G., Palomba, I., Richiedei, D., Trevisani, A.: Model updating in flexible-link multibody systems. In: Journal of Physics: Conference Series, vol. 744, p. 012073. IOP Publishing (2016)
2. Biagiotti, L., Melchiorri, C.: Trajectory Planning for Automatic Machines and Robots. Springer, Heidelberg (2008)
3. Boscariol, P., Gallina, P., Gasparetto, A., Giovagnoni, M., Scalera, L., Vidoni, R.: Evolution of a dynamic model for flexible multibody systems. In: Advances in Italian Mechanism Science, pp. 533–541. Springer (2017)
4. Boscariol, P., Richiedei, D.: Energy saving in redundant robotic cells: optimal trajectory planning. In: IFToMM Symposium on Mechanism Design for Robotics, pp. 268–275. Springer (2018)
5. Boschetti, G.: A picking strategy for circular conveyor tracking. J. Intell. Rob. Syst. **81**(2), 241–255 (2016)
6. Boschetti, G., Rosa, R., Trevisani, A.: Optimal robot positioning using task-dependent and direction-selective performance indexes: general definitions and application to a parallel robot. Robot. Comput. Integr. Manuf. **29**(2), 431–443 (2013)
7. Carabin, G., Vidoni, R., Wehrle, E.: Energy saving in mechatronic systems through optimal point-to-point trajectory generation via standard primitives. In: Second International Conference of IFToMM Italy, pp. 20–28. Springer (2018)
8. Carabin, G., Wehrle, E., Vidoni, R.: A review on energy-saving optimization methods for robotic and automatic systems. Robotics **6**(4), 39 (2017)
9. Chemnitz, M., Schreck, G., Krüger, J.: Analyzing energy consumption of industrial robots. In: 2011 IEEE 16th Conference on Emerging Technologies & Factory Automation (ETFA), pp. 1–4. IEEE (2011)
10. Goya, H., Matsusaka, K., Uemura, M., Nishioka, Y., Kawamura, S.: Realization of high-energy efficient pick-and-place tasks of SCARA robots by resonance. In: 2012 IEEE/RSJ International Conference on Intelligent Robots and Systems (IROS), pp. 2730–2735. IEEE (2012)

11. Ho, P.M., Uchiyama, N., Sano, S., Honda, Y., Kato, A., Yonezawa, T.: Simple motion trajectory generation for energy saving of industrial machines. SICE J. Control Meas. Syst. Integr. **7**(1), 29–34 (2014)
12. Inoue, K., Ogata, K., Kato, T.: An efficient induction motor drive method with a regenerative power storage system driven by an optimal torque. In: Power Electronics Specialists Conference, PESC 2008, pp. 359–364. IEEE (2008)
13. Lu, G., Kawamura, S., Uemura, M.: Proposal of an energy saving control method for SCARA robots. J. Robot. Mechatronics **24**(1), 115–122 (2012)
14. Paes, K., Dewulf, W., Vander Elst, K., Kellens, K., Slaets, P.: Energy efficient trajectories for an industrial ABB robot. Procedia Cirp **15**, 105–110 (2014)
15. Patel, S., Sobh, T.: Manipulator performance measures-a comprehensive literature survey. J. Intell. Rob. Syst. **77**(3–4), 547–570 (2015)
16. Richiedei, D., Trevisani, A.: Analytical computation of the energy-efficient optimal planning in rest-to-rest motion of constant inertia systems. Mechatronics **39**, 147–159 (2016)
17. Shiller, Z.: Time-energy optimal control of articulated systems with geometric path constraints. J. Dyn. Syst. Meas. Control **118**(1), 139–143 (1996)
18. Tanev, T., Stoyanov, B.: On the performance indexes for robot manipulators. Probl. Eng. Cybern. Robot. **49**, 64–71 (2000)
19. Trigatti, G., Boscariol, P., Scalera, L., Pillan, D., Gasparetto, A.: A new path-constrained trajectory planning strategy for spray painting robots. Int. J. Adv. Manuf. Technol. **98**(9–12), 2287–2296 (2018)
20. Vidoni, R., Gasparetto, A., Scalera, L.: 3-D ERLS based dynamic formulation for flexible-link robots: theoretical and numerical comparison between the finite element method and the component mode synthesis approaches. Int. J. Mech. Control (2018)

# 3D Printing with 6D of Freedom: Controlling Material Extrusion Speed

David Kraljić(✉), Matej Štefanič, and Roman Kamnik

Faculty of Electrical Engineering, Laboratory of Robotics,
University of Ljubljana, Tržaška cesta 25, Ljubljana, Slovenia
david.kraljic@gmail.com,
{matej.stefanic,roman.kamnik}@fe.uni-lj.si
http://robolab.si

**Abstract.** 3D printing is conventionally performed with manipulators with only three degrees of freedom (DOF), resulting in objects consisting of horizontal layers and inherent weakness in the vertical direction. This shortcoming is mitigated by printing along curved surfaces, which requires manipulators with more degrees of freedom, a new way of trajectory planning, and a dynamic control of material extrusion speed - the issue addressed in this paper. Our printing set-up, which we describe in the paper, includes an industrial 6 DOF manipulator. The manipulator has non-negligible inertia as well as other limiting constraints, and thus fails to achieve the programmed speed on many of the printing path segments leading to either overflow or underflow of material and poor object quality. We develop a simple empirical approach to predict the speed on problematic path segments and adjust the speed of extrusion, ensuring the correct amount of material is deposited.

**Keywords:** 3D printing · Curved surfaces · Constant speed · Industrial manipulators

## 1 Introduction

3D printing is a process in which material is deposited in layers to form a three dimensional object. The increasing popularity of this additive way of manufacturing is due to the fast and cheap way of building not only prototypes but also end-use products.

The shortcomings of conventional approaches to 3D printing is the use of robotic manipulators with only three degrees of freedom, which limit the deposition of material in horizontal layers. This leads to object weakness in the vertical direction and poor surface quality due to the 'stair-step' effect [1].

The issues above are mitigated by printing in curved layers [2,3], instead of horizontal ones. This requires the use of manipulators with more degrees of freedom and new methods of trajectory planning, which we presented in [4].

Our experimental set-up includes an industrial 6 DOF manipulator. The manipulator has heavier moving parts than conventional 3 DOF printers that


K. Berns and D. Görges (Eds.): RAAD 2019, AISC 980, pp. 170–178, 2020.
https://doi.org/10.1007/978-3-030-19648-6_20

only have a light printing head (printhead) moving. This results in the failure to attain the programmed speed of motion of the printhead on many path segments. Because the quality and strength of the printed objects depend crucially on the uniform deposition of material, the extrusion speed has to be adjusted for the actual motion speed of the printhead and not the programmed one.

Therefore, we develop an empirical approach to predicting the actual motion speed along the printing trajectory, such that the speed of material extrusion can be set correctly.

In Sect. 2 we describe our experimental set-up (both hardware and software), in Sect. 3 we present our approach to the control of the material extrusion speed, in Sect. 4 we illustrate the difference between objects printed with adjusted extrusion speeds and with fixed extrusion speeds, and in Sect. 5 we conclude.

## 2 Set-Up: Control System with Manipulator and Components

In this section, we present each individual component of our set-up. The building of the whole 3D printing system is ongoing and is part of a larger collaborative project.

**Robot Cell:** The robot cell for 3D printing will support the use of composite as well as natural materials, besides the standard plastics such as ABS, PLA, and NYLON. The robot cell is built as a mobile unit and fits the size of the KUKA KR 10 R900 Sixx manipulator.

The cell is designed as a heated chamber in order to reduce the difference between the extruded material and ambient temperature, thus enabling better print quality. The heat prevents uneven cooling of the extruded material which causes bending and internal stresses in the object. The cell includes a printing bed onto which the initial layers are printed - the printing bed is also heated to prevent warping and unsticking of the initial layers.

**Manipulator:** KUKA KR 10 R900 Sixx manipulator, one of standard industrial manipulators available on the market, is used together with the KR C4 Compact controller. The manipulator has 6 DOF enabling the access to the same point from multiple directions and printing on curved surfaces.

**Printing Head and Its Control Board:** The printing head incorporates two nozzles in a single head. The first nozzle supports material extrusion while the second nozzle supports the extrusion of carbon fibre filament as additional strengthening component. The deposit of material is controlled by controlling two stepper motors.

We use readily available open source 3D printing controller board[1], which is designed for conventional 3 DOF 3D printers. The board supports multiple

[1] http://www.geeetech.com/wiki/index.php/GT2560.

stepper motor drivers which are used for extrusion speed regulation of material. The board is also used for temperature regulation of two heating elements - one in the nozzle and one in the printing bed.

### 2.1 Software and Communication

The complete printing set-up includes a PC, which with its software represents the main data and control interface between the manipulator, the printing head, and the user.

We developed the trajectory planning module as well as the communication module [4] for communication between the PC, robot controller, and the printhead control board. The communication with the robot controller is implemented via Ethernet, and with the printhead control board via serial connection. The communication continuously provides the trajectory information and desired speed of motion to the robot controller as well as extrusion speed and temperature to the printhead. Figure 1 provides a schema of the set-up and Fig. 2 a sample output from our trajectory planning software.

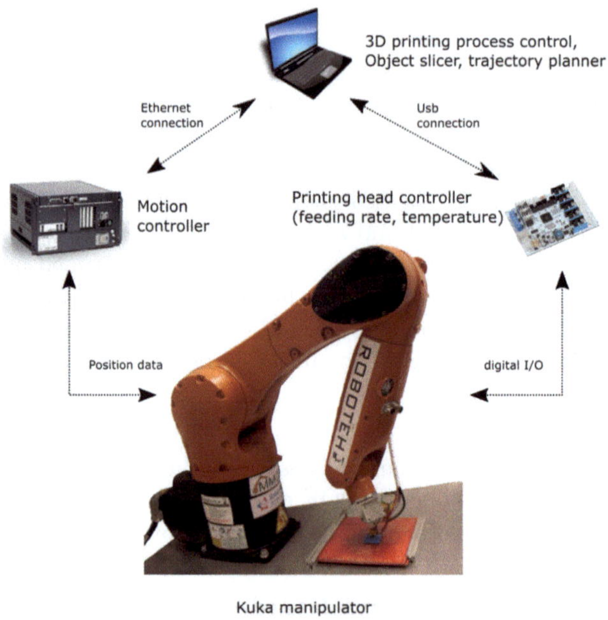


**Fig. 1.** Robot cell for 3D printing.

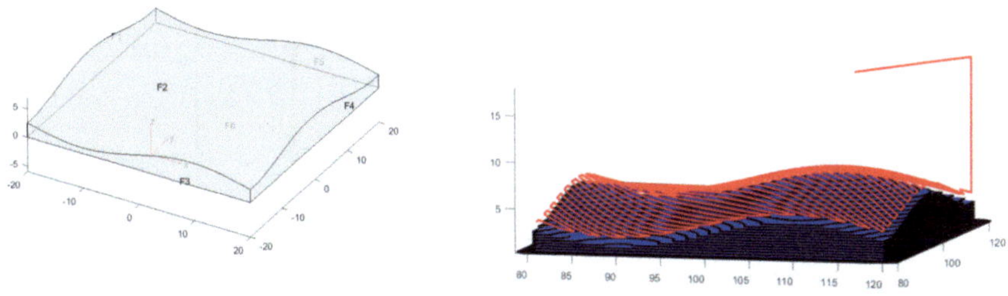

(a) (b)

**Fig. 2.** An example object (a) with printing head motion trajectory for curved surface printing (b)

## 3 Material Extrusion Control Based on Prediction of Motion Speed

The quality of printed object depends crucially on the uniform deposition of material. This is achieved only if the extrusion speed of the material from the printing head is synchronized with the motion speed of the printing head nozzle - the tool centre point (TCP). If the motion speed is too slow with respect to the extrusion speed, too much material is extruded leading to overflow and incorrect outer dimensions of the printed object. Conversely, if the motion speed is too fast, not enough material is extruded leading to lack of intertrack fusion, gaps, and weakness of the printed object.

For conventional 3D printers with only three axes, the main moving part is just the (light) printhead. There, the speed of the printhead and the speed of extrusion can be maintained constant along the printing trajectory. However, industrial manipulators with more degrees of freedom have moving parts with non-negligible inertia. This results in significant speed variation on and within the printing trajectory segments *despite* programming the robot controller to maintain constant speed along the trajectory. We illustrate this in Fig. 3.

Therefore, to achieve good print quality when printing with large and heavy manipulators, the actual speed of the tip of the nozzle has to be estimated in advance for the whole printing path, such that the extrusion of material can be adjusted before each trajectory segment begins. The prediction of actual speeds additionally enables a more accurate estimation of total printing times.

### 3.1 Predicting the Actual TCP Speed

The failure to attain the desired speed generally occurs in sharp corners, during large orientation changes, near robot singularities, and for very short trajectory segments. The reduction of speed from the desired one originates from the torque, acceleration, and axis-velocity limits, as well as particularities of singularity handling and positioning approximation.

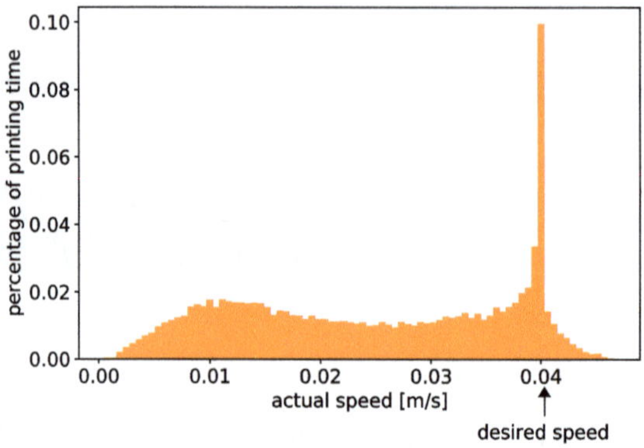


**Fig. 3.** The histogram of actual speeds that were attained during printing a sample object (in Fig. 2) with our set-up. The desired constant motion speed of 4 cm/s was set on the robot controller. Not accounting for the varying actual speed would lead to about 60% more of material extruded than needed. The slowest parts of the printing trajectory were on the curved surface which requires constant reorientation of the printhead.

One way to predict the actual speed along the printing trajectory, is to simulate all of the kinematics, dynamics, and specifics of the robot controller and manipulator for a particular trajectory.

We decided for a simpler and empirical solution - we measured the actual speed of the TCP along random trajectories, modelled the collected data, and used the models to predict speeds for future printing trajectories. The prediction is performed offline, before the printing starts. In this paper, we focus on demonstrating the viability of such an empirical approach, and not on the details of statistical modelling.

**Experimental Protocol.** We generated random trajectories in the workspace of the robot, with varying lengths of the path segments as well as varying orientations and speeds along the path. Such random trajectories ensure we efficiently sample the possible configurations the robot will encounter during the printing process.

The trajectories were generated by sampling relative coordinates for each trajectory point from a normal distribution of typical scale 6 mm. This represents typical lengths of printing path segments of objects we plan to print. The relative orientation angles along each trajectory segment were generated by sampling from a normal distribution of typical scale 10°. The programmed speeds along trajectories were set to 2 cm/s, 4 cm/s, and 6 cm/s, each for about 60 min, resulting in recorded data of about 68000 trajectory segments altogether. The robot controller we use enabled us to measure and record the actual speed as well as the desired speed along the trajectory. This gave us the input data (the trajectory) and the output data (the actual speed) to be used in statistical modelling.

We chose to predict the average speed on each path segment - relaying finer grained TCP speed within path segments could not be utilized in practice due to the communication delay between the robot controller and the printhead.

The actual TCP speed depends on the changes in the trajectory (such as sharp corners, orientation changes, ...). The controller in our experimental set-up implements positioning such that the executed trajectory consists of linear segments and parabolic blends between them. As a consequence, the information that determines the actual TCP speed needs to include not only the properties of the current path segment but also the previous and the next one.

Therefore, our input data to statistical models consist of the printhead coordinates, orientations, the path segment lengths, and the desired speed:

$(x, y, z, A, B, C, l, v)$, plus the same information for the previous and the next path segment along the trajectory. We illustrate this in Fig. 4a.

**Results.** The mapping between the input data (the trajectory) and the output data (the TCP speed) can be modelled in many ways. The relationship is expected to be non-linear, with complex, 'hidden', implementation of robot dynamics in-between. This set-up exactly corresponds to the architecture of neural networks (NN) [5]. Neural networks have been used for trajectory planning as well as for solving the inverse kinematics problem (see [6–8] and references in [9]). In this paper, however, we focus on using NN for predicting the *actual* TCP speed in the cases when the robot controller is programmed to keep the TCP speed constant.

We compare our non-linear NN models with a simple linear relation which serves as a baseline prediction between the input data and the output. We use 30 min of data for training and the rest for validation. The results are presented in Fig. 4b. With only 30 min of training data, we can predict the actual TCP speed to within 10 % for any future trajectory using our set-up. This way the printing head extrudes the correct amount of material on each trajectory segment, solving the issues with over and underflow.

For different experimental set-ups with different manipulators, controllers, or printing heads the approach we presented can be repeated; first collect data using random trajectories, then model this data giving a predictive model for actual TCP speeds and material extrusion speeds.

## 4 Printing Results: Comparison in Quality

We illustrate the improvements gained by our method by printing two identical objects (Fig. 5). A detailed study of material properties of objects printed by our set-up will be part of a separate publication.

The first object is printed by keeping the extrusion speed constant, such that the appropriate amount of material would have been extruded *if* the actual printhead motion speed had been as programmed on the robot controller. The second object is printed by dynamically adjusting the extrusion speed, as described in Sect. 3.

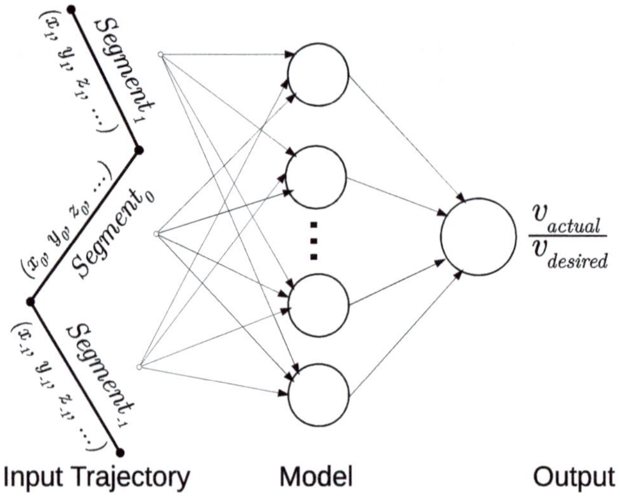


(a) Our modelling approach: The ratio between the actual average speed achieved and the programmed (desired speed) is predicted for the current path segment, taking as inputs the trajectory data from the current, previous, and next segments. Each path segment constitutes a single training sample. The architecture consists of a fully connected hidden layer of 100 logistic units trained with stochastic gradient descent.

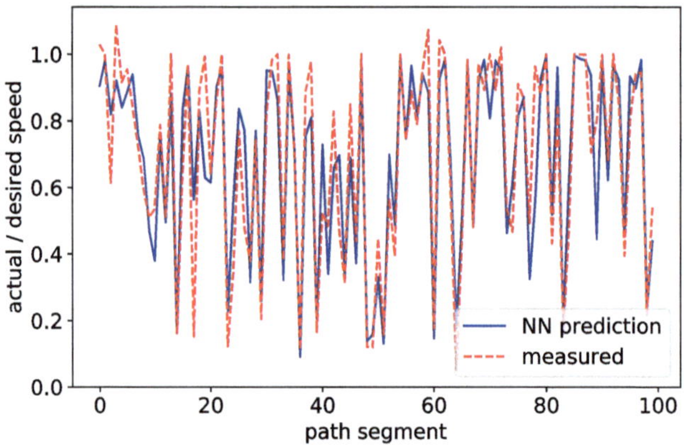


(b) Neural network predictor for the actual TCP speed given some trajectory. On the validation data set the average error between the predicted and true speed is 9.5 %. Note, that the baseline predictor (linear model) achieves the average error between the predicted and true speed of 21 %. The considerable reductions in the actual speed were observed for the path segments that include large re-orientations.

**Fig. 4.** Schema of our modelling approach (a) and predictions of the NN model (b). In (b) the vertical axis shows the ratio between the actual TCP speed achieved and the speed set on the robot controller for some path segment (which are plotted on the horizontal axis).

(a) 3D printed object with constant extrusion speed.

(b) 3D printed object with dynamically adjusting extrusion speed.

**Fig. 5.** Examples of different amount of extruded material, based on approaches to material extrusion control.

## 5 Conclusion

This paper deals with issues relating to 3D printing with 6 DOF industrial manipulators. We present our printing set-up consisting of the robot cell, manipulator, controller, printhead, and printhead controller.

In previous contribution, we dealt with new methods of trajectory planning. Here, we focus on the issue of material extrusion control in situations where the programmed motion speed of the printhead cannot be achieved due to limitations of the manipulator or controller. If the motion speed of the printhead and the extrusion speed of material are not synced, then an incorrect amount of material is deposited. This leads to poor quality of printed objects.

We demonstrate that a simple empirical method to predict the actual motion speed of the printhead is a viable approach to ensuring that the appropriate amount of material is extruded along the printing path. Firstly, we measure a number of random trajectories and speeds on them in the work space of our robot. This then serves as training data for our models which accept printing trajectory as input, while the actual speed of the printhead on each segment of the trajectory is the output of the models.

**Acknowledgements.** This work was supported by the European cohesion funds as part of the project Development of multiaxis robotic 3D printing of composite materials - MMO3D under the grant Interreg V-A Slovenia-Austria, ESRR No. SIAT73, and by the Slovenian Research Agency (ARRS) under research program Motion analysis and synthesis in man and machine (P20228).

## References

1. Dizon, J.R.C., Espera, A.H., Chen, Q., Advincula, R.C.: Mechanical characterization of 3D-printed polymers. Addit. Manuf. **20**, 44–67 (2018)
2. Allen, R.J.A., Trask, R.S.: An experimental demonstration of effective curved layer fused filament fabrication utilising a parallel deposition robot. Addit. Manuf. **8**, 78–87 (2015)

3. Lim, S., Buswell, R.A., Valentine, P.J., Piker, D., Austin, S.A., De Kestelier, X.: Modelling curved-layered printing paths for fabricating large-scale construction components. Addit. Manuf. **12**, 216–230 (2016)
4. Kraljić, D., Kamnik, R.: Trajectory Planning for additive manufacturing with a 6-DOF industrial robot. In: Proceedings of the 27th International Conference on Robotics in Alpe-Adria Danube Region (RAAD 2018). Advances in Service and Industrial Robotics, pp. 456–465, January 2019
5. Cheng, B., Titterington, D.M.: Neural networks: a review from a statistical perspective. Stat. Sci. **9**(1), 2–30 (1994)
6. Choi, B.B., Lawrence, C: Inverse kinematics problem in robotics using neural networks. NASA STI/Recon Technical Report N, 93, October 1992
7. Simon, D.: The application of neural networks to optimal robot trajectory planning. Robot. Auton. Syst. **11**(1), 23–34 (1993)
8. Hasan, A.T., Hamouda, A.M.S., Ismail, N., Al-Assadi, H.M.A.A.: An adaptive-learning algorithm to solve the inverse kinematics problem of a 6 d.o.f serial robot manipulator. Adv. Eng. Softw. **37**(7), 432–438 (2006)
9. Tang, S.H., Ang, C.K., Ariffin, M.M., Mashohor, S.: Predicting the motion of a robot manipulator with unknown trajectories based on an artificial neural network. Int. J. Adv. Robot. Syst. **11**(10), 176 (2014)

# Robotic Knitting in String Art as a Tool for Creative Design Processes

Marko Jovanović(✉), Marko Vučić, Bojan Tepavčević, Mirko Raković, and Jovica Tasevski

Faculty of Technical Sciences, University of Novi Sad,
Trg Dositeja Obradovića 6, Novi Sad, Serbia
{markojovanovic,vucic.marko,tepavcevicb,rakovicm,tasevski}@uns.ac.rs

**Abstract.** The application of industrial robots in creative industries has shown great potential in the last decade. The interpretation of images as tangible and artistic works of art is depicted with various elements and techniques. The usage of a continuous piece of string allows for the design process to explore different methods of achieving comprehensible images in the form of string art. With different computational tools and design techniques, it becomes possible to explore design thinking coupled with automated fabrication. In this paper, the emphasis is placed on generating a large scale string art pieces out of smaller tiles, that can be used as wall art. By using an integrated workflow it is possible to generate the program for robotic knitting of the string art piece, by adjusting the parameters that govern the entire process.

**Keywords:** Robotic knitting · Robots in creative industries · Tessellation · Strings per tile

## 1 Introduction

In recent years, robotic fabrication has been widely used as a creative tool in the process of design and art. The reason why industrial robots were adopted instantly as a powerful tool for the creation of interesting computer designed and artistic works lies in the fact that different end effectors can be used to achieve various visual and tangible properties of architectural objects or artwork pieces. Industrial robots have been used for variable carving and mould making [1], thermo-forming [2], milling [3], assembling [4], spatial wire cutting [5] and robotic knitting of fibre-based structures [6,7].

Robotic manufacturing has also been used in the area of fabricating tangible structures while utilizing an image as a driving factor for the artistic design process. By using a grayscale image and interpreting the pixels for various tangible modules depicting images from linear elements [8], and solid cubic modules [9] has been used. As opposed to using modular elements, novel approaches utilize a single element, a continuous piece of string which is spread over anchored pegs on a frame in order to generate string art.


K. Berns and D. Görges (Eds.): RAAD 2019, AISC 980, pp. 179–187, 2020.
https://doi.org/10.1007/978-3-030-19648-6_21

Novel computational design tools allow for such an approach to be made. Between 1–2 km of material is spread between 200 points on a circular frame measuring 70 cm in diameter which, with manual fabrication, represents an extremely time-consuming process [10]. Striving for an automated fabrication approach allows for the artwork pieces to become more efficient and interesting to design [11], with only 500 m of material being used and between 2000 and 3000 points on a frame 40 cm × 40 cm is size. By forming an integrated approach, it becomes possible to prepare the image to respond to the manufacturing process [12]. Based on the approaches done by Vrelis and Laarco, Birsak et al. used a mathematical approach to combine the two thus forming an automated fabrication approach of string art pieces on a circular frame with a small number of points [13]. The research in this paragraph is closely related to exploring relatively small artwork pieces.

In this research, the emphasis is placed on designing and fabricating large format string art pieces that can be applied as a part of interior wall art. The design and fabrication phases are used in an integrated approach, where the limitations of both are introduced at the start of the design process, especially the size of the robot's work area (Sect. 2.1). It is explored how the large string art area is tessellated with the string knitting process explained in Sect. 2.2. The number of strings in each part (Sect. 2.3) can influence the formation of an image in the eye of the observer, which can be seen after the fabrication process (Sect. 2.4).

## 2 Integrated Design Approach

An integrated design and fabrication approach is used, where all the parameters and limitations are taken at the start of the process. Large scale string art pertains to a planar area of a wall which is larger than 1 × 1 m. The workflow entitles that all the design, calculation, optimization and fabrication preparation is done in a single work environment of Rhinoceros modeling software with Grasshopper being used for parameterization. The first step is to determine the work area of the industrial robot, as an important limitation factor and parameter in the workflow.

### 2.1 Robot's Work Area

The information regarding the work area of the industrial robot is used as an input factor for the tessellation process and hence has to be determined early on. Each robot has a predefined work area according to its properties. However, the application of a specific end effector (tool) necessary for the process of string knitting, as well as the orientation of the frame with the points (Wobj) restricts the work area. The tool used in the fabrication process attaches such that the TCP, where the string goes through, represents the tool0 only translated along the vector z (Fig. 1a). The string bundle is locked on a handle located between the mounting part of the tool and the TCP. On the other hand, the frame with the targets has a horizontal orientation which requires the tool to rotate around the 5th axis and avoid causing a singularity when going between points (Fig. 1b).

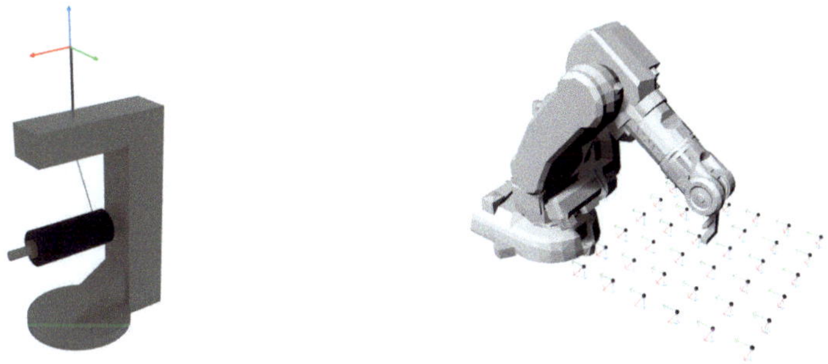

(a) (b)

**Fig. 1.** (a) The model of the tool used for the knitting process; (b) The orientation of the targets in reference to the robot

In order to determine the work area for such a tool and frame orientation, a calculation is performed by using an add-on for Grasshopper called Taco, which can check the target's reachability. The analysis is conducted by forming a horizontal grid of points and translating it vertically in order to check for the largest work area. The value of 1 m is chosen for the grid dimensions given that the available industrial robot is ABB IRB 140. A total of 1681 targets are generated in each horizontal grid. At a certain height, the targets are checked for reachability, culling the ones that are not within reach. Afterward, the same initial grid is moved vertically by 10 cm, for time efficiency, and the process is repeated. In the end, the remaining point cloud is used to generate a closed surface that determines the boundaries of the work area, which resemble a torus.

Figure 2a shows that the largest vertical cross-section of the torus work area (around 55 cm) is located at a height of 20 cm from the robot's GCS (Global Coordinate System) which is chosen as the height of the frame. The top view (Fig. 2b) shows that at that height, the horizontal cross-section of the work area can fit various shapes including:

- quads - from square $50 \times 50$ cm to rectangles $35 \times 100$ cm
- triangles - from equilateral triangles of 63 cm edges to isosceles triangles with basis of 125 cm and height of 54 cm and
- hexagons - that fit inside a $100 \times 50$ cm rectangle.

All shapes are placed symmetrically in reference to the x-axis of the robot's GCS. These 3 shapes are chosen since they can be used for regular tessellation of any planar surface. For this research, a planar surface of $300 \times 200$ cm is chosen. Once the largest shape of the work area is determined and the size of the string art, it is possible to tessellate the string art area following a condition that the tessellated tiles have to fit inside the robot's work area.

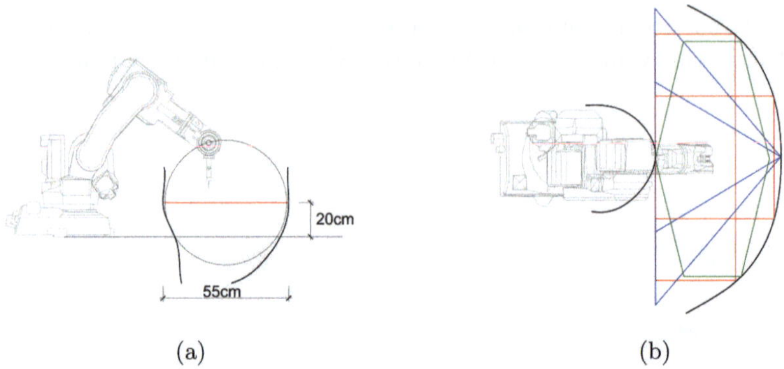


**Fig. 2.** (a) A vertical cross section of the robot's work area with the widest part located at a height of 20 cm; (b) the horizontal cross section of the robot's work area at a height of 20 cm showing different shapes that can fit inside - red quads, blue triangles and a green hexagon

## 2.2 The String Knitting Process

The method used for the generation of the string knitting process follows the method that Vrelis uses [10]. This means that in the design process a certain number of points is placed along the border edges of the shape which the string art piece is going to be created in. The string art is created on a white background with dark colored strings. Only a fraction of the image is used, the one that coincides within the area of the tile where the knitting is done (Fig. 3a). At the start of the process, the border edges are populated by a set of equally spaced points (Fig. 3b). The knitting is done by connecting one point to another in such a way that the resulting line (which represents a string) covers the darkest pixels of the image fraction. In order to determine the starting point, it would be necessary to test all the possible connections, which would be very time consuming to calculate. The premise of this research is that any point can be chosen for the starting point as the line originating from it would eventually connect to the point which the darkest string emanates from. Hence, once the starting point is chosen, all the lines connecting that point to the viable points are generated (points are viable if they are not located on the same edge as the starting point) (Fig. 3c). The lines are evaluated according to which pixels of the image they intersect. Each pixel has a certain grayscale value from 0 to 1, with 0 being black and 1 being white. Whichever line has the lowest sum of the crossed pixel values in reference to its length is chosen and its connecting point is used for the process iteratively (Fig. 3d). Every time this evaluation is performed, the image fraction is updated so that the intersected pixels get brighter by a certain amount (by adding 0.1 for example) (Fig. 3e) to prevent choosing the same line through consecutive iterations. For each tile, the process is repeated a desired number of times, after which the process is iteratively repeated on the next tile

(Fig. 3f). The number of strings within each of the tiles can have a great impact on the clarity of the image and needs to be addressed in more detail.

### 2.3 Forming the Image

The strings enhance the darker parts of the image and the more strings the art piece has, the more details it can show. However, a large number of strings can cause the end result to have low contrast due to the black color of the string overwhelming the white background. In order to test this, four tile shapes are selected - triangle, quad, triangle/quad combination and hexagon and the amounts of 100, 150, 200 and 250 strings per tile are generated following the process explained previously.

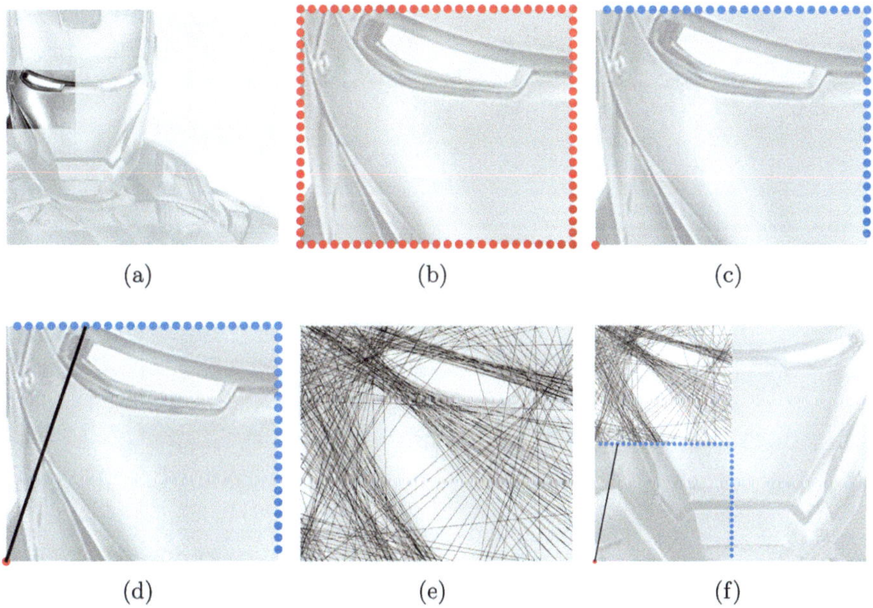

**Fig. 3.** (a) Referencing a fraction of the large scale image for tile knitting; (b) populating the frame with equally spaced points; (c) choosing the viable points according to a starting point; (d) Drawing a line representing a string which is the darkest portion of the image fraction in a certain iteration; (e) A finished portion of the image in a single tile; (f) The process started on a new tile

For the triangle shaped panels, a total of 18 properly sized tiles are chosen to cover the main parts of the reference image. As the (Fig. 4a) shows, the detailing around the white part of the eye is noticeable only after applying 200 strings per tile, while the rest shows a good resemblance to the original. The application of quad shaped panels requires 14 panels with all the necessary features being visible with 100 strings per tile and an increase in resemblance as the number

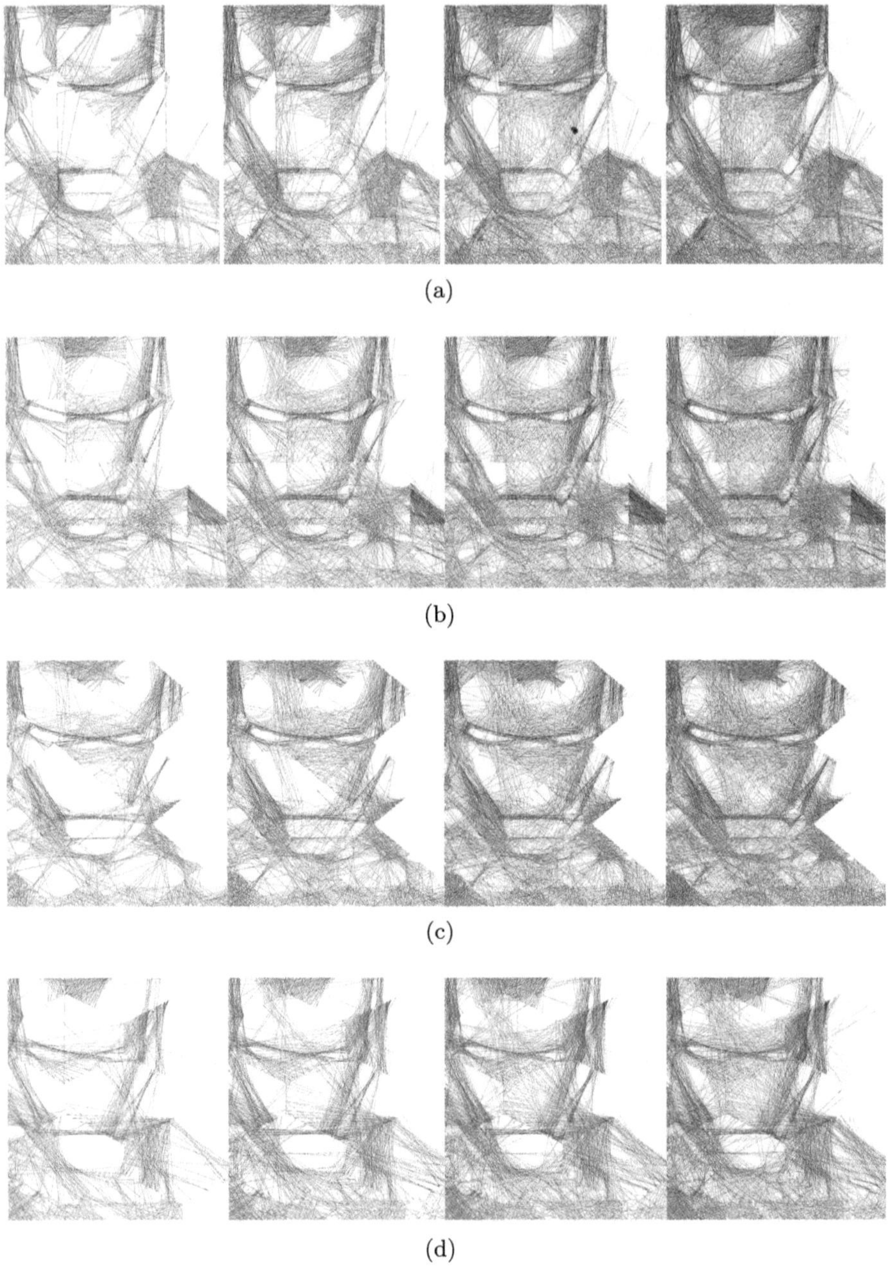

Fig. 4. Renderings of the string art models with 100, 150, 200 and 250 strings per tile showed from left to right and for (a) triangle, (b) quad, (c) triangle/quad combination, and (d) hexagon tile shape.

of tiles increases (Fig. 4b). Since around 80% of the shoulder tile depicts white pixels, the number of strings over-exceeds the number of required strings per unit area for proper depiction. The same problem occurs when using a combination of quad and triangle shapes as seen in (Fig. 4c), where 7 triangles and 8 diamond quad shapes are used. Due to their position to one another and their edges being angled in reference to the large scale image, the seam between the tiles is not as noticeable as in the quad case. Lastly, the hexagonal tessellation required only 11 tiles, due to its large area. In reference to the tile size, the number of strings is shown to be inadequate to depict the image properly, since the detailing on the left side of the head is overemphasized while lacking on the top of the head (Fig. 4d).

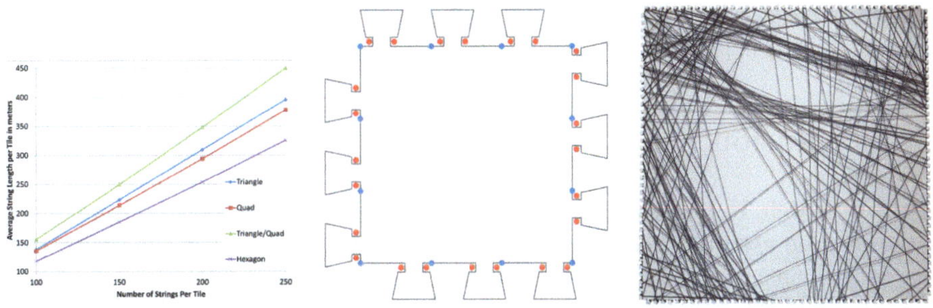


(a) (b) (c)

**Fig. 5.** (a) A graph showing the amount of strings used in reference to the average string length per tile; (b) The design and disposition of the teeth with the position of each shown in blue and the extra string space shown in red; (c) The finished tile with 150 strings knitted through the MDF board

In addition to this, it is tested how many meters of string is used per tile area in meters squared, which can be seen in (Fig. 5). It shows that hexagon shaped tiles use the least amount of string per area, while the quad shaped panels are second best. Combining the rendering appearance with the string usage, an optimal choice is found in quad panels, which is why they are chosen for the fabrication process.

### 2.4 The Fabrication Process

The fabrication process requires two things - a frame with anchored pegs for the string to hook through and a knitting path program that the robot can perform. The frame with anchored pegs consists of a flat surface, where the anchored pegs are oriented in the same plane as the board as serrated edge teeth. In this case a square shaped MDF board is used. The teeth are designed while keeping in mind two things. First design criteria is to allow sufficient space for the TCP tool to go between, as well as leave enough space for the connecting teeth of the other boards to fit. The second design criteria allows the string to remain in place, with a tapper connection of the peg to the board and additional space for the strings ((Fig. 5b) showed in red), which proved to be optimal for knitting and later board aligning. The teeth are positioned in such a manner that the connecting points of the knitting path (showed in blue on (Fig. 5b)) are located on the left side of the teeth in reference to the board, which ensures that the boards can align their nonadjacent edges. On the other hand, the knitting path represents a sequence of points that the robot can follow. Given that the string cannot be adequately knitted by simply going from one point to the other during the actual knitting process, each point is replaced with 4 points that go around each anchored peg in a square, enclosing the peg with a certain margin of distance. In order to avoid the collision of the tool and the teeth, the points are moved away from the frame, allowing the string to be moved around the peg. Such a sequence of points is converted into targets following the TCP definition in Sect. 2.1 and the fabrication is carried out accordingly (Fig. 5c).

## 3 Conclusion

In this research, it is tested how image tessellation, combined with robotic fabrication can be used for creating large string art pieces. First, it is concluded that the determination of the robot's work area is adamant in properly tessellating the large scale format of an image in order to use as few tiles as possible for the tessellation process. Second, using triangle, quad and hexagon shaped tiles with a variation in the number of strings per tile shows that quad tiles provide the best depiction of the large scale image in reference to the amount of strings used per unit area. The tessellation images for large scale string art, coupled with robotic fabrication enables additional potential for the fabrication of automated string art pieces. For future work, better edge detection recognition is considered for tessellation with application on different images.

**Acknowledgement.** This research was supported by the Serbian Ministry of Education and Science under contracts TR36042 and III44008; and by Provincial secretariat for science and technological development under contract 114-4512116/2011. The authors would like to thank the company ABB for providing the licenses for the RobotStudio software. The authors would like to thank collaborators in this project, Milana Mitrović and Tatjana Prčić.

## References

1. Clifford, B., Ekmekjian, N., Little, P., Manto, A.: Variable carving volume casting. In: Robotic Fabrication in Architecture, Art and Design, pp. 3–15. Springer, Cham (2014)
2. Weissenbck, R.: D-FORM. In: Robotic Fabrication in Architecture, Art and Design, pp. 249–260. Springer, Cham (2014)

3. Brell-Cokcan, S., Reis, M., Schmiedhofer, H., Braumann, J.: Digital design to digital production: flank milling with a 7-axis robot and parametric design. In: Computation: The New Realm of Architectural Design. 27th eCAADe Conference Proceedings, pp. 323–330 (2009)
4. Bonwetsch, T., Gramazio, F., Kohler, M.: Digitally fabricating non-standardised brick walls. In: ManuBuild, 1st International Conference, Rotterdam (2007)
5. Rust, R., Jenny, D., Gramazio, F., Kohler, M.: Spatial Wire Cutting: cooperative robotic cutting of non-ruled surface geometries for bespoke building components (2016)
6. Waimer, F., La Magna, R., Reichert, S., Schwinn, T., Menges, A., Knippers, J.: Integrated design methods for the simulation of fibre-based structures (2013)
7. Lu, X., Dounas, T., Spaeth, A.B., Bissoonauth, C., Galobardes, I.: Robotic simulation of textile as concrete reinforcement and formwork (2017)
8. Jovanović, M., Tasevski, J., Tepavčević, B., Raković, M., Mitov, D., Borovac, B.: Fabrication of digital anamorphic sculptures with industrial robot. In: International Conference on Robotics in Alpe-Adria Danube Region, pp. 568–575. Springer, Cham (2016)
9. Jovanović, M., Vesna, S., Tepavčević, B., Mitov, D., Bajšanski, I.: Generating an anamorphic image on a curved surface utilizing robotic fabrication process. In: Proceedings of 34th eCAADe 2016, August 2016, pp. 185–191. ISBN 9789491207105
10. Vrellis, P.: A New Way to Knit. http://artof01.com/vrellis/works/knit. Accessed 06 Feb 2018
11. Laarco: Laarco Studio—String Art. https://laarco.com/. Accessed 31 Jan 2019
12. Lindemeier, T., Pirk, S., Deussen, O.: Image stylization with a painting machine using semantic hints. Comput. Graph. **37**(5), 293–301 (2013)
13. Birsak, M., Rist, F., Wonka, P., Musialski, P.: String art: towards computational fabrication of string images. Comput. Graph. Forum **37**(2), 263–274 (2018)

# Towards Using Natural User Interfaces for Robotic Arm Manipulation

Florin Gîrbacia(✉), Cristian Postelnicu, and Gheorghe-Daniel Voinea

Transilvania University of Brasov, Eroilor 29, 500036 Brasov, Romania
garbacia@unitbv.ro

**Abstract.** In this paper, we propose a Natural User Interface (NUI) based on Leap Motion controller which enables intuitive manipulation of a 6-DOF Jaco robotic arm. By using this NUI interface, we aim to study the effects of hand movements and gestures for interaction with a robotic arm. We present a qualitative evaluation of the proposed NUI system by trials with subjects carrying out a standard robotic arm manipulation session using both the conventional joystick interface and the NUI based one. The results are discussed in the last part of the paper.

**Keywords:** Natural user interaction · Virtual Reality · Robot manipulation

## 1 Introduction

In order to mimic the way in which a person utters his/her ideas, new methods consisting in voice commands, hand movements and gesture recognition have been used to create intuitive human-computer interfaces that are part of the Natural User Interface (NUI) technologies area. The novelty of using NUI to human - robot interaction is the possibility of direct command and interaction using voice, gestures, touch and motion tracking, rather than instruct them by joystick or typing commands [10]. In this paper are explored the modes in which a NUI can be used to enhance robotic arm manipulation. We specifically focused on investigating the usability and ergonomics factors of hand movements and gestures for interacting with a robotic arm.

A number of applications have been identified where NUI has been successfully integrated [2, 4]. Previous works has shown that NUI technologies can be used to create human-robot interfaces for Unmanned Aerial Vehicle and drones [3, 7–9], underwater robots [5] or mobile robots [11]. Bassily et al. [1] proposed an algorithm that allows an optimum mapping between the user's hand movements, tracked by the Leap Motion controller, and the 6-DOF Jaco arm. The system is dedicated to people with upper limb problems, supporting them in performing some of the essential Activities of Daily Living. Recently, Sabattini et al. [10] developed a concept of NUI to allow people to interact with a generic robot. The proposed method is based on recognizing the users' forearm motion using a smartwatch to produce commands for the robotic system. In comparative tasks, they found that the proposed interaction system clearly shows advantages related to usability.


K. Berns and D. Görges (Eds.): RAAD 2019, AISC 980, pp. 188–193, 2020.
https://doi.org/10.1007/978-3-030-19648-6_22

The study presented in this paper attempts an analysis of possible answers to the question: Can hand movements and gestures replace the conventional joystick control for a robotic arm?

# 2 Materials and Methods

## 2.1 The Hardware and Software Architecture of the NUI Interface

The components of the developed system are presented in Fig. 1. It consists of the following subsystems:

- Robot used for testing the NUI interface: the developed NUI interface was developed for a serial six degrees of freedom Kinova JACO robotic arm (www.kinovarobotics.com). The robotic arm is equipped with a three fingers based gripper that is used to grasp the test object. In a conventional manner, the JACO arm is manually operated using a 7-button joystick. The user can send motion commands to the robotic arm by selecting three different joystick modes: (i) translate the robot arm along X, Y or Z axis, (ii) rotate the gripper around the X, Y or Z axis and (iii) open-close the gripper. A Kinova API is available to send motion controls and get sensors data for a specific task.
- The Leap Motion Controller: We chose the low-cost Leap Motion Controller (www.leapmotion.com) because, based on [12], it allows robust hand movements tracking and gestures recognition. The Leap Motion sensor can detect the position, orientation and velocity of hands along with the position, orientation and velocity of each component finger [6]. The operating principle of this sensor is simple: a beam is sent by infrared sources and when it is reflected by an obstacle this information is read using IR cameras. The sensor is sensitive to bright IR lighting conditions, but in those conditions is automatically set to activate the Robust Tracking mode which allows tracking to perform better.
- PC running the NUI controller software: Leap Motion Controller is connected to a PC running a software application developed by the authors, using C++ programming language based on Leap Motion SDK and Kinova API.

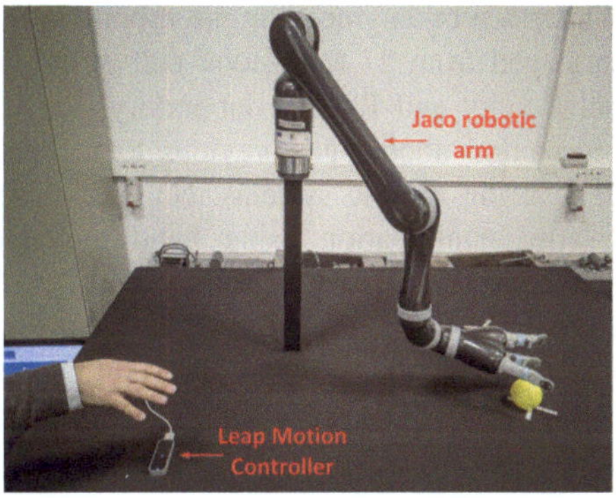


**Fig. 1.** The hardware architecture of the developed NUI interface

The user can send translation commands to the Jaco robotic arm in the Cartesian space by moving the right hand on the X, Y or Z axis. For the rotation of the robotic gripper, are used the right hand roll, pitch, and yaw angles. Because there are different coordinate system for the robot and the controller, the coordinates of the hand movements, received from the Leap Motion, were transformed in order to have a clear mapping between the hand and the robotic arm motion.

For each new frame acquired from the Leap Motion controller, the position and orientation of the right hand is compared with the previous one. If the difference is higher than a certain threshold, it is then transmitted as movement command to the Jaco robotic arm.

Grasping and releasing an object is handled by closing and opening the right hand (see Fig. 2). This implementation is based on the function "grabStrength" available in the Leap Motion SDK (developer.leapmotion.com) which specifies how close the hand is.

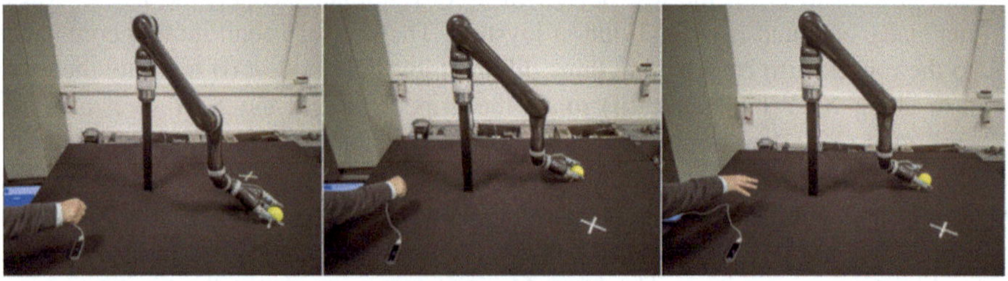

**Fig. 2.** Operating the JACO arm using developed NUI for a pick and place task

## 3 Evaluation of NUI Interface

The aim of the evaluation study presented in this section is to assess the way NUI interface can be used for robotic arm manipulation. We conducted an initial user study during which the users could familiarize with the robotic arm by controlling it former with the conventional joystick robotic arm manipulation, and later by hand movements and gestures. As the experimental task, we chose to use a pick and place operation. By using a subjective questionnaire we collected qualitative information about the acceptance of the NUI interface by engineers in the robotic arm manipulation process.

A number of 6 users aged from 27 to 38 took part at the experiment. They were members in our Virtual Reality and Robotic laboratory. Each user filled in a questionnaire providing information about age, previous experience of using NUI technology and experience in using robotic systems. None of the subjects had previous experience in robotic arm manipulation using hand movements and/or gestures. Instead, they had experience in using JACO joystick controller. At the beginning, each subject was informed about the purpose of the experiment and specific instructions were given. Before the test, each participant was allowed to understand and familiarize with the settings of NUI interface. The users had 10 min prior to the experiment, for practicing the with the NUI interface.

At the end of the experiment, the participants were asked to rate the NUI interface on a 7-point Likert scale: "Strongly agree" (7), "Agree" (6), "Slightly Agree" (5) "Neutral" (4), "Slightly Disagree" (3) "Disagree" (2) or "Strongly disagree" (1). Post-task scenario subjective measures included the following statements:

- As a first impression, I like the NUI multimodal interface (first impression).
- It is easy to interact with the application by using the NUI interface (interaction).
- The interface has all the functions and capabilities I would expect it to have for a robotic manipulation task (functionality).
- The NUI system is easy to use (easy to use).
- The interaction modalities are easy to learn (easy to learn).
- Overall, I am satisfied with the NUI interface (overall).

Figure 3 shows the results concerning the grades (Likert-scale) obtained by the NUI interface for each of the subjective criteria.

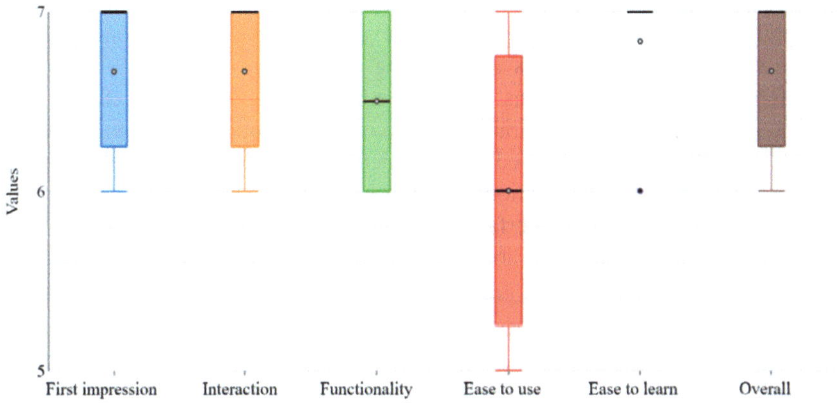


**Fig. 3.** Results for the subjective questionnaire

The research conducted led to the following findings:

(i) Interest: comparing with the conventional joystick controller, the users expressed a high interest for the use of the NUI technologies as an interface for robotic arm manipulation. The given reason relates to the more intuitive interaction and the simplification of the series of joystick commands needed for various operations. The users also appreciated the possibility to use the interaction modalities such as hand movements and gestures, which offers the possibility for a natural communication of commands.

(ii) Ergonomics: The proposed solution based on the NUI environment changes the workplace of the operator. Comparatively, the new workplace is in completely different architecture that affects dramatically the ergonomics of the work. During the use of the NUI interface large hand movements are involved, because of the spatial input and this makes the work tiring compared to the conventional joystick interface. The subjects indicated that the joystick interface generates a lower stress to the muscles, since only small movements are performed and, additionally, provide a comfortable support for the hand during the work. This conclusion demonstrates that if the future human-robot interfaces are to integrate the advantages of NUI, they need to provide a support for the hand which can reduce the hand fatigue. The hand movements and gestures are not the only alternative to replace the conventional joystick control. Future work will explore development of a multimodal interface that combines presented NUI with other human-machine communication channels in order to allow different ways of accomplishing a robotic task.

(iii) Difficulty, intuitiveness: the subjects considered the NUI commands easy to learn and understand, thanks to the natural and intuitive hand movements and gestures commands. In the conducted experiment there was observed an error in positioning the robot according to the user's input data caused by singularity. The Jaco robot has a built-in automatic singularity avoidance algorithm that reduces the number of degrees of freedom and causes a position error. The user's perception regarding the kinematic state of the robot is limited. The usage of haptic feedback (for example vibrotactile) could represent a solution in notifying the user when the robot is approaching a singularity situation. The joystick interface presents limitations that can be translated into mental efforts for the operator, especially at the training stage, when the operator has to make judgments on the commands used for a particular task, identify and select the right ones. Instead of using various joystick commands and consuming an extra mental effort, the NUI interface will allow the use of perceptual and mental systems at a normal level, so the user can focus on the task that needs to be solved.

## 4 Conclusions

This paper presents an initial NUI interface solution for robotic arm manipulation based on Leap Motion Controller and analyses the potential interest for replacement of conventional joystick controller. The user study shows that the NUI interface added value must be judged by combining positive and negative facts. Clearly there are some pros such as interest and simplification of commands but the negative ergonomic issues are quite important. More work is needed in order to reduce the weight of negative aspects and thus make it interesting for service and assistive robotics applications. The future work is related to the objective analysis of performances related to task completion times and accuracy with a larger number of participants.

**Acknowledgment.** This work was supported by a grant of the Ministry of National Education and Scientific Research, RDI Programme for Space Technology and Advanced Research - STAR, project number 566. We would also like to thank members of Transilvania University of Brasov for their participation in the experiment.

## References

1. Bassily, D., Georgoulas, C., Guettler, J., Linner, T., Bock, T.: Intuitive and adaptive robotic arm manipulation using the leap motion controller. In: Proceedings of 41st International Symposium on Robotics 2014, pp. 1–7. VDE, Munich (2014)
2. Cicirelli, G., Attolico, C., Guaragnella, C., D'Orazio, T.: A kinect-based gesture recognition approach for a natural human robot interface. Int. J. Adv. Rob. Syst. **12**(3), 1–11 (2015)
3. Chandarana, M., Trujillo, A., Shimada, K., Allen, B.D.: A natural interaction interface for UAVs using intuitive gesture recognition. In: Savage-Knepshield, P., Chen, J. (eds.) Advances in Human Factors in Robots and Unmanned Systems. AISC, vol. 499, pp. 387–398. Springer, Cham (2017)
4. Chen, S., Ma, H., Yang, C., Fu, M.: Hand gesture based robot control system using leap motion. In: Liu, H., Kubota, N., Zhu, X., Dillmann, R., Zhou, D. (eds.) Intelligent Robotics and Applications, ICIRA 2015. LNCS, vol. 9244, pp. 581–591. Springer, Cham (2015)
5. García, J.C., Patrão, B., Almeida, L., Pérez, J., Menezes, P., Dias, J., Sanz, P.J.: A natural interface for remote operation of underwater robots. IEEE Comput. Graph. Appl. **37**(1), 34–43 (2017)
6. Gîrbacia, F., Gîrbacia, T., Butnariu, S.: Design review of cad models using a NUI leap motion sensor. J. Ind. Des. Eng. Graph. **10**, 21–24 (2015)
7. Fernández, R.A.S., Sanchez-Lopez, J.L., Sampedro, C., Bavle, H., Molina, M., Campoy, P.: Natural user interfaces for human-drone multi-modal interaction. In: 2016 International Conference on Unmanned Aircraft Systems (ICUAS), pp. 1013–1022. IEEE, Arlington (2016)
8. Herrmann, R., Schmidt, L.: Design and evaluation of a natural user interface for piloting an unmanned aerial vehicle. i-com **17**(1), 15–24 (2018)
9. Peshkova, E., Hitz, M., Kaufmann, B.: Natural interaction techniques for an unmanned aerial vehicle system. IEEE Pervasive Comput. **16**(1), 34–42 (2017)
10. Sabattini, L., Villani, V., Secchi, C., Fantuzzi, C.: A general approach to natural human-robot interaction. In: Ficuciello, F., Ruggiero, F., Finzi, A. (eds.) Human Friendly Robotics, Springer Proceedings in Advanced Robotics, vol. 7, pp. 61–71. Springer, Cham (2019)
11. Villani, V., Sabattini, L., Riggio, G., Levratti, A., Secchi, C., Fantuzzi, C.: Interacting with a mobile robot with a natural infrastructure-less interface. IFAC-PapersOnLine **50**(1), 12753–12758 (2017)
12. Weichert, F., Bachmann, D., Rudak, B., Fisseler, D.: Analysis of the accuracy and robustness of the leap motion controller. Sensors **13**(5), 6380–6393 (2013)

# Grounding of Uncertain Force Parameters in Spoken Robot Commands

Kim Wölfel(✉) and Dominik Henrich

University of Bayreuth, 95440 Bayreuth, Germany
{kim.woelfel,dominik.henrich}@uni-bayreuth.de
http://www.robotics.uni-bayreuth.de

**Abstract.** Speech-based robot instruction is a promising field in private households and in small and medium-sized enterprises. It facilitates the use of robot systems for non-experts as well as experts, even while the user executes other tasks. Considering force-based robot motions, the common approach is to map verbs to robot motions depending on the tool and the work piece. While this method works well for a wide variety of applications, it limits the user to a fixed force for a given manipulation task and does not allow extensions like "hard" or "soft". To overcome this drawback, we contribute an approach for a defuzzification of uncertain force parameters to numerical robot motions. To proof the reliability of our approach, we apply it on a motion with varying material parameters.

**Keywords:** Fuzzy logic · Human-robot interfaces · Force control

## 1 Introduction

One long term goal in current robotic research is the development of robot systems, which have approximately the same cognitive, communicational, and handling abilities as humans. As part of this ongoing development, application domains for robot systems shall be expanded, from industrial settings with separated working cells, fixed object positions, and preprogrammed motions towards a flexible usage in small or medium-sized enterprises or private households.

An intuitive way to communicate with such a robot system is to instruct it via spoken commands ([1–4]), which is on the one hand useful in tele-robotics or for handicapped people, on the other hand this is challenging due to the fact that users often do not specify all necessary parameters or use uncertain descriptions of motion and object properties (see Fig. 1). Many approaches have been developed to manage these uncertain parameters and several styles of them have been studied in depth, except parameters regarding force [5].

Our contribution is a framework for transforming fuzzy input parameters from speech commands to force-based robot motions. Our approach is based on the idea that not only force parameters should be modelled using the fuzzy logic, but tool and object information as well. This enables us to flexibly generate force-based motions for a variety of tools and work piece materials.


K. Berns and D. Görges (Eds.): RAAD 2019, AISC 980, pp. 194–201, 2020.
https://doi.org/10.1007/978-3-030-19648-6_23

The remainder of this paper is organized as follows: Sect. 2 contains a brief summary of related work in the area of interpreting fuzzy voice commands. Our framework is presented in Sect. 3 and its evaluation in Sect. 4. Finally, Sect. 5 contains a summary of the scientific gain achieved by our approach and an outlook on future work.

**Fig. 1.** Example illustrating the problem of untold and uncertain parameters while using speech control.

## 2 Related Work

Since our contribution is mostly related to the interpretation of fuzzy voice commands, a brief overview is given in this section. The interpretation of uncertain parameters in spoken commands has been examined for many years and for a wide variety of parameter types. An extensive review of presented approaches is given in [5], which not only highlights the parameter types covered by the approaches but also that are not considered in detail until now (like force).

Research areas well examined are distance-related parameters, like *closer* [6–8], velocity-related parameters, like *faster* or *slower* in navigational applications [9,10], direction-related parameters, like *slightly left* [11], size-related parameters, like *big*, *large* or *smaller* while referring to objects [12,13] and joint configuration-related parameters when instructing a robot arm [14,15]. The interpretation of the uncertain information used in such commands is often done by either examining their use for a varying workspace (i.e. location of work pieces/ obstacles) or by giving the user the opportunity to adjust the motion in a post-processing step, also known as *coach-player-system*. The advantages and disadvantages are a faster execution of the robot but with a not that sufficient outcome for the first approach and a longer time to reach the goal but a maximized user satisfaction for the second approach.

What has not been investigated in depth until now are force-dependent parameters [5]. An approach to solve this problem is presented in the following sections.

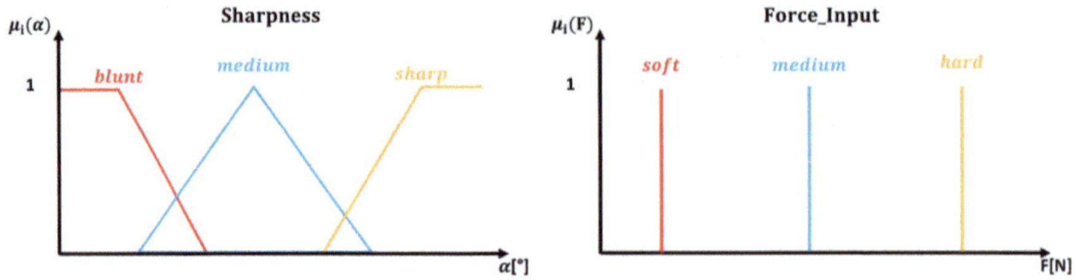


(a) Triangle membership functions for the linguistic variable Sharpness.

(b) Singleton membership functions for the linguistic variable Force_Input.

**Fig. 2.** Example membership functions for linguistic variables.

## 3 Interpreting Uncertain Force Parameters

### 3.1 Fuzzy Logic

In natural language usecases it is not always common to specify characteristics of a parameter using real values. For example, one would rather say "press it hard against the wall!" instead of "press it against the wall with a force of 10 N!". A well tested method for mapping this kind of uncertain sets to uncertain as well as crisp sets is the *Fuzzy Logic* [16]. It can even be used for control, where the main idea is to transform crisp input values into uncertain values (*fuzzification*), evaluate them using a fuzzy rule data base (*reasoning*) and then converting the uncertain output parameters into numerical or crisp output values (*defuzzification*).

Input and output parameters are modelled as a set of *linguistic variables* $V_l$, with different characteristics or *terms* $T_{\text{Param1}}$, e.g. $T_{\text{Param1}} = \{\text{low}, \text{medium}, \text{high}\}$. The mapping of a crisp value $x \in \mathbb{R}_0^+$ to a term is done using so called membership functions $\mu_i : \mathbb{R} \rightarrow [0, 1], i \in \{\text{low}, \text{medium}, \text{high}\}$ (see Fig. 2). For a measured value $x \in \mathbb{R}$ all membership functions are evaluated and stored in so called *sympathy vectors* $s(x) = (\mu_i(x))$, storing the evaluated membership functions from low to high.

In the reasoning step, predefined *fuzzy rules* are used to map input parameters to output parameters, and thereby identifying which characteristics of the output parameter are present. An example for an uncertain rule is:

$$\text{IF ParamInput} = \text{low THEN ParamOutput} = \text{high}, \tag{1}$$

where ParamInput and ParamOutput $\in V_l$ are linguistic variables representing the uncertain values of the input and output variables. The active output membership functions are then limited to the appropriate entries of $s(x)$, and merged for later computations.

In the defuzzication step, the uncertain output parameters gained in the previous step need to be transformed into crisp values. This can be done by using the maximum value of the combination of membership functions or using the centroid of the area beneath the combination of these functions.

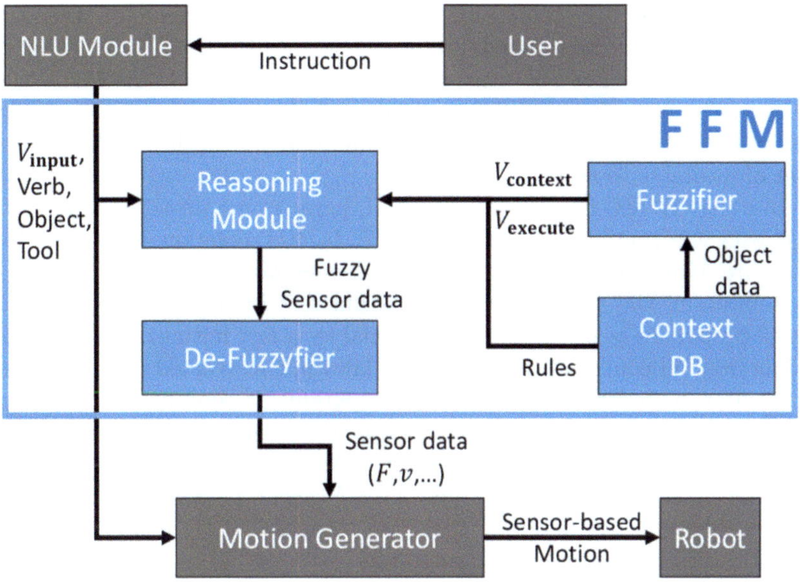


**Fig. 3.** Pipeline from an instruction to a sensor-based motion using our Fuzzy Force Model (FFM), a natural language understanding (NLU) module and a motion generator.

In our case, we want to investigate to which extend the fuzzy theory is applicable on force-based motion verbs like brushing, drawing, wiping, cutting, pressing, stamping, etc. The main challenge of interpreting uncertain parameters is, that besides the subjective perception of each user regarding phrases like *soft* or *hard*, the force during the execution depends on the material and geometry of the work piece and tool as well. To go one step closer to a solution of this, we present a model, which is capable of interpreting such uncertain parameters together with the work piece to manipulate and the tool used.

### 3.2 Fuzzy Force Model

Now we present our main contribution, the *Fuzzy Force Model* FFM (see Fig. 3). Based on fuzzy logic our model is able to transform a spoken command into a robot motion by not only considering the given force parameter by also uncertain tool and work piece properties.

Starting from the examplary command:

*Robot, wipe the desk softly with the cloth!*,

we extract a generic form of these type of instructions:

*Robot, MANIPULATE OBJECT FORCE_INPUT TOOL!*

*MANIPULATE* ∈ {cut, wipe, screw, ... } is an entry of the set of verbs representing (force-based) robot motions, *OBJECT* ∈ {cube, cake, desk, ... } is an entry of the set of work pieces, *FORCE_INPUT* ∈ {hardly, softly, fast, slowly, ... } is an entry of the set of force characteristics of the execution and *TOOL* ∈ {gripper, knife, scissors, cloth, ... } is an entry of the set of available tools.

While the OBJECT, VERB and TOOL parameters are here assumed to be precise except for ambiguaties, the *FORCE_INPUT* component is of an uncertain nature. One way to evaluate this parameter would be to create a look-up table containing the sensor values for parameters in combination with object and tool properties (see [17]). While this enables a fast initialization of a sensor-based motion, all combinations have to be known, constructed and stored before the execution. Our FFM instead, allows a parameter computation during run-time by using fuzzy rules and object parameters stored in a *context data base*, a *fuzzyfier* to transform parameters into uncertain values, a *reasoning module* for generating fuzzy output data and a *de-fuzzyfier* to generate sensor values for a sensor-based robot motion. The main components of our FFM are further explained in the following three paragraphs.

**Context Data Base.** In our case, the set of linguistic variables $V_l$ consists of three subsets: Specified input parameters $V_{\text{input}}$ that provided a sensor parameter like velocity or contact force of the robot, which is either specified by the user or set to *medium* if is not mentioned in the instruction, context parameters $V_{\text{context}}$ and execution parameters $V_{\text{execute}}$ that are not mentioned in the instruction but defined in a context data base, which holds the data for known materials, tools and robot parameters along with motion specific rules for the reasoning step. An example for these variables is:

- $V_{\text{input}} = \{\text{Force\_Input}, \text{Velocity\_Input}\}$
- $V_{\text{context}} = \{\text{Obj\_Mass}, \text{Tool\_Mass}, \dots\}$
- $V_{\text{execute}} = \{\text{Velocity\_Output}, \text{Force\_Output}, \dots\}$

While $V_{\text{input}}$ is already given as an uncertain value, it is modelled by a singleton-function (see Fig. 2b). The membership functions for $V_{\text{context}}$ and $V_{\text{execute}}$ should be modelled using appropriate functions, like triangle or trapezoid functions (see Fig. 2a), because they have to be evaluated during the interpretation of the user command.

**Reasoning Module.** We describe the remaining components on base of a multi-input single-output (MISO) framework, to highlight the force-based motions. A multi-input multi-output (MIMO) approach would be possible as well. First, the $V_{\text{execute}}$ is evaluated using the rules stored in the context data base. The following rules illustrate how this could look like for a cutting motion similar to Rule 1:

$$\begin{aligned}\text{IF Force\_Input} = \text{ soft AND Solidity} = \text{ soft AND Sharpness} = \text{ high}\\ \text{THEN Force\_Output} = \text{ soft}\end{aligned} \tag{2}$$

$$\begin{aligned}\text{IF Force\_Input} &= \text{soft AND Solidity} = \text{hard AND Sharpness} = \text{high} \\ &\text{THEN Force\_Output} = \text{medium}\end{aligned} \tag{3}$$

$$\begin{aligned}\text{IF Force\_Input} &= \text{soft AND Solidity} = \text{soft AND Sharpness} = \text{low} \\ &\text{THEN Force\_Output} = \text{medium}\end{aligned} \tag{4}$$

$$\begin{aligned}\text{IF Force\_Input} &= \text{soft AND Solidity} = \text{hard AND Sharpness} = \text{low} \\ &\text{THEN Force\_Output} = \text{hard}\end{aligned} \tag{5}$$

While in the Rules 2 through 5, the Force_Input is taken right from the instruction (or set to medium if nothing is specified), Solidity, Sharpness and Force_Output are computed using object data and the membership functions stored in our context data base.

Having found the activated output membership functions, each of them will be limited to the minimum value of the context values contained in the rule. This is done to avoid executions with an unnecessary high force.

**De-Fuzzyfier.** In the defuzzification step, we want to have a method that fulfils the following requirements: no maximum force or velocity while executing a motion (R1), no sensor noise (R2) and consider the whole output area (R3). After evaluating the defuzzification methods max-method, left-max-method, right- max-method, mean-max-method, centroid-method and modified centroid-method we come to the conclusion that the modified centroid-method is the best method regarding our application, because it fulfils R1–R3.

## 4 Prototype Evaluation

Our framework can be used to generate a variety of force-based robot motions for different material and motion properties (see Fig. 4). A brief example in form of a stamping motion is now described as a proof of concept. The materials in use are modelling clay and soap, which are described by their hardness. The assumed user instruction is: "Robot, stamp the object softly!".

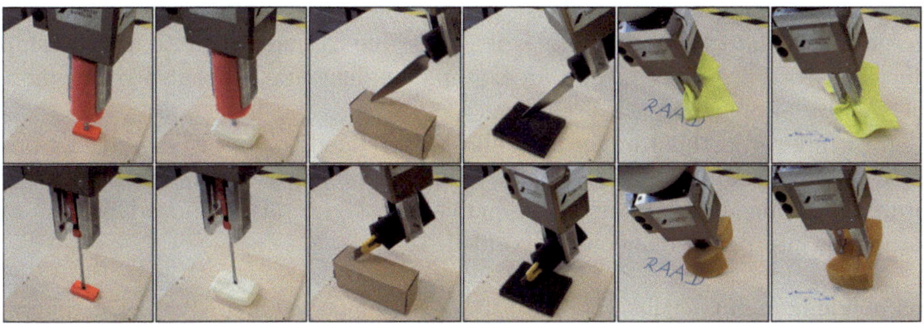

**Fig. 4.** Example combinations of tools and work pieces with different material properties in the use cases: stamping, cutting and wiping.

A natural language understanding component implemented with the *semvox* framework [18] extracts the verb "stamping", the work piece "object", the force input "softly" and detects the gripped tool "stamping tool". This information is then processed to the reasoning module and the motion generator.

The reasoning module requests the fuzzy rules for "stamping" and the membership functions together with the following object information for the linguistic variables from the context data base: Hardness and Force_Output. The membership functions for Hardness are modelled by using the shore hardness scale [19], the functions for Force_Output are modelled in a way, that the robot in use is capable of execute them (in our case 100 N). The following fuzzy rules are fulfilled for the modelling clay and the candle soap:

$$\begin{aligned} \text{IF Force\_Input } &= \text{ soft AND Hardness} = \text{ soft AND Tooltype} = \text{ solid} \\ &\text{THEN Force\_Output} = \text{ soft} \end{aligned} \tag{6}$$

$$\begin{aligned} \text{IF Force\_Input } &= \text{ soft AND Hardness} = \text{ medium AND ToolType} = \text{ solid} \\ &\text{THEN Force\_Output} = \text{ medium} \end{aligned} \tag{7}$$

The output force membership functions $\mu_{\text{soft}}$ respectively $\mu_{\text{medium}}$ are limited by the hardness values for each material. By using the modified centroid method the resulting forces $F_{\text{clay}} = 10$ N and $F_{\text{soap}} = 80$ N are computed for the sensor-based motions which produce sufficient results.

## 5 Conclusion

This paper tackles the problem of mapping uncertain force parameters in spoken commands to force-based robot motions depending on fuzzy logic. Our Fuzzy Force Model enables users to utilize robot motions via pre-defined membership functions and fuzzy rules in a flexible way after they have been specified by themselves or experts. While not only force parameters are modelled in a fuzzy way, specific motions do not have to be hard-coded for any possible combination of tools and work pieces as long as necessary properties are available in a data base or measurable at runtime. It is even possible to extend it to different sensor-based output data types, as long as it is possible to find adequate membership functions. As a proof of concept, an example implementation is also given where the validity of our approach is shown.

In future work, this framework could be extended to velocity-based motions as well as a combination of velocity-based and force-based motions. Also an intuitive graphical user interface to enable a fast generation and adaptation of parameter interpretations could be added.

**Acknowledgements.** This work has partly been supported by Deutsche Forschungsgemeinschaft (DFG) under grant agreement He2696-18.

## References


1. Knoll, A., et al.: Instructing cooperating assembly robots through situated dialogues in natural language. In: International Conference on Robotics and Automation (1997)
2. Norberto Pires, J.: Robot-by-voice: experiments on commanding an industrial robot using the human voice. Ind. Robot Int. J. **32**(6), 505–511 (2005)
3. Tellex, S., et al.: Understanding natural language commands for robotic navigation and mobile manipulation. In: AAAI, vol. 1 (2011)
4. Spangenberg, M., Henrich, D.: Towards an intuitive interface for instructing robots handling tasks based on verbalized physical effects. In: IEEE International Symposium on Robot and Human Interactive Communication (2014)
5. Muthugala, M.A.V.J., et al.: A review of service robots coping with uncertain information in natural language instructions. IEEE Access **6**, 12913–12928 (2018)
6. Schiffer, S., Ferrein, A., Lakemeyer, G.: Reasoning with qualitative positional information for domestic domains in the situation calculus. J. Intell. Robot. Syst. **66**(1–2), 273–300 (2012)
7. Jayasekara, A.G.B.P., Watanabe, K., Izumi, K.: Understanding user commands by evaluating fuzzy linguistic information based on visual attention. Artif. Life Robot. **14**(1), 48–52 (2009)
8. Muthugala, M.A.V.J., et al.: Enhancing human-robot interaction by interpreting uncertain information in navigational commands based on experience and environment. In: International Conference on Robotics and Automation (ICRA) (2016)
9. Pulasinghe, K., et al.: Modular fuzzy-neuro controller driven by spoken language commands. IEEE Trans. Syst. Man Cybern. **34**(1), 293–302 (2004)
10. Lin, C.-T., Kan, M.-C.: Adaptive fuzzy command acquisition with reinforcement learning. IEEE Trans. Fuzzy Syst. **6**(1), 102–121 (1998)
11. Skubic, M., et al.: Spatial language for human-robot dialogs. IEEE Trans. Syst. Man Cybern. **34**(2), 154–167 (2004)
12. Jayawardena, C., Watanabe, K., Izumi, K.: Learning of object identification by robots commanded by natural language. In: 9th International Conference on Intelligent Autonomous Systems (IAS) (2006)
13. Muthugala, M.A.V.J., Jayasekara, A.G.B.P.: Synthesizing fuzzy linguistic vocal responses by adapting perception of robot based on visual attention. In: 7th International Conference on Information and Automation for Sustainability (2014)
14. Jayawardena, C., Watanabe, K., Izumi, K.: Posture control of robot manipulators with fuzzy voice commands using a fuzzy coach-player system. Adv. Robot. **21**(3–4), 293–328 (2007)
15. Jayasekara, A.G.B.P., et al.: Interpreting fuzzy linguistic information by acquiring robot's experience based on internal rehearsal. J. Syst. Des. Dyn. **4**(2), 297–313 (2010)
16. Zadeh, L.A.: Fuzzy logic= computing with words. IEEE Trans. Fuzzy Syst. **4**(2), 103–111 (1996)
17. Wölfel, K., Henrich, D.: Grounding verbs for tool-dependent, sensor-based robot tasks. In: 27th IEEE International Symposium on Robot and Human Interactive Communication (RO-MAN) (2018)
18. Pfalzgraf, A., et al.: ODP-Ontology-based Dialogue Platform. Technical report (2008)
19. DIN EN ISO 868: Plastics and ebonite-Determination of indentation hardness by means of a durometer (Shore hardness) (2003)

# Integrated Palletizing Workstation with an Industrial Robot and a Cobot

Juraj Kováč[1(✉)], Robert Jenčík[1,2], Peter Andrejko[1], Mikuláš Hajduk[1], Zbigniew Pilat[3], Peter Tomči[2], Jozef Varga[1,2], and Martin Bezák[2]

[1] Faculty of Mechanical Engineering, Technical University of Košice, Letná 9, 04001 Košice, Slovakia
{juraj.kovac,mikulas.hajduk,jozef.2varga}@tuke.sk

[2] Manex spol. sr.o., Alvinczyho 12, POBOX E-7, 04001 Košice, Slovakia
robert.jencik@manex.sk

[3] Przemyslowy Instytut Automatyki I Pomiarów PIAP, Warsaw, Poland
zpilat@piap.pl

**Abstract.** It is anticipated that the field of industrial robot deployment will continue to grow in the coming years. However, an even higher growth rate is expected in deployment of cooperative robots. This means that the area of robotics will be one of the decisive factors in increasing competitiveness on the variable product market. It is logical that such an expansion of robotics cannot be based on solo robotic workplaces. Rather, ever increasing collaboration of different robot types is expected to take place. In co-operation with the SjF, MANEX sees the near future in implementing automation to its production, namely automation of lines and workplaces where industrial robots-cobots-men work together or in a complementary manner. The issue of this integration is relatively new and it is still difficult to find comprehensive methodologies of how to approach it.

The article presents a framework proposal for deployment of different robots and the man in the area of final palletizing and packaging. It goes without saying that newer and newer approaches and methodologies and practical experience will be garnered in the future in both, the academia and in practical applications.

**Keywords:** Palletizing · Industrial robot · Cobot

## 1 Introduction

Every production is subject to the rules and principles of the technological process. The desired shape, dimensions and quality of the processed product is achieved through technological operations. In addition to the technological process, the manufacturing system is also significantly affected by non-technological operations, without which the production itself could not run economically. Handling and transport tasks are done at every stage of the technological process. Material preparation, production, inspection, assembly, storage and transportation of the manufactured product require the processes be arranged rationally. Planning, material handling and transport at all stages of the product processing nowadays posits the greatest opportunity for innovation in production and it significantly contributes to savings. At the same time, the costs of


K. Berns and D. Görges (Eds.): RAAD 2019, AISC 980, pp. 202–209, 2020.
https://doi.org/10.1007/978-3-030-19648-6_24

logistical or material handling processes are increasing. This spurs an increase in investments into modernization and automation of storage, transport and handling. One of the chosen paths emerged from the use of industrial robots. Since about 1990, industrial robots have begun to replace not only manual palletizing but also traditional palletizing machines. This trend was triggered by the need for greater flexibility of palletizing workplaces due to dynamic development of different palletizing item variations and different production volumes. Relatively high investments that the palletizing workplaces require make their implementation a priority as their utilization is great and, as a rule, because each flow of finished products passes through this functional section. These are the main reasons why the incentives for automating the palletization should be paid attention to.

## 2 Specification of Handling and Palletizing Activities of the Production Process for Robot Deployment

The basis for palletization is the concentration of material into a compact unit, the so-called handling packet. The handling packet create consists of the goods (palletized items) stocked on a pallet which can be handled as a single unit. It is determined by its size (length × width × height), the weight and volume of the material stacked on it", Fig. 1.

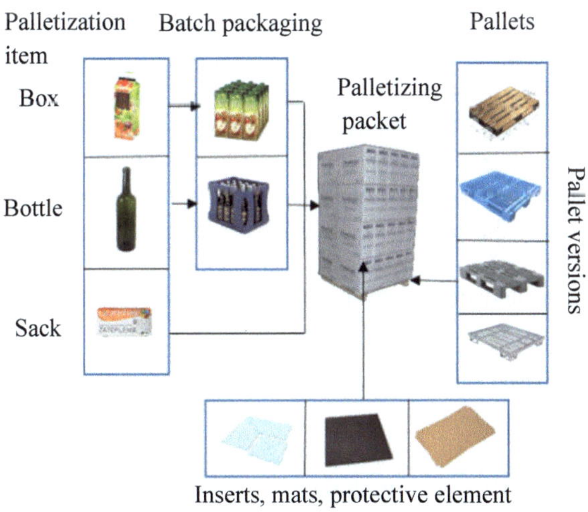


**Fig. 1.** Creating palletizing packets

In order to achieve automation of the logistic chain, the area of robot use is mainly in the so-called initial manufacture i.e. the storage of palletized items in the so-called transport packages or secondary packaging. This is the creation of the so-called groups of items in one "packaging" e.g. cartons, crates, etc., Fig. 2.

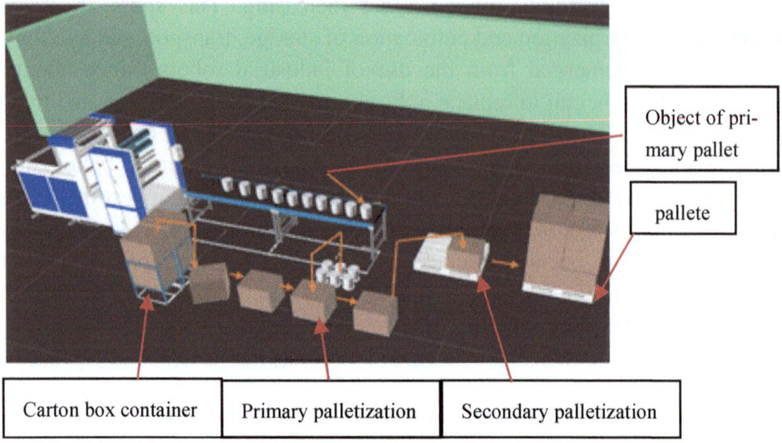


**Fig. 2.** The material flow of an item of the production logistic chain

The second area of palletization is the so-called palletization using transport pallets. The decisive criterion for robot deployment for these activities is the flexibility requirement of the palletizing workplace.

# 3 Options of Robot Deployment to Palletizing

Currently, there are quite a few devices for automatic palletizing available, ranging from simple palletizers to highly flexible robots and, most recently, cobots. If the palletizing workstation is to be effective, it is not easy to choose the most appropriate type of a palletizer, a robot or a cobot. It is important to note that palletization is not just a palletizer or a robot. It is a system of cooperating devices ensuring the entry of items to be palletized and of the pallets or other objects such as partitions, stabilizing elements of the palletized packet and many others [1].

Robot-executed palletization may nowadays be divided into three categories:

- palletization by means of palletizers
- industrial, robot-based palletization
- collaborative robot - human.

## 3.1 Palletization by Means of Palletizers

These are essentially automatic devices that palletize the items designated for palletization in layers. Prior to palletization, the items are assembled into a layer and then placed onto a pallet. Their advantage is that they palletize quickly and are also designed to handle heavier objects.

The principle of palletizer's activity lies in creating the entire layer and then grabbing it with the gripper and moving it onto the pallet as a whole.

The palletizers may be applied mainly to processes with items to be palletized being the same. Their main disadvantage is very low flexibility when it comes to objects of different shapes and dimensions. The advantage of palletizers is low investment costs and high palletizing speed.

### 3.2 Palletization with Industrial Robots

Current automation of the palletizing process requires a highly flexible structure of the palletizing workplace, with highly adaptive material flows. Flexible structure of the palletizing workplace is only possible to achieve with robotic technology. The robot becomes the dominant element of the palletizing workplace. It is highly adaptive to change in handling movements. Its flexibility in changing the items to be palletized is achieved by abandoning the concept of the gripper versatility, which is replaced with rapid automatic griper swapping. Other advantages of robot-executed palletization is high precision of repeated handling, a wide-ranging robot capacity, stable position of the robot's desired working space, high speed of handling motions, and high operation reliability and safety.

The robot's application capabilities in palletizing processes are very broad and the structures of their workplace vary considerably [2].

### 3.3 Palletizing Cobot

Man-cooperating cobots are a simple and easy way to palletize or to package the products into the secondary packaging Fig. 3. The palletizing or the cobot-man packaging workplace is a compact, rapidly executable and operable site, highly reconfigurable and easily reprogrammable.

**Fig. 3.** Palletizing cobot – human

The cell is of a plug and play type. The cobot-man collaboration is safe and does not require a safety fence. The collaborative workplace offers the possibility of several modifications such as changing the position of the robot, the conveyors, positioning of the robot on different supports, etc. The palletizing patterns of how items are positioned on the pallet are stored in the system memory and can be immediately retrieved to make changes in the layout in seconds.

The benefits of the collaborative palletizing workplace are as follows:

- Space saving: the palletizing robot requires, on average, 60% smaller space than the traditional industrial robot or palletizer. This is important when you do not have enough space on the shop floor [4].
- Cost reduction: Capital costs are much lower and return on investment faster than those of palletizing workplaces with industrial robots or automatic palletizers.
- Reduction in both, physical strain and worker injuries: Cobots are safe. The slightest touch of the robot stops its operation or, if safety sensors are used, the robot slows its movements to a safe speed. Its speed of motion is suitable for most applications: It is true that the cobot is not as fast as a robot that is in a fully enclosed guard system. However, with intelligent security features, its movements can compete with classical palletizing robot. However, most applications are slower.
- In the common workplace of the man and the collaborative robot, the man is not restricted in motion by a safety fence.
- Operating costs are low, with 220 V power supply

The cobots are designed to be used within a defined workspace for collaboration with human workers. Cobots usually perform tasks that are repetitive or more challenging for the man, while a human worker performs simpler manual tasks that involve more intellectual or decision-making skills.

Essentially, the cobots mimic human activities and perform these tasks at similar speed to men, but with higher payloads, reach, and higher accuracy. The robot's application capabilities in palletizing processes are very broad and the structures of their workplace vary considerably (Fig. 4).

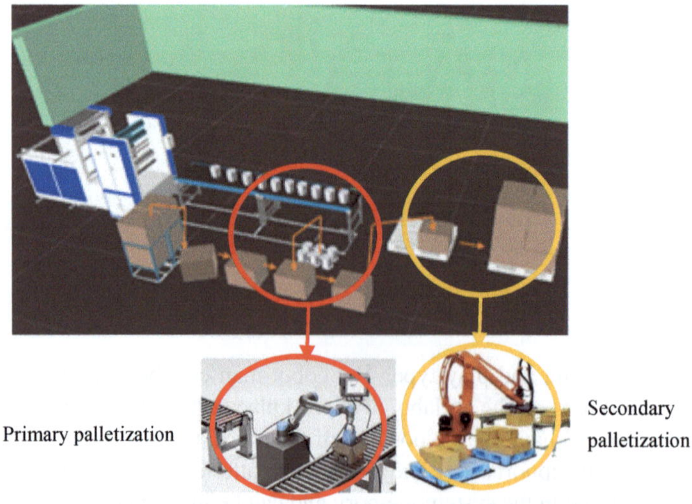


**Fig. 4.** Industrial robot and Cobot paletting options

# 4 Methodological Framework of the Design of Mixed Palletizing Workplaces for Robots

In order to be able to decide correctly which robot to use for the automation - palletization of the production logistics chain, we must in particular:

- Have a detailed knowledge of handling activities and of items handled in the process of manufacturing
- Be aware of the possibilities and differences of industrial robots.

As seen in Fig. 5, in manufacturing e.g., in the food production process, we have to distinguish between several material flows, items that are handled (product in primary packaging, products in secondary packaging, product group, cartons, etc.) of the outputs. We need to make a detailed analysis of these handling flows, items to be handled and their movements, to be able to specify the handling activities for robot-executed automation-palletization. The analysis focuses on what is handled, how often, from where to where, on the handling times and more.

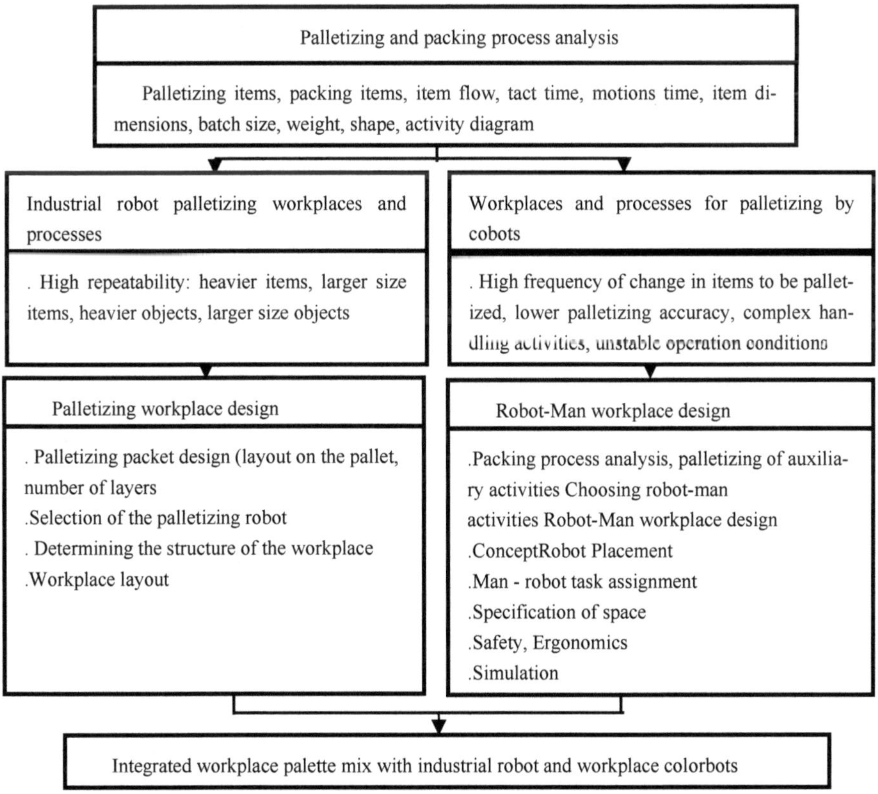


**Fig. 5.** Methodical steps of a mix of robotized palletizing workplace

Step 2 follows the detailed analysis, in which we put processes into one of two groups, namely:

- Processes suited to an industrial robot
- Processes which enable the robot-man collaboration.

So far, the industrial robot has been working in a cage, the workplace structure is relatively stable, the robot is fast and accurate. From the productivity viewpoint, the industrial robot workplace is suitable for relatively stable handling tasks and items to be handled.

The Cobot is lightweight, its dimensions are smaller, its speed is lower, as is its precision, but can be relocated easily, allowing workplaces to be varied. It is suitable for frequent rotation of items to be handled. The items designated for palletization are most often production objects in primary packaging of different shapes, dimensions and weights. It is true that the cobots are also suitable for handling and palletizing heavier and larger objects, but decisive are always the economic indicators and the advantage of their use, or their simplicity, reliability and safety of operations involving the man.

Step 3 deals with the workplace design, either with an industrial robot or with a cobot and the man. It should be emphasized here that the priority of the final workplace with the industrial robot is the efficiency and productivity. Workplaces with a cobot are designed in a manner and in the number that ensures the production process continuity, and that the man is relieved of heavier and more complicated handling tasks.

The last step consists of designing how these robot workplaces can be integrated.

## 5 Conclusion

For a number of small and medium-sized enterprises, cobots have brought many benefits that could not have been achieved with industrial robots. The cobots allowed these enterprises to automate such processes to such an extent that they could have not dreamed of until recently. And it finally enabled them to gain some competitive advantages of automation that have so far been achieved only by larger manufacturers. While cobots cannot offer all the benefits of complete industrial automation, they can be an effective first step to increasing the degree of automation.

There are many questions that need to be addressed when deciding whether an industrial robot or a cobot is the most suitable for your project. What are your automation goals and how much space do you have for the automation device, are your processes fixed or are they changing regularly?

While in doing the same operation a robot is certainly faster than the man, this is not crucial.

"What is more important is that the cobot relieves the man of an activity that he does not want to do."

This article was created by implementation of the grant project APVV-17-0258 Digital engineering elements application in innovation and optimization of production flows.

## References


1. Cheng, H., Xu, Y., Liu, Y., Hao, L.: Design of palletizing algorithm based on palletizing robot workstation. In: Conference on Real-time Computing and Robotics, Angkor Wat, Cambodja (2016)
2. Penington, R.A., Tanchoco, J.M.A.: Robotic palletization of multiple box sizes, pp. 95–105. Received 01 Jan 1987, Published Online 16 May 2007
3. Moura, M., Silva, F.: ICARSC 2018 - 18th IEEE International Conference on Autonomous Robot Systems and Competitions (2018)
4. Kravec, D., Staš, O.: Creation of control algorithms for automatic palletization STU Scripts in Bratislava. Publishing House STU (2014). ISBN 9788022741781
5. Kováč, J., Rudy, V., Kováč, J.: Automation of Production. Edition of Scientific and Professional Literature. SjF TU Košice (2016). ISBN 978-80-553-2311-4
6. Gołębski, R.: Parametric Programming of CNC Machine Tools (2016). https://doi.org/10.1051/matecconf/20179407004
7. Kanuch, J., Girovský, P.: The device to measure the load angle for the salient-field synchronous machine in the educational laboratory. Measurement **116**, 49–55 (2018)
8. Kanuch, J., Girovský, P.: Motor for direct drive of electric wheelchair. Int. J. Eng. Res. Afr. **31**, 94–103 (2017)

# A General Approach for Automating Teleoperated Construction Machines

Hyung Joo Lee[1(✉)], Sigrid Brell-Cokcan[1], and Katharina Schmitz[2]

[1] Chair of Individualized Production (IP), RWTH Aachen University, Schinkelstr. 1, 52062 Aachen, Germany
office@ip.rwth-aachen.de

[2] Institute for Fluid Power Drives and System (IFAS), RWTH Aachen University, Campus-Boulevard 30, 52074 Aachen, Germany
katharina.schmitz@ifas.rwth-aachen.de

**Abstract.** Despite enormous research advances in robotics, most of repetitive and dangerous tasks on construction sites are still manually performed. The challenge to robotic automation lies in the nature of construction site. Robots must be able to handle large payloads, outdoor conditions, dirty and dynamic environment. Therefore, attempting to develop a new robot for construction sites is a formidable task. In this paper, a generic method is introduced to adapt existing stable hydraulically driven construction machinery so that proven robots can be obtained at a significant reduction in effort. Fist, the bus system of the machinery is utilized to communicate with the user's computer. Then, a control framework is developed which is based on the closed-loop inverse kinematic (CLIK). The method is experimentally validated on a demolition machine Brokk 170, that is originally designed to be controlled with a remote controller.

**Keywords:** Construction robotic · Closed-loop inverse kinematics · CAN bus · Demolition robotics · Path planning

## 1 Introduction

Integrating robots into construction sites and automating the construction work is an attractive paradigm, as it relieves workers from heavy physical exertion and repetitive work meeting societal need for safe, inexpensive and sustainable construction. Many approaches have been proposed in the literature to integrate robots into construction sites [1–6].

The majority of approaches proposed customize classical industrial robots so that they can partially handle the unstructured nature of the construction site. In [3] and [4], a robot system with a customized ABB IRB 4600 robot arm is introduced. The system is capable of driving on non-flat or soft ground and performs multiple tasks. However, the 40 kg payload and the indoor design of the industrial robot limit the usage of the system to certain types of tasks.


K. Berns and D. Görges (Eds.): RAAD 2019, AISC 980, pp. 210–219, 2020.
https://doi.org/10.1007/978-3-030-19648-6_25

The tasks on the construction sites often require high payloads. Correspondingly, the classical electrically actuated robot arms without compliant elements will always suffer from a low payload/weight ratio and their susceptibility to vibration and impact damage.

Another approach is to develop new robots according to desired requirements. In [1] and [2], a system for teleoperated decommissioning applications is described. Two 7-DOF hydraulically actuated robot arms are developed and attached to the end of a Brokk demolition machine. This approach allows the authors to semi-automate selected decommissioning tasks such as cutting pipes.

In these works, the introduced approaches are restricted to a laboratory indoor environment or to their custom robots. Implementation on the real construction sites and acceptance from the construction industry still remain open. In the context of works on the construction site, it is also necessary to take into account the ease of control and the reliability. The highly specified custom designs often lack of these. Hence, the research described in this paper aims to introduce a generic method that can adapt existing construction machines rather than developing entirely new systems.

On construction sites, hydraulic actuators form a robust and central element for managing heavy loads and are controlled by an interaction of valves and pumps. In automated machines, a precise movement is a key part, however, there are several factors hindering such a control in hydraulic systems, which have to be considered. A major challenge are the abundance of nonlinearities due to the design of pumps, valves and the interaction of different actuators, see e.g. [7]. A closed-loop control is required to compensate outer forces and prevent unstable modes due to oscillations. New designs of the machines hydraulic circuit offer improvements regarding energy efficiency, as it was demonstrated in [8] or the usage of independent metering enables a high adaptability of the systems [9]. Overall, the introduction of electro-hydraulic components and a variety of sensors can support the automation, as in the design of a forestry crane in [10].

Hence the aim of this work is too answer the question of whether proven construction machinery can be adapted so that one can program the motion as if it were a robot. This question is the foundation of future steps and will open up limitless opportunities for automation in different fields. Instead of developing new robots for each specific task, machines that have been already used on the fields for years are adapted so that humans can program their movement and also reasoning systems. We believe that adapted machines would have a significant impact on building construction in the near future.

## 2 Communication Between PC and Machine

To enable the PC-based control of a construction machine, the machine must be able to communicate with the PC. For this purpose, we utilize the control bus of the machine. Establishing the new communication link can be described with the following steps: (1) Build a CAN data gateway (2) Analyze the CAN bus communication (3) Filter out and replace commands.

First we build our communication gateway between the computer and the machine using CAN bus. The CAN bus interface can be built with an arduino micro-controller and an CAN adapter. It's worth mentioning that the messages from the original control device have to be blocked first, before sending messages from the PC to prevent messages from the original control device overwriting those of the PC. Therefore, the CAN bus interface consists of two CAN controllers, as depicted in Fig. 1. Once, the communication gateway is built, it is necessary to analyze how the machine behaves to each specific CAN command. In the CAN bus system, each control unit has one unique identifier (ID) on which it broadcasts messages. In other words, specific tasks always use the same message on the same ID. One way to gain this information can be from the manufacturer of the system or technical documentation.

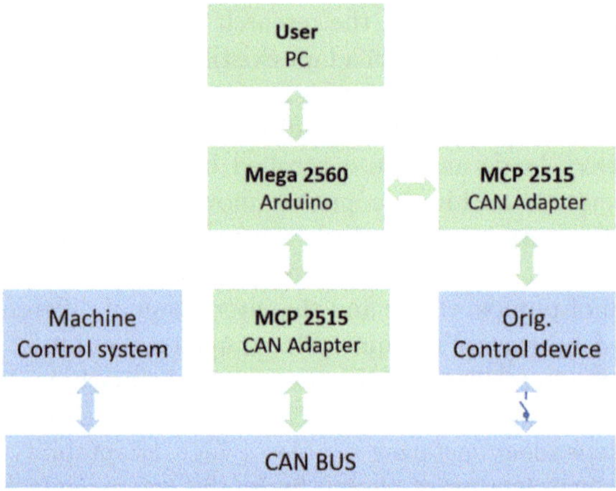


**Fig. 1.** Illustration of wiring concept. To filter out the messages from the original control device, the connection to CAN bus is separated. The messages transmitted by the original control device are filtered out and only select messages are combined with the messages from the user's computer. The combined messages are sent to the control system via the bus system.

Another way is to write an Arduino script which can record the complete machine communication while for instance executing a particular operation. After having recorded the relevant CAN messages, it is necessary to examine the message data bytes in detail to determine the required CAN data (e.g. CAN message, CAN ID, cycle time and length of the message) to execute that particular operation. This step is repeated for every operation that must be simulated later. In case the machine is equipped with sensors, it is also possible to read out their values from the bus system, thereby gaining valuable information for control loops in later work.

After having identified the relevant CAN data to operate the machine, these CAN data from the control device are filtered out and replaced with the custom

CAN data. This way, it is not necessary to emulate the whole control device, but only the parts that the user wants to control. Having performed the steps above, it is possible to setup the desired operations for the machine system.

## 3 Real-Time Cartesian Control

A mapping algorithm based on closed-loop inverse kinematic (CLIK) solver is used to find the corresponding joint configuration of the machine for the defined trajectory. The calculated joint angle values are changed into corresponding CAN data and forwarded to the machine via the Arduino. The block diagram of the control method is illustrated in Fig. 2.

In the following sections, it is described how the mapping between the trajectory defined by the user and the corresponding machine's motion can be achieved.

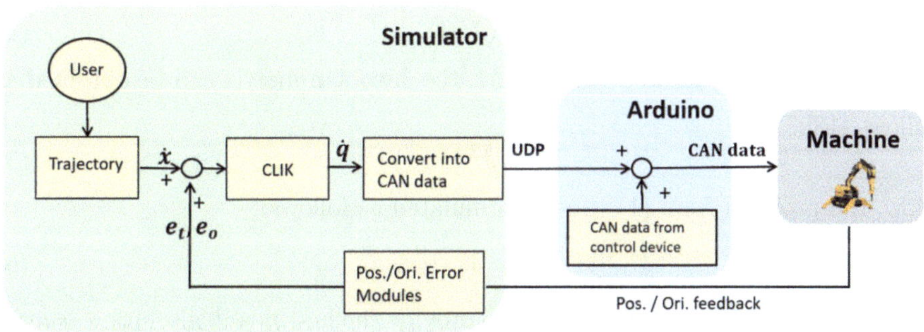


**Fig. 2.** System overview for the Cartesian control.

### 3.1 Kinematic Model

Cartesian control in robotics can be formulated as finding proper joint space configurations $\mathbf{q} \in \mathbb{R}^N$ of a robot given an operational space trajectory $\mathbf{x} \in \mathbb{R}^M$, where N is the number of degree of freedoms (DoF) of the robot and M is the operation dimension. The mapping between the joint space velocities and operational space velocities can be obtained as,

$$\dot{\mathbf{x}} = \mathbf{J}(\mathbf{q})\dot{\mathbf{q}} \tag{1}$$

where $\mathbf{J} \in \mathbb{R}^{M \times N}$ is the Jacobian matrix. The Jacobian matrix can be decomposed to its rotational $\mathbf{J_o}$ and translational $\mathbf{J_t}$ components.

$$\mathbf{J} = \begin{pmatrix} \mathbf{J}_t \\ \mathbf{J}_o \end{pmatrix} \tag{2}$$

### 3.2 Cartesian Control

To obtain the $\dot{\mathbf{q}}$ values out of the relation in (1), one can analytically solve the Eq. (1). There also exist a numerical ways to solve the problem [11], however they are based on iterative methods. For real-time control the number of iterations must be limited and therefore they may fail to provide a solution even if one exists. For this reason, iterative methods are not treated in this work.

The Cartesian control approach used here is based on a control method known as CLIK. Following the determination of the inverted Jacobian matrix is described for different cases:

*Case 1*: In the case of non-redundant problem $N = M$, the Jacobian matrix can be directly inverted to obtain the joint space velocity $\dot{\mathbf{q}}$ [12]:

$$\mathbf{J}^{+} = \mathbf{J}^{-1} \tag{3}$$

*Case 2*: For manipulators with $N < M$, the Jacobian matrix can be obtained as follows [12]:

$$\mathbf{J}^{+} = (\mathbf{J}^{T}\mathbf{J})^{-1}\mathbf{J}^{T} \tag{4}$$

*Case 3*: For manipulators with $N > M$, the Jacobian matrix can be obtained as follows [13]:

$$\mathbf{J}^{+} = \mathbf{J}^{T}(\mathbf{J}\mathbf{J}^{T})^{-1} \tag{5}$$

Now, the problem from (1) can be formulated as follows:

$$\dot{\mathbf{q}} = \mathbf{J}^{+}(\mathbf{q})\dot{\mathbf{x}} \tag{6}$$

The pseudo-inverse has the property to provide the best possible solution according to the equation $\mathbf{J}\dot{\mathbf{q}} = \dot{\mathbf{x}}$. If $\dot{\mathbf{x}}$ is in the range of $\mathbf{J}$, $\dot{\mathbf{q}}$ is unique vector of smallest magnitude which satisfies $\mathbf{J}\dot{\mathbf{q}} = \dot{\mathbf{x}}$. If $\dot{\mathbf{x}}$ is not in the range of $\mathbf{J}$, an exact value of $\dot{\mathbf{q}}$ is not available. However, the provided $\dot{\mathbf{q}}$ still minimizes the magnitude difference of $\mathbf{J}\dot{\mathbf{q}} = \dot{\mathbf{x}}$ [16].

The Eq. (6) can be extended to (7) by adding a feedback correction term $\mathbf{Ke}$, where $\mathbf{K} \in \mathbb{R}^{M\times M}$ is a positive definite gain matrix and $\mathbf{e} \in \mathbb{R}^{M\times 1}$ is the remaining [14]. If $M = 6$ the $\mathbf{e} = [\mathbf{e}_t\mathbf{e}_o]^T$ can be decomposed to the translation error $\mathbf{e}_t \in \mathbb{R}^{3\times 1}$ and the orientation error $\mathbf{e}_o \in \mathbb{R}^{3\times 1}$ components.

$$\dot{\mathbf{q}} = \mathbf{J}^{+}(\mathbf{q})(\dot{\mathbf{x}} + \mathbf{Ke}) \tag{7}$$

The error $\mathbf{e}$ between the desired and actual motion occurs for instance due to model inaccuracies. The translation error can be simply defined as $\mathbf{e}_t = \mathbf{x}_d - \mathbf{x}_a$, where $\mathbf{x}_d$ and $\mathbf{x}_a$ represent the desired and the actual operation space position, respectively. To obtain the orientation error, first the desired operation frame $\mathbf{T}_d$ and the actual operation frame $\mathbf{T}_a$ are estimated. The orientation error can then be calculated with the desired $(\mathbf{n}_d; \mathbf{s}_d; \mathbf{a}_d)$ and the actual unit orientation vectors $(\mathbf{n}_a; \mathbf{s}_a; \mathbf{a}_a)$, respectively [15]. The functional expression of the orientation error is then:

$$\mathbf{e}_o = \frac{1}{2}(\mathbf{n}_a \times \mathbf{n}_d + \mathbf{s}_a \times \mathbf{s}_d + \mathbf{a}_a \times \mathbf{a}_d) \tag{8}$$

Using the Eq. (7) both the desired velocity and position can be obtained. The discrepancy between the desired and actual value will be numerically eliminated with a rate depending on the eigenvalues of K. The value of K is experimentally tuned. The CLIK algorithm above described is used to control the machine to follow a trajectory which is defined by the user.

### 3.3 Kinematic Constraints

A machine system typically has kinematic constraints which must be properly handled. Typical industrial machines may not have a sufficient number of redundant DoF to enforce these constraints in the nullspace. In this work, the kinematic constraint problem is therefore handled without using the redundancy.

1. *Kinematic singularity problem*
   Recall the inverse kinematics problem, where a proper set of joint space velocities of a robot is estimated given a set of operational space velocities, as defined in Sect. 3.2. However, a matrix is only invertible if it has full row rank. Thus, the Eqs. (4) and (5) will have stability problems in the neighborhoods of singularities where the Jacobian matrix $\mathbf{J}$ is about to lose it's row rank [16]. In other words, a small movement near the singular configuration can cause large joint angle changes, such that the inverse solution becomes impractical.
   The most widely used solution for the robust usage of a pseudo-inverse matrix is the damped least squares method (DLS), which is introduced in [16,17]. The basic idea of DLS is to damp the robot's movement using a damping factor $\lambda$, whenever the robot is near the singularity. For manipulators with $N > M$ the functional expression is then:

$$\mathbf{J}^{+} = \mathbf{J}^T(\mathbf{J}\mathbf{J}^T + \lambda^2\mathbf{I})^{-1} \tag{9}$$

   If $\lambda$ in (9) is chosen too large, the machine's movement will be dampened too much. Despite of $\dot{\mathbf{x}}$, the corresponding joint velocity $\dot{\mathbf{q}}$ will be small. Using the manipulability measure $m$, the damping factor $\lambda$ can be dynamically estimated depending on the current configuration [18]:

$$\lambda = \begin{cases} k(1 - \frac{m}{m_t}), & \text{if}\, m < m_t \\ 0, & \text{elsewhere} \end{cases} \tag{10}$$

   where $k$ is a constant, m $= \sqrt{det(JJ^T)}$ and $m_t$ is a threshold value.
2. *Joint limit problem*
   The joint limit avoidance is achieved based on the Weighted Least-Norm (WLN) method proposed in [17]. The basic idea of WLN is similar to the above DLS method. The motion of a joint is penalized, whenever a joint is near it's limit. In order to penalize the motion of a joint over the others, the gradient of the performance criterion $H(q)$ is defined as follows:

$$\frac{\partial H(q)}{\partial q_i} = \frac{(q_{i,max} - q_{i,min})^2(2q_i - q_{i,max} - q_{i,min})}{4(q_{i,max} - q_i)^2(q_i - q_{i,min})^2}, \tag{11}$$

where $q_i$ is the $i$th joint angle, $q_{i,max}$ and $q_{i,min}$ are the upper and lower limits respectively. The gradient $\frac{\partial H(q)}{\partial q_i}$ is equal to zero if the $i$th joint is at the middle of its limit and goes to infinity at either limit. Correspondingly, the elements of the $n \times n$ diagonal matrix $W$ are defined by,

$$w_i = \begin{cases} 1 + |\frac{\partial H(q)}{\partial q_i}|, \text{ if } \triangle |\frac{\partial H(q)}{\partial q_i}| \geq 0 \\ 0, \qquad\qquad \text{elsewhere} \end{cases} \tag{12}$$

From (12), it can be seen that the weighting factor $w$ is equal to 1 if the value of the gradient decreases. In other words, if the joint moves away from either of its limits the value of the gradient decreases and consequently the joint freely moves without any penalty. However, if the value of the gradient increases, it indicates that the joint is moving toward its limit. Correspondingly, the joint limit is guaranteed by the large weighting factor. The calculated weighting matrix $W$ can be integrated into (9) as follows:

$$\mathbf{J}^{+} = \mathbf{W}^{-1}\mathbf{J}^{T}(\mathbf{J}\mathbf{W}^{-1}\mathbf{J}^{T} + \lambda^{2}\mathbf{I})^{-1} \tag{13}$$

## 4 Experimental Setup

The developed method is evaluated on a demolition machine (Brokk170), which has proven itself to be highly reliable even in hazardous environments such as Fukushima nuclear power plants [19]. As these machine are not equipped with any axis sensors, the implemented cartesian control method can't be implemented as a closed loop system. Therefore the aim of the experiment is to show the feasibility of the proposed method and not a measurement of achievable accuracy.

To integrate the capabilities of adapted machines into an construction workflow, making the machines accessible to architects and civil engineers, the CAD-tool Rhino/Grasshopper is used in to define the machine's trajectory. However, any software with a UDP interface can be used. The machine is originally designed to be controlled with a remote control and was adapted as described in Sect. 2. In this experiment, the machine is programmed to follow a virtual red ball in Rhino/Grasshopper. To visualize the discrepancy between the desired and actual motion, a 3D model of Brokk170 is also controlled with the desired motion values. A discrepancy exists, however these errors are inherent to the missing sensor system. Nevertheless, a reasonable PC-based Cartesian control of Brokk170 could be achieved, as presented in Fig. 3.

Although these experiments are still within a simple framework, they show the feasibility of our method for real-time cartesian control of construction machinery. A key advantage of the proposed approach is the significant reduction in effort to automate a proven machine into a robot suited for construction sites and other hazardous environments.

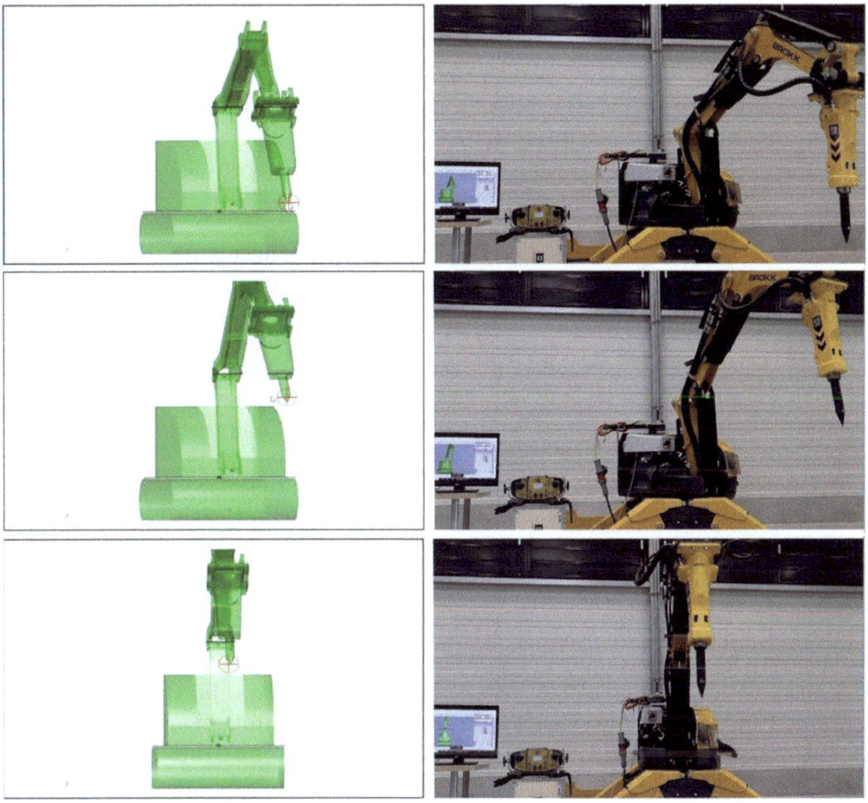

**Fig. 3.** Frames from the video capturing the experiment. In this experiment, both the 3D model and Brokk170 are programmed to follow the red point in Rhinoceros.

## 5 Conclusion

In this paper, motivated by the need for robust mobile robots on construction sites, a new method is presented, which can adapt existing construction machines. For this purpose, we have analyzed the existing CAN communication between the control device and the machine. Based on these results, we have implemented a Cartesian controller. In this way, robust robots can be obtained at a significant reduction in efforts compared to existing methods, such as developing a complete new robotic system. The first results are showcased in this paper. In future work, we plan to integrate sensor systems to inform our Cartesian controller with feedback from the machine.

## References


1. Montazeri, A., Udo, E.J.: Development of dynamic model of a 7DOF hydraulically actuated tele-operated robot for decommissioning applications. In: American Control Conference (ACC), pp. 1209–1214 (2016)
2. Taylor, C.J., Chotai, A., Robertson, D.: State dependent control of a robotic manipulator used for nuclear decommissioning activities. In: Intelligent Robots and Systems, pp. 2413–2418 (2010)
3. Giftthaler, M., Sandy, T., Dörfler, K., Brooks, I., Buckingham, M., Rey, G., Kohler, M., Gramazio, F., Buchli, J.: Mobile robotic fabrication at 1:1 scale: the In situ Fabricator: system, experiences and current developments. Constr. Robot. **1**, 3–14 (2017)
4. Lussi, M., Sandy, T., Doerfler, K., Hack, N., Gramazio, F., Kohler, M., Buchli, J.: Accurate and adaptive in situ fabrication of an undulated wall using an on-board visual sensing system. In: IEEE International Conference on Robotics and Automation, pp. 3532–3539 (2018)
5. Stumm, S., Braumann, J., Brell-Cokcan, S.: A human-machine interaction for intuitive programming of assembly tasks in construction. In: Conference on Assembly Technologies and Systems, vol. 44, pp. 269–274 (2016)
6. Lublasser, E., Hildebrand, L., Vollpracht, A., Brell-Cokcan, S.: Robot assisted deconstruction of multi-layered façade constructions on the example of external thermal insulation composite systems. Constr. Robot. **1**, 39–47 (2017)
7. Lee, S., Chang, P.: Control of a heavy-duty robotic excavator using time delay control with switching action with integral sliding surface. In: Proceedings of the 2001 IEEE International Conference on Robotics & Automation (2001)
8. Vukovic, M.: Hydraulic hybrid systems for excavators, Ph.D. thesis RWTH Aachen University, Shaker (2017)
9. Vukovic, M., Murrenhoff, H.: Single edge meter out control for mobile machinery. In: Proceedings of the ASME/BATH 2014 Symposium on Fluid Power & Motion Control (2014)
10. Wydra, M., Geimer, M., Weiss, B.: An approach to combine an independent metering system with an electro-hydraulic flow-on-demand hybrid-system. In: The 15th Scandinavian International Conference on Fluid Power (2017)
11. Korein, J.U., Balder, N.I.: Techniques for generating the goal-directed motion of articulated structures. IEEE Comput. Graph. Appl. **2**, 71–81 (1982)
12. Goldenberg, A., Benhabib, B., Fenton, R.: A complete generalized solution to the inverse kinematics of robots. IEEE J. Robot. Autom. **1**, 14–20 (1985)
13. Buss, S.R.: Introduction to inverse kinematics with Jacobian transpose, pseudoinverse and damped least squares methods. IEEE Trans. Robot. Autom. **17**, 16 (2004)
14. Dariush, B., Gienger, M., Jian, B., Goerick, C., Fujimura, K.: Whole body humanoid control from human motion descriptors. In: IEEE International Conference on Robotics and Automation, pp. 2677–2684 (2008)
15. Deo, A.S., Walker, I.D.: Adaptive non-linear least squares for inverse kinematics. In: Proceedings IEEE International Conference on Robotics and Automation, vol. 1, pp. 186–193 (1993)
16. Nakamura, Y., Hanafusa, H.: Inverse kinematics solutions with singularity robustness for robot manipulator control. J. Dyn. Syst. Meas. Control **108**, 163–171 (1986)

17. Chan, T.F., Dubey, R.V.: A weighted least-norm solution based scheme for avoiding joint limits for redundant joint manipulators. IEEE Trans. Robot. Autom. **11**, 286–292 (1995)
18. Yoshikawa, T.: Foundations of Robotics, Analysis and Control. The MIT Press, Cambridge (1990)
19. Osumi, H.: Application of Robot Technologies to the Disaster Sites, Report of JSME Research Committee on the Great East Japan Earthquake Disaster, pp. 58–74 (2014)

# Path Planning and Navigation

# Automatic Path Planning for Unmanned Ground Vehicle Using UAV Imagery

Jurgen Zoto[1,3(✉)], Maria Angela Musci[2,3], Aleem Khaliq[1,3], Marcello Chiaberge[1,3], and Irene Aicardi[2,3]

[1] DET, Politecnico di Torino, Turin, Italy
{jurgen.zoto,aleem.khaliq,marcello.chiaberge}@polito.it
[2] DIATI, Politecnico di Torino, Turin, Italy
{mariaangela.musci,irene.aicardi}@polito.it
[3] PIC4SeR, Politecnico di Torino Interdepartmental Centre for Service Robotics, Turin, Italy

**Abstract.** Field machines play an important role in the management of agricultural environments. Increasing use of automated machines in precision agriculture has gained significant attention of farmers and industries to minimize human work load to perform tasks such as land preparation, seeding, fertilizing, plant health monitoring and harvesting. Path planning is considered as a fundamental step for agricultural machines equipped with autonomous navigation system. For mountain vineyards, path planning is a big challenge due to terrain morphology and unstructured vineyards.

This paper proposes a workflow to generate an automatic coverage path plan for unmanned ground vehicles (UGVs) using georeferenced imagery taken by an unmanned aerial vehicle (UAV). First, image acquisition is performed over a vineyard to generate an orthomosaic and a digital surface model, which are then used to identify the vine rows and inter-row terrain. This information is then used by the algorithm to generate a path plan for UGV.

**Keywords:** Agricultural field machines · Automatic coverage path planning · UAV imagery · Unmanned ground vehicle · Photogrammetry

## 1 Introduction

Increase in use of the automated machines in agricultural environment has gained significant attention of farmers and industries to minimize human work load to perform related tasks. Automated vehicles need to have an automatic navigation system and they have to be able to autonomously select the path according to the specific area and obstacles. To increase agricultural productivity, path planning techniques have been implemented to cover the crop fields. In many cases, the crop field plots are planned with the knowledge of on-field expert


K. Berns and D. Görges (Eds.): RAAD 2019, AISC 980, pp. 223–230, 2020.
https://doi.org/10.1007/978-3-030-19648-6_26

for autonomous agricultural machines. Several advancements in path planning methods for agricultural productions using different optimization techniques for single or multiple coordinated vehicles which span the entire field, are discussed in [1]. For agricultural field vehicles, the greedy algorithms can be used to find an efficient route considering the required condition of covering the whole field [2].

For agricultural vehicles, path planning for known environments usually involves the partition of the area to cover in small polygonal regions called parcel, to find the best path for each parcel and then aggregate the results. Several optimization techniques have been implemented to find an approximate solution that solves the coverage path planning problem, such as neural networks [3] and genetic algorithms [4]. Due to widely available optimization techniques, selection of a suitable algorithm is an important step, for both path planning and modeling the environment.

Using high spatial resolution imagery (HSRI), the plantation, boundaries and inter-row terrains become distinguishable, providing capabilities to discriminate and characterize the area [5]. However, considering the capability of HSRI with automatic processes needs new image processing methods, allowing the analysis and classification of vegetative and non-vegetative pixels from textured image. Fast Fourier Transform (FFT) and Gabor Filtering have been discussed in [5–7], which permits to delimit vineyards and to determine inter-row width and row orientation. The main problem of this method is related to boundary precision in particular when missing plants are present.

This work proposes a workflow that automatically generates coverage path plan for UGV using high resolution imagery acquired from UAV. Study area and data acquisition related details are discussed in Sects. 2 and 3 respectively. Data processing, mask generation and algorithms used in the study are discussed from Sect. 4 to 8 followed by the conclusion and future work.

## 2 The Case Study

The survey was conducted in a rural area near the town of Baldichieri d'Asti (AT), in Piedmont. This hilly area, called "Basso Monferrato Astigiano", is dominated by forests and vineyards [8]. The area of interest, property of the "Azienda Agricola Ciabot", is around 2 ha. It is not flat and it is characterized by the presence of obstacles inside and outside the area. Indeed, in the area there are few small houses and it is rounded by woods.

Image acquisition and 3D model were generated for the whole area, instead the path planning was focused in a small area with only one obstacle, to test, as first step, the path algorithm in a simple real case (Fig. 2).

## 3 Fieldwork and Data Acquisition for Base Map Generation

The data acquisition was carried out in October 2018. Due to the specific exposure of the area that causes a lot of shadows, two campaigns of acquisition were carried out in the same day (at 10.30 am and at 5.00 pm). Due to the lack of natural points, 16 plastic targets were placed on the ground, Ground Control Points (GCPs), to allow the model georeferencing. The coordinate of each target were measured using a Network Real Time Kinematic (NRTK) Global Navigation Satellite System (GNSS) technique [9]. The NRTK survey was performed by a SP80 Trimble GNSS receiver, using the real time correction of the permanent GNSS station of Canelli.

The UAV was chosen according to the size of study area, flight time and expected output of the survey. The used platform was a DJI Phantom 4 Pro, equipped with a RGB camera with focal length 8.8 mm, CMOS sensors 13.2 × 8.8 mm, pixel size of 2.4 m. The flights were planned using the open source software Mission Planner that connects the platform to the ground station. This tool was also used to set all the parameters of the flight plan, configuration of the flight (nadir images in North-South direction), height of the flight (33 m to have a Ground Sample Distance, GSD, of 8 m) and the overlap between different images (80% longitudinal and 60% transverse overlap).

The duration of each flight was about 15 min and a total of 261 images for the first flight and 209 images for the second flight were acquired.

## 4 Data Processing and Validation

The data processing consists on the 3D model generation to extract the orthophoto, the Dense Digital Surface Model (DDSM), and the Dense Digital Terrain Model (DDTM) [10]. The data processing was performed with the commercial software Agisoft PhotoScan 1.2.6. For each flight, all the images acquired were processed in a single block, following these steps: alignment and orientation of the cameras, images georeferencing by GCPs (horizontal and elevation accuracy are less than 2 cm), point cloud densification (about 80 million points) [11]. From the dense points cloud, the standard workflow of the software allows to generate the mesh and the DDSM. However, in this case a further step was made to generate also the DDTM. The software, by the "*Classify ground point*" tool, allows to perform a unsupervised classification of the point cloud to detect only the points on the terrain surface [12]. The classification was performed based on three main parameters:

- max angle: maximum slope of the ground within the scene (20°);
- max distance between the point above the ground and terrain model (0.05 m);
- cell size of the largest area that does not contain any ground points (1 $m^2$).

Each parameter was chosen according to the features of the area described in Sect. 2. The outputs of the whole procedure are DDTM, DDSM and an orthophoto in raster format with a cell size of 0.02 m, in WGS84 UTM 32N coordinate system.

## 5 Mask Generation for Path Planning

This step aims at the definition of a binary image where nonzero-value pixels represent vineyard lines, building and all the potential obstacles for the UGV path. A methodology based only on a 3D geometric information was developed. The map with potential obstacles and vineyard lines was carried out by an approach that considers the differences between DDSM and DDTM. This step was performed with the commercial software ArcMap using the "*Raster math-minus*". After that, the final map was obtained classifying the differences into two classes (greater than 0.50 cm, and smaller than 0.50 cm). Using the "*Raster Reclass*" tool, non-zero value was assigned to the differences greater than 0.50 cm which represent the presence of obstacles and zero value to the differences smaller than 0.50 cm which represent the ground. The process was applied to the data of both flights and the masks were merged to obtain a complete information.

## 6 Steps to Generate Path Plan

In the study area, there are three vineyard fields (Fig. 2), hereafter called parcels. Proposed algorithm automatically generates a path plan, which the UGV has to follow to cover a parcel.

A generic approach has been followed to develop the algorithm for kind of crops having considerable inter-row spacing for the UGV movement. The flowchart in Fig. 1 summarizes the adopted approach to develop the algorithm.

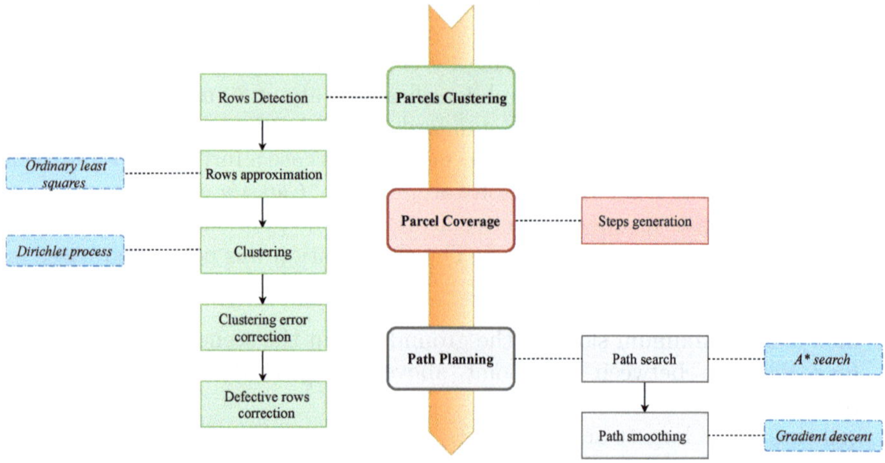


**Fig. 1.** Workflow.

## 7 Parcel Recognition and Rows Clustering

The first step of the algorithm deals with the detection of the rows that compose the vegetation. Each group of interconnected black pixels represents one canopy row cluster. The algorithm is based on the idea of visiting all the mask (Sect. 5) using a sliding window: starting from the first pixel in the upper left to the last pixel on the bottom right. The sliding window is defined with a proper width called $\rho$, which depends on the type of vegetation density. To take into account the possibility of having different distances in accordance to the plantation type a threshold value for $\rho$ is required: for example, maize rows are closer to each other with respect to vineyards, where vine rows are more distant. In the environment under investigation a $\rho = 7$ pixels has been used. As the acquisition was performed after the harvesting season of the vineyards, therefore, some missing part of the vine rows were detected during the clustering and mask generation process.

When dealing with more than one parcel per map, it may occur that two of them are separated by a narrow path where the farmer or the vehicle have the possibility to pass. For this reason, the problem is twofold and it is necessary to recombine only the segments that effectively belong to the same row. The idea is to perform the linear regression of every row, finding the line that best fits each cluster. Ordinary Least Squares ($OLS$) method has been chosen to make the parameterization. It determines the parameters of a linear function given a set of variables: it minimizes the sum of the squares of the differences between the observed dependent variable in the given dataset and those predicted by the linear function.

As the canopy rows are, generally, disposed as straight lines, the most recurrent form is a first order polynomial. Due to the limitation of the model, if the points to be approximated are distributed along a vertical line and the variance occurs mainly in the horizontal direction, the $OLS$ will not perform a good parametrization of the model. The problem occurs mainly because the $OLS$ is designed to fit points with varancies in the vertical direction. To solve this problem, before projecting the data points into the $OLS$ space, a rotation of 90° has been applied to the clusters which presented such *verticality*. Once all the rows have been identified, they need to be separated and grouped in parcels.

It is necessary to find a criterion which can be used to properly correlate only the rows of the same parcel, avoiding as much as possible errors.

Analyzing the literature [13,14], the biggest problem encountered is related to the fact that the number of parcels within a field is not known a priori, unless the final user provides this information to the algorithm. For this reason the simplest algorithms or very hardware-demanding, such as *K-means* or *DBSCAN*, have been discarded [15,16], deciding to implement clustering using the *Dirichlet Process* model.

Considering parcels separated by a free space of higher dimension with respect to the inter-row distance belonging to the same parcel, it is possible to use their *slope* to identify them uniquely. The result will be a collection of slopes close to each other if they belong to the same parcel. Due to the presence of different slopes, the resulting statistical distribution of angular coefficients will be in general multimodal. The assumption that all the data are generated from a mixture of finite number of distributions with unknown parameters allows to exploit the *Dirichlet Process*. At first hand it seems that slope feature is not enough restrictive to bind together the right rows. To solve this problem a further feature must be included, which can correlate the rows according to their mutual position: from each row the coordinates of its mean point have been picked, in such a way that distant mean points are unlikely to be recognized as part of the same cluster.

A drawback of this method derives from the fact that labelling errors can occur, affecting small piece of land where rows are not contiguous. The labelling correction has been performed by considering the majority of row's neighbors. This procedure is repeated recursively until no more label changes are present. The connection of two contiguous rows is done by extending the approximating line from the row's extremes until the neighbour is reached (Fig. 2).

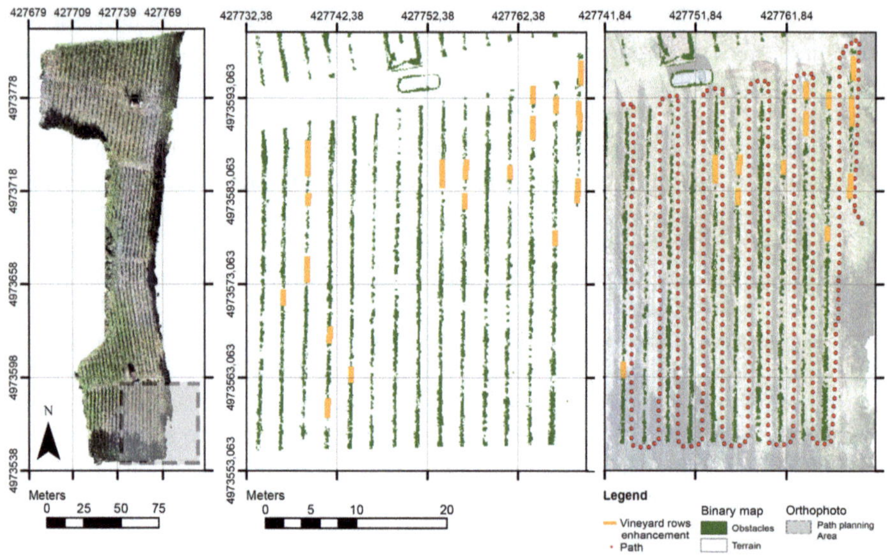


**Fig. 2.** Case Study. From left to right: overview, vineyard enhancement, path plan.

## 8 Steps Generation and Path Planning

To guarantee the coverage, the UGV needs to receive some nodes along the path between the starting position and the goal.

A possible solution to this problem is to exploit the line which best approximates the row and the distance between two close rows. With these information a grid that contains the parcel can be created, whose dimensions must be such that the robot can navigate free of incident. Hence a safety distance must be included when it is built. The rows are extended in both directions of at least the dimensions of robot in such a way that it can perform the curve. Then the average point is computed, meaning that when the robot will end to perform the curve, it will lie exactly in the middle of the two rows.

It is still necessary to define the order in which the robot must cover them. A sorting function has been developed: it minimizes the *euclidean* distance between the current position and the following one. According to the needs it can be used either to cover all the parcels which compose the map or only the single parcel.

The planner is then developed by implementing the $A^*$ algorithm [17], which uses as starting and goal nodes the points generated previously, and determines the optimal path between them avoiding obstacles. Then a smoothing of the $A^*$ trajectory is performed with the first-order iterative optimization *Gradient Descent* algorithm.

## 9 Conclusion and Future Work

In this work, an automatic coverage path planning for UGV for hilly vineyard environment is proposed. UAV imagery is used to obtain a DDTM and a DDSM, which are used to build the mask for the path planning. The experimental results show that the work as a whole presents some significant contribution in coverage path planning for UGV in the challenging environment like hilly vineyards that can be useful for the farmers to manage agricultural tasks.

However, we found some weaknesses when dealing with environments that deviate significantly from one parcel to another in the considered study area. This can be dealt by considering one parcel at a time as we did in this work. In future, the same approach can be applied and tested with different vineyards fields and by considering the acquisition campaigns in accordance with the peak phenological stage of the vineyard.

**Acknowledgement.** This work has been developed with the contribution of the Politecnico di Torino Interdepartmental Centre for Service Robotics PIC4SeR, https://pic4ser.polito.it and Azienda Agricola Ciabot, https://www.aziendaagricolaciabot.com/.

## References

1. Galceran, E., Carreras, M.: A survey on coverage path planning for robotics. Robot. Auton. Syst. **61**(12), 1258–1276 (2013)
2. Oksanen, T., Visala, A.: Coverage path planning algorithms for agricultural field machines. J. Field Robot. **26**(8), 651–668 (2009)
3. Yang, S.X., Luo, C.: A neural network approach to complete coverage path planning. IEEE Trans. Syst. Man Cybern. Part B (Cybern.) **34**(1), 718–724 (2004)
4. Hameed, I.A., Bochtis, D., Sørensen, C.A.: An optimized field coverage planning approach for navigation of agricultural robots in fields involving obstacle areas. Int. J. Adv. Robot. Syst. **10**(5), 231 (2013)

5. Da Costa, J.P., et al.: Delineation of vine parcels by segmentation of high resolution remote sensed images. Precis. Agric. **8**(1–2), 95–110 (2007)
6. Comba, L., et al.: Vineyard detection from unmanned aerial systems images. Comput. Electron. Agric. **114**, 78–87 (2015)
7. Comba, L., et al.: Unsupervised detection of vineyards by 3D point-cloud UAV photogrammetry for precision agriculture. Comput. Electron. Agric. **155**, 84–95 (2018)
8. SistemaPiemonte. Carta dei paesaggi agrari e forestali 1:250.000. http://www.sistemapiemonte.it/eXoRisorse/dwd/servizi/Agricoltura/ServiziGeografici/note.pdf
9. Cina, A., et al.: Network real time kinematic (NRTK) positioning description, architectures and performances. In: Satellite Positioning-Methods, Models and Applications. InTech (2015)
10. Vivalda, C., et al.: Forest wildfire risk mapping and the influence of the weather and geo-morphological input data, p. 171, June 2017. https://doi.org/10.1201/9781315210469-145
11. Aicardi, I., et al.: Recent trends in cultural heritage 3D survey: the photogrammetric computer vision approach. J. Cult. Herit. **32**, 257–266 (2018)
12. Grilli, E., Menna, F., Remondino, F.: A review of point clouds segmentation and classification algorithms. Int. Arch. Photogramm. Remote. Sens. Spat. Inf. Sci. **42**, 339 (2017)
13. Antoniak, C.E.: Mixtures of Dirichlet processes with applications to Bayesian nonparametric problems. Ann. Stat. **2**, 1152–1174 (1974)
14. Kallenberg, O.: Foundations of Modern Probability. Springer, New York (2006)
15. Wagstaff, K., et al.: Constrained k-means clustering with background knowledge. In: ICML, vol. 1, pp. 577–584 (2001)
16. Hartigan, J.A., Wong, M.A.: Algorithm AS 136: a k-means clustering algorithm. J. R. Stat. Soc. Ser. C (Appl. Stat.) **28**(1), 100–108 (1979)
17. Hart, P.E., Nilsson, N.J., Raphael, B.: A formal basis for the heuristic determination of minimum cost paths. IEEE Trans. Syst. Sci. Cybern. **4**(2), 100–107 (1968)

# Indoor Navigation Using Existing Infrastructure for Professional Service Robots

Denis Chikurtev[1(✉)], Kaloyan Yovchev[2], Nayden Chivarov[3], and Ivaylo Rangelov[3]

[1] Institute of Information and Communication Technologies, Bulgarian Academy of Sciences, Sofia, Bulgaria
dchikurtev@gmail.com
[2] Faculty of Mathematics and Informatics, Sofia University, Sofia, Bulgaria
k.yovchev@fmi.uni-sofia.bg
[3] European Polytechnical University, Pernik, Bulgaria
nshivarov@code.bg, rangelov.ivaylo@yahoo.com

**Abstract.** The share of service robots is increasing. Large number of those robots are the professional service robots such as customer service and logistics robots. One of the main requirements of these robots is the ability to navigate in an environment where a GPS system cannot be used. This article investigates existing methods for indoor localization and navigation. It proposes a new complete approach. This approach does not require the integration of a new indoor navigation system. Instead, the existing 24/7 video surveillance infrastructure, as well as the capabilities of the service robot, are considered. The proposed approach can reduce both the initial cost of integrating the robotic system as well as the maintenance costs. Subsequent maintenance can be done entirely remotely. The suggested approach is validated experimentally on a mobile robot.

**Keywords:** Indoor localization · Indoor navigation · Service robots · Vision-based navigation approach · Customer service robots

## 1 Introduction

Today's service robots are used to support day-to-day activities performed by people. These robots operate in a non-industrial setting and must be tailored and designed to their environment. Unlike industrial robots that work in a controlled and specifically-designed work environment, service mobile robots typically work in a dynamic and volatile environment.

Nowadays, the service robots' market is experiencing rapid development. According to forecasts of the International Robotics World Robotics, it is expected their market to reach around $ 37 billion between 2019 and 2021 [1]. One of the basic requirements for professional service robots is that they can work fully or semi-autonomously and have some degree of mobility. Some robots might have direct interaction with people. Robots that operate in a commercial, hospital, or storage


K. Berns and D. Görges (Eds.): RAAD 2019, AISC 980, pp. 231–239, 2020.
https://doi.org/10.1007/978-3-030-19648-6_27

environment usually have a direct interaction with humans. While those who perform dangerous tasks such as demolition or work in a hazardous environment must be able to perform their assignments remotely. They should be operating in an autonomous, semi-autonomous or manual mode of control.

This study will look at customer service and logistics robots. Customer service robots need to be able to communicate with a customer. They should help him/her find the product he/she is looking for or guide him/her through the surrounding environment. Using them would result in lower staff costs and a more enjoyable customer experience. Logistics robots are used to automate the process of packaging and delivering certain goods. They mostly work autonomously in a warehouse environment.

One of the basic requirements for these types of robots is the ability to navigate autonomously in their workspace. Typically, they operate in a closed environment where GPS cannot be used for robot localization and navigation [2, 3]. This necessitates the development of additional systems and methods for navigation localization in enclosed spaces [4, 5]. Currently, there are many such systems based on Bluetooth beacon technology, RFID or ultrasonic sensor systems [5–7]. All of them require additional equipment in the work environment with the necessary devices in order to use the technology. This increases the investment required, and also increases the cost of maintaining the embedded robotic system.

In paper [8], indoor navigation is realized based on a view-based approach using a camera mounted on the head of humanoid robot. The authors present a method to generate a stable view sequence is proposed based on the detection of optical flows.

In other researches, the authors propose navigation system that allowed a robot to complete 26.2 miles of autonomous navigation in a real office environment. They present the methods required to achieve this level of robustness, including an efficient Voxel-based 3D mapping algorithm that explicitly models unknown space [9]. Another research represents vision-based SLAM method to improve the navigation performance of mobile robot. It consists of a set of vision sensors and encoders which are equipped on mobile robot's wheels [10]. More papers for SLAM processes represents methods and systems for real-time localization [11, 12].

This research will offer a new complete approach for localization and navigation of a mobile service robot. The proposed approach does not require the deployment and installation of any additional indoor navigation system [13, 14]. Instead, the already available and maintained infrastructure in the work environment will be used. For the customer service and logistics robots it can be assumed that they work in an environment where 24/7 video surveillance system is installed. Camera footage can be processed in real time and used to locate and navigate the mobile service robot.

## 2 Case Study

For the purpose of this research, we will consider a customer service robot. This robot will have to operate in a hotel environment. Its main task will be to deliver orders from the restaurant to the hotel rooms. In other words, it should be a room service robot.

Most hotels are equipped with 24/7 video surveillance in their hallways. The robot will have to be localized and navigated through this available infrastructure.

Robot management will be implemented on two levels. The first level involves localizing, trajectory generating, and navigation of the robot. The second level includes motion control, recognition and avoidance of obstacles (Fig. 1).

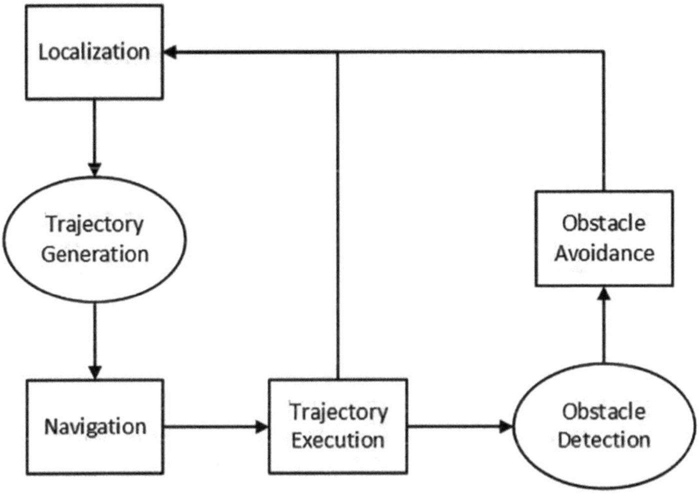


**Fig. 1.** Algorithm of the proposed method.

On the basis of the presented algorithm, a system will be built to allow successful and correct implementation of the set goals.

## 3 A New Approach to Indoor Navigation Based on Already Existing Infrastructure

The proposed new approach is to locate and navigate indoors using the already available 24/7 video surveillance infrastructure. This approach must take into account the particularities of video surveillance systems. Their most important deficiency is that the video stream will have some delay during processing, which may even be more than 1 s. In this situation, it is important for the mobile robot to be equipped with additional sensors to take into account the presence of obstacles and avoid collision with other objects. Corridor walls geometry and detection can also be taken into consideration by the robot control system in order to execute straight-line movements.

The robot must be guided by commands given by the navigation system but cannot rely on them to avoid obstacles. For this, the robot must be equipped with at least distance measurement sensors - ultrasonic or infrared. Also, in order to determine the robot orientation, it is imperative that the robot is equipped with an orientation sensor. Through it, the robot will monitor its orientation in real time and execute correctly the movement commands. The robot must also have feedback from the drive motors. In this way, it could perform the planned trajectory autonomously without relying on the external navigation system in real time. This results in a clear separation between the trajectory planning and the trajectory execution process. This allows the method proposed below to be responsible for locating, planning, and submitting a trajectory. This trajectory will then be autonomously executed by the robot. During execution, the navigation system will monitor for an unacceptable deviation from the trajectory and may send trajectory corrections to the robot. This will not have to be done in real time, because the robot will be responsible for detecting and avoiding immediate obstacles. The architecture of the proposed approach is shown in Fig. 2.

### 3.1 Preliminary Preparation for Use of the Available Infrastructure for Robot Localization and Navigation

In order to be used for localization and navigation, the available infrastructure needs to be initially prepared. A two-dimensional workspace map must be created. From the video stream of each available surveillance camera, a region of interest (ROI) should be selected and matched to a particular area of the two-dimensional map. Then, the perspective of the selected area must be corrected and oriented towards the zero direction of the orientation sensor mounted on the robot. This will now allow a robot positioning in a given video stream with a correct orientation in the workspace.

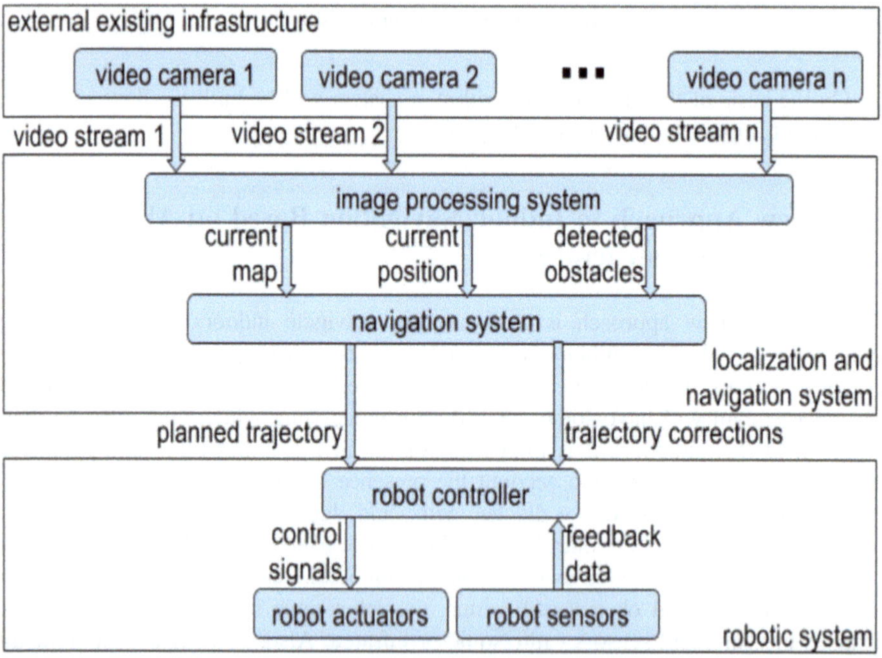


**Fig. 2.** Architecture of the proposed approach

For the subsequent location of the robot, it is necessary to take from each video stream a zero frame that does not contain neither the robot, nor any other objects or obstacles.

In other similar navigation and localization methods, it is also necessary to create a map beforehand, and in some techniques, it is even required for the robot to physically crawl the whole space and create a calibrated map itself [15, 16]. With new suggestions, the map can be created without the need for the robot to crawl through space.

With the proposed method, the map can also be created fully remotely, and subsequent maintenance and correction can also be done remotely. In other methods, damage to a sensor would require physical replacement, and perhaps even additional on-site calibration.

### 3.2 Initial Localization of a Robot

For the initial robot localization, it is necessary to review the current frames of all video streams and then compare those frames to the preliminary captured zero frames. The robot will be located in the frame with the most significant difference. By finding the difference position in both frames, the exact location of the robot can then be determined. In addition, the shape, color and size of objects will be checked if there are more than one different area in the image. Because if there are multiple objects appearing in the new frame, we need additional recognition criteria to determine which of them is the robot. This will allow us to recognize pre-set obstacles, which will help in trajectory generation.

### 3.3 Planning a Trajectory

The trajectory can be represented as a sequence of commands, with the simplest possible forward/reverse/left/right rotation (allowing you to adhere to a certain distance from the wall when there is no obstruction). Trajectory planning itself is trivial when the starting position, end position, and geometric features of the map are known [17]. Planning process can also take into consideration any detected and known obstacles.

### 3.4 Detecting Fixed and Movable Obstacles

Fixed obstacles can be easily detected by finding the difference between zero and current frame from each video stream. Since the CCTV system works with some delay, the robot cannot rely on it to detect movable obstacles in real time. This must be done through the robot's sensors. When a particular obstacle is detected, the robot must avoid it. This is most likely to happen if the robot stops and waits for a correction of the planned trajectory from the navigation system.

### 3.5 Monitoring and Correction of the Trajectory Execution

The monitoring is carried out using the following method: Depending on the current position of the robot, the video stream corresponding to that position as well as the video streams of the neighboring zones are monitored. It can also be used to correct the already planned trajectory for any newly detected obstacles. The monitoring also allows for any deviations from the desired trajectory to be corrected.

### 3.6 Overlapping of the Observed Areas

It is worth noting that it is not strictly necessary for the video surveillance to cover the entire map. This is because the robot is equipped with enough sensors to allow it to make a transition between two observed areas. However, the robot will have to autonomously avoid obstacles in the invisible areas. Additionally, both the start and endpoints of the desired trajectory must be within the area covered by the cameras.

The proposed approach is suitable for use by almost all available mobile platforms and does not require the use of costly sensors and an indoor navigation system.

## 4 Experiments and Results

The proposed approach has been validated experimentally. For this purpose, a mobile robot equipped with compass and distance sensors was used. The work area is monitored by two video cameras.

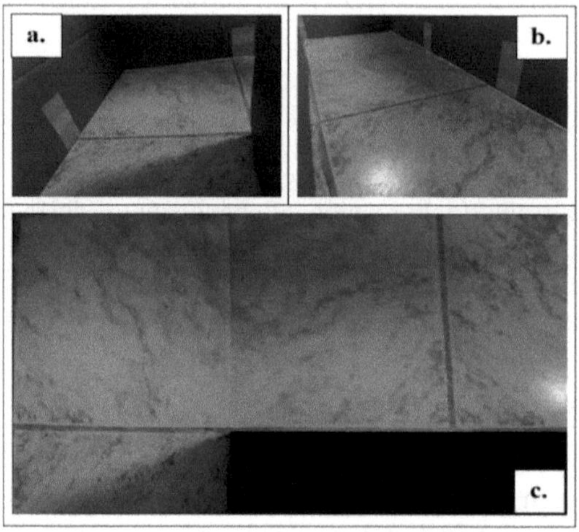


**Fig. 3.** Preliminary preparation: a. zero frame from first camera; b. zero frame from second camera; c. two-dimensional map matched to the corresponding camera frames.

Zero frames from the video streams are shown in Fig. 3a and b. The corresponding map created after selecting specific regions of interests and applying perspective transform on them is shown in Fig. 3c.

The next Fig. 4 shows the robot's starting position and its location on the map. Tests were also carried out in which sudden obstacles appeared. In all trials, the mobile platform managed to successfully reach the desired end position.

Due to delay in the video streams the robot was not able to execute precisely the planned trajectory (Fig. 5). However, with the proposed navigation method the robot was able to reach the goal position and to avoid collision with any obstacles.

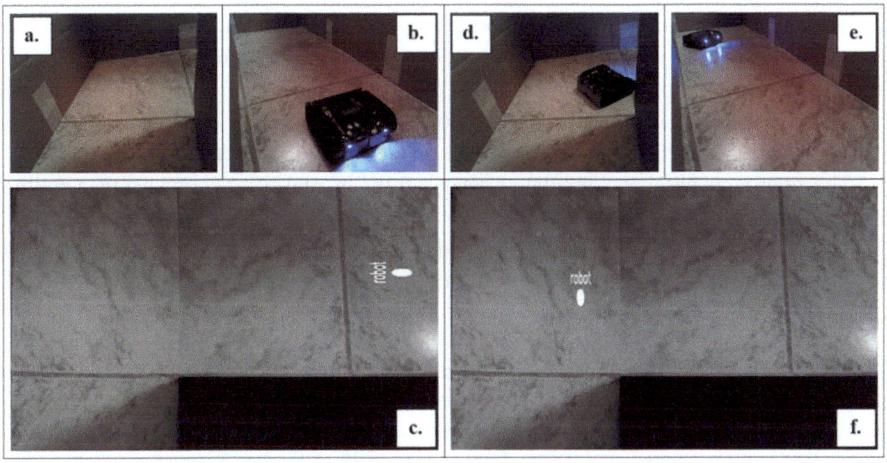


**Fig. 4.** Starting position: a. video frame captured from first camera; b. video frame captured from second camera; c. the computed starting position of the mobile robot is marked with white circle on the map; Intermediate position: d. frame captured from first camera; e. frame captured from second camera; f. the computed intermediate position is marked with white circle on the map.

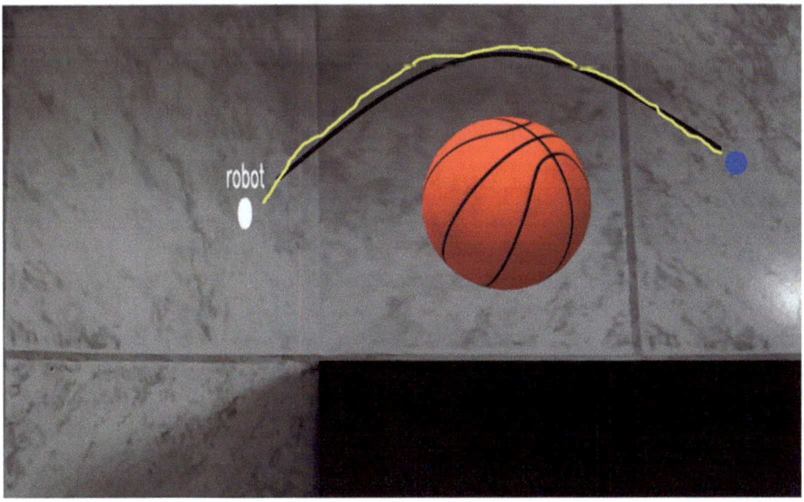


**Fig. 5.** Planned (black curve) and executed (yellow curve) trajectory for collision avoidance.

## 5 Conclusion

This publication offers a new complete approach for solving the problem of robots localization and navigation in an indoor area where GPS can not be used. The proposed approach does not require the installation and subsequent maintenance of additional navigation systems. Instead, it is supposed to use the already established system for 24/7 video surveillance. The paper describes the necessary requirements for the service robot that allow the approach to be used. A great advantage of this method is the ability for a remote maintenance of the localization and navigation system. This would also significantly reduce the cost of subsequent service robotic system maintenance.

The approach has been validated successfully in real-life testing using a mobile robot.

The results obtained are very good and the approach can be successfully applied into the control of a professional service robot.

**Acknowledgement.** The research presented in this paper was partially supported by the Bulgarian National Science Fund under the contract No. DN 07/23 – 15.12.2016 "Telecontrolled Service Robots for Increasing the Quality of Life of Elderly and Disabled" and partially by contract № KP-06-M27/1 – 04.12.2019 "Research and Development of Innovative, Intelligent Information and Communication Technologies for Control of Service Robots".

## References

1. IFR Press Conference, 18 October 2018, Tokyo (2018). https://ifr.org/downloads/press2018/WR_Presentation_Industry_and_Service_Robots_rev_5_12_18.pdf
2. Maaref, H., Barret, C.: Sensor-based navigation of a mobile robot in an indoor environment. Robot. Auton. Syst. **38**(1), 1–18 (2002). https://doi.org/10.1016/S0921-8890(01)00165-8. ISSN 0921-8890
3. Ivanov, V., Garvanov, I.: Implementation of jumping average window (JAW) in FPGA devices. Technics. Technologies. Education. Safety. 30.5–02.06.2018 Veliko Tarnovo, 2, Scientific technical union of mechanical engineering "Industry-4.0" (2018). ISSN:2535-0315, 194–198
4. Biswas, J., Veloso, M.: WiFi localization and navigation for autonomous indoor mobile robots. In: 2010 IEEE International Conference on Robotics and Automation, Anchorage, AK, pp. 4379–4384 (2010). https://doi.org/10.1109/robot.2010.5509842
5. Mester, G., Rodić, A.: Sensor-based intelligent mobile robot navigation in unknown environments. Int. J. Electr. Comput. Eng. Syst. **1**(2), 55–62 (2010)
6. Park, S., Hashimoto, S.: Autonomous mobile robot navigation using passive RFID in indoor environment. IEEE Trans. Industr. Electron. **56**(7), 2366–2373 (2009). https://doi.org/10.1109/TIE.2009.2013690
7. Kulyukin, V., Gharpure, C., Nicholson, J., Pavithran, S.: RFID in robot-assisted indoor navigation for the visually impaired. In: 2004 IEEE/RSJ International Conference on Intelligent Robots and Systems (IROS) (IEEE Cat. No. 04CH37566), Sendai, vol. 2, pp. 1979–1984 (2004). https://doi.org/10.1109/iros.2004.1389688
8. Ido, J., Shimizu, Y., Matsumoto, Y., Ogasawara, T.: Indoor navigation for a humanoid robot using a view sequence. Int. J. Robot. Res. **28**(2), 315–325 (2009). https://doi.org/10.1177/0278364908095841

9. Marder-Eppstein, E., Berger, E., Foote, T., Gerkey, B., Konolige, K.: The office marathon: robust navigation in an indoor office environment. In: 2010 IEEE International Conference on Robotics and Automation, Anchorage, AK, pp. 300–307 (2010). https://doi.org/10.1109/robot.2010.5509725
10. Lee, Y.J., Sung, S.: Vision based SLAM for mobile robot navigation using distributed filters. In: Barrera, A. (ed.) Mobile Robots Navigation. InTech (2010)
11. Labbé, M., Michaud, F.: Long-term online multi-session graph-based SPLAM with memory management. Auton. Robots **42**(6), 1133–1150 (2018). https://doi.org/10.1007/s10514-017-9682-5
12. Labbé, M., Michaud, F.: Online global loop closure detection for large-scale multi-session graph-based SLAM. In: 2014 IEEE/RSJ International Conference on Intelligent Robots and Systems, Chicago, IL, pp. 2661–2666 (2014). https://doi.org/10.1109/iros.2014.6942926
13. Blanc, G., Mezouar, Y., Martinet, P.: Indoor navigation of a wheeled mobile robot along visual routes. In: Proceedings of the 2005 IEEE International Conference on Robotics and Automation, Barcelona, Spain, pp. 3354–3359 (2005). https://doi.org/10.1109/robot.2005.157062
14. Tomono, M., Yuta, S.: Mobile robot navigation in indoor environments using object and character recognition. In: Proceedings 2000 ICRA. Millennium Conference. IEEE International Conference on Robotics and Automation. Symposia Proceedings (Cat. No. 00CH37065), San Francisco, CA, USA, vol. 1, pp. 313–320 (2000). https://doi.org/10.1109/robot.2000.844076
15. Se, S., Lowe, D., Little, J.: Vision-based mobile robot localization and mapping using scale-invariant features. In: Proceedings 2001 ICRA. IEEE International Conference on Robotics and Automation (Cat. No. 01CH37164), Seoul, South Korea, vol. 2, pp. 2051–2058 (2001)
16. Zaman, S., Slany, W., Steinbauer, G.: ROS-based mapping, localization and autonomous navigation using a Pioneer 3-DX robot and their relevant issues. In: 2011 Saudi International Electronics, Communications and Photonics Conference (SIECPC), Riyadh, pp. 1–5 (2011)
17. Nagarajan, U., Kantor, G., Hollis, R.L.: Trajectory planning and control of an underactuated dynamically stable single spherical wheeled mobile robot. In: 2009 IEEE International Conference on Robotics and Automation, Kobe, pp. 3743–3748 (2009)

# Online Trajectory Planning and Collision Avoidance for a Group of Robots Using Distributed Model Predictive Control

Giuliano Costantini(✉), Ramin Rostami, and Daniel Görges

Department of Electrical and Computer Engineering, University of Kaiserslautern,
Erwin-Schrödinger-Straße 12, 67663 Kaiserslautern, Germany
{costantini,rostami,goerges}@eit.uni-kl.de

**Abstract.** In this paper the trajectory planning problem for unicycle robots is studied. This problem is particularly difficult to solve online due to its inherently nonlinear and nonconvex structure. The first aim of the paper is to overcome these difficulties using model predictive control (MPC) when a single robot is considered. Several requirements must be fulfilled by the optimal trajectory, such as respecting physical limitations as well as track constraints. The second aim of the paper is to extend the results to a group of robots using distributed MPC providing formation control and inter-robot collision avoidance.

**Keywords:** Autonomous vehicles · Unicycle robots · Trajectory planning · Collision avoidance · Distributed model predictive control

## 1 Introduction

Trajectory planning is a fundamental task for autonomous mobile robots. In the industry an increasing number of applications require robots to rapidly generate an optimal trajectory while moving in an unknown environment. Planning a trajectory basically means to find the set of future control actions which transfers the robot between two different configurations. This control problem has been a long-standing research topic because it poses many challenges [1]. In particular Brockett's conditions prove that there is a whole class of systems including nonholomonic mobile robots for which no smooth time-invariant feedback law exists able to regulate the tracking error dynamics to zero. Different methods can be used to overcome this weak controllability limitation including time-varying feedback control [2] or discontinuous and piecewise continuous control [3–5] which are mostly based on Lyapunov techniques. Recently model predictive control (MPC) has been proven to be a good candidate for solving the motion planning problem [6,7]. MPC is an advanced control method which combines the advantages of open loop optimization with the power of feedback control, particularly robustness against disturbances and uncertainty. The controller uses a discrete-time model to predict the behaviour of the system over a finite time horizon.


K. Berns and D. Görges (Eds.): RAAD 2019, AISC 980, pp. 240–249, 2020.
https://doi.org/10.1007/978-3-030-19648-6_28

An optimization is performed at each discrete time step yielding effectively a static discontinuous nonlinear feedback law. The major advantage of this approach is the possibility to add different kind of constraints in the optimization phase. This is particularly important when the robot is moving in an unknown environment where the planned trajectory must avoid colliding with obstacles.

The first aim of this paper is to find a minimal set of constraints which ensure the control target to be achieved while keeping the prediction horizon small, thus only using a small number of optimization variables. This is actually a challenge because the typical constraints for autonomous robots are nonlinear and most critically nonconvex. The main contribution of this paper is the design of an iterative model predictive control strategy in which a sequence of quadratic optimization problems is solved and the resulting trajectory is refined at each step. Commonly only a small number of iterations is required for convergence.

The second contribution of the paper is the practical solution of a frequent tracking problem which requires the robot to move on a curved track. In a typical industry application robots are employed for goods transportation in a closed spaced where safety tracks have been marked on the ground. It is assumed that by means of onboard sensors the robot is able to detect in its field of view the lines delineating the track. This information is conveyed to the control algorithm where the iterative methods imposes tighter constraints to keep the robot on track. Similar situations arise e.g. in autonomous driving.

The final part of this work extends the consideration above to the case of multiple robots moving in the same environment. Problems such as coordination [8] and flocking [9] are of great interest. The most popular methods here are the use of potential functions and consensus techniques. Recently the distributed version of MPC has been proposed to solve robot coordination tasks in [10,11]. Distributed MPC has been developed to control a network of dynamical agents. Here the global optimization problem is composed of simple local problems coupled together through the objective or constraints [12]. The most common strategy to solve this problem is to split the optimization over the network and the agents' task is to alternatively optimize and exchange information until a consensus is reached. This is clearly an iterative method and can be easily embedded in the iterative MPC developed in this work for a single robot. The third contribution of this paper is a tailored distributed model predictive control method for mobile robots.

The paper is organized as follows. The remainder of Sect. 1 gives an introduction of the model being used. The main algorithm is developed in Sect. 2 for a single robot and is extended to multiple robots in Sect. 3. Conclusions are finally drawn in Sect. 4. In each of the main sections several examples and simulation results are given to illustrate the theoretical derivations.

### 1.1 Robot Dynamics

For the modeling of the robot, the *unicycle model* is employed which provides simplicity and yet captures the typical nonlinear dynamics of a mobile robot. Furthermore this model can be easily linearized via feedback linearization.

$$\underline{\text{nonlinear model}} \qquad\qquad\qquad\qquad \underline{\text{linear model}}$$

$$\begin{array}{l} \dot{x} = v\cos(\theta) \\ \dot{y} = v\sin(\theta) \\ \dot{\theta} = \omega \\ \dot{v} = \alpha \end{array} \quad \xleftrightarrow[\text{Feedback Linearization}]{} \quad \begin{array}{l} \dot{x} \;\, = v_x \\ \dot{y} \;\, = v_y \\ \dot{v}_x = a_x \\ \dot{v}_y = a_y \end{array} \tag{1}$$

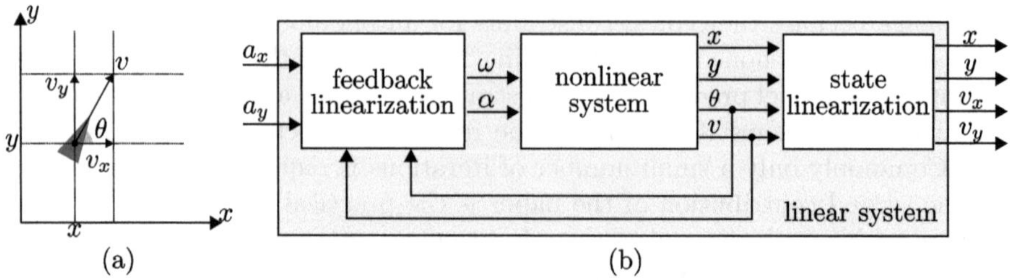


**Fig. 1.** (a) Robot configuration (b) Feedback linearization

As shown in Fig. 1(a), the robot configuration is completely described by the quadruple $(x, y, \theta, v)$ indicating position, orientation and velocity, respectively. The inputs to the system $(\omega, \alpha)$ are the angular velocity and the linear acceleration respectively. Equivalently, when $v \neq 0$, it is possible to describe the robot configuration through the linear state $(x, y, v_x, v_y)$ and input $(a_x, a_y)$. The relation between these two descriptions is given by the transformation

$$\begin{array}{ll} v_x = v\cos\theta & \quad a_x = \cos\theta\,\alpha - v\sin\theta\,\omega \\ v_y = v\sin\theta & \quad a_y = \sin\theta\,\alpha + v\cos\theta\,\omega \end{array} \tag{2}$$

which can be inverted when $v \neq 0$, yielding to the feedback linearization law

$$\begin{array}{l} \omega = \frac{1}{v}(-\sin\theta\,a_x + \cos\theta\,a_y) \\ \alpha = \cos\theta\,a_x + \sin\theta\,a_y. \end{array} \tag{3}$$

The final structure of the linearized system is shown in Fig. 1(b).

In order to employ a model predictive control scheme, the linear system $\dot{\xi} = A_c\xi + B_c u$ in (1) with $\xi = [x\ y\ v_x\ v_y]^T$ and $u = [a_x\ a_y]^T$ needs to be approximated by a discrete time model $\xi_{k+1} = A\xi_k + Bu_k$. In the following a first-order approximation is performed, i.e. $A = I + hA_c$ and $B = (hI + \frac{h^2}{2}A_c)B_c$ with $h$ as fixed step size.

$$\begin{array}{ll} x_{k+1} & = x_k \quad + hv_{x,k} + \frac{1}{2}h^2 a_{x,k} \\ y_{k+1} & = y_k \quad + hv_{y,k} + \frac{1}{2}h^2 a_{y,k} \\ v_{x,k+1} & = v_{x,k} + ha_{x,k} \\ v_{y,k+1} & = v_{y,k} + ha_{y,k} \end{array} \tag{4}$$

To ease the notation, in what follows the equations in (4) are rewritten in a compact form by defining the vectors $p = [x\ y]^T$, and $q = [v_x\ v_y]^T$

$$\begin{aligned} p_{k+1} &= p_k + hq_k + \tfrac{1}{2}h^2 u_k \\ q_{k+1} &= q_k + hu_k \end{aligned} \tag{5}$$

## 1.2 Model Constraints

The controller is required to fulfil a certain number of constraints imposed by physical limitations such as actuator saturation or keeping the robot within the track boundaries. Furthermore it is also the responsibility of the controller to avoid the singularity condition $v = 0$ which prevents the feedback linearization of the system. Additionally it is also required that the robot plans its trajectory in the forward direction which means that a minimal positive velocity is desired. Most of these requirements are defined directly on the nonlinear system model and for this reason they are inherently nonlinear and possibly nonconvex. In order to keep the underlying MPC optimization problem simple it is paramount to redefine such constraints as linear ones. The following box constraints are considered here on the inputs $|\alpha| \leq \bar{\alpha}$, $|\omega| \leq \bar{\omega}$ and velocity $v_{\text{min}} \leq v \leq v_{\text{max}}$. From the last equation in (3) it results $|\alpha| \leq \sqrt{a_x^2 + a_y^2} = \|u\|$, and thus it suffices to bound the norm of the linear system input. This is equivalent to a quadratic constraint which can be approximated by a set of hyperplanes as depicted in Fig. 2(a) using 16 hyperplanes. The angular velocity can be indirectly bounded by restricting the velocity vector at next time step $q_{k+1}$ according to the following proposition.

**Proposition 1.** *The constraint* $|\omega_k| \leq \bar{\omega}$ *is equivalent to* $A_{q_k} q_{k+1} \leq b_{q_k}$ *where*

$$A_{q_k} = A_{q_k}(\theta_k) = \begin{bmatrix} \sin\theta_k & \cos\theta_k \\ \sin\theta_k & -\cos\theta_k \end{bmatrix} \qquad b_{q_k} = b_{q_k}(v_k) = hv_k\,\bar{\omega}.$$

*Proof.* From (3) it results $\omega_k = \frac{1}{v_k}[-\sin\theta_k\ \cos\theta_k]u_k$. Substituting $u_k = \frac{q_{k+1}-q_k}{h}$ from (5) yields $\omega_k = \frac{1}{hv_k}[-\sin\theta_k\ \cos\theta_k]q_{k+1}$ since $q_k = v_k[\cos\theta_k\ \sin\theta_k]^T$. The constraint inequality $|\omega_k| \leq \bar{\omega}$ finally implies two hyperplanes for the upper and lower bound, i.e. the polytope $A_{q_k} q_{k+1} \leq b_{q_k}$. □

Graphically this constraint corresponds to a tube running in the direction of the current velocity $q_k$ as shown in Fig. 2(b). Notice that the higher the velocity $v_k$ the tighter this constraint is, corresponding to the fact that at higher velocity it is harder to steer the vehicle. Finally in order to enforce the velocity constraint $v_{\text{min}} \leq v_k \leq v_{\text{max}}$ the norm of $q_{k+1}$ is bound similar to the input $u_k$ and further a new hyperplane orthogonal to the driving direction and with some offset is used to maintain a minimum velocity.

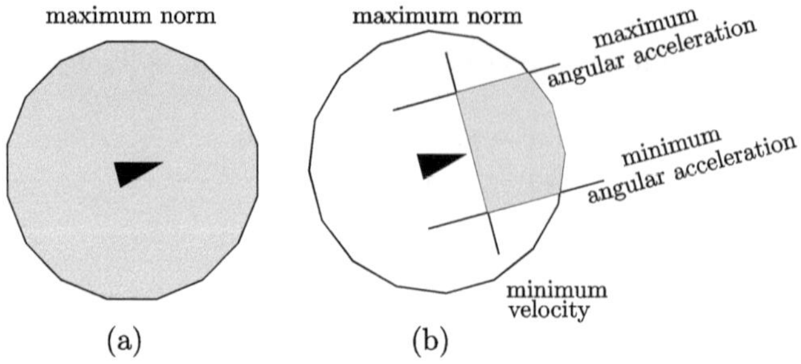


**Fig. 2.** (a) Input $u_k$ constraint (b) Velocity $q_{k+1}$ constraint

## 2 Trajectory Planning

In the this section an iterative model predictive control scheme is presented which aims to provide an optimal trajectory for a typical robot tracking problem. First it is considered that the robot is able to move anywhere in the plane. Later this assumption is removed and the robot is restricted to move along a track.

### 2.1 Model Predictive Control

In model predictive control (MPC) a finite horizon optimal control problem is solved at each time instant and the first element of the optimal solution is applied to the system. The optimization problem depends on the current system state $\xi_k$ and therefore this control strategy provides in general a nonlinear feedback control law $u_k = f(\xi_k)$. The control objective is encoded into a cost function which should be minimized. Although there exist many MPC schemes, in the present paper a linear MPC is employed where the underlying optimization is recast as a quadratic programming problem for which well-established methods exists. The problem to be solved at each time instant is given by

$$\begin{array}{lll} \underset{\xi_{\tau+1},\, u_\tau}{\text{minimize}} & \sum_{\tau=0}^{N-1} l_\tau(\xi_{\tau+1}, u_\tau) & \\ \text{subject to} & \xi_0 = \bar{\xi} & \\ & \xi_{\tau+1} = A\xi_\tau + Bu_\tau, & \tau = 0 \ldots N-1 \\ & \xi_{\tau+1} \in \mathscr{X}_{\tau+1} \quad u_\tau \in \mathscr{U}_\tau, & \tau = 0 \ldots N-1 \end{array} \tag{6}$$

where $N$ is the prediction horizon and $\bar{\xi} = \xi_k$ is the current measured or estimated state. Note that $(\xi_k, u_k)$ denote the state and input at real time $t = hk$ and the $(\xi_\tau, u_\tau)$ denote the predicted state and input which are the optimization variables (with a slight abuse of notation). The function $l_\tau$ is called stage cost while the sets $\mathscr{X}_\tau, \mathscr{U}_\tau$ are the stage constraints. In order for problem (6) to represent a convex quadratic program, it is required that $l_\tau$ is a convex quadratic function and that $\mathscr{X}_\tau, \mathscr{U}_\tau$ are convex polytopic sets. In the following these quantities are described in detail for the unicycle model.

### 2.2 Target Tracking

A tracking problem is defined by a reference configuration $r_1 = [r_x \; r_y \; r_{v_x} \; r_{v_y}]$ which it is desired to be reached. It is assumed that the reference velocity is larger than zero to avoid the singularity condition. A reference trajectory is easily generated from this configuration using the linear model with zero input $r_{\tau+1} = A \; r_\tau$ for the whole horizon $N$. The cost function is simply the weighted squared norm difference of the planned and reference trajectory

$$l_{\tau,1} = \|\xi_\tau - r_\tau\|^2_{Q_k} + \|u_\tau\|^2_{R_k}$$

where $Q_k$ and $R_k$ are positive definite matrices which penalize the state deviation and the input. In order to keep the horizon small a second term is added which is used to attract the trajectory onto the hyperplane where the reference trajectory lies as depicted in Fig. 3(a) which is given by $l_{\tau,2} = \|a_r^T \xi_\tau - b_r\|^2$.

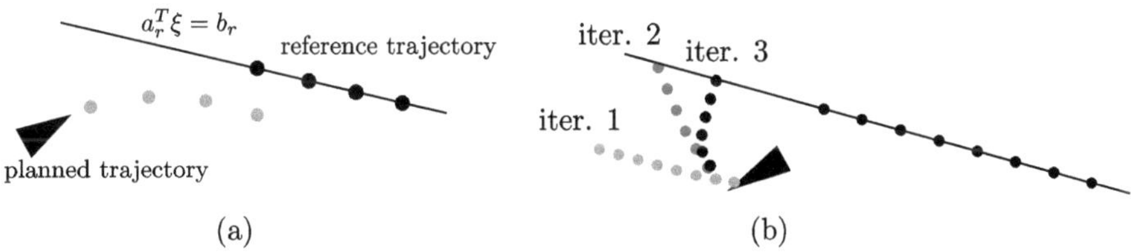


**Fig. 3.** (a) Trajectory planning (b) Iterative MPC

### 2.3 Constraints

The nonlinear constraints of the unicycle robot have been already recast into simple polytopes. The exact description results as

$$A^u_\tau u_\tau \leq b^u_\tau, \qquad A^\xi_\tau \xi_\tau \leq b^\xi_\tau$$

where $(A^u_\tau, b^u_\tau)$ describes the linear approximation of the quadratic norm bound while $(A^\xi_\tau, b^\xi_\tau)$ contains different terms including the minimum velocity, a quadratic norm bound approximation, and the angular velocity. The latter poses some difficulties because at each prediction step $\tau + 1$ the hyperplane depends nonlinearly on the velocity vector at the previous prediction step $\tau$ (cf. Preposition 1).

To cope with such complication, an iterative MPC algorithm has been devised which attempts at each step to improve on the previous trajectory plan as shown in Fig. 3(b). The following pseudocode describes this iterative procedure.

**Algorithm - Iterative MPC**

**Input:** Current velocity vector $\bar{q} = [v_{x,k},\ v_{y,k}]^T$
Define $q_\tau = \bar{q} \qquad \tau = 0 \ldots N-1$
**Loop**
  Define the hyperplanes $(A_{q_\tau}, b_{q_\tau})$ as in Proposition 1
  Solve the linear MPC problem (6)
  **Stop** if trajectory has not changed significantly
  Redefine $q_\tau$ according to the new MPC trajectory plan
**end**
**Output:** Optimal input $u^*$

It can be readily checked that each new optimization problem contains the previous solution as a feasible decision vector. This means that the objective can only improve at each step and since the problem is strictly convex it must converge to a solution. Moreover it is possible to warm start each iteration for a faster convergence of the underlying solver. The number of iterations of this procedure depends on the horizon length $N$, but typically converges in few steps.

In Fig. 4 the resulting final trajectory is shown for two different scenarios where the horizon $N = 8$ has been selected.

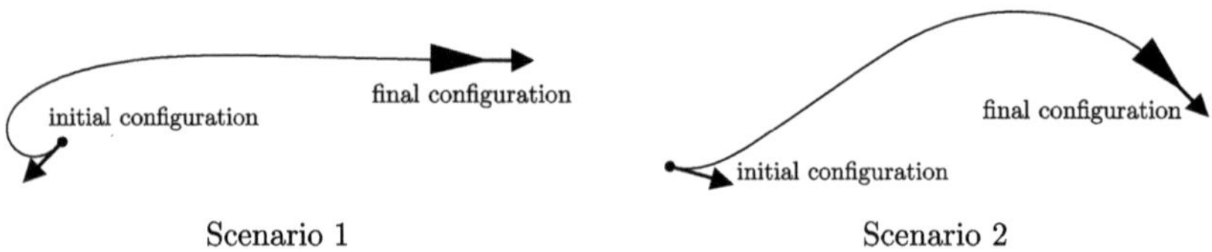


**Fig. 4.** Planned trajectory for two different scenarios

## 2.4 Track Boundaries

This section introduces a new challenge given by the presence of boundaries in which the robot is constrained to move. This information is provided to the controller as a set of coordinates $(x, y)$ which is typically quite huge. Including all this data in the MPC optimization problem is unviable because of the enormous amount of computation which would be required. Instead, an iterative procedure is again devised which can be directly embedded in the previously described algorithm. Basically the first iteration is run without considering the track boundary. This step gives some trajectory which might get off the track. Next for each point in the planned trajectory, the closest point on the track is searched and a corresponding separating hyperplane is defined $a_\tau^T p_\tau \leq b_\tau$ where $p_\tau = [x_\tau\ y_\tau]^T$ is the robot position vector. The affine factor $b_\tau$ can be made smaller (tightening) in order to give some margin from the real boundary. The hyperplane norm is obtained by numerically differentiating the track boundaries using the central difference method. Next the problem is solved again with only $N$ new constraints which are then directly replaced in the successive iterations. A simple example which illustrates this procedure is shown in Fig. 5(a). Simulation results show that if the curve is sufficiently smooth and the angle between the curve and the first planned trajectory is small (typically $\leq \frac{\pi}{6}$), then after a few iterations the procedure converges to a final trajectory which remains within the track boundaries. A complete example where the robot drives through a circuit is shown in Fig. 5(b). Here it is assumed that some points are defined as target checkpoints to avoid deadlock situations. These points might be provided for instance by a GPS navigation system. The set of separating hyperplanes calculation is conducted directly inside the loop of the *Iterative MPC* algorithm after the MPC problem is solved.

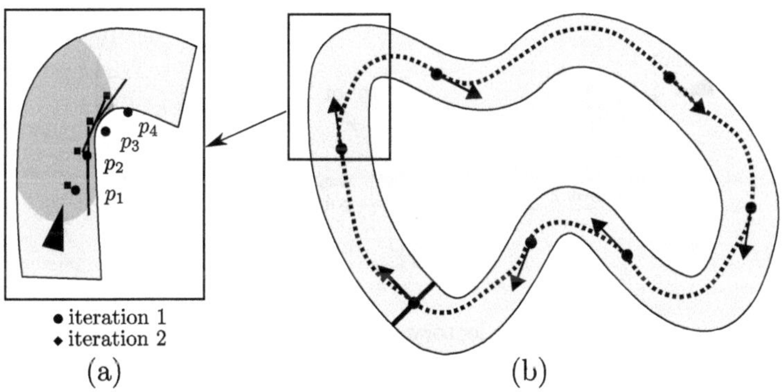


**Fig. 5.** (a) Separating hyperplanes and track boundaries (the shaded area shows what the robot detects) (b) Circuit scenario with some checkpoints provided to the controller

## 3 Distributed Model Predictive Control

In this section the control algorithm is extended to a group of mobile robots. The objective of the controllers is to keep the vehicles in a close formation while preventing inter-robot collision. The distributed version of MPC offers a natural choice to accomplish this task. In distributed MPC the optimization problems of the involved agents are typically coupled and cannot be entirely solved locally. If the agents are able to communicate over a network, then distributed optimization methods can be employed to find the optimal global solution. These are iterative procedures in which the agents solve a local optimization problem and communicate some part of the result to their neighbours. This information is then used to adjust the optimization problem so that at the end a consensus is reached. There are many distributed optimization methods which are mostly based on dual decomposition like the Alternating Direction Method of Multipliers (ADMM). Although these can be successfully employed for the problem at hand, a new method is proposed in this paper which is tailored to the particular structure considered and results to be faster and simpler to implement.

The objective of the robots is augmented with a third term which tries to minimize the distance between the robots

$$l^i_{3,\tau} = \sum_{\substack{j=1 \\ j \neq i}}^{M} \Delta_{ij} \|\xi_{i,\tau} - \xi_{j,\tau}\|^2$$

where $M$ is the number of robots and the index $i$ refers to the current agent. The parameters $\Delta_{ij}$ can be used to describe the shape of the formation.

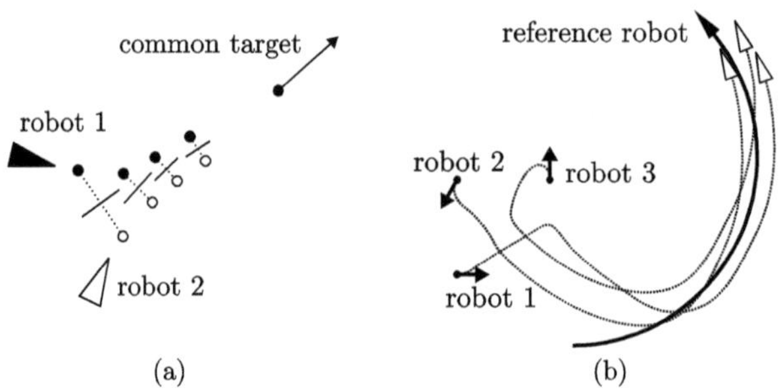


**Fig. 6.** (a) Distributed trajectory planning (b) Simple circle tracking

The proposed method consists in approximating the terms $\xi_{j,\tau}$ with the values at the previous algorithm iteration. The first iteration is run without such information and the agents selfishly decide their optimal solution. These trajectory predictions are then communicated over the network and the augmented cost function is formed. To avoid collision a separating hyperplane with some offset is also defined for each received position $p_{j,\tau}$. This procedure is illustrated in Fig. 6(a). The proposed strategy is particularly simple to implement and it avoids the need of augmenting the local decision variables with neighbour copies as in consensus ADMM. Furthermore since each local controller is itself already using an iterative method, it is straightforward to embed this new procedure as a further step in the main control loop. A simple simulation result is shown in Fig. 6(b), where each robot tracks a common target which moves in circle.

## 4 Conclusion

A novel iterative MPC strategy to solve the trajectory planning problem of mobile robots has been introduced. The iterative nature of this technique was motivated by the presence of constraints which couple the decision variables nonlinearly between successive prediction steps. The same approach has been also successfully used to plan motion within nonconvex track boundaries and finally to achieve coordination between a group of robots.

## References

1. Oriolo, G., Luca, A.D., Vendittelli, M.: WMR control via dynamic feedback linearization: design, implementation, and experimental validation. IEEE Trans. Control Syst. Technol. **10**(6), 835–852 (2002)
2. M'Closkey, R.T., Murray, R.M.: Exponential stabilization of driftless nonlinear control systems using homogeneous feedback. IEEE Trans. Autom. Control **42**(5), 614–628 (1997)
3. Astolfi, A.: Discontinuous control of nonholonomic systems. Syst. Control Lett. **27**(1), 37–45 (1996)
4. Pourboghrat, F.: Exponential stabilization of nonholonomic mobile robots. Comput. Electr. Eng. **28**(5), 349–359 (2002)
5. Lee, T.C., Song, K.T., Lee, C.H., Teng, C.C.: Tracking control of unicycle-modeled mobile robots using a saturation feedback controller. IEEE Trans. Control Syst. Technol. **9**(2), 305–318 (2001)
6. Gu, D., Hu, H.: A stabilizing receding horizon regulator for nonholonomic mobile robots. IEEE Trans. Robot. **21**(5), 1022–1028 (2005)
7. Klančar, G., Škrjanc, I.: Tracking-error model-based predictive control for mobile robots in real time. Robot. Auton. Syst. **55**(6), 460–469 (2007)
8. Cao, Z., Xie, L., Zhang, B., Wang, S., Tan, M.: Formation constrained multi-robot system in unknown environments. In: Proceedings of the 2003 IEEE International Conference Robotics and Automation, vol. 1, pp. 735–740. IEEE (2003)
9. Olfati-Saber, R.: Flocking for multi-agent dynamic systems: algorithms and theory. IEEE Trans. Autom. Control **51**(3), 401–420 (2006)
10. Perizzato, A., Farina, M., Scattolini, R.: Formation control and collision avoidance of unicycle robots with distributed predictive control. IFAC-PapersOnLine **48**(23), 260–265 (2015)
11. Farina, M., Perizzato, A., Scattolini, R.: Application of distributed predictive control to motion and coordination problems for unicycle autonomous robots. Robot. Auton. Syst. **72**, 248–260 (2015)
12. Negenborn, R.R., Maestre, J.: Distributed model predictive control: an overview and roadmap of future research opportunities. IEEE Control Syst. Mag. **34**(4), 87–97 (2014)

# Whole-Body Planning for Obstacle Traversal with Autonomous Mobile Ground Robots

Martin Oehler(✉), Stefan Kohlbrecher, and Oskar von Stryk

Technische Universität Darmstadt, Darmstadt, Germany
{oehler,kohlbrecher,stryk}@sim.tu-darmstadt.de

**Abstract.** A common challenge for autonomous mobile ground robots in unstructured environments is the traversal of obstacles without risking to tip over. Previous research on prevention of vehicle tip-over is mostly limited to basic mobility systems with only few degrees of freedom (DOF). In this paper, a novel whole-body motion planning approach is presented. Based on a 3D world model and a given planned path, the trajectories of all joints are optimized to maximize robot stability. The resulting motion plan allows the robot to cross obstacles without tipping over. Compared to existing approaches, the proposed approach considers environment- and self-collisions during planning. Few assumptions about the robot configuration are made which enables the adoption to different mobile platforms. This approach is evaluated for a simulated and a real robot. The platform is a tracked vehicle with adjustable flippers and a five DOF manipulator arm. In several test scenarios, it is shown that the proposed approach effectively prevents tip-over and increases robot stability.

**Keywords:** Stability optimization · Unstructured environment

## 1 Introduction

A common challenge for autonomous robots operating in unstructured environments is the traversal of obstacles. To effectively increase its stability the robot can use arm joints to shift the center of mass (COM) and flipper joints to keep contact with the ground.

We propose a division into two sub-problems: While *pre-planning* computes a trajectory for all joints based on a map of the environment, *reactive behavior* corrects for disturbances during execution, such as unstable ground or slippage.

This work focuses on *pre-planning* and presents a novel whole-body planning approach that utilizes an online-generated 3D map of the environment to optimize joint trajectories along a given path. Vehicle tip-over is prevented by assessing joint configurations with a stability metric. The resulting motion plan allows the robot to safely cross generic obstacles. The proposed method is applicable to a wide range of wheeled or tracked robot platforms. Performance is evaluated in various scenarios in simulation and on the real robot.


K. Berns and D. Görges (Eds.): RAAD 2019, AISC 980, pp. 250–258, 2020.
https://doi.org/10.1007/978-3-030-19648-6_29

The evaluation platform is the *Hector Tracker* robot. The mobile platform features adjustable flippers, a five DOF arm and multiple sensors. An environment map is generated in real-time using a rotating Velodyne VLP-16 Lidar located at the rear.

# 2 Related Work

## 2.1 Stability Margin

Postural stability is critical for a vehicle's ability to traverse uneven terrain. Various analytical stability margins have been proposed that predict the probability of tip-over given the vehicle location. Commonly, the calculation is based on the location of the COM and the support polygon which is defined as the convex hull of all ground contact points.

For the analysis of quadruped walking gaits McGhee and Frank [6] proposed a stability margin defined as the shortest distance from the COM, projected onto the plane of the support polygon, to any point on the boundary of the support polygon. An energy-based formulation was proposed by Messuri *et al.* [7]. Their stability margin is defined as the minimum impact energy which can be sustained by the vehicle without tipping over. Unlike the work by McGhee *et al.*, this formulation considers COM-height changes. The force-angle stability margin was proposed by Papadopoulos *et al.* [11] and has been used in numerous works [2–4,8]. It combines the distance of the COM to the support polygon edges with the angles of the gravity vector to the edges. The least stable axis determines the stability of the whole system. A positive value indicates a stable position. Similar to the energy-based formulation by Messuri *et al.* [7], the measure is sensitive to height changes of the COM. Furthermore, the evaluation is computationally cheap. For these reasons, we utilize the force-angle stability margin.

## 2.2 Stability Control

We categorized the existing work on the prevention of vehicle tip-over on uneven terrain into two classes: *Pre-planning* and *reactive behavior*.

**Reactive Behavior.** Grand *et al.* [4] optimized stability and traction of the wheel-legged robot *Hylos* by adjusting its posture. A different approach for the same robot was proposed by Besseron *et al.* [3]. They exploit existing redundancies of robot kinematics by decoupling control of posture and trajectory. Stability is optimized with a potential field formulation. In contrast, Ohno *et al.* [9] performed online reconfigurations of the flipper on a tracked vehicle to prevent tip-over around the roll axis. Ground contacts are approximated dependent on sub-track contact.

**Pre-Planning.** Norouzi *et al.* [8] proposed a path planning method that also generates optimal configurations of the flipper and 1-DOF arm of the tracked *iRobot Packbot*. The optimization is embedded into the $A^*$ search algorithm and considers visibility, traction, energy consumption and stability. The physics engine Open Dynamics Engine (ODE) is used together with a 3D model of the environment to predict contact points. The work of Beck *et al.* [2] was implemented on the same platform. Given a path, the approach optimizes flipper and arm position by considering stability, equal distribution of contact forces, low energy consumption and operation within nominal joint positions. The prediction of contact points is idealized and depends on the flipper position and terrain slope.

The presented methods of stability control are unsuitable for adoption on the Hector Tracker. With five DOF its arm kinematics is much more complex than previous evaluation platforms. Furthermore, these works did not address important aspects like environment- and self-collision avoidance.

## 3 Whole-Body Planning

The whole-body planner generates a motion plan to safely cross obstacles. The planner takes a path $\mathbf{s}$, defined by $N_s$ equidistant waypoints, as input. Each waypoint $s_i = (x, y, \psi)^T, i = 1, ..., N_s$ is defined by a 2D-position $(x, y)$ and a heading $\psi$ in the xy-plane. The remaining pose components height $z$, roll $\phi$ and pitch $\theta$ are constrained by the ground geometry and estimated as part of the approach.

### 3.1 Cost Function

The goal of optimization is the maximization of robot stability by finding an optimal robot posture $\mathbf{p}_i^*$ at each point on the path $s_i$. The parameter vector $\mathbf{p} = (q_1, ..., q_{N_p})$ specifies the joint angles of the robot, where $N_p$ is the number of available DOF. It is evaluated using an appropriate objective function $\varphi(\mathbf{p})$, which has to be minimized:

$$\mathbf{p}^* = \arg\min_{\mathbf{p}} \varphi(\mathbf{p}) \tag{1}$$

**Stability Criterion.** The main component of the objective function is the stability criterion. It is evaluated by applying a cost function $w(\cdot)$ to the force-angle stability margin $\beta_i$ of every axis of the support polygon:

$$\varphi(\mathbf{p}) = \frac{1}{N_s} \sum_{i}^{N_s} w(\beta_i(\mathbf{p})) \tag{2}$$

where $N_s$ specifies the number of support polygon edges. The ground contact points depend on $\mathbf{p}$ and are predicted by an appropriate contact estimation using a model of the world.

We propose to use an exponential weighting function:

$$w(x) = ae^{-bx+c} \tag{3}$$

Compared to a quadratic-inverse formulation as for example used in [4], it also punishes negative stabilities as would be the case for unstable postures.

**Movement Penalty.** To prevent large movements between waypoints that increase the stability only slightly, a penalty term $r(\cdot)$ on joint movement is added to the objective:

$$\varphi(\mathbf{p}) = \frac{\sum_i^{N_s} w(\beta_j(\mathbf{p}))}{N_s} + K_m \sum_j^{N_p} r(p_j - q_j) \tag{4}$$

where $\mathbf{q}$ is the previous joint configuration and $K_m$ is a factor that determines the trade-off between stability and trajectory execution time. It is reasonable to do large motions if it is coupled with a meaningful stability increase. The Lorentzian function (Eq. 5) fulfills this requirement because the applied penalty is capped.

$$r_{lor}(x) = \ln\left(1 + \frac{x^2}{2\sigma^2}\right) \tag{5}$$

### 3.2 Optimization Constraints

To prevent damage to the robot, collisions with the environment and itself have to be avoided. This is achieved by modeling collisions as constraints of the optimization problem.

**Environment Collision Avoidance.** The world representation used to formulate environment collision constraints is an Euclidean Signed Distance Field (ESDF) [10]. It consists of a 3-dimensional uniform grid with each cell containing the distance to the closest surface. Environmental collisions are only modeled for the manipulator arm since the flipper is intended to be in contact with the ground. Each link of the arm is approximated with multiple spheres (Fig. 1a). One constraint per sphere is added to the optimization problem:

$$b_{i,j}(\mathbf{p}) = \Phi({}^W c_{i,j}) - r_{i,j} > 0 \tag{6}$$

The function $\Phi(\cdot)$ evaluates the ESDF at the sphere center ${}^W c_{i,j}$ with link index $i$ and sphere index $j$ to get the distance to the closest obstacle. Subsequently, the radius $r_{i,j}$ of the respective sphere is subtracted to check for a collision.

**Self-collision Avoidance.** To check for self-collisions efficiently, we distinguish between dynamic and static links. The position of dynamic links depends on the optimization parameter $\mathbf{p}$ whereas the static part is rigidly attached to the robot base. Analogous to the environment collision avoidance, the dynamic links of the robot are sampled with spheres (Fig. 1a). One constraint per unique pair of spheres is added to the problem. The number of constraints is reduced by using the Allowed Collision Matrix (ACM). To prevent a collision, the distance between the sphere centers minus their radii must be greater than zero:

$$b_{i,j;m,n}(\mathbf{p}) = \|{}^B c_{i,j} - {}^B c_{m,n}\|_2 - r_{i,j} - r_{m,n} > 0; \quad i \neq m \tag{7}$$

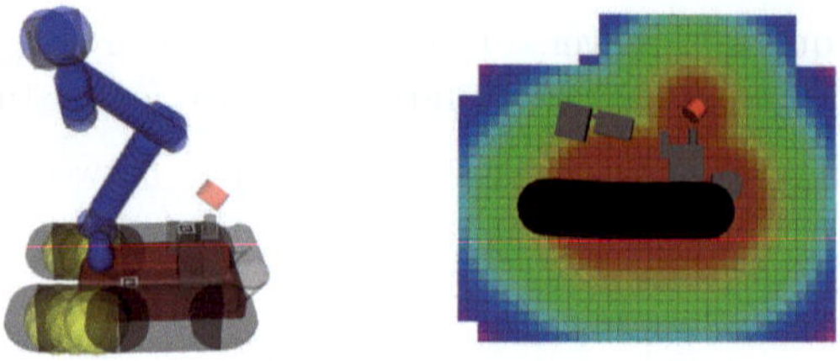

(a) (b)

**Fig. 1.** Models used for collision avoidance. 1a shows the sphere-decomposition. Blue spheres model the arm and are used for environment and self-collision. The flippers, visualized with yellow spheres, are only considered for self-collision. 1b shows a slice of the robot ESDF used for self-collision checking against static links.

The first link is indexed with $i$ and its spheres with $j$, the second link is indexed with $m$ and its spheres with $n$.

To reduce computational cost, static links are represented by an ESDF. It is generated by computing the distances to the links in a grid around each link (Fig. 1b). Collision checks are formulated analogous to environment collision checks (Eq. 6).

### 3.3 Optimization Process

As gradient-based optimization techniques can be faster and more efficient than gradient-free ones, a differentiable objective function is desired. In the original definition of the force-angle stability margin [11], the sign of the angle $\theta_i$ between force vector $\hat{f}_i$ and axis normal $\hat{l}_i$ is given by the piecewise-defined function $\sigma_i$. As the discontinuity is not differentiable, a signed-angle formulation is used instead:

$$\theta_i = \operatorname{atan2}((\hat{f}_i \times \hat{l}_i) \cdot \hat{a}_i,\ \hat{l}_i \cdot \hat{f}_i) \tag{8}$$

Computation of the stability margin relies on the contact estimation to determine robot pose and support polygon. This function space is highly complex and discontinuous. To find feasible solutions, the optimization process is separated into two consecutive phases. In the first phase, only the contact geometry is optimized by adjusting the flipper position, eliminating the need for collision constraints. As the contact estimation is non-differentiable, only gradient-free solvers are suitable. Because we do not optimize directly for stability in this phase, an adjusted objective function is used that maximizes the area $A$ of the support polygon:

$$\varphi(p_f) = -A - K_t \frac{C_t}{N_t} + K_c \frac{C_c}{N_c} \tag{9}$$

Additionally, two terms for assessment of ground geometry are introduced. The first term rewards contact points $C_t$ with the tracks, the second term punishes contact points $C_c$ with the chassis. The total number of sampled points on the tracks is labeled $N_t$ and $N_c$ for the chassis respectively.

In the next phase, the COM is shifted by varying the arm joint angles. The contact geometry is fixed and given by the first phase. Since the contact estimation is no longer involved, efficient gradient-based optimization algorithms can be applied that also include the presented collision constraints.

If the final optimization solution $\mathbf{p}^*$ is instable, the procedure stops as a safe traversal is not possible.

## 4 Evaluation

The proposed whole-body planning method was implemented using ROS[1] as middleware. The platform used for evaluation in simulation and reality is the *Hector Tracker* robot. To assess the effectiveness of the approach, the robot has to cross different obstacles without tip-over. The commanded path is a straight line and defined with a resolution of 0.02 m. The optimized robot postures $\mathbf{p}_i^*, i = 1, ..., N_s$ at each waypoint are connected by a joint trajectory which is executed in synchronization with vehicle movement. The environment map is represented by an ESDF. It was generated online with *Voxblox*[2] and captured using the VLP-16 Lidar and a depth camera. The contact estimation operates directly on the ESDF and is based on iterative optimization. The parameters used for evaluation were determined experimentally and are listed in Table 1. Derivatives of objective and constraint functions were determined with automatic differentiation using RBDL [5] and Ceres [1]. During trajectory execution, the actual stability margin is measured and compared to the predicted stability during pre-planning. In simulation, the ground truth is used as robot pose. On the real robot, a SLAM estimate based on fusing track odometry, IMU and Lidar data is used.

**Table 1.** Parameters used for evaluation in simulation and on the real robot.

| Exponential | | Flipper objective | | Penalty flipper | | Penalty arm | |
|---|---|---|---|---|---|---|---|
| a | 14 | $K_t$ | 4 | $K_m$ | 0.2 | $K_m$ | 0.0025 |
| b | −2.6 | $K_c$ | 3 | $\sigma$ | 0.2 | $\sigma$ | 0.2 |
| c | 0.3 | | | | | | |

**Simulation.** Experiments in simulation were performed in the open-source robotics simulator Gazebo[3]. Four different testing scenarios were chosen to cover a variety of obstacles: A step with a height of 0.15 m, a double ramp with an inclination angle of 40°, a cinder block with a height of 0.14 m and an asymmetrical step with the same height as the normal step (Fig. 2).

The results are compared to a scripted behavior which switches hand-tuned joint configurations based on IMU feedback.

[1] http://www.ros.org/.
[2] https://github.com/ethz-asl/voxblox.
[3] http://gazebosim.org/.

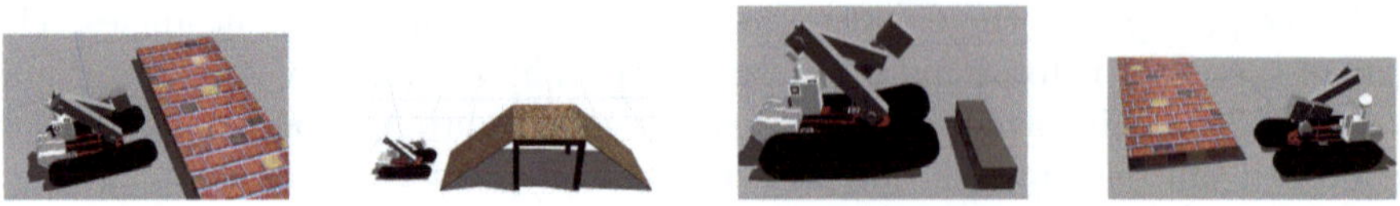

(a) Step (b) Ramp (c) Cinder Block (d) Asym. Step

**Fig. 2.** The four evaluation scenarios in Gazebo.

The minimum and average stabilities for each scenario are summarized in Table 2. Using whole-body planning, the robot managed to cross all obstacles successfully, as can be seen by the positive minimum stability. The reference behavior failed to traverse the ramp and fell during descend on the downwards slope. Overall, the proposed approach significantly outperformed the reference behavior in all scenarios in terms of average stability. Comparing the minimum stability, the reference behavior performed only slightly better in the cinder block task. Differences between measured and predicted stabilities are explained by approximation effects in map data and contact estimation. A detailed visualization of the generated motion plan in the ramp scenario can be seen in Fig. 3.

**Table 2.** Comparison of stability margin results. *Plan* labels the predicted stability during planning, *Measured* refers to the actual stability during execution. Measured values during execution of the reference behavior are given in the last column.

| | Plan | | Measured | | Reference | |
|---|---|---|---|---|---|---|
| | avg | min | avg | min | avg | min |
| Step | 2.26 | 1.8 | 2.23 | 0.96 | 1.7 | 0.28 |
| Ramp | 1.98 | 0.63 | 1.75 | 0.04 | 1.16 | −1.65 |
| Cinder block | 2.3 | 1.78 | 1.92 | 0.02 | 1.68 | 0.13 |
| Asymmetrical step | 1.68 | 0.87 | 1.57 | 0.07 | 0.57 | 0.05 |
| Step (Real Robot) | 2.26 | 1.8 | 2.23 | 0.96 | - | - |

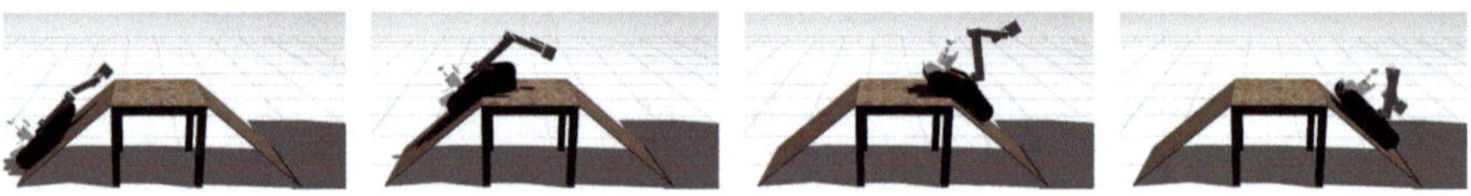

(a) $x = 1\,\text{m}$ (b) $x = 2\,\text{m}$ (c) $x = 3.25\,\text{m}$ (d) $x = 3.8\,\text{m}$

**Fig. 3.** Visualization of the whole-body motion plan in the ramp scenario.

**Real Robot.** To demonstrate, that the simulation results transfer to reality, we also evaluated our method on the real robot. Compared to simulation, the planner has to deal with inaccuracies of the robot model. The test was conducted using a single step with a height of 0.18 m. The robot successfully managed to climb the step as can be seen in Fig. 4 and Table 2. Even with model inaccuracies, the difference between average expected stability and measured stability is only 0.03, showing that the method successfully extends to real robots. A full video of the experiment is available online[4].

**Fig. 4.** Visualization of the whole-body motion plan on the real robot.

## 5 Conclusion

In this paper, we present a novel whole-body planning approach for traversal of obstacles with autonomous mobile ground robots. Based on a 3D map of the environment and a given path, optimal joint trajectories for manipulator arm and flippers are computed to maximize stability and safely cross obstacles. The method generalizes to arbitrary robot platforms using wheeled or tracked locomotion, as long as basic sensor requirements are satisfied. The effectiveness of the approach was shown by successfully crossing diverse obstacles in simulation and with a real robot. The proposed method consistently outperformed the hand-tuned reference in terms of average stability.

## References


1. Agarwal, S., Mierle, K., et al.: Ceres solver. http://ceres-solver.org
2. Beck, C., Miró, J.V., Dissanayake, G.: Trajectory optimisation for increased stability of mobile robots operating in uneven terrains. In: IEEE ICCA, pp. 1913–1919 (2009)
3. Besseron, G., Grand, C., Amar, F.B., Bidaud, P.: Decoupled control of the high mobility robot hylos based on a dynamic stability margin. In: IEEE/RSJ IROS, pp. 2435–2440 (2008)
4. Grand, C., Benamar, F., Plumet, F., Bidaud, P.: Stability and traction optimization of a reconfigurable wheel-legged robot. Int. J. Rob. Res. **23**(10–11), 1041–1058 (2004)
5. Kudruss, M., Manns, P., Kirches, C.: Efficient derivative evaluation for rigid-body dynamics based on recursive algorithms subject to kinematic and loop constraints. Optimization Online Preprint (2019). http://www.optimization-online.org
6. McGhee, R.B., Frank, A.A.: On the stability properties of quadruped creeping gaits. Math. Biosci. **3**, 331–351 (1968)
7. Messuri, D., Klein, C.: Automatic body regulation for maintaining stability of a legged vehicle during rough-terrain locomotion. IEEE J-RA **1**(3), 132–141 (1985)


[4] https://youtu.be/dVLi2w4l3Lg.

8. Norouzi, M., Miro, J.V., Dissanayake, G.: Planning stable and efficient paths for reconfigurable robots on uneven terrain. J. Intell. Rob. Syst. **87**(2), 291–312 (2017)
9. Ohno, K., Takeuchi, E., Chun, V., Tadokoro, S., Yuzawa, T., Yoshida, T., Koyanagi, E.: Rollover avoidance using a stability margin for a tracked vehicle with sub-tracks. In: IEEE SSRR, pp. 1–6 (2009)
10. Oleynikova, H., Taylor, Z., Fehr, M., Siegwart, R., Nieto, J.: Voxblox: incremental 3D euclidean signed distance fields for on-board MAV planning. In: IEEE/RSJ IROS (2017)
11. Papadopoulos, E., Rey, D.A.: The force-angle measure of tipover stability margin for mobile manipulators. Veh. Syst. Dyn. **33**(1), 29–48 (2000)

# Connections for Path Planning in Mobile Robotics

Mircea Nitulescu(✉) and Mircea Ivanescu

Department of Mechatronics and Robotics,
University of Craiova, Craiova, Romania
{nitulescu, ivanescu}@robotics.ucv.com

**Abstract.** This work presents the theoretical solutions that are possible for connecting the linear segments of the global trajectories developed for mobile robots in their operating scenes. Generated most often in a purely polygonal form by the global planner of trajectories, they cannot be executed directly than by mobile robots that have the pivotal capability around their axis of vertical symmetry. For all the other constructive structures of the mobile robots and whatever type of their control, e.g. with the trajectory stored in memory or with the wired trajectory mapped on the ground, these techniques are needed. This paper presents three techniques for connecting the polygonal paths using only pure circular arcs, circular arcs and splines curves, respectively arcs of clothoid. Theoretical support of each method, examples and the assessment of specific influences on the behavior of a mobile robot at changing sections on trajectory are included.

**Keywords:** Mobile robots · Global trajectories · Paths planning · Methods of connecting linear sections

## 1 Introduction

Global path planning of mobile robots operating in their scenes leads to polygonal trajectories, closed (if they are cyclical) or open. Alternatively, the overall trajectory can be obtained in the form of a sequence of positions imposed in the operating scene. However, a purely polygonal trajectory, like that in Fig. 1, cannot be tracked with the accuracy required only by mobile robots whose locomotion system provides pivoting along its vertical axis of symmetry. Thus, in the significant points of the trajectory, denoted in Fig. 1 with *A, B, C,* it is necessary to stop the robot and to make a rotational movement of the pivotal type with the angles $\alpha_A, \alpha_B, \alpha_C$ that characterize the next trajectory section.

Another solution is to connect all linear sections of the initial polygonal trajectory planned through judiciously chosen curves. This can be an additional feature for mobile robots with pivotal capability, but it is mandatory for other constructive mobile robots. Using the connections between the linear sections of a polygon type trajectory will ultimately lead to:

- Eliminating the need to stop the robot from changing sections.
- Optimization of the overall time of the trajectory.


K. Berns and D. Görges (Eds.): RAAD 2019, AISC 980, pp. 259–266, 2020.
https://doi.org/10.1007/978-3-030-19648-6_30

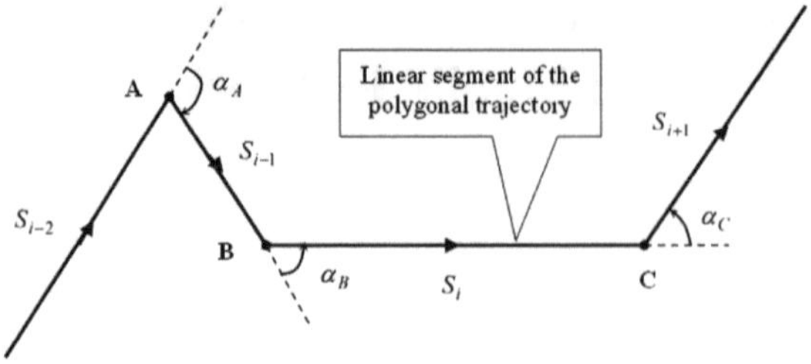


**Fig. 1.** Trajectory planned through an open polygonal structure.

- Elimination of discontinuities in the evolution of kinematic and dynamic parameters (linear velocity, angular velocity, forces, couples) of the mobile robot along the trajectory to be accomplished by it.

In principle, connecting a global trajectory composed exclusively of linear sections can be done both off-line and on-line [1, 2]. The off-line solution has the purpose of finalizing the final trajectory that is to be memorized by the robot's control system, or it can then be materialized on the ground of the operating scene in the form of fixed map for filo guiding, opto guiding or video guiding mobile robots [3]. The on-line solution can be used when the robot's control system calculates the necessary connections even while it is moving, using the database that includes the entire polygon path or only the coordinates of its peaks.

There are several possible techniques for connecting the linear sections of a trajectory and which are suitable for the intended purpose. They take into account to a varying extent the functional limits introduced by the electric and mechanical parameters of the mobile robot, thus directly influencing the final performances obtained. In order of complexity, these are the followers.

## 2 Connecting the Linear Sections of the Global Trajectories

### 2.1 Connection Through Arcs of Circle

We will consider a portion of the global trajectory consisting of two linear and adjacent segments $S_{i-1}$ and $S_i$ as shown in Fig. 2. The two segments achieve the known and horizontal angles $\theta_{i-1}$, $\theta_i$ and intersect at the point $O'(x', y')$, the peak of the polygon trajectory provided by the global trajectory planner. Let $R$ be the length required for the radius of the circular connecting arc. It is desirable to determine the position of the connection center $C(x_C, y_c)$. The angle made by the axis that will contain the horizontal connection center can be obtained immediately:

$$\alpha = \frac{\theta_{i-1} + \theta_i}{2} \quad (1)$$

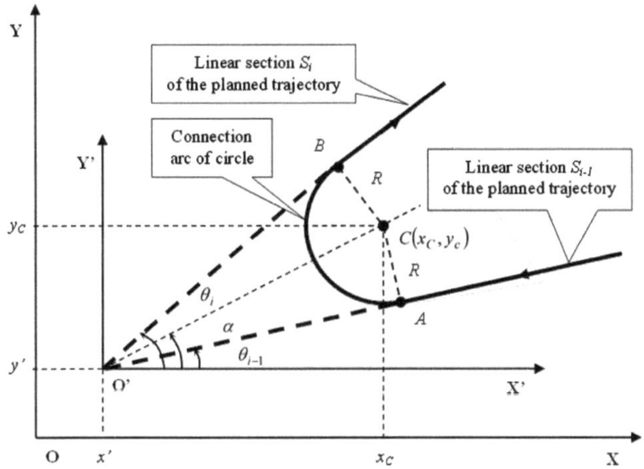


**Fig. 2.** Connection of the linear segments of the trajectory planned through circle arcs.

The trajectory should be routinely reported to the global reference system attached to the operating scene or configuration space. The final coordinates of the connection center $C$ are:

$$x_C = x' + R \cdot \left[ (\cos \theta_{i-1}) \cdot \left( ctg \frac{\theta_{i-1} + \theta_i}{2} \right) - \sin \theta_{i-1} \right] \tag{2}$$

$$y_C = y' + R \cdot \left[ \left( ctg \frac{\theta_{i-1} + \theta_i}{2} \right) \cdot (\sin \theta_{i-1}) - \frac{\sin^2 \theta_{i-1}}{\cos \theta_{i-1}} \right] \tag{3}$$

Although straightforward, connecting the linear sections of a global trajectory with circular arcs introduces curvature discontinuities at all points of a linear segment connected with a circular arc. Assuming, for example, that the robot's mass center wants to run the arc at the same speed as the linear sections connected thereto, the different rays on which the two drive wheels of an mobile robot move will cause discontinuities of the angular speed of the control, which requires motor couples infinite due to the robust inertia of the robot. As this is not practically possible, it means that the mobile robot will not be able to accurately track the prescribed trajectory. As a result, there will be lateral errors of the actual trajectory in relation to the prescribed one, and consequently will result in a considerable decrease in the performance of the robot in terms of tracking the prescribed trajectory, whether it is mapped in its operating space or stored in its control system memory.

### 2.2 Connection Through Arcs of Circle and Splines Curves

We will consider a portion of the global trajectory consisting of three linear and adjacent segments $S_{i-1}$, $S_i$ and $S_{i+1}$ as in Fig. 3. It is desirable to transform this purely polygonal trajectory into a curvilinear trajectory, which will use only circle arcs and spline arcs for connections.

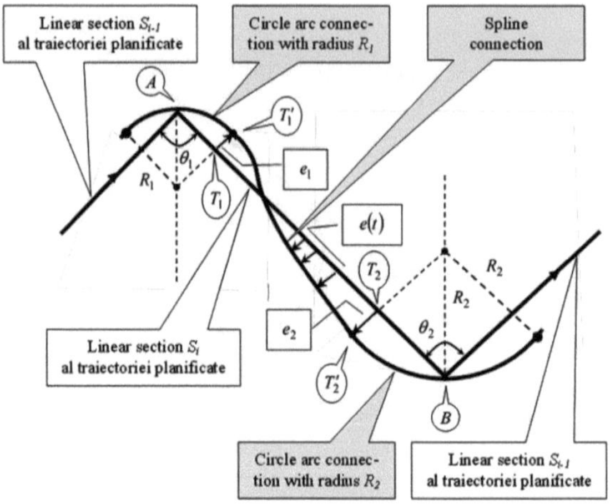


**Fig. 3.** Connection of trajectory linear segments through circle arcs and spline curves.

Between the tip $A$ of the prescribed polygonal trajectory and a point $T_1'$ of the linear section $S_i$ of the same trajectory, the mobile robot will describe a circle arc trajectory. It passes through point $A$ and is chosen to satisfy the following conditions:

- The tangent at the arc at the point $T_1'$ is parallel to the linear section $S_i$.
- The center of the arc belongs to the bisector of the angle formed by the linear section $S_i$ with the previous one $S_{i-1}$ of the initially prescribed polygonal trajectory.

Since it is desirable for the deviation between the real path of the mobile robot and the originally planned polygonal trajectory to be less than a required boundary [5], we will introduce the function of the side error (or the offset) to the prescribed trajectory for its estimation $e_C(t)$.

Let be $\theta_1$ the angle between the current $S_i$ and the preceding $S_{i-1}$ linear segment and $R_1$ the radius of the circle arc used for connecting these segments on the actual trajectory. Figure 3 shows the following relationships:

$$|AT_1| = R_1 \cdot \cos\frac{\theta_1}{2} \quad |T_1T_1'| = e_C(t_1) = R_1 \cdot \left(1 - \sin\frac{\theta_1}{2}\right) \tag{4}$$

where $t_1$ corresponds to the time at which the mobile robot is on the circle arc trajectory at the point $T_1'$, or equivalent, it would ideally be placed on the reference polygonal trajectory at the point $T_1$ on the linear section $S_i$.

The general form of the function that expresses the evolution of the lateral error of the robot relative to its reference trajectory is:

$$|e_C(t_1)| = \sqrt{R_1^2 - \left(R_1 \cdot \cos\frac{\theta_1}{2} - t\right)^2} - R_1 \cdot \sin\frac{\theta_1}{2} \tag{5}$$

For the execution of the circle arc trajectory, a series of additional restrictions on side error may be imposed, such as:

- $R_1 > R_{\min}$ for reasons related to the minimum radius of rotation that the mobile robot can perform.
- $e_C(t) < e_{C_{\max}}$ for controlling the width of the passage that is physically available to the robot displacement around the polygonal trajectory initially prescribed.

In view of these restrictions, finally results:

$$R_{\min} < R_1 < \frac{e_{C_{\max}}}{1 - \sin\frac{\theta_1}{2}} \tag{6}$$

In conclusion, the Eq. (5) in which the radius $R_1$ follows the restriction expressed by the inequality (6) will describe the circle arc replacing the linear section $\overline{AT_1}$. Similarly, replace the linear section $\overline{T_2B}$ with a new radius circle arc $R_2$. The procedure can continue for all the peaks of the reference polygonal trajectory. Between the points $T_1$ and $T_2$ along $S_i$, the lateral error function used for the circle arc $e_C(t)$ is replaced by a third-order spline function $e_S(t)$ with the following form [4]:

$$e_S(t) = a \cdot t^3 + b \cdot t^2 + c \cdot t + d \tag{7}$$

To ensure continuity of positioning and orientation of the mobile robot at junction points with circular arcs, the restrictions must be met simultaneously:

$$e_S(t_1) = e_C(t_1) \quad e_S(t_1) = e_C(t_1) \quad e'_S(t_1) = 0 \quad e'_S(t_2) = 0 \tag{8}$$

which finally leads to the need for the four polynomial coefficients a, b, c, d to be dimensioned so as to verify the equations resulting from the following matrix form:

$$\begin{bmatrix} t_1^3 & t_1^2 & t_1 & 1 \\ t_2^3 & t_2^2 & t_2 & 1 \\ 3 \cdot t_1^2 & 2 \cdot t_1 & 1 & 0 \\ 3 \cdot t_2^2 & 2 \cdot t_2 & 1 & 0 \end{bmatrix} \cdot \begin{bmatrix} a \\ b \\ c \\ d \end{bmatrix} = \begin{bmatrix} e_C(t_1) \\ e_C(t_2) \\ 0 \\ 0 \end{bmatrix} \tag{9}$$

Although more laborious, this strategy eliminates discontinuities in the speed of the mobile robot, but the discontinuities in accelerations are maintained due to the existence of the curvature discontinuities at the junctions circle arcs - spline curves.

## 2.3 Connection Through Arcs of Clothoid

The equations of the mathematical model of a mobile robot are nonlinear. There are possibilities of linearization based essentially on the assumption that the robot orientation varies extremely slowly, staying in the vicinity of a constant, which allows approximation of sinus and cosine trigonometric functions. The aim is to considerably simplify the theoretical study of the robot's actions. But these assumptions do not remain valid if the trajectory has significant curvature or if the curvature is not constant during short time intervals. Another approach to the problem can be made by requiring the accelerations of the two left and right drive mobile robot wheels, $a_S$ respectively $a_D$, to remain constant along the connected trajectory. If the accelerations of the two drive wheels are opposite as a sign, and if its initial linear velocity is not zero, the trajectory of the mobile robot is a clothoid type, i.e. a curve whose curvilinear abscissa is proportional to its curvature. It starts with a null curve and tends to a null radius [6].

The parametric equations of a curve of clothoid are as follows:

$$x(\sigma) = v \cdot \int_{t_i}^{t_f} \cos\left(\sigma \cdot \frac{t^2}{2} + \omega_0 \cdot t\right) \cdot dt \quad y(\sigma) = v \cdot \int_{t_i}^{t_f} \sin\left(\sigma \cdot \frac{t^2}{2} + \omega_0 \cdot t\right) \cdot dt \tag{10}$$

The rotation function that characterizes this type of curve can be expressed by:

$$\Psi(t) = \frac{\sigma}{2} \cdot t^2 + \omega_0 \cdot t \tag{11}$$

and the function expressing the variation in curvature of a clothoid is as follows:

$$k(\sigma, \lambda) = \frac{\sigma}{v^2} \cdot \lambda \tag{12}$$

In previous relations (10)–(12), several notations have the following meanings:

- $\sigma$, the curvature parameter of the clothoid, given by the relation:

$$\sigma = \frac{2 \cdot a}{l_A} \tag{13}$$

  where $l_A$ is the length of the distance between the centers of two coaxial driving wheels of a mobile robot, and $a$ represents the acceleration of the outer wheel (left or right) in the turn.
- $\lambda$, curvilinear abscissa, defined by:

$$\lambda = v \cdot \left(t_f - t_i\right) \tag{14}$$

- $\omega_o$ is the initial angular velocity.
- $v$ is the linear velocity, constant along the arc of clothoid.
- $\Psi(t)$ is the function expressing the rotation of a clothoid.
- $k(\sigma, \lambda)$ is the function that expresses the variation in the curvature of a clothoid.

A clothoid arc can be used as a connection between two trajectory sections with different curves, such as a straight segment and a circle arc, as shown in Fig. 4a. Using two clothoid arcs, the connection can be achieved between the two linear segments of a trajectory, as can be seen in Fig. 4b.

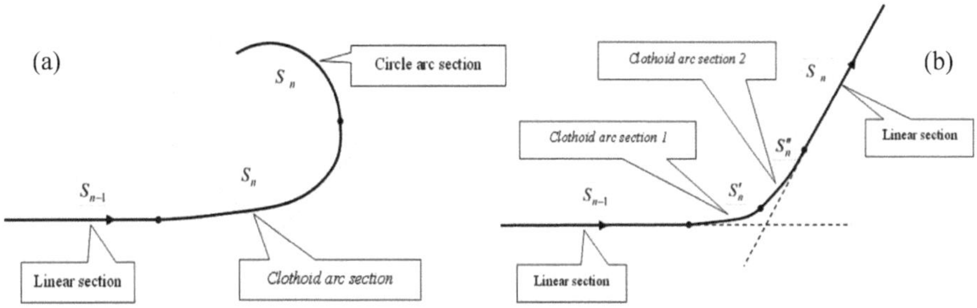


**Fig. 4.** Use of a clothoid arc to connect a linear section and a circle arc section (a); use of two clothoid arcs to connect two linear sections of the trajectory.

As can be seen from the Eqs. (12)–(13), the curvature of a clothoid varies in a continuous manner. Using this special class of curves to connect successive linear or circular segments of an initially planned trajectory will make it possible to obtain a global trajectory without curvature discontinuities, that is, with a continuous evolution of acceleration. The continuity of the curvature is important because a real mobile robot with a nonzero mass, and therefore inertia that cannot be neglected, cannot achieve discontinuities in speed and acceleration along a prescribed trajectory. Consequently, there will be errors tracking the trajectory, errors that can alter the entire operation if they lead to the impossibility of retrieving it. Discontinuities in the speed prescribed to a mobile robot will require to the mobile robot actuators an infinite force so that it can achieve an infinite linear acceleration. Discontinuities in the acceleration prescribed to a mobile robot will require an infinite torque from the drive system to assure an infinite angular acceleration when its angular velocity is not null. It is therefore possible that a global polygon trajectory is transformed into a global trajectory connected with clothoid arcs as in Fig. 5, thus obtaining the minimization of trajectory tracking errors. In Fig. 5, the clothoid arcs $AA'$ and $A'B$ are chosen such that:

- Their curvatures at the connection points $A$ and $B$ with the linear segments $S_{n-2}$ and respectively $S_{n-1}$ are null.
- Their curvatures at the mutual coupling point $A'$ have equal values.

Similarly, the $CC'D$ and $EE'F$ arcs were also introduced in Fig. 5a. A mobile robot with no pivoting ability will now be able to traverse the planned path between the initial point $IP$ and the final point $FP$ without stopping, as there are no bending discontinuities along the planned trajectory. Similarly to the previously discussed connection method, the clothoid connection arcs must be entered in a lane fixed with the desired width $l_{\max}$. The width of each corridor can be defined for each top of the overall trajectory, but it should be taken into account to avoid collisions with obstacles present in the robot operating scene, as shown in Fig. 5b.

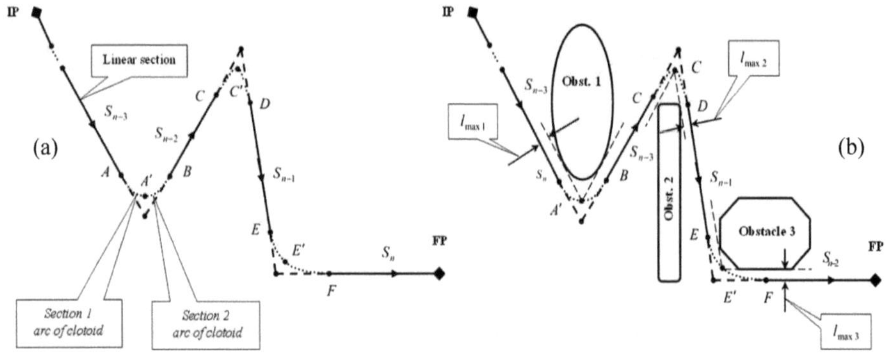


**Fig. 5.** Using two arcs of clothoids to eliminate curvature discontinuities (a); physical limitations by imposing the width of the connecting corridors (b).

## 3 Conclusions

This paper evaluates three possible solutions for connecting the linear segments of a two-wheel drive robot mobile robot trajectory that was originally planned only by linear segments. Although straightforward, connecting through circular arcs introduces curvature discontinuities in all connection points and requires impossible motor couples infinite to eliminate the discontinuities in speeds and accelerations. The second strategy, based on connection through arcs of circle and splines curves, eliminates discontinuities in the speed, but the discontinuities in accelerations are maintained due to the existence of the curvature discontinuities at the junctions points circle arcs - spline curves. Although more laborious, the third strategy based on connection through arcs of clothoid will make it possible to obtain a global trajectory without curvature discontinuities, that is, with a continuous evolution also for the acceleration. The continuity of the curvature is important because a real mobile robot has a nonzero mass, and therefore an inertia that cannot be neglected. Discontinuities in the speed prescribed to a mobile robot will require to the mobile robot actuators an infinite force so that it can achieve an infinite linear acceleration, while discontinuities in the acceleration prescribed to a mobile robot will require an infinite torque from the drive system to assure an infinite angular acceleration when its angular velocity is not null.

## References


1. Dolgov, D., Thrun, S., Montemerlo, M., Diebel, J.: Path planning for autonomous vehicles in unknown semi-structured environments. Int. J. Robot. Res. **29**(5), 485–501 (2010)
2. Laumond, J.P., Jacobs, P.E., Taix, M., Murray, R.M.: A motion planner for nonholonomic mobile robots. IEEE Trans. Robot. Autom. **10**(5), 577–593 (1994)
3. Kanayama, Y., Hartman, B.I.: Autonomous Robot Vehicles. Springer, New York (1990)
4. Sprunk, C.: Planning Motion Trajectories for Mobile Robots Using Splines. ALUF (2008)
5. Latombe, J.C.: Robot Motion Planning. Springer, US (1991)
6. Clothoid. https://en.wikipedia.org/wiki/Euler_spiral. Accessed 17 Nov 2018

# Safe and Efficient Navigation of an Autonomous Shuttle in a Pedestrian Zone

Qazi Hamza Jan(✉), Sascha Klein, and Karsten Berns

Robotics Research Lab, Technische Universität Kaiserslautern,
67663 Kaiserslautern, Germany
{hamza,sascha.klein,berns}@cs.uni-kl.de

**Abstract.** This paper summarizes the navigation of autonomous shuttle in a pedestrian area using a behavior-based architecture with Pedestrian Interaction System (PIS). For pedestrians to accept autonomous shuttles in their vicinity, it is important to have a safe distance between the shuttle and the pedestrians. The PIS determines the scheme of interaction with pedestrians based on their behavior. It also keeps the pace of the shuttle by recognizing safe conditions in a pedestrian zone. The experiments are performed in a simulated environment. It shows that the shuttle moves efficiently on a predefined track, keeping the safety of the environment when PIS is active.

**Keywords:** Autonomous navigation · Pedestrian · Collision avoidance · Social environment · Autonomous shuttle · Human-machine interaction

## 1 Introduction

Autonomous shuttles traversing in a pedestrian area have become popular in the transport industry. Due to the limited width of pedestrian areas, shuttles can only be used for such purposes. Self-driving shuttles become meaningful to be used here. The first mile and last mile travel also comes in this context where it is required to carry passengers in and out from the last bus stop and busy streets. Major concern while driving in a pedestrian area is about the safety of the people due to the fact that people have random behavior. For autonomous shuttles moving in such environments, it is efficient to keep their pace while avoiding any collision with pedestrians that can have irregular motions. There should always be a safe distance between the shuttle and the crowd when maneuvering through them as described in [1]. This means that the vehicle should gradually move along the path so that the humans do not get intimidated.

Generally, there are two major problems associated with this topic. Firstly, the navigation of autonomous vehicle in a limited space and secondly, the pedestrians. The major focus of the work presented here is on the interaction of


K. Berns and D. Görges (Eds.): RAAD 2019, AISC 980, pp. 267–274, 2020.
https://doi.org/10.1007/978-3-030-19648-6_31

the autonomous shuttle with pedestrians. Interaction includes how the shuttle should react to different number and behaviors of the pedestrian to keep its smooth motion. Work in this regard is considered in the literature; however, the focus is often on improving the motion model. We propose a different strategy in this regards, i.e., instead of improving the motion model, a PIS is developed to interact with pedestrians on different levels and adapt the shuttle control accordingly. This PIS also considers different types of pedestrians such as retired persons, adults, childs and invalids with different characteristics in order to use specific interaction level.

Usually, the major problem which occurs in this scenario is the so-called freezing vehicle problem and it cannot be solved even with exactly known human trajectories [2,3]. Previously to encounter this problem was by improving the navigation within crowded areas. Most of the planning algorithms do not consider humans as dynamic obstacles. A similar application is given in [4] where the robot is moving among the crowd in a museum. It uses the extended version of the Dynamic Window Algorithm (DWA) which controls the velocity of the robot for obstacle avoidance. It considers humans as static obstacles and re-plans its motion. Bauer et al. [5] uses simultaneous localization and mapping (SLAM) for navigation and treats pedestrian as a static obstacle. It uses behavior selection at every update to avoid an obstacle. Some approaches such as [6] uses blob tracker for pedestrians tracking. Recently, [7] has used POMDP with pedestrian intentions but uses a simple straight line motion model for the pedestrian. It ignores the situations where pedestrians ignore the vehicle.

The experiments in this paper were carried out in a simulated environment of the campus of the TUK with a simulated autonomous shuttle. The Simulation was created using the Unreal Engine[1]. In this work, the characteristics of the pedestrians are predefined for testing different scenarios. Section 2 describes the behavior-based architecture which is used for basic navigation of the vehicle and how this is modified to be used for shuttle navigating in the pedestrian area. Section 3 discusses in detail about the PIS which uses Social Collision Avoidance (SCA) and Tentacle-based Evasion. We describe the experiments in Sect. 4 and the conclusion is given at the end.

## 2 Architecture

PIS is included in a behavior-based architecture which is employed for off-road navigation [8]. This control architecture provides basic functionalities such as hardware interfaces, navigation, pilot and low-level control. Extension of this architecture by PIS adapts it to navigate the autonomous shuttle in a pedestrian area. PIS provides flexibility to this architecture in pedestrian areas to avoid jolting motion and keep its speed when pedestrians are well aware of the shuttle. An attempt is made to avoid the pedestrians with a tentacle approach [9], whereby the costs are calculated and, if the costs are too high, the behavior is switched to SCA designed in the PIS.

[1] https://www.unrealengine.com/.

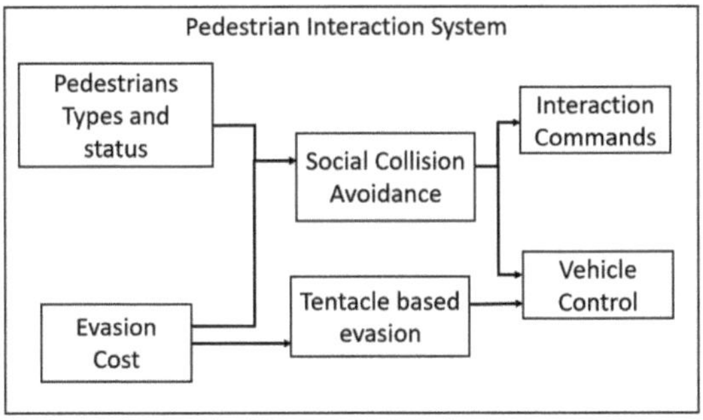


**Fig. 1.** General Block diagram of PIS

# 3 Pedestrian Interaction System

PIS is responsible for the interaction process between the vehicle and the pedestrians. The PIS contains two different behaviors that are activated based on pedestrians in the detection area. PIS takes a lot of inputs such as pedestrians, their characteristics, their type, and their states and decides a feasible navigation and interaction strategy. This section introduces the two basic behaviors of PIS, SCA and Tentacle-based Evasion. These two behaviors form the basis for collision avoidance in this paper. If humans are in the detection area of the vehicle, one of the behaviors must become active and navigate through humans. The desired behavior of the autonomous shuttle is described in detail in Table 1.

## 3.1 Social Collision Avoidance

SCA is inspired by [2,3], where the freezing robot problem and possible solutions are discussed. The SCA behavior assumes that humans also take responsibility to avoid collision with the vehicle. In general, they try to avoid colliding with the shuttle. Therefore, SCA assumes that pedestrian evades and the vehicle retains its trajectory and adjusts only its speed. This speed is changed according to the Eq. 1. Therefore the pedestrian is segmented in three sets.

- A: Pedestrians, where SCA can be used, even if an alternative route exists
- B: Pedestrians, where SCA could use when no alternative route
- C: Pedestrians, where SCA cannot be used

$$f = \begin{cases} max((f_{sr}^{|B|} * min(1.0, \frac{d_s}{d_{min}})), f_{es}) & \text{for } |A| \neq \emptyset \text{ , } |B| \neq \emptyset \text{ and } |C| = \emptyset \\ max(min(1.0, \frac{d_s}{d_{min}}), f_{es}) & \text{for } |A| \neq \emptyset \text{ , } |B| = \emptyset \text{ and } |C| = \emptyset \\ max((f_{sr}^{|B|}), f_{es}) & \text{for } |A| = \emptyset \text{ , } |B| \neq \emptyset \text{ and } |C| = \emptyset \\ 1 & \text{for } |A| = \emptyset \text{ , } |B| = \emptyset \text{ and } |C| = \emptyset \\ 0 & \text{for } |C| \neq \emptyset \end{cases} \quad (1)$$

**Table 1.** Desired behaviors of the autonomous shuttle

| Characteristic | Shuttle recognized | Human behavior | Social collision avoidance | Perform interaction |
|---|---|---|---|---|
| Child | Yes | arbitrary | Only if no alternative route exists | |
| Child | No | arbitrary | No | If there is no alternative: contact the person and ask him/her to clear the lane |
| Retired person | Yes | arbitrary | Only if no alternative route exists | |
| Retired person | No | arbitrary | No | If there is no alternative: contact the person and ask him/her to clear the lane with increased volume |
| Invalid | Yes | arbitrary | Only if no alternative route exists | |
| Invalid | No | arbitrary | No | If there is no alternative: contact the person and ask him/her to clear the lane |
| Adult | No | arbitrary | No | If there is no alternative: honk, contact the person and ask him/her to clear the lane |
| Adult | Yes | Standing on the phone | Only if no alternative route exists | |
| Adult | Yes | Walking and telephoning | Yes, with lower speed depending on the distance to the vehicle | |
| Adult | Yes | Walking | Yes, with lower speed depending on the distance to the vehicle | |
| Adult | Yes | Sitting | No | If there is no alternative: honk, contact the person and ask him/her to clear the lane |
| Adult | Yes | Lying | No | If there is no alternative: honk, contact the person and ask him/her to clear the lane |
| Adult | Yes | Kneeling | No | If there is no alternative: honk, contact the person and ask him/her to clear the lane |
| Adult | Yes | Talking | Only if there is no alternative, use the horn | |
| Adult | Yes | Screaming | No | If there is no alternative: contact the person and ask him/her to clear the lane |

Equation 1 distinguishes five different cases with which the velocity reduction factor $f$ is calculated. The factor $f_{sr}((1,0])$ defines the speed reduction by pedestrians from the set B. The factor $f_{es}$ describes the speed reduction that is necessary to be able to brake the vehicle in an emergency situation without driving over a person. In the first case, pedestrians from sets A and B are in the detection zone. The speed is reduced by the factor $f_{sr}$ for all pedestrians from set B. This is due to the fact that the risk of an accident rises rapidly due to an increase in the number of people in the detection area from set B. Furthermore, the distance to the next pedestrian from Set A ($d_s$) is used to reduce the speed based on this distance. For this, the parameter $d_{min}$ was defined, from which the reduction should take place proportional to the distance. The calculated speed reduction factor is limited downwards by the factor $f_{es}$. The next two equations contain the necessary parts of the first equation depending on the presence of pedestrians. The fourth equation lets the vehicle drive without speed reduction because no pedestrians were detected. The last equation describes the case where pedestrians are in the detection area for which SCA is not allowed. The vehicle is decelerated for this purpose.

The desired behavior of the autonomous shuttle depends on the characteristic of the pedestrian. The behavior of the pedestrians plays a major role in the decision about the motion of the shuttle. The type of behavior of the shuttle in SCA is highly correlated to the type of pedestrians. For instance, children are more active than adults. Due to this reason, it becomes important to classify pedestrians based on their ages. A distinction is made between the characteristics, i.e., child, retired person, invalid and adult. For a child, retired person and invalid pedestrian types, SCA generally can only be used at low speed due to their unpredictable and restricted movement. If Tentacle-based Evasion for the shuttle is not possible then the person can be asked to clear the way by using interaction schemes such as voice commands, horns or flashing lights. The entire behavior should be as friendly as possible to the pedestrians and try not to frighten the pedestrians. Furthermore, the direction of the person's gaze is analyzed and conclusions are drawn as to whether the shuttle is recognized or not. If the shuttle cannot be seen due to the direction of the pedestrian's gaze, it must, of course, be proceeded more cautiously and, if necessary, appropriate aids must be used to attract attention.

### 3.2 Tentacle-Based Evasion

Tentacle-based Evasion and the associated cost calculation are based on the approach [9]. In the original version, this approach was based on an obstacle map. The implementation has been modified so, that instead of working on the obstacle map, the data of the pedestrian detector of the simulation can be used directly. The costs for the Tentacle-based Evasion are calculated and based on this it is decided whether a Tentacle-based Evasion or the SCA is used.

## 4 Experiments and Results

Experiments are performed with a varying number of pedestrians and behaviors to test the shuttle with and without PIS. Results for one of the trials are illustrated here. The simulated autonomous shuttle maneuvers through randomly walking pedestrians from left to right for 80 m. Simulated pedestrians avoid collision with the shuttle by turning ±90°. Due to the randomness of the pedestrian, the average of three trials was taken for the same scene.

### 4.1 Trajectory and Activities

Figure 2 shows the entire trajectory of the shuttle and the trajectory of pedestrians within a radius of 7 m of the shuttle by means of a test run. The shuttle starts at the green dot and passes the finish line at the red cross. The color of each trajectory represents the time axis. In the beginning, two pedestrians walk towards the shuttle. Both pedestrians recognize the approaching shuttle and turn to avoid a collision. From the activity plot Fig. 3(a), it can be seen that an attempt was made to avoid pedestrians with Tentacle-based Evasion. Shortly thereafter, due to the high cost of Tentacle-based Evasion, the SCA becomes active. After 12 s, when the pedestrian moves, the shuttle accelerates. Shortly thereafter, Tentacle-based Evasion takes place because of the second pedestrian. At approx 35 s a pedestrian approaches the shuttle again and the shuttle first avoids the pedestrian and then changes to SCA for approx 3 s and 2 s until it passes the pedestrian. The pedestrian has also recognized the shuttle and has evaded to the left due to an approaching collision. The shuttle then uses the default behavior Tentacle-based Evasion until the target point, which passes the control signals without any changes.

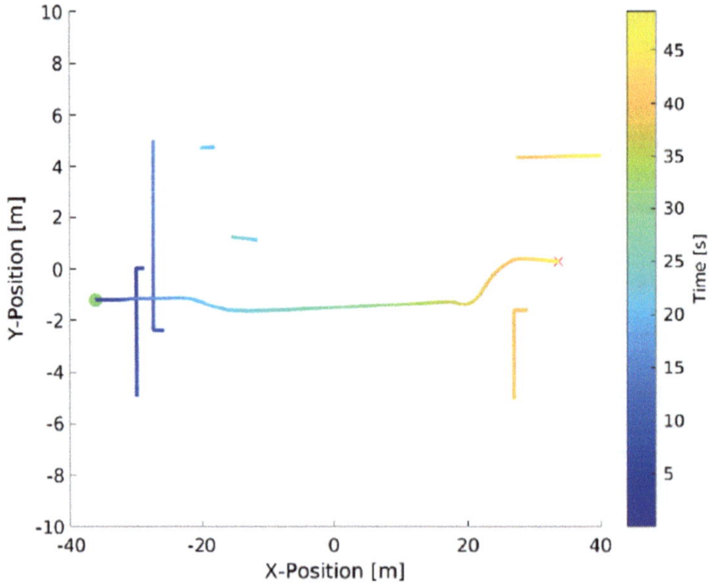


**Fig. 2.** Example trajectory of an experiment, where the robot trajectory and the trajectories of the pedestrians are represented in the detection area of the robot.

### 4.2 Travel Time

To examine the efficiency of the PIS, the travel time of the shuttle is observed. The travel time is taken from the starting point to the goal point of the shuttle. Due to non-deterministic pedestrian motion, the average value from three trials is taken for each scene. Four scenes are considered with 5, 10, 15 and 20 pedestrians. All experiments were performed with and without PIS. The graph Fig. 3(b), shows that the travel time with PIS switched on is shorter than PIS switched off. Only with 15 pedestrians, the travel time is almost the same. Altogether, with PIS switched on, 9.5% time was saved.

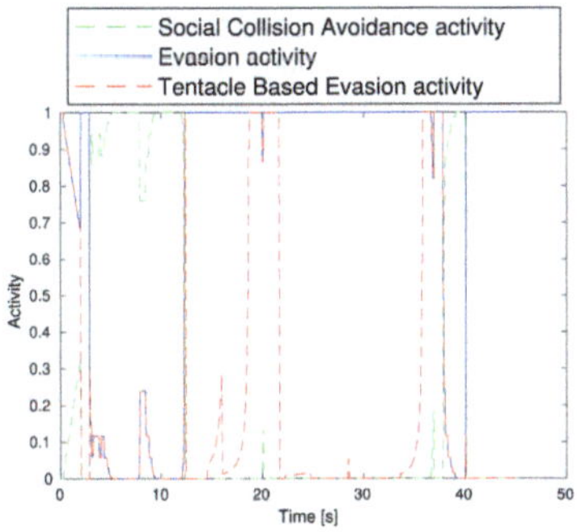


(a) Activities of experiment

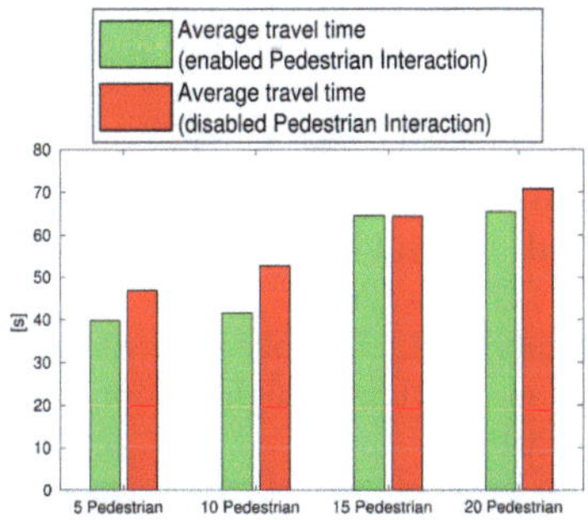


(b) Average travel time based on the three experiments per Scenario

**Fig. 3.** Travel time for the experiment

## 5 Conclusion

When the PIS is activated, the shuttle shows better handling and saves an average of 9.5% of the driving time. So far, the results have been based on simulation and in the near future, we are planning to experiment it on a real shuttle prototype. To move to a real autonomous shuttle, suitable pedestrian detectors must be used, which also make it possible to classify pedestrians and define its state. The classifications do not necessarily have to correspond to the classification of this work, but can easily be adapted accordingly. Deep learning approaches are conceivable for detection in practice. Due to limitations in the simulation environment, such as the unrealistic avoidance behavior of the simulated pedestrians, even better results are expected on the real machine. It should also be noted that the detection accuracy of the persons will play a major role. For these reasons, tests on the real shuttle are absolutely necessary.

## References


1. Bera, A., Randhavane, T., Kubin, E., Wang, A., Manocha, D., Gray, K.: Classifying group emotions for socially-aware autonomous vehicle navigation. In: Proceedings of the IEEE Conference on Computer Vision and Pattern Recognition Workshops, pp. 1039–1047 (2018)
2. Trautman, P., Krause, A.: Unfreezing the robot: navigation in dense, interacting crowds. In: 2010 IEEE/RSJ International Conference on Intelligent Robots and Systems (IROS), pp. 797–803. IEEE (2010)
3. Trautman, P., Ma, J., Murray, R.M., Krause, A.: Robot navigation in dense human crowds: statistical models and experimental studies of human-robot cooperation. Int. J. Robot. Res. **34**(3), 335–356 (2015)
4. Burgard, W., Cremers, A.B., Fox, A.B., Hähnel, D., Lakemeyer, G., Schulz, D., Steiner, W., Thrun, S.: The interactive museum tour-guide robot. In: AAAI/IAAI, pp. 11–18 (1998)
5. Bauer, A., Klasing, K., Lidoris, G., Mühlbauer, Q., Rohrmüller, F., Sosnowski, S., Xu, T., Kühnlenz, K., Wollherr, D., Buss, M.: The autonomous city explorer: towards natural human-robot interaction in urban environments. Int. J. Soc. Robot. **1**(2), 127–140 (2009)
6. Kümmerle, R., Ruhnke, M., Steder, B., Stachniss, C., Burgard, W.: Autonomous robot navigation in highly populated pedestrian zones. J. Field Robot. **32**(4), 565–589 (2015)
7. Bai, H., Cai, S., Ye, N., Hsu, D., Lee, W.S.: Intention-aware online pomdp planning for autonomous driving in a crowd. In: 2015 IEEE International Conference on Robotics and Automation (ICRA), pp. 454–460. IEEE (2015)
8. Wolf, P., Ropertz, T., Berns, K., Thul, M., Wetzel, P., Vogt, A.: Behavior-based control for safe and robust navigation of an unimog in off-road environments. In: Berns, K., Dressler, K., Fleischmann, P., Görges, D., Kalmar, R., Sauer, B., Stephan, N., Teutsch, R., Thul, M. (eds.) Commercial Vehicle Technology 2018. Proceedings of the 5th Commercial Vehicle Technology Symposium – CVT 2018, Commercial Vehicle Alliance Kaiserslautern (CVA), pp. 63–76. Springer Vieweg, Kaiserslautern, March 13–15 2018. iSBN 978-3-658-21299-5. https://doi.org/10.1007/978-3-658-21300-8, https://link.springer.com/content/pdf/10.1007%2F978-3-658-21300-8.pdf
9. Ropertz, T., Wolf, P., Berns, K.: Behavior-based low-level control for (semi-) autonomous vehicles in rough terrain. In: Proceedings of ISR 2018: 50th International Symposium on Robotics, pp. 386–393. VDE VERLAG GMBH, Munich, Germany, June 20–21 2018. iSBN: 978-3-8007-4699-6

# Task Dependent Trajectory Learning from Multiple Demonstrations Using Movement Primitives

Josip Vidaković(✉), Bojan Jerbić, Bojan Šekoranja, Marko Švaco, and Filip Šuligoj

Department of Robotics and Production System Automation, Faculty of Mechanical Engineering and Naval Architecture, University of Zagreb, Zagreb, Croatia
{josip.vidakovic,bojan.jerbic,bojan.sekoranja, marko.svaco,filip.suligoj}@fsb.hr

**Abstract.** We propose a model for learning robot task constrained movements from a finite number of observed human demonstrations. The model uses the variation between demonstrations to extract important parts of the movements and reproduce trajectories accordingly. Regions with low variability are reproduced in a constrained manner, while regions with higher variability are approximated more loosely to achieve shorter trajectories. The demonstrations are sampled into states and an initial state sequence is chosen by a minimum distance criterion. Then, a method for state variation analysis is proposed that weights the states according to its similarity to all the other states. A custom function is constructed based on the state-variability information. The time function is then coupled with a state driven dynamical system to reproduce the trajectories. We test the approach on typical two-dimensional task constrained trajectories with constrains on the beginning, in the middle and the end of the movement. The approach is further compared with the case of using a standard exponentially decayed time function.

**Keywords:** Robot learning from demonstration · Dynamical movement primitives · Constrained trajectory

## 1 Introduction

As robots perform more complex tasks and humans trying to optimize the time being necessary to program them, robot learning methods are becoming more and more interesting. Learning can have different goals and forms. It ranges from learning a collision free trajectory and other trajectory constrains until learning to fulfill an entire task. Learning information can be collected by self-exploration or by encoding human demonstrations into useful learning data. The second one usually requires less time and computation resources and is therefore a promising concept for various industrial robotic applications. The field is known as Robot learning by demonstration (LfD) or Robot programming by demonstration (PbD).


K. Berns and D. Görges (Eds.): RAAD 2019, AISC 980, pp. 275–282, 2020.
https://doi.org/10.1007/978-3-030-19648-6_32

Here, methods for trajectory modeling are divided into statistical methods where generalized trajectories are modeled from a number of demonstrations, and dynamical systems mostly suitable for "one-shot" learning. Hybrid approaches exist where the shape of the trajectory is generated by the dynamical system or some other model but the parameters are learned by statistical methods [1] or reinforcement learning algorithms [2–5].

Statistical methods often use the combination of statistical models for encoding the demonstration data e.g. Gaussian mixture model (GMM) and regression methods as Linear Weighted regression (LWR) and Gaussian mixture regression (GMR) to obtain trajectory data [6]. It is shown that such generalized trajectories can represent movements more accurately than trajectories extracted from a single demonstration [7]. Statistical variability of demonstrations can also be used for force based controlled robots to encode stiffness information based on the variability of demonstrations as shown in [8]. A widely used framework for trajectory modeling with dynamical systems are the Dynamic movement primitives – DMP, which will be covered more detailed in the rest of the paper.

Trajectories encoded from demonstrations can also be task dependent through the use of learning mechanisms. Most approaches utilize a set of demonstrations with low variability to learn task specific parameters for the execution trajectories. Tasks executions are therefore not well generalizable to different starting positions and spatial variations. Some approaches are modeling the shape parameter of a trajectory with respect to a task parameter [9–11] which finally leads to a multi-dimensional approximation of the forcing term [12].

In this paper, we present an approach for task dependent modeling of trajectories based on a set of variable demonstrations. We assume the validity of the demonstration data, and define the goal of the executed movement as to reproduce the most important parts of the demonstrations in a natural and robust way. For this we, use a state representation of demonstrated trajectories and weight the states according to a variability analysis, where states with lower variability gain higher importance and vice versa. Such approach is useful for modeling of constrained spatial movements where multiple trajectories can be demonstrated. It can also generalize trajectories from the entire state space which means it is not sensitive to new initial positions.

For driving the state dependent DMP and achieving the already mentioned importance dependency of the trajectory, a custom time function is proposed that substitutes the standard canonical system of the DMP. It is constructed based on a variability analysis of each extracted state and drives the state DMP through the attractor points by giving higher time priority to states with lower demonstration variations.

## 2 Methods

### 2.1 Learning Task Redundancies

It is assumed that $M$ examples of a task are demonstrated from highly different initial positions. Each demonstration consists of position data $x$ for every degree of freedom of the demonstrated trajectories. The demonstration data is resampled for a fixed time step and states are extracted evenly across the trajectories based on linear time spacing and a manually chosen number of states $N$. This way the number of data points is reduced significantly and the data matrix $T_{i,j} = \left\{ \{x_j\}_{j=1}^{N} \right\}_{i=1}^{M}$ is constructed. Depending on the complexity of the demonstrated trajectories, the number of states should be high enough in order to capture important transitions.

Prior to trajectory reproduction, an initial state sequence is chosen from the sampled demonstrations for every reproduction attempt individually, depending on the starting state (initial position) of the rollout. The state sequence is chosen based on a minimal distance criterion, meaning that the next state is always chosen by looking for the closest following state. This results in a state vector $\mu_x$ whose elements are chosen by using (1) where $z$ is the current state position and $y$ represents a vector with all possible next states.

$$\mu_i = \left\{ y \rightarrow \min_{\substack{y \in T_{i+1} \\ z \in T_i}} \|y - z\| \right\}_{i=1}^{N} \tag{1}$$

The importance of each state in the sequence is determined by calculating the variability with respect to other states of the demonstrations. For a two-dimensional setup which is presented in this paper, this is done by calculating the Euclidian distance from the current state to all the other states from the demonstrations, $d_{i,j} = \left\| \mu_i - T_{i,j} \right\|$ and linearly normalizing the distances between values zero and one:

$$d_{i,j} = \frac{d_{i,j} - \min(d)}{\max(d) - \min(d)} \tag{2}$$

To achieve a better quantitative measure of state density which is here considered as the inverse to variability, the normalized data is additionally mapped to an scaled exponential decay function $xf = 0.9^k$ with a manually set base of 0.9, and $k \in \{1, \ldots, 100\}$. This weights the distance vector so that the closer states gain higher values then the states further away. A new distance vector $d^e$ is calculated. Finally, variability of the state is then calculated as the sum of all the components of the weighted distance vector $d^e$, $\sigma_i = sum(d^e)$. This calculation is conducted for every state in the state sequence $\mu$ and the entire variability vector $\sigma$ is linearly normalized once more across its values to obtain more relevant variability information. This results in a variability vector that contains a single scalar value for every state of the initial state sequence as the measure of importance. This will be used as relevant data in the reproduction phase.

## 2.2 Reproduction

Dynamic movement primitives were firstly presented in [13] and extended in [14, 15]. In a classical DMP system, a proportional-derivative i.e. transformed system is driven by a canonical system from the starting point to the goal point. The transformed system has a forcing term which shapes the initial linear trajectory in a certain desired way.

While the canonical system has the form of an exponential decay function which diminishes the influence of the forcing term as the transformed system reaches its goal.

This concept is widely used in imitation learning as the forcing term can be learned from demonstrated trajectories with various regression methods and can be altered to achieve collision avoidance. A drawback of DMPs in some applications can be that it is not state driven. This means that, by default, it is not possible to navigate a DMP through a certain state without ending the motion. This is a problem when creating a task constrained movement.

Here, we use a variation of DMP in order to achieve following of the state vector $\mu$ and consequently generating a trajectory. The forcing term is substituted by a time-state dependency. This is achieved by shifting the goal of the dynamical system by the means of a time probability to be in a state (4). This approach is originally presented in [8], for encoding stiffness information into a torque controlled robot trajectory and is here used without the originally proposed stiffness matrix. The acceleration of the system is calculated based on the sum of $N$ proportional-derivative systems where $N$ represents the number of states, $y$ and $\dot{y}$ the current position and velocity of the system and $a_y$ and $b_y$ are the gain parameters.

$$\hat{\ddot{y}} = \sum\nolimits_{i=1}^{N} h_i(t) \cdot a_y \left[ b_y(\mu_i - y) - \dot{y} \right] \tag{3}$$

The already mentioned time-state dependency is realized by using an exponential decay function $\dot{t} = -\alpha_x \cdot t$ as the time function, where $t$ is initialized with $t = 1$ on the beginning of the reproduction phase and converges to zero by a fixed time-step by means of the gain parameter $\alpha_x$. The time function is coupled with the dynamics of the system through a set of $N$ Gaussians $\mathcal{N}\left(t; \mu_i^t, \Sigma_i\right)$ with equally distributed centers over the period from zero to one where each of the Gaussians corresponds to a state from $\mu$. The time dependent probability to be in a state is given by:

$$h_i(t) = \frac{\mathcal{N}\left(t; \mu_i^t, \Sigma_i\right)}{\sum_{k=1}^{N} \mathcal{N}\left(t; \mu_k^t, \Sigma_k\right)} \tag{4}$$

By using this model, sequential state following can be achieved.

It can be noted that the shape of the time function is determining how much time will be assigned to each state. When using a standard exponential decay term initialized at 1, the function has higher falling rates on the beginning, indicating that the states occurring in this period will have less importance then the states on the end of the movement, as the decay term slowly converges to zero and the time is not falling with a high rate.

In order to achieve importance based trajectory following we are modifying the standard decay function in a way that variable amounts of time can be assigned to each of the states. State redundancy/variability information from the variability vector $\sigma$ presented in Sect. 2.1, is used to construct a custom decayed time function.

As already mentioned, to achieve the state coupling with the time function, the execution time i.e. the total time period of the decay term is linearly sampled by the already defined number of states. Each state $\mu_i$, is represented with a time parameter $\mu_i^t$ according to the sequential order where the first state corresponds with the time parameter 1 ($\mu_i^t = 1$). Then, a vector for each state is formed with the time parameters as its values, where the length of the vector is proportionally depending on the variability of the state $\theta_i = \{x_j\}_{j=1}^{N}$, $n = f(\sigma_i), x_j \in \mu_i^t$. The full time vector is then obtained by concatenating the individual state vectors into one single vector $\theta = \{\theta_1, \theta_2, \ldots, \theta_N\}$. This results in time function where the amount of time-steps assigned to a state depends on the variability with respect to all other demonstrations (Fig. 1).

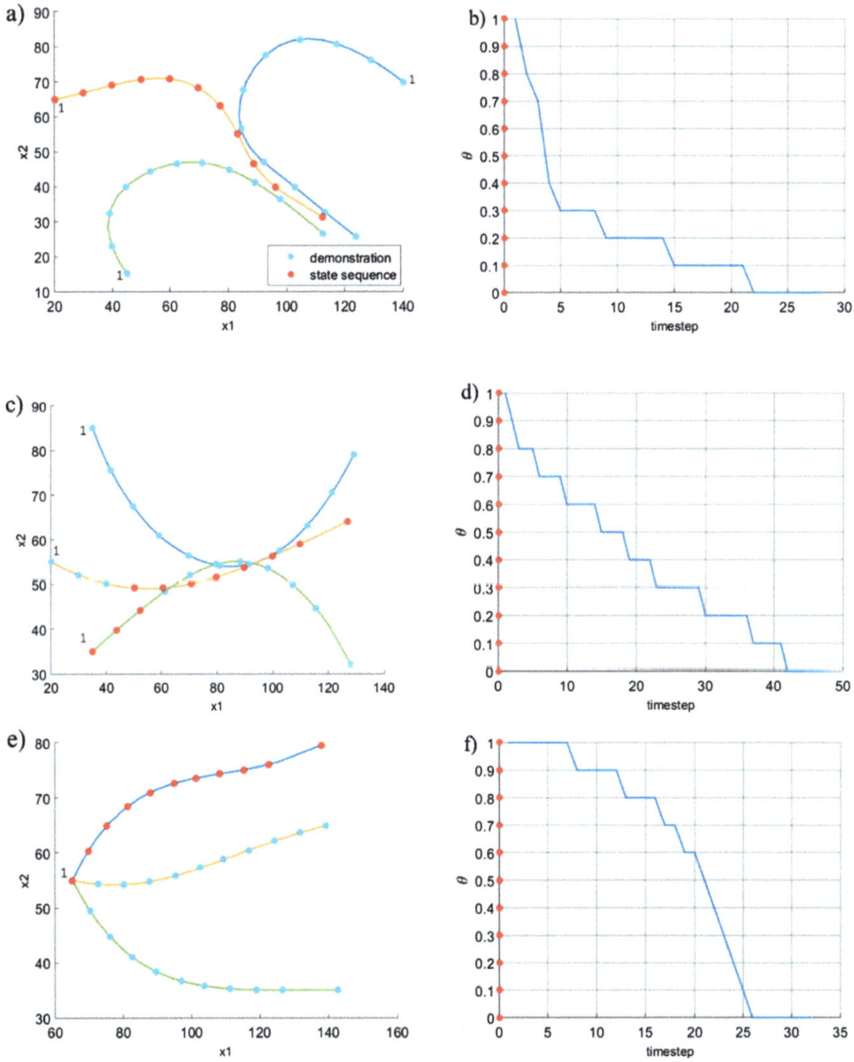


**Fig. 1.** Two dimensional trajectory demonstrations in Cartesian space (a, c, e) with the chosen trajectory sequence consisting of 11 states represented with red dots. Corresponding time functions are given in (b, d, f) where the red dots represent the assigned states in the time domain. The starting positions of the demonstrations are marked with the index 1. It can be seen that states with lower variability are assigned with more time-steps in the decay function.

## 3 Results

The presented system is tested for two-dimensional trajectory modeling in Cartesian space. Three most general situations will be covered: moving towards a target, passing through an obstacle, reproduction from a mutual starting position. Three demonstrations are arbitrarily simulated to represent each of the test situations (Fig. 2), and the demonstration data saved as position coordinates in a vector form. State selection and variability analysis has been performed as described in Sect. 2.1. Trajectories are

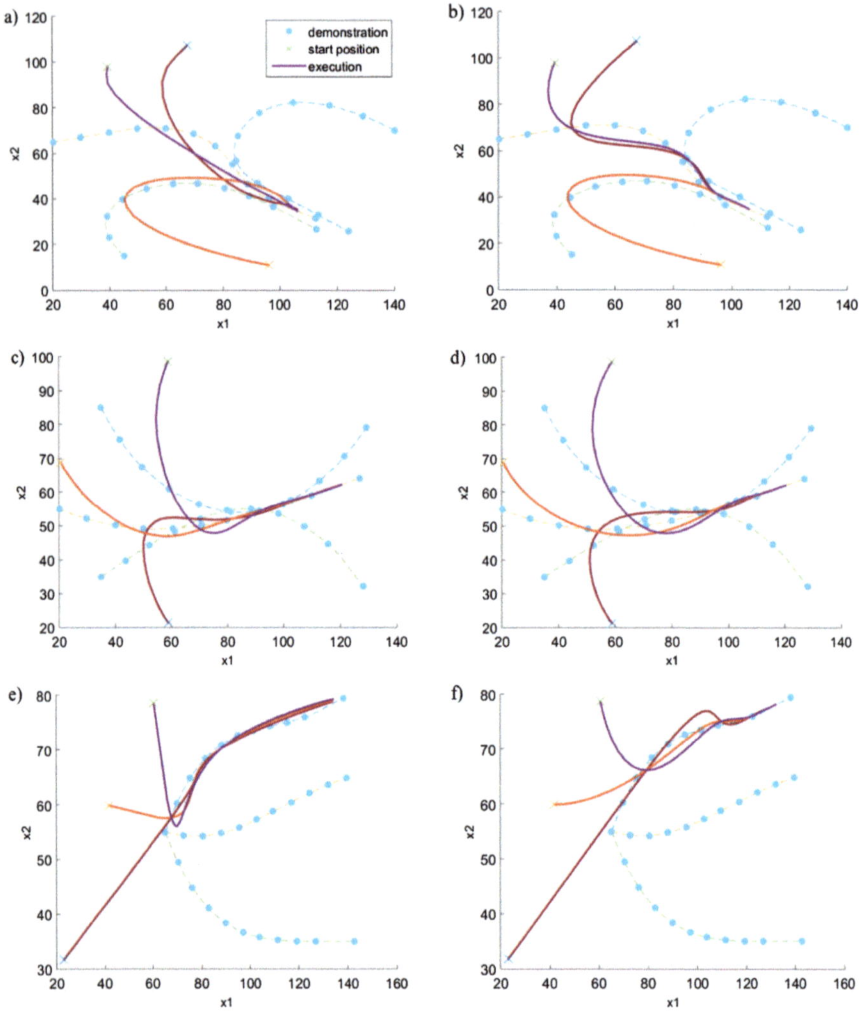


**Fig. 2.** Trajectory reproduction for three different test scenarios. Three demonstration are provided for each. Trajectory reproduction from three different initial positions is shown in full lines. Reproduction with the proposed modified time function is shown in (a), (c), (e). Reproduction with the standard exponential decay time function is shown in (b), (d), (f).

reproduced using the proportional-derivative system described in Sect. 2.2 from three different starting positions each. For comparison, trajectories for the same demonstrations and starting positions are also modeled with the same dynamical system but using a standardly used exponential time decay function and shown in the right part of Fig. 2.

Figure 2 shows that the variability analysis and the modified decay function enable the DMP based system to follow important parts of the demonstrations while loosely approximating demonstration parts with high variability. On the other hand, DMP based system with a standard decay time function is giving the highest importance to the goal state of the movement while neglecting the importance of other states. The proposed approach has the ability of closely interpolating the more important parts of the demonstrations which is an advantage over the standard approach. The other benefit is that the obtained trajectories are shorter and with less overshot.

## 4 Discussion and Future Work

In this work, we used the assumption of all the demonstrations being valid, which is common in imitation learning. The variability analysis presented is based on a Euclidian distance calculation. This is suitable for two and three-dimensional trajectory learning but when increasing the dimensionality by including the orientation information, distance can be calculated based on the norm difference between full homogenous matrices. For demonstrations including additional degrees of freedom, other similarity measures can be specifically developed.

Substituting the standard time function with the variability dependent time function for the reproduction of the state DMP showed to give good results in terms of task dependent generalization. The trajectories can be reproduced from different starting positions while preserving important parts of the demonstrations and reaching the goal of a demonstration in an approximate manner.

Future work will include testing the approach for higher dimensional problems with real robot demonstrations extracted through kinesthetic teaching. The final goal is to add extrapolation capabilities to the model and compare its performance with other learning frameworks e.g. reinforcement learning when learning specific robot tasks.

## 5 Conclusion

We proposed an approach for learning from multiple demonstrations. It is based on a variability analysis to determine the importance of demonstrated states and a modified version of Dynamic Movement Primitives based on sequenced state approximation for the reproduction of trajectories. We demonstrated the feasibility of the approach on two-dimensional trajectory problems with high demonstration variability. The reproduced trajectories showed to generalize from different starting positions in a continuous way by approximating demonstrated states, while giving higher priority to states with lower variability.

**Acknowledgments.** Authors would like to acknowledge the Croatian Scientific Foundation through the "Young researchers' career development project – training of new doctoral students".

## References


1. Calinon, S., Li, Z., Alizadeh, T., Tsagarakis, G.N., Caldwell, G.D.: Statistical dynamical system for skill acquisition in humanoids. In: 12th IEEE-RAS International Conference on Humanoid Robots, pp. 323–329. IEEE, Osaka (2012)
2. Mülling, K., Kober, J., Kroemer, O., Peters, J.: Learning to select and generalize striking movements in robot table tennis. Int. J. Robot. Res. **32**(3), 263–279 (2013)
3. Theodorou, E., Buchli, J., Schaal, S.: Learning policy improvements with path integrals. In: PMLR, Proceedings of the Thirteenth International Conference on Artificial Intelligence and Statistics, Sardinia, Italy, vol. 9, pp. 828–835 (2010)
4. Nair, A., McGrew, B., Andrychowicz, M., Zaremba, W., Abbeel, P.: Overcoming exploration in reinforcement learning with demonstrations. In: 2018 IEEE International Conference on Robotics and Automation (ICRA), Brisbane, Australia (2018). https://doi.org/10.1109/icra.2018.8463162
5. Švaco, M., Jerbić, B., Polančec, M., Šuligoj, F.: A reinforcement learning based algorithm for robot action planning. In: Proceedings of the 27th International Conference on Robotics in Alpe-Adria Danube Region, RAAD (2018)
6. Calinon, S., Guenter, F., Billard, A.: On learning, representing, and generalizing a task in a humanoid robot. IEEE Trans. Syst. Man Cybern. Part B (Cybernetics) **37**(2), 286–298 (2007)
7. Hewitt, A., Yang, C., Li, Y., Cui, R.: DMP and GMR based teaching by demonstration for a KUKA LBR robot. In: 2017 23rd International Conference on Automation and Computing (ICAC), pp. 1–6. IEEE, Huddersfield (2017)
8. Calinon, S., Sardellitti, I., Caldwell, D.G.: Learning-based control strategy for safe human-robot interaction exploiting task and robot redundancies. In: IEEE/RSJ International Conference on Intelligent Robots and Systems, Taipei (2010)
9. Stulp, F., Raiola, G., Hoarau, A., Ivaldi, S., Sigaud, O.: Learning compact parameterized skills with a single regression. In: 13th IEEE-RAS International Conference on Humanoid Robots (Humanoids), pp. 417–422. IEEE, Atlanta (2013)
10. Forte, D., Gams, A., Morimoto, J., Ude, A.: On-line motion synthesis and adaptation using a trajectory database. Robot. Auton. Syst. **60**(10), 1327–1339 (2012)
11. Figueroa, N., Pais Ureche, A.L., Billard, A.: Learning complex sequential tasks from demonstration: A pizza dough rolling case study. In: The Eleventh ACM/IEEE International Conference on Human Robot Interaction, pp. 611–612. IEEE, Christchurch (2016)
12. Pervez, A., Lee, D.: Learning task-parameterized dynamic movement primitives using mixture of GMMs. Intel. Serv. Robot. **11**(1), 61–78 (2018)
13. Ijspeert, J.A., Nakanishi, J., Schaal, S.G.: Trajectory formation for imitation with nonlinear dynamical systems. In: International Conference on Intelligent Robots and Systems (IROS), pp. 752–757. IEEE, Maui (2001)
14. Schaal, S., Mohajerian, P., Ijspeert, J.: Dynamics systems vs. optimal control a unifying view. Prog. Brain Res. **165**, 425–445 (2007)
15. Ijspeert, A., Nakanishi, J., Hoffmann, H., Pastor, P., Schaal, S.: Dynamical movement primitives: learning attractor models for motor behaviors. Neural Comput. **25**(2), 328–373 (2013)

# Advances in Human-Robot Interaction

# Bounded Self-motion of Functional Redundant Robots

Leon Žlajpah(✉) and Tadej Petrič

Jožef Stefan Institute, Ljubljana, Slovenia
{leon.zlajpah,tadej.petric}@ijs.si

**Abstract.** In this paper, we consider a problem how to exploit a task space motion for lower-priority tasks when the end-effector motion allows some deviation of the motion for the primary task. Using a common redundancy resolution methods, the self-motion is only possible in the null-space. Therefore, we propose a novel combination of controllers in two spaces: in the task space and in the reduced task space, where DOFs corresponding to spatial directions allowing the deviations are excluded. The motion generated by the controller in the reduced task space is mapped into the main task and by properly selecting the controller parameters the resulting motion does not violate the constraints. To demonstrate the effectiveness of the proposed control we show simulation examples where motion in a constraint region is used to avoid joint limits.

**Keywords:** Redundant robots · Functional redundancy · Null-space control · Tracking relaxation

## 1 Introduction

Robot tasks are usually defined as a motion of the robot's end-effector. Typically, the motion of the end-effector is restricted by different constraints arising from the task itself or the working environment. In general, to perform the task without violating any constraints the robot has to be redundant. In this case, there are infinite joint configurations which result in the desired task-space pose of the end-effector. The problem is how to find such joint motions that realize the desired end-effector motion while guaranteeing that no constraints are violated.

Typically, the redundancy resolution is done by selecting a suitable objective function and then a feasible solution is found by using a local or a global optimization method. In most cases, it is assumed that the robot is kinematically redundant. However, there are tasks like arc welding or spray painting, where not all six Cartesian dimensions are important to accomplish the task. These unused DOFs can then contribute to the degree-of-redundancy (DOR) of the

This work was supported by EU Horizon 2020 RIA Programme grant 820767 CoLLaboratE and by Slovenian Research Agency grant P2-0076.


K. Berns and D. Görges (Eds.): RAAD 2019, AISC 980, pp. 285–292, 2020.
https://doi.org/10.1007/978-3-030-19648-6_33

robot and they can be exploited for the self-motion. This type of redundancy has been recognized as the *functional redundancy* [2,5,8]. To control robots with functional redundancy, Baron [2] proposed to add a virtual joint to the robot and used the usual redundancy resolution with the extended Jacobian. A more general version has been proposed in [7], where a robot has been virtually extended by three prismatic joints and three rotational joints at the end-effector allowing flexible redundancy control when the task is not fully constraint in the Cartesian space. Huo and Baron [5] proposed the orthogonal decomposition of the task-space without considering the null-space of the Jacobian matrix. In [9] the use of the extended Jacobian is proposed for tasks where axis-symmetric tools are used (like a spray-painting gun) and the orientation of the tool along one axis is not important. For 5 DOF tasks, a redundancy resolution algorithm based on sequential quadratic programming is proposed in [6] and based on convex optimization in [4].

These approaches do not consider the possibility that tasks may allow some freedom regarding the motion of controlled DOF, i.e. the task can be performed with a set of possible end-effector positions and/or orientations. For example, for spray-painting and welding tasks it is common to assume that the tool is axis-symmetric and hence, it has one additional rotational DOR. However, the task can be performed also when the orientation of the tool-axis is not exactly oriented but in some region around the object. In [3] a concept of Task Space Regions is proposed, where the idea is that the end-effector pose is not fixed but can belong to a region of admissible positions and/or orientations. These regions are then used for motion planning. In [1] a method for optimizing the trajectories considering allowed deviations around the path is presented.

All path planning algorithms require apriori knowledge of constraints. If this is not the case, a common approach is to perform redundancy resolution in the control. The simplest approach to solve the kinematic redundancy is to resolve it on the velocity level, where the differential kinematic equation is solved by using a kind of generalized inverse, and the projection of an arbitrary vector onto the null-space of the Jacobian is used to exploit the self-motion of the robot. When the motion in one spatial direction has to be controlled, but the task allows some deviation in this direction then we can treat this freedom-of-motion as *bounded functional redundancy*, i.e. this DOF belongs to the task space and allows simultaneously some motion which can be used to perform lower priority tasks.

This paper aims at contributing to this field by presenting efficient kinematic control strategies for bounded functional redundancy. The main goal is to improve the capabilities of robots using intrinsic and functional redundancy in the same framework. For that, we propose a control framework which allows exploiting freedom-of-motion in the task DOFs for lower priority tasks like joint limit avoidance or obstacle avoidance.

## 2 Modelling

We deal with robot manipulators, which can be represented by a serial kinematic chain. Let the configuration of the robot manipulator be represented by the $n$-dimensional vector $\boldsymbol{q}$ of joint positions. The robot's end-effector pose $\boldsymbol{x} \in SE(3)$, the six-dimensional Cartesian space, is represented by the position $\boldsymbol{p} \in \mathbb{R}^3$ and orientation $\mathcal{Q} \in SO(3)$ (quaternion representation), which can be expressed as a function of joint coordinates using the direct kinematic equations, $\boldsymbol{p} = \boldsymbol{p}(\boldsymbol{q})$ and $\mathcal{Q} = \mathcal{Q}(\boldsymbol{q})$.

In general, the spatial end-effector velocity $\boldsymbol{v}$ is expressed as

$$\boldsymbol{v} = [\dot{\boldsymbol{p}}^T \boldsymbol{\omega}^T]^T , \tag{1}$$

where $\dot{\boldsymbol{p}} \in \mathbb{R}^3$ and $\boldsymbol{\omega} \in \mathbb{R}^3$ are the linear and angular velocity of the end-effector, respectively. The relation between joint velocities $\dot{\boldsymbol{q}}$ and end-effector velocity $\boldsymbol{v}$ is represented by the differential forward kinematics in the form

$$\boldsymbol{v} = \begin{bmatrix} \dot{\boldsymbol{p}} \\ \boldsymbol{\omega} \end{bmatrix} = \mathbf{J}(\boldsymbol{q})\dot{\boldsymbol{q}} . \tag{2}$$

where $\mathbf{J}(\boldsymbol{q}) \in \mathbb{R}^{6\times n}$ is a geometric Jacobian matrix.

## 3 Redundancy

A redundant robot has more DOFs than it is necessary to perform a task. This implies that the redundancy is not only a feature of the robot structure but depends also on the task. To identify different types of redundancy, it is beneficial to define the following spaces:

- the *configuration space* $\mathcal{C} \in \mathbb{R}^n$ where the joint variables $\boldsymbol{q}$ are defined;
- the *operational space* $\mathcal{O} \triangleq SE(3)$ is the 6-dimensional Cartesian space, where the positions/orientations of the robot end-effector are defined; and
- the *task space* $\mathcal{T}$ is a Cartesian subspace where the task is defined, $\mathcal{T} \subseteq \mathcal{O}$.

Serial robots for which $\dim(\mathcal{O}) < \dim(\mathcal{C})$ are *intrinsic redundant*. When the task does not require that all 6 spatial DOFs are controlled, i.e. $\dim(\mathcal{T}) < \dim(\mathcal{O})$, the robot becomes *functional redundant*.

The task can be viewed as constraints of the end-effector pose $\boldsymbol{x}$ along the path. In some cases, the task allows that the constraints for one or more DOFs are relaxed, i.e. the end-effector can be in some region $\mathcal{A} \subset \mathcal{T}$ around the nominal path. Let the task be defined as a path in task space $\mathcal{T}$ as

$$\boldsymbol{x}_d = \boldsymbol{x}_d(s) \in \mathcal{T} , \tag{3}$$

where $s$ is a strictly increasing path parameter, $s \subseteq [0, 1]$ and let $\boldsymbol{\xi} \in \mathbb{R}^k$ be a set of coordinates in $\mathcal{T}$ corresponding to spatial directions in which the task allows some free motion. Then the free-motion region $\mathcal{A}(s)$ is defined as a subspace of $\mathcal{T}$ spanned by $\boldsymbol{\xi}$ with origin in $\boldsymbol{x}_d(s)$ and with bounds

$$B_i = [\underline{\boldsymbol{\xi}}_i(s,t), \overline{\boldsymbol{\xi}}_i(s,t)], \qquad i = 1, \ldots, k \,, \tag{4}$$

where $\underline{\boldsymbol{\xi}}_i(s)$ and $\overline{\boldsymbol{\xi}}_i(s)$ are minimal and maximal deviations from $\boldsymbol{x}_d(s)$ in spatial direction $\boldsymbol{\xi}_i$, respectively. The region $\mathcal{A}$ can change along the path or in time. In [3] these regions are termed Task Space Regions (TSRs) and are used for manipulation planning. In the following, we will present how the motion in $\mathcal{A}$ can be included in the redundancy resolution at the control level. We denote that when $\mathcal{A}$ exists, the robot becomes *bounded functional redundant.*

## 4 Kinematic Control

To this day many different control approaches for robots have been proposed. As robot manipulators are highly nonlinear dynamical systems, most of the proposed control strategies use some kind of inner-loop inverse dynamic control like computed torque technique or operational space control to compensate nonlinearities. Here, we focus on the redundancy resolution at the velocity level and we assume that the inner-loop controller already compensates the nonlinear dynamics of the robot.

The main idea of the redundancy resolution at the velocity level is to compute the necessary outer-loop control velocity $\dot{\boldsymbol{q}}_c$ by inverting Eq. (2)

$$\dot{\boldsymbol{q}}_c = \mathbf{J}^{\#}(\boldsymbol{q})\dot{\boldsymbol{x}}_c + \left(\mathbf{I} - \mathbf{J}^{\#}(\boldsymbol{q})\mathbf{J}(\boldsymbol{q})\right)\dot{\boldsymbol{q}}_n \,, \tag{5}$$

where $\dot{\boldsymbol{x}}_c$ represents the desired task-space control velocity and $\dot{\boldsymbol{q}}_n$ is an arbitrary joint velocity, which is projected into the null-space of $\mathbf{J}$ and can be used to perform some additional lower priority subtasks.

It is common to select the task-space control velocity $\dot{\boldsymbol{x}}_c$ in (5) as

$$\dot{\boldsymbol{x}}_c = \boldsymbol{v}_d + \mathbf{K}_p \boldsymbol{e} \,, \tag{6}$$

where $\boldsymbol{v}_d$ is the desired end-effector position/rotation velocity, $\mathbf{K}_p$ is the diagonal gain matrix, which define the close-loop behavior, and $\boldsymbol{e}$ is the end-effector position/rotation error. It is defined as

$$\boldsymbol{e} = \begin{bmatrix} \boldsymbol{p}_d - \boldsymbol{p} \\ 2\log(\mathcal{Q}_d \mathcal{Q}^{-1}) \end{bmatrix} , \tag{7}$$

where $\boldsymbol{x}_d = \{\boldsymbol{p}_d, \mathcal{Q}_d\}$ represents the desired pose of the end-effector. Note that the velocity $\dot{\boldsymbol{x}}_c$ and errors $\boldsymbol{e}$ can be expressed in any task space as long as the Jacobian matrix $\mathbf{J}$ relates the robot configuration space and the selected task space $\mathcal{T}$.

The task space $\mathcal{T}$ and consequently the Jacobian matrix $\mathbf{J}$ are defined considering intrinsic and/or functional redundancy of the robot. However, when bounded functional redundancy is present, then (5) does not make possible to use the motion in $\mathcal{A}$ for additional lower priority subtasks. As $\mathcal{A} \subset \mathcal{T}$ the spatial directions $\boldsymbol{\xi}$ are also included in $\mathcal{T}$. Therefore, projecting $\dot{\boldsymbol{q}}_n$ into the null-space of $\mathbf{J}$ using $(\mathbf{I}-\mathbf{J}^{\#}\mathbf{J})$ does not produce any motion in $\mathcal{T}$, meaning that the available motion in $\mathcal{A}$ cannot be exploited.

To exploit the motion in $\mathcal{A}$ for secondary tasks, it is necessary that spatial directions $\boldsymbol{\xi}$ are excluded from $\mathcal{T}$ and included in null-space projector $(\mathbf{I}-\mathbf{J}^{\#}\mathbf{J})$. Let denote $\mathcal{T}_{\mathcal{A}}$ the task space spanned over spatial directions excluding directions $\boldsymbol{\xi}$ and the corresponding Jacobian matrix as $\mathbf{J}_{\mathcal{A}}$. Then, projected velocities $(\mathbf{I}-\mathbf{J}^{\#}_{\mathcal{A}}\mathbf{J}_{\mathcal{A}})\dot{\boldsymbol{q}}_n$ have components also in directions $\boldsymbol{\xi}$. However, using $\mathbf{J}_{\mathcal{A}}$ in (5) is not allowed as this would allow moving along the task path. To overcome this problem, we propose to use the following kinematic control, which combines redundancy resolution in space $\mathcal{T}$ and in task space $\mathcal{T}_{\mathcal{A}}$.

The affordance that free-motion region around the path exists can be expressed in (6) by applying dead-zones to the components of $\boldsymbol{e}$ that are corresponding to $\boldsymbol{\xi}$

$$\widehat{\boldsymbol{e}}_i = \begin{cases} (\boldsymbol{e}_i - \overline{\boldsymbol{e}}_i) & \text{for } \boldsymbol{e}_i > \overline{\boldsymbol{e}}_i \\ 0 & \text{otherwise} \\ (\boldsymbol{e}_i - \underline{\boldsymbol{e}}_i) & \text{for } \boldsymbol{e}_i < \underline{\boldsymbol{e}}_i \end{cases} \quad \text{and} \quad i \in \boldsymbol{\xi} \tag{8}$$

Next, we augment the control (6) with a term corresponding to the task space velocities generated by the projection of $\dot{\boldsymbol{q}}_n$ in the null-space of $\mathbf{J}_{\mathcal{A}}$

$$\dot{\boldsymbol{x}}_{\mathcal{A}} = \mathbf{J}(\mathbf{I}-\mathbf{J}^{\#}_{\mathcal{A}}\mathbf{J}_{\mathcal{A}})\dot{\boldsymbol{q}}_n \tag{9}$$

yielding

$$\dot{\boldsymbol{x}}_c = \boldsymbol{v}_d + \mathbf{K}_p\widehat{\boldsymbol{e}} + \mathrm{sat}(\dot{\boldsymbol{x}}_{\mathcal{A}}, \overline{\dot{\boldsymbol{x}}}_i, \underline{\dot{\boldsymbol{x}}}_i)\,, \tag{10}$$

where $\mathrm{sat}(a)$ is the saturation function

$$\mathrm{sat}(a, a_{max}, a_{min}) = \begin{cases} a_{max} & \text{for } a > a_{max} \\ a & \text{for } a_{min} < a < a_{max} \\ a_{min} & \text{for } a < a_{min} \end{cases}. \tag{11}$$

To prevent the violation of constraints defined by $\mathcal{A}$, the dead-zone parameters in (8) are selected as

$$\overline{\boldsymbol{e}}_i = \overline{\boldsymbol{\xi}}_i - \frac{1}{\mathbf{K}_{p,i}}\,\overline{\dot{\boldsymbol{x}}}_i \ \text{and} \ \underline{\boldsymbol{e}}_i = \underline{\boldsymbol{\xi}}_i - \frac{1}{\mathbf{K}_{p,i}}\,\underline{\dot{\boldsymbol{x}}}_i\,, \tag{12}$$

where $\overline{\dot{\boldsymbol{x}}}_i$ and $\underline{\dot{\boldsymbol{x}}}_i$ are the limits for the task velocities generated by (9), respectively.

## 5 Case Study

The application of the proposed control scheme does not depend on the complexity of the motion the robot has to perform. It solely allows some freedom-of-motion around the task trajectory, which depends on the task requirements. Hence, to illustrate the usefulness of the proposed control scheme we have selected a simple line tracking task with a 7-DOF robot KULA LWR. The task was to move along line from the point $P_1$, $\boldsymbol{p}_1 = [0.25,\ -0.5,\ 0.4]^T$, to the point $P_2$, $\boldsymbol{p}_2 = [0.25,\ 0.5,\ 0.4]^T$ with a constant orientation $\mathcal{Q} = \mathcal{Q}(R_y(\pi))$ (rotation around $y$-axis) and with velocity $\boldsymbol{v}_d = [0, 0.125,\ 0,\ 0,\ 0,\ 0]^T$ (see Fig. 1a). The robot has one redundant DOF, which has been used to avoid joint limits. For that, we propose to use the following null-space velocity control

$$\dot{q}_{n,i} = K_{n,i} \begin{cases} (a\, q_{i,max} - q_i) & \text{for } q_i > a\, q_{i,max} \\ 0 & \text{otherwise} \\ (a\, q_{i,min} - q_i) & \text{for } q_i < a\, q_{i,min} \end{cases} \qquad i = 1, \ldots, n, \tag{13}$$

where $q_{i,max}$ and $q_{i,min}$ are the joint limits and $a$ has been used to select the critical distance to the joint limits.

The path was intentionally selected so, that the typical robot control algorithm could not move along the path without violating constraints. Figure 1 shows the response of the system with control (5). We can see that the end-effector tracks the desired path but during the motion one joint has reached the limit ($q_4$) (see Fig. 1b) and hence, the robot has stopped and aborted the task execution.

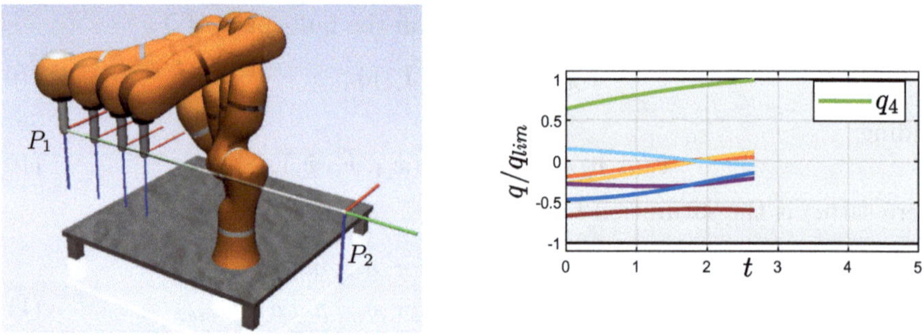


a) 3D path tracking

b) Normalized joint positions (regarding joint limits)

**Fig. 1.** KUKA LWR robot moving along straight line without exploiting bounded functional redundancy; $\mathbf{K}_p = 100$, $a = 0.9$.

If some deviations from the desired path are allowed, then the task can be completed. Here we assumed that when moving along the path a rotation around $y$-axis is allowed in a bounded region. Note that no deviations are allowed at the end of the path at $P_2$, where the robot has to be in the desired pose. This free rotation forms a one dimensional subspace $\mathcal{A}_1$ spanned over $\boldsymbol{\xi} = [r_y]$ (we denote with $r_a$ the axis of spatial rotation around $a$-axis). The bounds of $\mathcal{A}$ have been selected as $\overline{\boldsymbol{\xi}} = -\underline{\boldsymbol{\xi}} = \min(2\|\boldsymbol{p}(s) - \boldsymbol{p}(1)\|, 0.4)$rad. Figure 2 shows the simulation results. We can see that using motion in $\mathcal{A}$ enables the robot to complete the

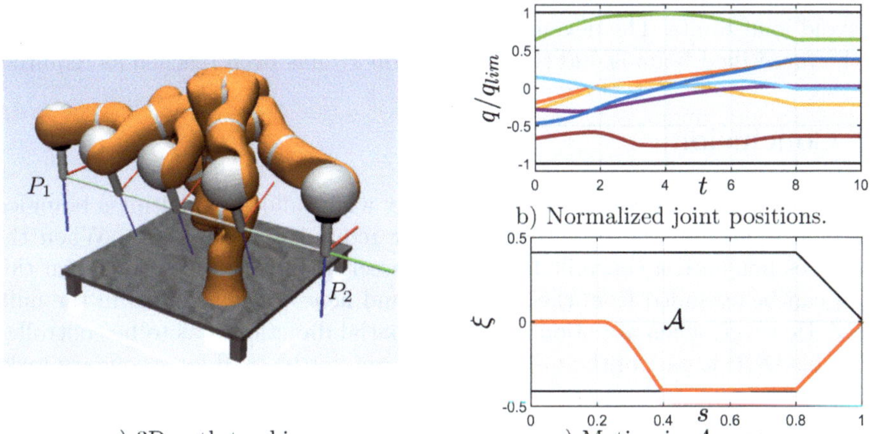


a) 3D path tracking

b) Normalized joint positions.

c) Motion in $\mathcal{A}$ space.

**Fig. 2.** KUKA LWR robot moving along straight line exploiting bounded functional redundancy (rotation around $y$-axis); $\mathbf{K}_p = 100$, $a = 0.9$.

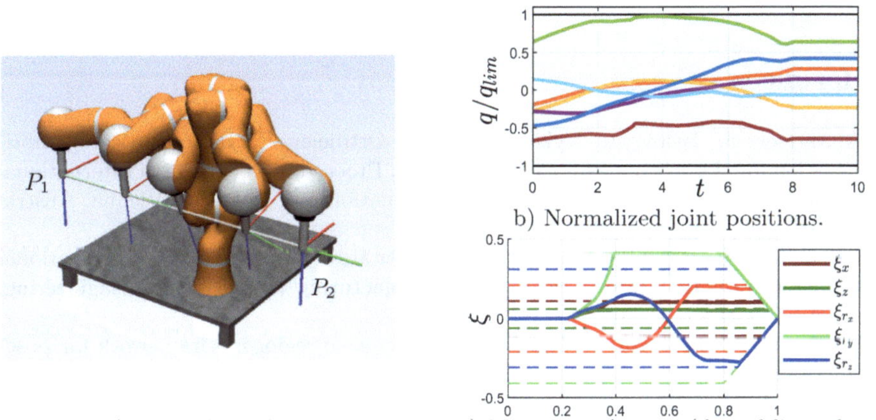


a) 3D path tracking

b) Normalized joint positions.

c) Motion in $\mathcal{A}$ space (dotted lines show bounds of $\mathcal{A}$).

**Fig. 3.** KUKA LWR robot moving along straight line exploiting bounded functional redundancy (rotation around all axes); $\mathbf{K}_p = 100$, $a = 0.9$.

task. When $q_4$ is close to the joint limit, the null-space control uses motion in direction $\boldsymbol{\xi}_{r_y}$ to avoid joint limit.

Finally, let assume that path tracking allows deviations in orientation and positions, except in the direction toward the goal point $P_2$. In this case, the subspace $\mathcal{A}$ has been spanned over $\boldsymbol{\xi} = [x, z, r_x, r_y, r_z]^T$ and the bounds have been $\overline{\boldsymbol{\xi}} = -\underline{\boldsymbol{\xi}} = \min(2\|\boldsymbol{p}(s) - \boldsymbol{p}(1)\|, [0.1\text{m}, 0.05\text{m}, 0.2\text{rad}, 0.4\text{rad}, 0.6\text{rad}]^T)$. The resulting motion is shown in Fig. 3. We can see that the robot uses motion in $\mathcal{A}$

to avoid joint limits. The deviations from the path during the motion have been in the prescribed bounds and the final pose in $P_2$ has been reached as required.

## 6 Conclusion

In this paper introduced a novel methodology which allows exploiting a bounded free-motion around the prescribed path for redundancy resolution. When the task does not need a controlled motion in a certain spatial direction, then this DOF can be excluded from the task space and hence, it is included in the null-space. However, when a motion in certain spatial directions has to be controlled then this DOF is part of the task space and can not be used for secondary tasks using common redundancy resolution methods. The proposed strategy combines controllers in two task spaces: in main task space $\mathcal{T}$, where DOFs with allowed deviations are included, and in the second task space $\mathcal{T}_{\mathcal{A}}$, where these DOFs are excluded. The motion generated by the controller in $\mathcal{T}_{\mathcal{A}}$ is mapped into $\mathcal{T}$ and by properly selecting the saturation parameters of this controller, the resulting motion does not violate the constraints in $\mathcal{T}$. This is also illustrated by simulation examples.

## References

1. Alatartsev, S., Belov, A., Nykolaychuk, M., Ortmeier, F.: Robot trajectory optimization for the relaxed end-effector path. In: Proceedings of the 11th International Conference on Informatics in Control, Automation and Robotics, Vienna, Austria pp. 385–390, September 2014
2. Baron, L.: A Joint-limits avoidance strategy for arc-welding robots. In: International Conference on Integrated Design and Manufacturing in Mechanical Engineering, Montreal, Canada (2000)
3. Berenson, D., Srinivasa, S., Kuffner, J.: Task space regions: a framework for pose-constrained manipulation planning. Int. J. Robot. Res. **30**(12), 1435–1460 (2011)
4. From, P.J., Gravdahl, J.T.: A real-time algorithm for determining the optimal paint gun orientation in spray paint applications. IEEE Trans. Autom. Sci. Eng. **7**(4), 803–816 (2010)
5. Huo, L., Baron, L.: Kinematic inversion of functionally-redundant serial manipulators: application to arc-welding. Trans. Can. Soc. Mech. Eng. **29**(4), 679–690 (2005)
6. Leger, J., Angeles, J.: A redundancy-resolution algorithm for five-degree-of-freedom tasks via sequential quadratic programming. In: TrC-IFToMM Symposium on Theory of Machines and Mechanisms, Izmir (2015)
7. Sanderud, A.R., Thomessen, T., Hashimoto, H., Osumi, H., Niitsuma, M.: An approach to path planning and real-time redundancy control for human-robot collaboration. In: IEEE/ASME International Conference on Advanced Intelligent Mechatronics, AIM, pp. 1018–1023 (2014)
8. Sciavicco, L., Siciliano, B.: Modelling and Control of Robot Manipulators. Advanced Textbooks in Control and Signal Processing, 2nd edn. Springer, London (2000)
9. Zanchettin, A.M., Rocco, P.: On the use of functional redundancy in industrial robotic manipulators for optimal spray painting. In: IFAC Proceedings Volumes (IFAC-PapersOnline), vol. 18, pp. 11495–11500 (2011)

# On-line Adaption of Virtual Guides Through Physical Interaction

Tadej Petrič(✉) and Leon Žlajpah

Jožef Stefan Institute, Ljubljana, Slovenia
{tadej.petric,leon.zlajpah}@ijs.si

**Abstract.** Virtual guide framework allows efficient learning and control of complex robot behaviors in human-robot interaction scenarios. The framework can help to guide users to move in a predefined direction or prevent them to enter a forbidden-region. As such, the framework also allows efficient modulation of regions by changing of parameters. In this paper, we introduce and evaluate the means of adapting path parameters through physical interaction. The main goal was to introduce an algorithm into a virtual guide framework which can partially modify the path trajectories. The path updates are based on physical interaction and allow human intervention to improve the task performance. This enables to update the path trajectory only where needed and hence, to bypass the need to re-learn the whole trajectory from scratch. Since virtual guides are also active while learning, the required effort from the user is lower compared to the required effort when the user is teaching the robot with kinesthetic guidance. The effectiveness of the proposed algorithm has been demonstrated with simulation results and experiments on a KUKA LWR robot.

**Keywords:** Human-robot physical cooperation · On-line path adaptation · Virtual guides

## 1 Introduction

A common strategy for defining a task on industrial robots is to involve experts to program them. This is usually a very time-consuming and a costly process. One of the possibilities to decrease learning time is to use the Programing-by-Demonstration (PbD). This also has an advantage since not only robotic experts but also skilfully personnel that are familiar with the task can program the robots. By utilizing kinesthetic teaching, users can easily demonstrate the task to the robot, what would otherwise require many programing hours by the robotic expert [2]. Programming methods that enable easy and intuitive robot programming, enhance the application of robots in many domains ranging from small and medium-sized enterprises to our daily home-environments.

This work was supported by IJS Director's found grand CoBoTaT.


K. Berns and D. Görges (Eds.): RAAD 2019, AISC 980, pp. 293–300, 2020.
https://doi.org/10.1007/978-3-030-19648-6_34

Human-robot interaction can be further improved, by exploiting the mechanical capabilities of robots and combining them with cognitive and perception capabilities of humans [14]. In such setup, the main task for the robot is to follow the human actions and improve his/her accuracy and payload capabilities. Typical examples of such collaboration are cobots [4,9,14] and telemanipulators [8,12]. It is common for both types that the operator controls the motion through the direct (cobots) or indirect contact (telemanipulators). In both cases, the aim of the robot is to provide support for the operator by guiding or constraining the motion by using *virtual guides* introduced in [17] as *virtual fixtures*. Virtual guides are virtual constraints implemented in an algorithm having similar functionality as real mechanical constraints [1,3].

The kinesthetic PbD relies on the interaction forces between the operator and the robot and can either be implemented using admittance control or impedance control [18]. In general, the impedance control is more suitable in situations when the robot is in contact with the stiff environment and the admittance control has its advantages when the environment is more compliant [13]. This distinction is significant when used in Physical Human-Robot Interaction (PHRI) framework. Namely, admittance based control prevents motion induced by external forces while the impedance-based constraints generate forces to nullify the motion. To ensure safe operation the characteristics of human arm need to be considered when defining the control parameters of either impedance or admittance controller for the robotic system [5,15].

However, not many paper deal with the on-line adaption of trajectories based on human feedback. For example, an adaption based on supervised learning combined with voice commands were used in [7]. In [11] human feedback was used to learn high-level task representations rather than motor skills. In [16] an approach based on verbal instructions was used to modify movements obtained by human demonstration. Kinesthetic teaching based on iterative updates was proposed in [10]. In [6] a method for altering robot behaviors based on human coaching gestures was proposed.

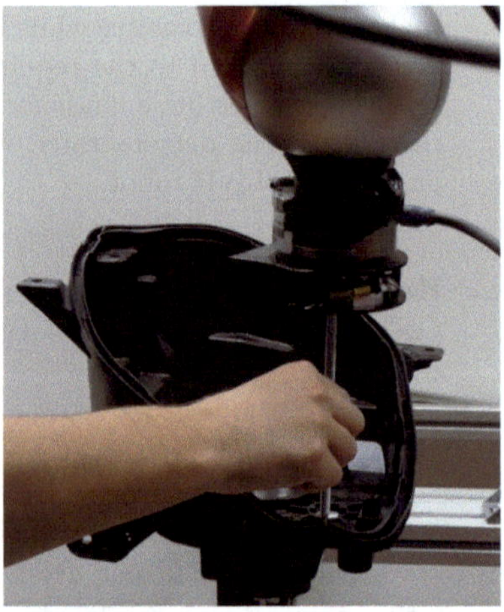

**Fig. 1.** Experimental setup of physical human robot interaction on KUKA LWR robot.

In this paper, we combine our work on coaching algorithms [6] and our recent work, where the framework for virtual guides using admittance control for the redundant robot was introduced [20]. The novelty in this paper is an algorithm for enabling on-line learning and adaption of the desired path based on the physical interaction. In Sect. 2 we introduce and explain the algorithm for on-line adaption of paths based on physical interaction. In Sect. 3 we show experimental results on KUKA LWR robot arm adapting to path based on human physical interaction as shown in Fig. 1.

## 2 Adaptation of Virtual Guides

The unified virtual guides framework based on admittance controller was introduced in [20]. We first give a brief overview of the admittance controller and then we describe how the path can be adapted on-line through physical interaction. On a velocity level, we assume that the virtual robot is defined as an integrator with the desired velocity as input and the inverse kinematics can be given as

$$\dot{\boldsymbol{q}}_d = \mathbf{J}^{\#}\boldsymbol{v}_c + (\mathbf{I} - \mathbf{J}^{\#}\mathbf{J})\dot{\boldsymbol{q}}_n \,. \tag{1}$$

The admittance controller is defined as

$$\boldsymbol{v}_c = \boldsymbol{v}_d + \mathbf{D}^{-1}(\mathbf{K}\boldsymbol{e}_x + \boldsymbol{F}_{\text{ext}}) \,, \tag{2}$$

where the parameters $\mathbf{K}$ and $\mathbf{D}$ define the dynamics of the virtual robot. For simplicity and w.l.o.g. we have limited the derivation and description of the proposed method only to the Cartesian position. Therefore, the end-effector error has been defined as

$$\boldsymbol{e}_x = \boldsymbol{p}_d - \boldsymbol{p} \,, \tag{3}$$

where $\boldsymbol{p}_d$ is the desired end-effector position. The details of the parametric description for the orientations are given in [19]. Let the path $\boldsymbol{f}$ be defined as a parametric curve depending on the path parameter $s$

$$\boldsymbol{p} = \boldsymbol{f}(s) \tag{4}$$

$$\boldsymbol{v} = \mathbf{J}_s\dot{s} \tag{5}$$

$$\dot{\boldsymbol{v}} = \mathbf{J}_s\ddot{s} + \dot{\mathbf{J}}_s\dot{s}^2 \,, \tag{6}$$

where $\mathbf{J}_s \in \mathbb{R}^{3\times 1}$ is the path Jacobian and represents the path direction at $s$. A good candidate for $s$ is a normalized path length, $s \in [0,\, 1]$. We assume that $\boldsymbol{f}(s)$ is a continuous function for all $s$.

When moving along the path, we assume first order dynamics given as

$$\dot{s} = k_s\mathbf{J}_s^T\boldsymbol{F}_{\text{ext}}, \tag{7}$$

where $k_s$ is the force scaling constant and $\boldsymbol{F}_{\text{ext}}$ is the humena interaction force, respectively. As it can be seen from (7), the projection of the external force $\boldsymbol{F}_{\text{ext}}$ on the path tangent has been used to guide the robot along the path.

In general a path can be encoded using parametric description given as

$$\boldsymbol{f}(s) = \frac{\sum_{i=1}^{m} \boldsymbol{w}_i\Psi_i(s)}{\sum_{i=1}^{m} \Psi_i(s)} \tag{8}$$

where $\boldsymbol{w}_i \in \mathbb{R}^{3xm}$ are weights that define the path. Here the radial basis kernel functions $\Psi_i$ are given as

$$\Psi_i = \exp\left(-\frac{(s-c_i)^2}{2h_i}\right). \tag{9}$$

where $c_i$ and $h_i$ define the centers and the widths of kernel functions, respectively. If not stated otherwise, we use $c_i$, $i = 1, \ldots, m$, that are equally spread along the path, $c_i \in [0,\ 1]$. The number of kernel functions $m$ depends on the required accuracy of the path description.

By defining $\mathbf{\Phi}(s) \in \mathbb{R}^m$ as a row vector with components

$$\mathbf{\Phi}_i(s) = \frac{\Psi_i(s)}{\sum_{i=1}^{m} \Psi_i(s)} \tag{10}$$

and using it in (8) yields

$$\boldsymbol{f}(s) = \mathbf{\Phi}(s)\boldsymbol{w} \tag{11}$$

Assuming that the values of $\boldsymbol{f}(s)$ are known along the path, the corresponding wights $\boldsymbol{w}$ can be computed by solving (11) in a least square sense as

$$\boldsymbol{w} = (\hat{\mathbf{\Phi}}^T\hat{\mathbf{\Phi}})^{-1}\hat{\mathbf{\Phi}}^T\hat{\boldsymbol{f}}, \tag{12}$$

where $\hat{\boldsymbol{f}} \in \mathbb{R}^{r\times\dim(\boldsymbol{f})}$ are path values and $\hat{\mathbf{\Phi}} \in \mathbb{R}^{r\times m}$ basis vectors for all path steps, respectively, and $r$ is the number of path steps.

Note that using this method multiple trajectories, i.e. a library of the demonstrated movements that represent the same path, can be used to determine the weights $\boldsymbol{w}$.

For the on-line adaptation of trajectory using physical interaction, we assume that the initial trajectory is given in advance, i.e. the corresponding weights $\boldsymbol{\omega}_0$ are known. This is required since we need to determine the path parameter $s$ (see Eq. 7) at the point where we would like to change the trajectory. Note that the path parameter $s$ is bounded to the interval $s \in [0,\ 1]$, and without knowing the initial trajectory in advance we can not integrate Eq. 7. Therefore, we propose to use the initial path in (7) yielding

$$\dot{s} = k_s \mathbf{J}_{s0} \boldsymbol{F}_{\text{ext}}, \tag{13}$$

where the path Jacobian $\mathbf{J}_{s0}$ depends on the initial weights $\boldsymbol{\omega}_0$.

To enable the on-line adaptation of the trajectory using the physical intercalation we need to transform the least square problem from Eq. 12 into a locally weighted recursive problem. The locally weighted recursive updates for on-line adaption is governed by

$$\boldsymbol{\omega}_{n+1} = \boldsymbol{\omega}_n + \mathbf{\Phi}_i(s)\boldsymbol{e}_u^T, \tag{14}$$

where the updating factor $\boldsymbol{e_u}$ was governed by

$$\boldsymbol{e_u} = k_u(\mathbf{I} - \mathbf{J}_s\mathbf{J}_s^T)\boldsymbol{F}_{\text{ext}}. \tag{15}$$

Here, $k_u$ is the force scaling constant and $\mathbf{J}_s$ is the path Jacobian and the projection of the external force $\boldsymbol{F}_{\text{ext}}$ perpendicular on the path was used to update the trajectory. Note that once the learning was finished, the new weights become the initial weights when learning was re-activated again.

## 3 Results

We first demonstrate the performance of the algorithm for an on-line adaption of trajectory using simulation of one dimensional case. The initial trajectory was set to a strait line with all weights equal to 0.1, i.e. $\boldsymbol{\omega}_i = 0.1$ for all $i = 0...N$, where $N = 30$. The simulation results are shown in Fig. 2, were we introduce a virtual force at path location $s = 0.5$ such that updating factor $\boldsymbol{e_u}$ pulls the path to the value $x(0.5) = 0.8$. On the right-hand side of the Fig. 2 the update rate dependent on the iteration is shown, where $i$ indicates the number of iterations. We can see that the proposed algorithm gradually approaches towards new desired point at $x(0.5) = 0.8$. In this particular case $\sim 20$ iterations were needed to fully update the trajectory at $s = 0.5$ to reach the desired point. On the left-hand side and middle plots we can also see the influence of a single point on the whole trajectory. For example, here the with of the kernel function was set to $h_i = 0.375$ and the influence of a single point is spanned over 4-5 neighboring kernel functions. With a given parameter set, this results in a change of trajectory on a interval $s \approx (0.4, 0.6)$. Note that if the width of the kernel functions is small (large) then the influence of a single point change on the trajectory is small (large) as well. Last but not least, we can also see that the change of the trajectory is continuous, even when discontinuous coaching points are introduced. This makes system easy applicable and robust even when used with noisy sensors, as force/torque sensors typically are.

To demonstrate the performance of the on-line adaption of trajectory for virtual guides framework control concept we have selected a task, where the tip of the end-effector has to follow the groove of the car light housing. This is a typical path following task, which would be difficult to program by hand. To reduce the programming effort, a virtual guides framework control concept was used with on-line trajectory adaption, based on human physical interaction, as shown in Fig. 1. There are two major advantages of such trajectory learning approach. First of all, it reduces the human mental and physical effort compared to typical kinesthetic guidance. Using both approaches the user grasps and moves the robot. However, the main advantage or the proposed approach is that the robot guides the operator along the already learned path and the human just corrects the path. Note that here the path can be already finalized in a single iteration. The difference between kinesthetic guidance and the proposed control is also that here the applied force is decoupled to the tangential and perpendicular component regarding the path direction. This allows changing the trajectory only when force component perpendicular to the path is present. If the user applies a force in path direction then the robot moves just along the path. Another important advantage is that the proposed approach is designed to work on-top of an admittance control framework. Therefore, the proposed approach can be applied to

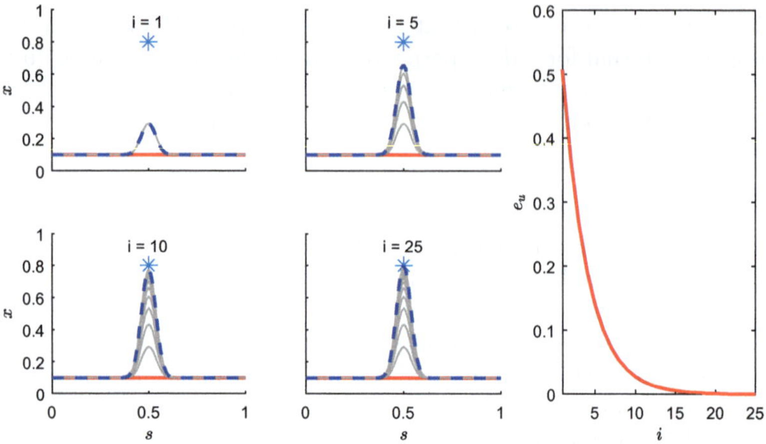


**Fig. 2.** A simulation example of a one-dimensional case. Left-hand side and middle plots show how the trajectory is adapted on-line for a different number of iterations. Red line denotes the initial trajectory, blue doted-line denotes the final trajectory and gray lines denoted the intermediate trajectories. The desired point for adaption is denoted with the blue star. The right-hand side plot shows the adaption speed.

a robot even if it is very stiff (as most of the industrial are) as long as it is equipped with a force/torque sensor at the end-effector.

The results of on-line adaption based on physical human interaction for learning trajectory of the car light-housing experiment are shown in Fig. 3. We can see the initial, the intermediate and the final robot path. Note that the orientation of the robot was kept constant along the path, as shown in Fig. 1. To demonstrate the effectiveness of the algorithm we have chosen an ellipsoid as

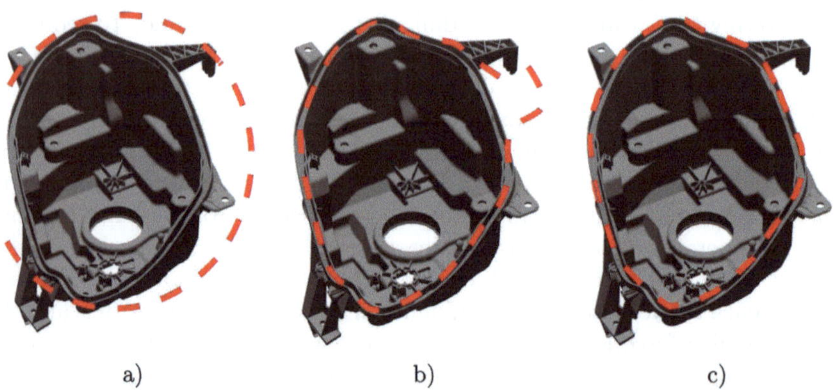


**Fig. 3.** Car light-housing experiment: (a) initial trajectory, (b) intermediate trajectory, (c) final trajectory

the initial path. Figure 4 shows the evolution of the path in 3D Cartesian space. We can see form Figs. 3 and 4 that the user could easily adapt the path to fit into the groove.

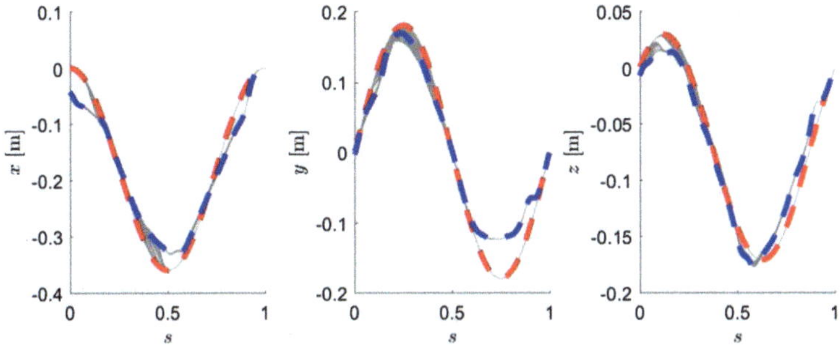


**Fig. 4.** The adaption of the Cartesian coordinates for the car light-housing experiment. The red doted line denotes the initial trajectory, blue doted line denotes the final trajectory and gray lines denoted the intermediate trajectory during learning phase.

## 4 Conclusion

In this paper, we introduced an algorithm for the on-line adaption of the trajectory based on human physical iteration. The proposed algorithm upgrades the virtual guides framework and as such allows on-line learning and adaption of complex paths. The proposed algorithm was evaluated using simulations and with experiments on a KUKA LWR robot, which has been assisting a human to do a path following task. Experimental results have shown, that virtual guides combined with the on-line trajectory adaption algorithm based on physical inter action can be successfully used to acquire complex paths. In the future, we plan to conduct research on how to further improve the on-line learning for tasks where the robot is in contact with the environment.

## References

1. Abbott, J.J., Marayong, P., Okamura, A.M.: Haptic virtual fixtures for robot-assisted manipulation. In: Robotics Research, pp. 49–64 (2007)
2. Billard, A., Calinon, S., Dillmann, R., Schaal, S.: Robot Programming by Demonstration, pp. 1371–1394. Springer, Heidelberg (2008). https://doi.org/10.1007/978-3-540-30301-5_60
3. Bowyer, S.A., Davies, B.L., Rodriguez y Baena, F.: Active constraints/virtual fixtures: a survey. IEEE Trans. Robot. **30**(1), 138–157 (2014)
4. Colgate, J., Peshkin, M., Klostermeyer, S.: Intelligent assist devices in industrial applications: a review. In: Proceedings 2003 IEEE/RSJ IROS, vol. 3, pp. 2516–2521, October 2003

5. Ficuciello, F., Villani, L., Siciliano, B.: Variable impedance control of redundant manipulators for intuitive human-robot physical interaction. IEEE Trans. Robot. **31**(4), 850–863 (2015)
6. Gams, A., Petrič, T., Do, M., Nemec, B., Morimoto, J., Asfour, T., Ude, A.: Adaptation and coaching of periodic motion primitives through physical and visual interaction. Robot. Auton. Syst. (2015). http://linkinghub.elsevier.com/retrieve/pii/S0921889015001992
7. Gruebler, A., Berenz, V., Suzuki, K.: Coaching robot behavior using continuous physiological affective feedback. In: 2011 11th IEEE-RAS International Conference on Humanoid Robots (Humanoids), Bled, Slovenia, pp. 466–471 (2011)
8. Kragic, D., Marayong, P., Li, M., Okamura, A.M., Hager, G.D.: Human-machine collaborative systems for microsurgical applications. Int. J. Robot. Res. **24**(9), 731–741 (2005)
9. Lecours, A., Mayer-St-Onge, B., Gosselin, C.: Variable admittance control of a four-degree-of-freedom intelligent assist device. In: Proceedings - IEEE International Conference on Robotics and Automation, vol. 2, pp. 3903–3908 (2012)
10. Lee, D., Ott, C.: Incremental kinesthetic teaching of motion primitives using the motion refinement tube. Auton. Robots **31**(2–3), 115–131 (2011)
11. Nicolescu, M.N., Mataric, M.J.: Natural methods for robot task learning: instructive demonstrations, generalization and practice. In: Proceedings of the Second International Joint Conference on Autonomous Agents And Multiagent Systems, pp. 241–248 (2003)
12. Niemeyer, G., Preusche, C., Hirzinger, G.: Telerobotics. In: Springer Handbook of Robotics, pp. 741–757. Springer, Heidelberg (2008)
13. Ott, C., Mukherjee, R., Nakamura, Y.: Unified impedance and admittance control. In: Proceedings - IEEE International Conference on Robotics and Automation, pp. 554–561 (2010)
14. Petrič, T., Cevzar, M., Babič, J.: Utilizing speed-accuracy trade-off models for human-robot coadaptation during cooperative groove fitting task. In: 2017 IEEE-RAS 17th International Conference on Humanoid Robotics (Humanoids), pp. 107–112, November 2017
15. Ranatunga, I., Lewis, F., Popa, D.O., Tousif, S.M.: Adaptive admittance control for human-robot interaction using model reference design and adaptive inverse filtering. IEEE Trans. Control Syst. Technol. **00**, 1–10 (2016)
16. Riley, M., Ude, A., Atkeson, C., Cheng, G.: Coaching: an approach to efficiently and intuitively create humanoid robot behaviors. In: 2006 6th IEEE-RAS International Conference on Humanoid Robots (Humanoids), Genoa, Italy, pp. 567–574 (2006)
17. Rosenberg, L.: Virtual fixtures: Perceptual tools for telerobotic manipulation. In: Proceedings of IEEE Virtual Reality Annual International Symposium, pp. 76–82 (1993)
18. Siciliano, B., Sciavicco, L., Villani, L., Oriolo, G.: Robotics - Modelling. Planning and Control. Springer, London (2009)
19. Ude, A., Nemec, B., Petrič, T., Morimoto, J.: Orientation in cartesian space dynamic movement primitives. In: 2014 IEEE International Conference on Robotics and Automation (ICRA), pp. 2997–3004, May 2014
20. Žlajpah, L., Petrič, T.: Virtual guides for redundant robots using admittance control for path tracking tasks. In: Aspragathos, N.A., Koustoumpardis, P.N., Moulianitis, V.C. (eds.) Advances in Service and Industrial Robotics: Proceedings of the 27th International Conference on Robotics in Alpe-Adria Danube Region (RAAD 2018), pp. 13–23. Springer, Cham (2019)

# A Comparison of Policy Search in Joint Space and Cartesian Space for Refinement of Skills

Alexander Fabisch(✉)

DFKI GmbH, Robotics Innovation Center,
Robert-Hooke-Str. 1, 28359 Bremen, Germany
alexander.fabisch@dfki.de

**Abstract.** Imitation learning is a way to teach robots skills that are demonstrated by humans. Transfering skills between these different kinematic structures seems to be straightforward in Cartesian space. Because of the correspondence problem, however, the result will most likely not be identical. This is why refinement is required, for example, by policy search. Policy search in Cartesian space is prone to reachability problems when using conventional inverse kinematic solvers. We propose a configurable approximate inverse kinematic solver and show that it can accelerate the refinement process considerably. We also compare empirically refinement in Cartesian space and refinement in joint space.

**Keywords:** Learning from demonstration · Imitation learning · Reinforcement learning · Policy search · Inverse kinematics

## 1 Refinement of Demonstrated Skills

Autonomous, mobile robots with manipulators can be benificial in many different environments, for example, deep sea [7] or household and factories [13]. In order to create these robots for real, dynamic environments they have to be able to learn. Skill learning frameworks for robots often combine various approaches to leverage intuitive knowledge from humans [3]. Most of these fall into the categories imitation learning and reinforcement learning (RL). A standard approach to learning skills is to initialize with a demonstrated movement and then refine the skill with policy search [2,3,9]. Both methods are complementary because on the one hand policy search methods are often local optimizers that require good initialization and on the other hand imitation learning usually does not produce a perfect skill from the start because of the correspondence problem [3], that is, different kinematic and dynamic properties of the target system and the demonstrator result in different outcomes. We argue that it is sometimes better to do policy search directly in task space because demonstration and main objectives of the learning problem are given in task space, for example, end-effector poses in Cartesian space. When policy search methods are used to


K. Berns and D. Görges (Eds.): RAAD 2019, AISC 980, pp. 301–309, 2020.
https://doi.org/10.1007/978-3-030-19648-6_35

learn end-effector trajectories in task space it often occurs that a trajectory is not completely in the robot's workspace. This results in reward landscapes that are difficult to optimize. We will address this problem with an approximation of inverse kinematics (IK) and investigate when to use it.

## 2 Related Work

We will present the setting in which our problem is embedded and give a brief overview of inverse kinematics that are related to our work.

Transfering skills from humans to robots is prone to the correspondence problem. Consider, for example, a reaching trajectory that is recorded from a human and transfered to the robot's end-effector. Because the robot probably has a different hand shape, it cannot grasp the object with exactly the same end-effector trajectory. In this case, policy search in Cartesian space can be used to refine the initial trajectory. Transfering trajectories directly in joint space is even more difficult because the segment lengths of a robot arm and the human arm are most likely different. Examples for skills that have been demonstrated in task space and transfered to robots are pancake flipping (learned from kinesthetic teaching; [11]), peg in hole (learned from tele-operation; [15]), and serving water (kinesthetic teaching with a similar arm; [18]). There are several reasons for learning trajectories in task space. Learning in Cartesian space is often easier when the main objectives are defined in Cartesian space. For example, learning to grasp an object in joint space might result in complicated Cartesian trajectories. When we want to transfer a skill from one robot to another it is easier to go over end-effector poses instead of joint angles because of different kinematic structures [18] because part of the correspondence problem is solved by forward and inverse kinematics. All of the mentioned works use dynamic movement primitives (DMP) as underlying trajectory representation. In our work, we will use a DMP based on quaternions [23] to generate end-effector poses.

Let $\boldsymbol{q}_t$ be the joint angles of a kinematic chain at time $t$. The forward kinematics of the chain is given by $f(\boldsymbol{q}_t) = \boldsymbol{p}_t$, where $\boldsymbol{p}_t$ denotes the end-effector's pose. An exact solution to the inverse kinematics (IK) problem would be $f^{-1}(\boldsymbol{p}_t) = \boldsymbol{q}_t$. However, $f^{-1}$ is usually not a function. Many joint configurations might result in the same end-effector pose. Numerical solutions to inverse kinematics handle redundant kinematic chains that can have many solutions or complex analytical solutions efficiently. Widely used approaches are based on the Jacobian pseudo-inverse or Jacobian transpose [17]. However, sequential quadratic programming has been shown to outperform these in terms of joint limit handling and computation time [1]. Indirect formulations of the form

$$\arg\min_{\boldsymbol{q}_t \in \mathbb{R}^n} (\boldsymbol{q}_{t-1} - \boldsymbol{q}_t)^T (\boldsymbol{q}_{t-1} - \boldsymbol{q}_t), \quad \text{s.t. } g_i(\boldsymbol{q}_t) \leq b_i, \quad i = 1, \ldots, m$$

where the constraints include the Euclidean distance error, the angular distance error, and joint limits (e.g. [12,14]), have a lower success rate than direct formulations of the form

$$\arg\min_{\boldsymbol{q}_t \in \mathbb{R}^n} \; d(f(\boldsymbol{q}_t), \boldsymbol{p}_t^{\text{des}})) \tag{1}$$

$$s.t. \; g_i(\boldsymbol{q}_t) \leq b_i, \quad i = 1, \ldots, m \tag{2}$$

where $d$ is a pose distance metric, $\boldsymbol{p}_t^{\text{des}}$ is the desired end-effector pose, and the inequality constraints only consist of the joint limits [1]. Few works investigate approximate IK solutions. However, in cases where the desired end-effector pose cannot be reached exactly, we would like to have at least the closest possible solution. Unreachable poses might be a problem of robots that have a low number of joints [6] or at the borders of workspaces as this can be seen for example in visualizations of capability maps [24]. Traditional methods based on the Jacobian pseudoinverse tend to be instable and run into local minima [6,17].

# 3 Configurable Approximate Inverse Kinematics

A disadvantage of previous approaches is that they do not allow to configure the approximation. In the domain of robot skill learning it is often not required to reach each pose in a trajectory exactly. Consider the problem of learning to grasp an object. At the beginning of a reaching movement orientation of the end-effector is not relevant. It only becomes important at the end. That is the reason why we develop a configurable IK solver based on [1]. In the IK formulation in Eqs. 1 and 2 we can use a weighted distance metric of the form

$$\begin{aligned} d(\boldsymbol{p}_1, \boldsymbol{p}_2) = w_{\text{pos}} ||\boldsymbol{p}_{1,1:3} - \boldsymbol{p}_{2,1:3}||_2^2 \; + w_{\text{rot}} \big[ \min(|| \log(\boldsymbol{p}_{1,3:7} * \overline{\boldsymbol{p}}_{2,3:7})||_2, 2\pi \\ - || \log(\boldsymbol{p}_{1,3:7} * \boldsymbol{p}_{2,3:7})||_2) \big]^2, \end{aligned}$$

where $\boldsymbol{p}_{1:3}$ represents the position of the end-effector and $\boldsymbol{p}_{3:7}$ is a quaternion that represents the orientation of the end-effector. It consists of a position distance metric and a rotation distance metric. There are other distance metrics for rotations that we could use [8]. This allows us to set different weights on position and rotation because it is often much more important to reach a desired position than the desired rotation. We could extend this pose metric so that we can weight each of the six degrees of freedom individually.

# 4 Experiments

## 4.1 Methods

We use DMPs as policies because they are well known and stable trajectory representations that can be used for imitation and RL. In the following experiments we will only learn the weights of the DMPs. Metaparameters are constant.

Many policy search algorithms [5,10,16,19,21,22] are based on similar principles: a Gaussian distribution is learned from which we sample policy parameters. We will use Covariance Matrix Adaption Evolution Strategy (CMA-ES; [4]) to learn the policy because it has few critical hyperparameters. Even when those are not set perfectly CMA-ES is still reliable. The most important hyperparameter is the initial step size $\sigma$. It determines the width of the initial search distribution. If it is not large enough, the algorithm will first have to increase the step size for several generations before it can converge. If it is set too large, convergence will take longer. For our comparison it is important to select the correct step size ratio between joint space and Cartesian space so that the results will not be distorted by a wrong choice of this parameter. In our experiments, we use the Kuka LBR IIWA 14 R820 robot arm with 7 DOF. We determined empirically that the initial step size in joint space must be 2 to 3 times higher than in Cartesian space to achieve similar effects on the robot's end-effector. In the first two experiments, the optimum solutions are close to the border of the workspace so that not every orientation can be reached. In all experiments we compare learning weights of DMPs in joint space, in Cartesian space with the proposed approximate IK, and in Cartesian space with an "exact" IK. The term exact throughout this paper means a numerical IK based on the pseudoinverse of the Jacobian that does not move the end-effector if no valid solution has been found. We will execute 30 runs per configuration with different random seeds for CMA-ES. We use the results to plot learning curves with the mean and standard error of the maximum reward obtained so far. Each DMP in our setup has 50 weights per joint space or task space dimension.

### 4.2 Problems and Results

The following paragraphs will summarize the results obtained for several problems that have different reward surfaces. Implementations for most experiments are available at https://github.com/rock-learning/approxik.

In the viapoint problem, the end-effector should pass through intermediate positions (viapoints). A simpler problem has been used to compare various policy search methods [10,20]. In this version we want the end-effector to pass through five Cartesian points (see Fig. 1(d)) while minimizing joint velocities and accelerations. The reward is $R = -10 \sum_{(t_\nu, \boldsymbol{\nu})} ||f(\boldsymbol{q}_{t_\nu}) - \boldsymbol{\nu}||_2 - 10^{-3} \sum_{t,j} |\dot{\boldsymbol{q}}_{t,j}| - 10^{-5} \sum_{t,j} |\ddot{\boldsymbol{q}}_{t,j}|$, where $t \in \{1, \ldots, 101\}$ represents the step, $j$ are joint indices and $(t_\nu, \boldsymbol{\nu})$ represents a position $\boldsymbol{\nu}$ that has to be reached at time $t_\nu$. The results are shown in Fig. 1(a). Learning in Cartesian space is more sample-efficient because the primary objective is defined in Cartesian space. The exact IK performs worse because the approximate IK usually generates more smooth trajectories. Viapoint problems belong to the most simple class of problems for policy search methods because there are usually no flat regions or abrupt changes in the reward surface so that it is easy to determine the direction of improvement. We can get a rough impression of the reward surface of the problem from Fig. 1(g). To generate a projection of the reward surface on one DMP weight dimension, we learn a DMP for 1000 episodes, keep the best policy, modify the 50th weight by adding

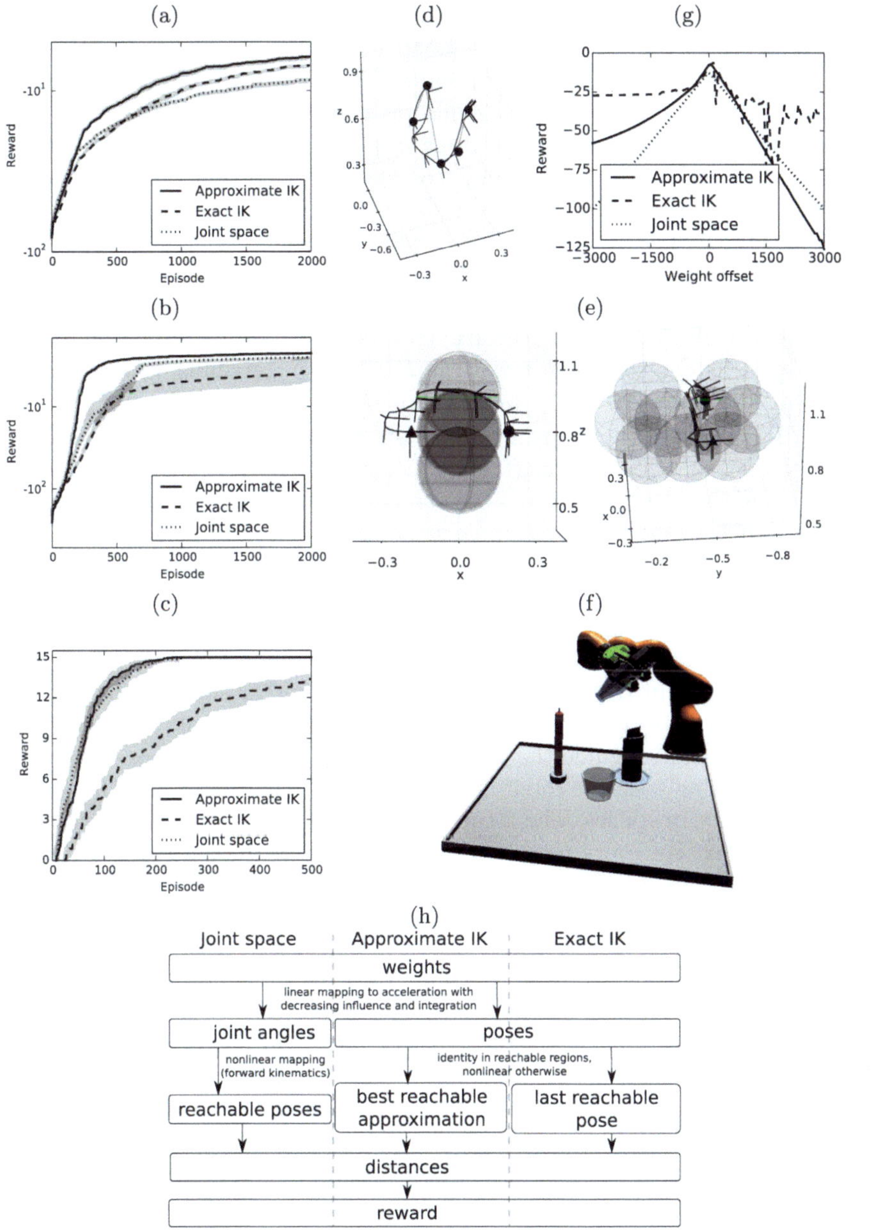


**Fig. 1.** Learning curves for (a) viapoint, (b) obstacle avoidance, and (c) pouring problem. Illustrations of environments: (d) Viapoint: 5 viapoints (circles) have to be reached. A line shows the temporal order. An additional line indicates an optimized trajectory (rotations are indicated by small coordinate frames). (e) Obstacle avoidance: ball-shaped obstacles have to be avoided between start (triangle) and goal (circle). An example of an optimized trajectory is displayed. (f) Pouring. (g) Projection of the reward surface of the viapoint problem on one axis. (h) Mapping from weights to corresponding reward in the viapoint problem. For each setup different mappings are involved to compute the reward for a weight vector.

an offset, and measure the corresponding reward. We can see that for both the joint space DMP and the Cartesian DMP with an approximate IK the reward surface is smooth with only one maximum. With the exact IK, the reward surface is rough and abruptly changing in some regions, which makes learning more difficult for CMA-ES. However, learning in joint space is worse than learning in Cartesian space because of the nonlinear mapping from weights to positions in Cartesian space through the forward kinematics of the robot arm. This results in more complex interrelations between parameters.

In the obstacle avoidance problem, the end-effector has to avoid several obstacles represented by spheres. It is an artificial problem because we only consider end-effector collisions and not the arm to which it is attached. The environment is displayed in Fig. 1(e). The reward is $R = -10 \sum_{\rho} p\left(\min_{\boldsymbol{q}_t} ||f(\boldsymbol{q}_t) - \boldsymbol{\rho}||_2\right) - 100||f(\boldsymbol{q}_T) - \boldsymbol{g}||_2 - 10^{-2} \sum_{t,j} |\dot{q}_{t,j}| - 10^{-5} \sum_{t,j} |\ddot{q}_{t,j}|$, where $t \in \{1, \ldots, T\}$, $T = 101$ represents the step in time, $j$ are the joint indices, $\boldsymbol{g}$ is the desired goal position, and $p(d) = \max(0, 1 - \frac{d}{0.17})$ is greater than zero when the end-effector is in the vicinity of one of the obstacles. The result is shown in Fig. 1(b). We see that learning in Cartesian space is again more sample-efficient because the primary objective is defined in Cartesian space. The exact IK performs worse because the approximate IK and learning in joint space results in more smooth trajectories. The difficulty of this task and its reward surface are similar to those of the viapoint task. The difference is that it is less smooth because the negative penalty for being in the vicinity of obstacles abruptly vanishes when the end-effector is outside of the radius of the spheres. In this case only the velocity and acceleration penalties guide the search to a better solution.

In the pouring task, the robot fills a glass with 15 marbles from a bottle while avoiding nearby obstacles. The setup is displayed in Fig. 1(f). The reward function is complex. For every marble that is inside of the glass we add 1. For every marble that is outside of the glass but on the table the negative squared distance to the center of the glass is given. For every marble that is still in the bottle we give a reward of −1. For collisions with obstacles we give a reward of −100 and abort the episode. Marbles that fall down the table also finish the episode and a reward of −1000 is given. Results are displayed in Fig. 1(c). Learning in joint space and learning in Cartesian space results in nearly identical learning curves. Using an exact IK is again worse than using an approximate solution. The mapping from weights to reward in this problem is complex, nonlinear, and certainly not smooth. There are abrupt changes in the reward function when a small change of one weight results in a marble falling down the table or the arm touching an obstacle. There are flat regions, for example, when all marbles stay in the bottle because the arm does not turn the bottle upside down, or all marbles miss the table. Therefore, the structure of the reward surface is complex independent of the space in which we describe DMPs, hence, learning in joint space and learning in Cartesian space work similarly well.

### 4.3 Discussion

The artificial RL problems that we presented vary in difficulty, that is, indirection and smoothness of the reward function. Weights in a DMP are bound to a specific radial basis function. They only influence one specific joint or pose dimension and only affect accelerations of the arm locally in time. They will affect the positions in all following time steps but with a decreasing influence because the weight of the learnable forcing term converges to zero. So it is possible to design a reward function so that the optimization problem becomes partially separable and, hence, easy to solve. For example, in the viapoint and obstacle avoidance problems there is a direct relation between weights of the Cartesian DMP and obtained reward (see Fig. 1(h)). With the approximate IK we use the weights to generate a pose trajectory. An acceleration is computed based on the linear forcing term that includes the weights and decays exponentially over time. The acceleration is integrated to obtain a trajectory of end-effector poses. This trajectory might not be perfectly executable so that in unreachable regions the pose is mapped to the closest reachable pose. For each viapoint we compute the distance to the corresponding poses from the trajectory. Hence, the mapping from specific weights to the reward in the workspace of the robot is straightforward and the reward function is partially separable. Joint space DMPs result in a nonlinear mapping with coupling between dimensions of the weight space because of the forward kinematics of the robot. This results in a difficult optimization problem that is not partially separable any more with probably multiple global and local optima. The more complex the relation between weights and reward becomes, the less significant is the difference between learning in joint space and learning in Cartesian space. An example is the pouring problem. It has an almost flat reward surface where the arm collides with an obstacle or a marble falls down. Small weight changes can make a difference between a marble staying within the glass and falling down, hence, the reward will abruptly change in the corresponding regions of the weight space. The relation between the DMP weights and a successful behavior is complex, highly nonlinear, and non-separable. This eradicates the advantage of learning in Cartesian space.

## 5 Conclusion

We develop a configurable, approximate variation of a state-of-the-art numerical IK solver and show that it accelerates policy search for movement primitives in Cartesian space in comparison to a conventional IK solver. Using inverse kinematics can be regarded as a way to include knowledge about the target system in the learning process. Learning in Cartesian space is not always the best option. It will be advantageous if the main objective has to be solved in Cartesian space and the reward function is almost separable. It is not always obvious when this is the case though. The advantage of learning in Cartesian space vanishes for more complex, nonlinear, non-separable reward functions. Policy search works best in the space in which the primary objectives are defined directly. Generating smooth trajectories is simpler in joint space.

**Acknowledgements.** We thank Manuel Meder for support in setting up simulation environments. This work was supported through grants of the German Federal Ministry of Economics and Technology (BMWi, FKZ 50RA1216 and 50RA1217), the German Federal Ministry for Economic Affairs and Energy (BMWi, FKZ 50RA1701), and the European Union's Horizon 2020 research and innovation program (No H2020-FOF 2016 723853).

## References

1. Beeson, P., Ames, B.: TRAC-IK: an open-source library for improved solving of generic inverse kinematics. In: Humanoids, pp. 928–935 (2015)
2. Deisenroth, M.P., Neumann, G., Peters, J.: A survey on policy search for robotics. Found. Trends Robot. **2**(1–2), 328–373 (2013)
3. Gutzeit, L., Fabisch, A., Otto, M., Metzen, J.H., Hansen, J., Kirchner, F., Kirchner, E.A.: The besman learning platform for automated robot skill learning. Front. Roboti. AI **5**, 43 (2018)
4. Hansen, N., Ostermeier, A.: Completely derandomized self-adaptation in evolution strategies. Evol. Comput. **9**, 159–195 (2001)
5. Heidrich-Meisner, V., Igel, C.: Evolution strategies for direct policy search. In: Parallel Problem Solving from Nature PPSN X, pp. 428–437 (2008)
6. Henning, A.: Approximate inverse kinematics using a database. Technical report, Worcester Polytechnic Institute (2014)
7. Hildebrandt, M., Albiez, J., Kirchner, F.: Computer-based control of deep-sea manipulators. In: OCEANS 2008, pp. 1–6 (2008)
8. Huynh, D.Q.: Metrics for 3D rotations: comparison and analysis. J. Math. Imaging Vis. **35**(2), 155–164 (2009)
9. Kober, J., Bagnell, J.A., Peters, J.: Reinforcement learning in robotics: a survey. Int. J. Robot. Res. **32**(11), 1238–1274 (2013)
10. Kober, J., Peters, J.: Policy search for motor primitives in robotics. Mach. Learn. (2010)
11. Kormushev, P., Calinon, S., Caldwell, D.G.: Robot motor skill coordination with EM-based reinforcement learning. In: IROS, pp. 3232–3237 (2010)
12. Kumar, S., Sukavanam, N., Raman, B.: An optimization approach to solve the inverse kinematics of redundant manipulator. Int. J. Inf. Syst. Sci. **6**, 414–423 (2010)
13. Lemburg, J., de Gea Fernndez, J., Eich, M., Mronga, D., Kampmann, P., Vogt, A., Aggarwal, A., Shi, Y., Kirchner, F.: AILA - design of an autonomous mobile dual-arm robot. In: ICRA, pp. 5147–5153 (2011)
14. Fallon, M., et al.: An architecture for online affordance-based perception and whole-body planning. J. Field Robot. **32**(2), 229–254 (2015)
15. Krüger, N., et al.: Technologies for the fast set-up of automated assembly processes. Künstliche Intelligenz **28**(4), 305–313 (2014)
16. Neumann, G.: Variational inference for policy search in changing situations. In: ICML, pp. 817–824 (2011)
17. Nilsson, R.: Inverse kinematics. Master's thesis, Luleå University of Technology (2009)
18. Pastor, P., Hoffmann, H., Asfour, T., Schaal, S.: Learning and generalization of motor skills by learning from demonstration. In: Robotics and Automation (ICRA), pp. 763–768 (2009)

19. Peters, J., Mülling, K., Altun, Y.: Relative entropy policy search. In: AAAI (2010)
20. Peters, J., Schaal, S.: Policy gradient methods for robotics. In: IROS (2006)
21. Peters, J., Schaal, S.: Reinforcement learning by reward-weighted regression for operational space control. In: ICML, pp. 745–750 (2007)
22. Theodorou, E., Buchli, J., Schaal, S.: A generalized path integral control approach to reinforcement learning. J. Mach. Learn. Res. **11**, 3137–3181 (2010)
23. Ude, A., Nemec, B., Petri, T., Morimoto, J.: Orientation in Cartesian space dynamic movement primitives. In: Robotics and Automation (ICRA), pp. 2997–3004 (2014)
24. Zacharias, F., Borst, C., Hirzinger, G.: Capturing robot workspace structure: representing robot capabilities. In: IROS, pp. 3229–3236 (2007)

# KUKA LWR Robot Cartesian Stiffness Control Based on Kinematic Redundancy

Branko Lukić[1(✉)], Tadej Petrič[2], Leon Žlajpah[2], and Kosta Jovanović[1]

[1] School of Electrical Engineering, University of Belgrade, Belgrade, Serbia
{branko,kostaj}@etf.bg.ac.rs
[2] Jožef Stefan Institute, Ljubljana, Slovenia
{tadej.petric,leon.zlajpah}@ijs.si

**Abstract.** This paper is dealing with one important topic for physical human-robot interaction, and that is achieving and/or optimizing of an arbitrary Cartesian stiffness of robot's end-effector (EE). The focus is given on redundant compliant robots with serial elastic actuators with fixed joint stiffness, but can reconfigure without changing the EE position. The work presented in this paper is an approach where the robot redundancy is exploited to achieve the desired or at least some optimal Cartesian stiffness of robots EE. Robot tasks can be divided into primary and secondary tasks. In our case, the primary task is to track the Cartesian position reference and the secondary task is to optimize EE Cartesian stiffness behavior while keeping the desired EE position. This means that the EE position is a constraint in the robot Cartesian stiffness optimization. The algorithm for the Cartesian stiffness optimization has been initially tested using the simulation, and then evaluated on the 7-DOFs KUKA light weight robot.

**Keywords:** Cartesian stiffness control · Robot redundancy · Null space · Physical human-robot interaction

## 1 Introduction

First industrial robots were design to help and to substitute humans in tasks that were dangerous, physically demanding, and to prevent work related injuries. Typically, the robots were introduced in repetitive tasks prevailing in precision and productivity. The industrial development requires that humans work in the robot proximity and that they collaborate with robots [1]. To enable physical interaction between humans and robots, it is necessary to make the collaboration inherently safe. One of the first approach toward the safe interaction was introduced through the impedance control [2], where the robot end-effector (EE) behaves as a mechanical impedance.

This paper was partly funded by the Ministry of Education, Science and Technological development of the Republic of Serbia, under contracts TR-35003, and bilateral project ARRS BI-RS/18-19-053.


K. Berns and D. Görges (Eds.): RAAD 2019, AISC 980, pp. 310–318, 2020.
https://doi.org/10.1007/978-3-030-19648-6_36

A challenge in the impedance control are sudden impacts with high-force amplitude, since an active controller cannot react timely to prevent damage, nor it can secure a safe robot interaction with the environment [3]. These shortcomings can be overcome with the use of compliant actuators. Therefore, the traditional premise for a good robot design "Stiffer is better" is no longer valid. Compliant actuators are often designed as serial elastic actuators (SEA) with constant actuator stiffness [4, 5], or as variable stiff actuators (VSAs). VSAs have variety of realization such as adjustable mechanical compliance [6], structure-controlled stiffness [7] or antagonistic controlled stiffness [8]. Beside the position, actuator output stiffness is an additional control state in VSAs. Thus, VSAs are systems with two outputs: position and stiffness. Different approaches to the mechanical design and realization of VSAs are pointed out in [9], while VSA's main properties are described in [10, 11].

Since achieving safe physical human-robot interaction is essential for wider use of robots in industrial and home environment, control of robot EE mechanical impedance is an attractive research topic. Advances in this field will lead to an increased human (and robot) safety [12]; possibility for safe human-robot collaboration [13]; and enable new breakthroughs in healthcare domain [14]. There are different approaches in achieving or optimizing a safe (compliant) robot behavior, i.e. a desired mechanical impedance of robot EE. The first one is adjusting individual joint stiffness which yield the desired Cartesian stiffness, and the second one is the exploitation of the robot kinematic redundancy to reconfigure the robot to an optimal configuration regarding the EE compliance.

The first approach is applicable to robots with VSAs regardless of their realization. In this case, the stiffness in robot joints is adjusted to satisfy the desired Cartesian stiffness of robot EE [15]. Symbiosis of a passive stiffness and an active stiffness is exploited when the desired Cartesian stiffness of EE is hardly obtained by adjusting the joint stiffness only [16]. One solution, applied in [15, 16], to compute an optimal joint stiffness is the Karush-Kuhn-Tucker (KKT) theorem or the KKT condition for the optimization with nonlinear constraints [17]. A detailed overview of optimization methods is described in [18, 19].

The second approach is to exploit robot redundancy, i.e. when there are more DOFs than the task requires. By changing the robot configuration, it is possible to influence the EE stiffness without changing EE position. This approach can be applied on robot driven by VSA or SEA. However, if the robot is driven by SEAs, this approach is of greater importance since there are no additional control states which are present in robots using VSAs.

The reconfiguration of a robot posture for EE stiffness optimization in Cartesian space is the main topic of this paper. The robot posture reconfiguration problem for EE stiffness optimization can be analyzed as a nonlinear optimization with constraints [15, 16]. In this paper, the paradigm is changed and the reconfiguration is observed as a control problem. Therefore, joint movements in the null-space are used for the Cartesian stiffness optimization.

In Sect. 2 we introduce a control design for achieving the optimal Cartesian stiffness using the robot motion in the null-space. Simulation and experimental results are presented in Sect. 3, while conclusions and final remarks are in Sect. 4.

## 2 Cartesian Control Design Through the Null-Space Optimization

The Cartesian stiffness matrix $K_C$ for the robot's EE is a function of joint's stiffness $K_j$ and the robot configuration reflected in the Jacobean matrix $J$, which relates joint velocities and EE velocities. It is given by

$$K_C = \left(J(q)K_j^{-1}J(q)^T\right)^{-1}, \tag{1}$$

where $K_C$ is a symmetric $m \times m$ Cartesian stiffness matrix, $K_j$ is $n \times n$ a diagonal joint stiffness matrix, $J$ is a $m \times n$ Jacobean matrix, $x$ is $m$-dimensional EE Cartesian position vector and $q$ is $n$-dimensional joint positions vector. Parameters $m$ and $n$ are the dimensions of the task and the joint space, respectively.

The performance index $H$ for the optimality of the Cartesian stiffness matrix $K_C$ with respect to the desired one $K_{CD}$ is defined as

$$H = \left\|K_{CD} - K_C\left(K_j, q\right)\right\|. \tag{2}$$

The problem of preserving the Cartesian stiffness while tacking a path can be observed as the minimization of $H$. Using the Euclidean norm, $H$ can be computed as

$$H = \sum\nolimits_{i,j} W_{ij}\left(K_{CD_{ij}} - K_C\left(K_j, q\right)_{ij}\right)^2, \tag{3}$$

where $W$ is a $m \times m$ weighting matrix, $i$ and $j$ denote the elements of the corresponding matrix. The role of the weighting matrix is the prioritization of multiple elements which are optimized simultaneously. When this is not necessary, the weighting matrix can be omitted.

The first order differential kinematics of a robot is defined as

$$\dot{x} = J\dot{q}, \tag{4}$$

and the inverse kinematics has a general solution [20] in following form

$$\dot{q} = J^{\dagger}\dot{x} + \left(I - J^{\dagger}J\right)\dot{q}_N. \tag{5}$$

Here, $\dot{q}_N$ is an arbitrary joint-space velocity and $J^{\dagger}$ is general inverse of $J$. In the following we will use the Moore-Penrose pseudo inverse defined as $J^{\dagger} = J^T\left(JJ^T\right)^{-1}$.

For redundant robots an arbitrary joint space velocity $\dot{q}_N$ can be selected to optimize an additional requirement since there are $n - m$ redundant DOFs not required for the primary task execution. For the implementation of an additional task related to the performance index $H$, $\dot{q}_N$ is calculated using the gradient of the optimization function [20]

$$\dot{q}_N = -K_n \nabla H(q), \tag{6}$$

where $K_n$ is scalar gain and $\nabla H(q)$ is the gradient of $H$.

We have used a basic Cartesian controller at the velocity level given by

$$\dot{x} = K(x_d - x), \tag{7}$$

where $K$ is positive constant gain matrix, and $x_d$ are the desired Cartesian positions.

Combining (5), (6) and (7) we obtain the Cartesian controller with Cartesian stiffness matrix optimization in the form

$$\dot{q} = J^{\dagger}(x_d - x)K - K_n\left(I - J^{\dagger}J\right)\nabla H(q). \tag{8}$$

Thus, the desired joint positions $q_d$ are computed as

$$q_d = q_0 + \int \dot{q}dt = q_0 + \int \left(J^{\dagger}(x_d - x)K - K_n\left(I - J^{\dagger}J\right)\nabla H(q)\right)dt, \tag{9}$$

where $q_0$ is a $n$-dimensional vector of initial joint positions. Since this algorithm uses the gradient method, converging to global optimum cannot be guaranteed. Also, selection of $K$ and $K_n$ should be considered carefully since their values affect the algorithm convergence, oscillations and stability.

We have validated the algorithm on KUKA LWR robot arm that does not have passive compliant joints. However, compliant behavior has been achieved as virtually passive through the PD torque control, where joint torques $\tau$ are computed as

$$\tau = K_p(q_d - q) + K_d(\dot{q}_d - \dot{q}) + g(q). \tag{10}$$

Proportional and differential gains, $K_p$ and $K_d$, are tuned to secured desired joint behavior so: $K_p = K_j$ and $K_d = 2\sqrt{K_p}$, while the term $g(q)$ is representing the gravity compensation.

Equation (9) is an algorithm for computing desired joint positions, while (10) is proposed torque controller. Together they make a set of equations required for the algorithm implementation.

## 3 Simulation and Experimental Results

In the experiment, task space comprises only the Cartesian position, (the orientation is not considered) in the robot coordinate frame. Presented results are obtained for the proposed control (10) using algorithm for planning generalized coordinates (9). Commanded joint stiffness in each joint is $K_j = 200\,\mathrm{Nm/rad}$, while EE position is $(x, y, z) = (-0.4\,\mathrm{m}, 0\,\mathrm{m}, 0.4\,\mathrm{m})$.

Minimal configuration requirements of a robot, where the redundant DOFs are used for the position control and to achieve the Cartesian stiffness matrix, are discussed in [15]. Summarizing, if $m$ is the dimension of the task space, then number of variables in

Cartesian stiffness matrix is $m(m+1)/2$. Thus, total number of variables is $m+m(m+1)/2 = m(m+3)/2$. Consequently, for a full control in 6-dimensional task space a minimum of 27 DOFs is required, while 9 DOFs are required for a full stiffness matrix and position control in a 3-dimensional task space.

In the presented experiment, the robot has 7 joints, but because EE position is located at the surface of the last link only first 6 joint affect the results, i.e. the EE position and the EE stiffness. Thus, the robot has 6 DOFs ($n = 6$) while the task requires 3 DOF ($m = 3$), giving 3 redundant DOFs. Considering that stiffness matrix for a 3 DOF task space has 6 variables and the position has 3 variables, not all of them can achieve the desired values simultaneously with only 3 DOFs of redundancy. Therefore, the optimization of one element, i.e. the stiffness along $y$ axis, has been applied.

The experiment contains measurements and model-based estimations of achieved stiffness along an axis for two different desired stiffness levels $S_{low} = 700\text{N/m}$ and $S_{high} = 900\text{N/m}$. EE position has been located at the surface of the last link (location where the sensor with a hook has been attached to the robot). Different external force levels have been applied by adding weights, one by one, on the hook attached to the force sensor (6 weights, 0.5 kg each). The external force has been measured with a 6-axis force/torque sensor. The Cartesian displacement resulted from the applied external force has been computed from the joint position measurements. The estimated stiffness along $y$-axis has been computed as the ratio between the applied force and the displacement along $y$ axis.

Graphical simulation of KUKA LWR robot with Cartesian axis orientation realized in the robot coordinate frame, for two different poses (initial position and working position) in MuJoCo graphical environment as illustrated in Fig. 1. Since the robot is mounted on the pedestal, robot frame in the laboratory setup is different from the world frame. All measurements are recorded respectively to the robot frame in the laboratory setup. An accurate dynamic model of LWR KUKA robot arm developed at Jožef Stefan Institute (Department for Automation, Biocybernetics and Robotics) has been used for the simulation experiments and for the comparison with the results obtained from experiments.

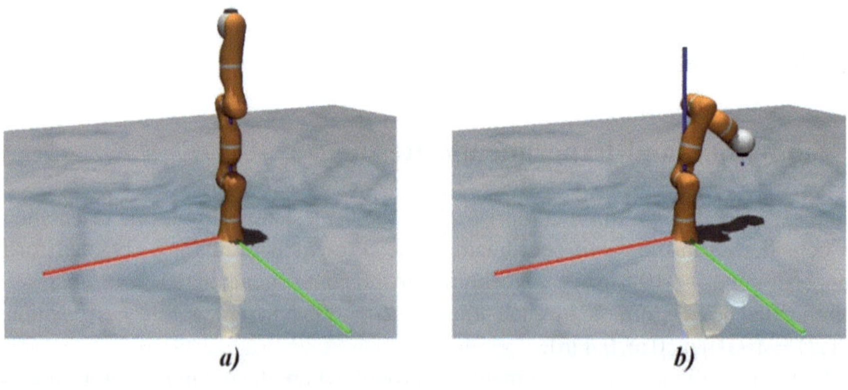


**Fig. 1.** Robot configuration for two different position where $x$ axis is labeled with red line, $y$ axis with green line and $z$ axis with blue line: (*a*) initial position $q_i = 0$ for $i = 1, \ldots 7.$, (*b*) robot in the working position.

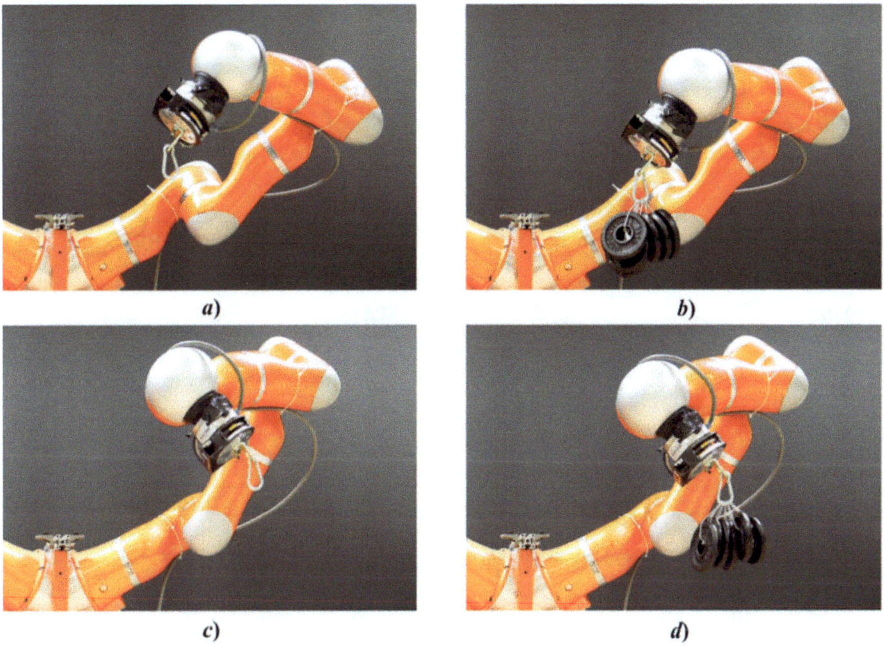

**Fig. 2.** Two different robot configuration for same EE position and different EE Cartesian stiffness along $y$ axis with and without payload: ($a$) lower stiffness value $S_{low}$ without payload; ($b$) lower stiffness value $S_{low}$ with 3 kg payload; ($c$) higher stiffness value $S_{high}$ without payload; ($d$) higher stiffness value $S_{high}$ with 3 kg payload.

Figure 2 shows the experimental setup, where we can see two different configurations for same EE position, both with low $(S_{low})$ and high commanded stiffness $(S_{high})$, with and without a payload.

Results obtained for these two robot configurations with same EE position and same joint stiffnesses which result in different EE Cartesian stiffness are presented in Fig. 3. For $S_{low}$, the force applied on robot's EE is shown in Fig. 3$a$. The displacement influenced by this force on the simulation model (labeled with Sim.) and on the experimental setup KUKA LWR robot arm (labeled with Exp.) are shown in Fig. 3$b$. The comparison of the stiffness obtained in simulation and in experiment with the theoretical stiffness are shown in Fig. 3$c$. Same results, but for a higher stiffness value $S_{high}$, are shown in Fig. 3$d$–$f$, respectively.

Results are obtained in a steady state when the transient effects are elapsed. For $S_{low}$, the experimentally estimated stiffness is within ±12% of the theoretically calculated value, while for $S_{high}$ the value is within ±6% of the theoretically calculated value.

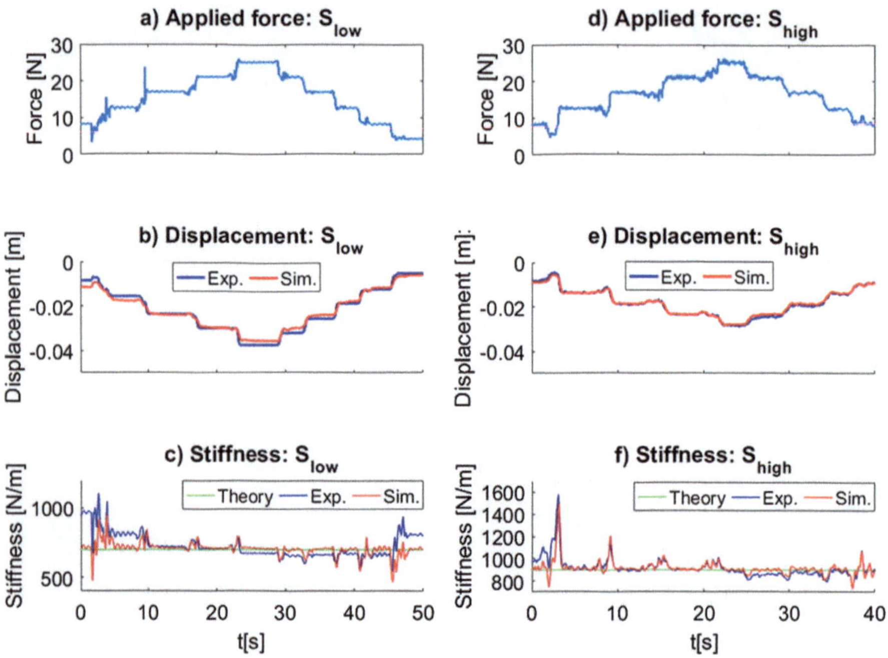


**Fig. 3.** Experimental stiffness estimation. For the first robot configuration resulted in lower Cartesian stiffness $S_{low} = 700\text{N/m}$: (*a*) Robots EE displacement along *y* axis; (*b*) Applied force on robot EE along *y* axis; (*c*) Theoretical (green) and experimental (blue) and simulation (red) stiffness values along *y* axis. For the second robot configuration resulted in higher Cartesian stiffness $S_{high} = 900\text{N/m}$: (*d*) Robots EE displacement along *y* axis; (*e*) Applied force on robot EE along *y* axis; (*f*) Theoretical (green) and experimental (blue) and simulation (red) stiffness values along *y* axis.

## 4 Conclusion

This paper presents a novel approach of the robot's EE Cartesian stiffness control which is accomplished through solving the optimization problem. In this control approach, a paradigm is changed from the optimization to the control problem utilizing the robot's motion in the null-space. This result is particularly important for robots with SEAs that do not have additional control states introduced by VSAs to adjust EE stiffness. Although, it is not possible in practice to control all elements in the stiffness matrix (due to insufficient degree-of-kinematic-redundancy of most robots), control of individual targeted elements in the stiffness matrix can be achieved. The presented control algorithm has been successfully implemented and tested in simulation and on a real KUKA LWR robot arm demonstrating the Cartesian stiffness shaping through the robot reconfiguration.

## References

1. Pedrocchi, N., Vicentini, F., Malosio, M., Tosatti, L.M.: Safe human-robot cooperation in an industrial environment. Int. J. Adv. Robot. Syst. **10**(1), 27 (2013)
2. Hogan, N.: Impedance Control: An approach to manipulation: Part II—Implementation. J. Dyn. Syst. Measur. Control **107**(1), 8–16 (1985)
3. Haddadin, S., Albu-Schäffer, A., Hirzinger, G.: Safe physical human-robot interaction: measurements, analysis and new insights. In: Springer Tracts in Advanced Robotics, vol. 66, pp. 395–407 (2010)
4. Pratt, G.A., Williamson, M.M.: Series elastic actuators. In: Proceedings of 1995 IEEE/RSJ International Conference on Intelligent Robots and Systems 95. Human Robot Interaction and Cooperative Robots, vol. 1, pp. 399–406 (1995)
5. Robinson, D.W., Pratt, J.E., Paluska, D.J., Pratt, G.A.: Series elastic actuator development for a biomimetic walking robot. In: Proceedings of 1999 IEEE/ASME International Conference on Advanced Intelligent Mechatronics, pp. 561–568 (1999)
6. Van Ham, R., Vanderborght, B., Van Damme, M., Verrelst, B., Lefeber, D.: MACCEPA, the mechanically adjustable compliance and controllable equilibrium position actuator: design and implementation in a biped robot. Robot. Auton. Syst. **55**(10), 761–768 (2007)
7. De, A., Tasch, U.: A two-DOF manipulator with adjustable compliance capabilities and comparison with the human finger. J. Robot. Syst. **13**(1), 25–34 (1996)
8. Migliore, S.A., Brown, E.A., DeWeerth, S.P.: Biologically inspired joint stiffness control. In: Proceedings - IEEE International Conference on Robotics and Automation 2005, pp. 4508–4513 (2005)
9. Vanderborght, B., Albu-Schaeffer, A., Bicchi, A., Burdet, E., Caldwell, D.G., Carloni, R., Catalano, M., Eiberger, O., Friedl, W., Ganesh, G., Garabini, M., Grebenstein, M., Grioli, G., Haddadin, S., Hoppner, H., Jafari, A., Laffranchi, M., Lefeber, D., Wolf, S.: Variable impedance actuators: a review. Robot. Auton. Syst. **61**(12), 1601–1614 (2013)
10. Grioli, G., Wolf, S., Garabini, M., Catalano, M., Burdet, E., Caldwell, D., Carloni, R., Friedl, W., Grebenstein, M., Laffranchi, M., Lefeber, D., Stramigioli, S., Tsagarakis, N., VanDamme, M., Vanderborght, B., Albu-Schaeffer, A., Bicchi, A.: Variable stiffness actuators: the user's point of view. Int. J. Robot. Res. **34**(6), 727–743 (2015)
11. Van Ham, R., Sugar, T.G., Vanderborght, B., Hollander, K.W., Lefeber, D.: Compliant actuator designs. Robot. Autom. Mag. **16**(3), 81–94 (2009)
12. De Santis, A., Siciliano, B., De Luca, A., Bicchi, A.: An atlas of physical human–robot interaction. Mech. Mach. Theory **43**(3), 253–270 (2008)
13. Edsinger, A., Kemp, C.C.: Human-robot interaction for cooperative manipulation: handing objects to one another. In: Proceedings - IEEE International Workshop on Robot and Human Interactive Communication, pp. 1167–1172 (2007)
14. Yu, H., Huang, S., Chen, G., Pan, Y., Guo, Z.: Human-robot interaction control of rehabilitation robots with series elastic actuators. IEEE Trans. Robot. **31**(5), 1089–1100 (2015)
15. Albu-Schaffer, A., Fischer, M., Schreiber, G., Schoeppe, F., Hirzinger, G.: Soft robotics: what cartesian stiffness can obtain with passively compliant, uncoupled joints? In: Proceedings of International Conference on Intelligent Robots and Systems, 2004 (IROS 2004), vol. 4, pp. 3295–3301 (2004)
16. Petit, F., Albu-Schäffer, A.: Cartesian impedance control for a variable stiffness robot arm. In: Proceedings of IEEE/RSJ International Conference on Intelligent Robots and Systems (IROS), pp. 4180–4186 (2011)

17. Wu, H.-C.: The Karush–Kuhn–Tucker optimality conditions in an optimization problem with interval-valued objective function. Eur. J. Oper. Res. **176**(1), 46–59 (2007)
18. Momoh, J.A., El-Hawary, M.E., Adapa, R.: A review of selected optimal power flow literature to 1993. I. Nonlinear and quadratic programming approaches. IEEE Trans. Power Syst. **14**(1), 96–104 (1999)
19. Momoh, J.A., El-Hawary, M.E., Adapa, R.: A review of selected optimal power flow literature to 1993. II. Newton, linear programming and interior point methods. IEEE Trans. Power Syst. **14**(1), 105–111 (1999)
20. Chiaverini, S., Oriolo, G., Maciejewski, A.A.: Redundant robots. In: Springer Handbook of Robotics, 2nd edn., pp. 221–242. Springer, Cham (2016)

# Influence of Unmodelled External Forces on the Quality of Collision Detection

Zaviša Gordić(✉) and Kosta Jovanović

School of Electrical Engineering, University of Belgrade, 11000 Belgrade, Serbia
zavisa@etf.rs

**Abstract.** Physical Human-Robot Interaction requires collision detection to enable a safe sharing of workspace between humans and robots, mostly using model based algorithms. Majority of robot tasks involve physical interaction with the environment, and consequently the forces occurring during the interaction. This paper presents an experimental testing of influence of unmodelled but intentional forces on the quality of collision detection algorithms. Results from testing of different manipulation and assembly tasks are shown and discussed in terms of their significance to collision detection and similarity with real collisions. Presented results and conclusions from this paper may serve as guidelines for future collision detection related work by offering a better understanding and insight on the relevance of intentional external forces.

**Keywords:** Physical Human-Robot Interaction · Collision detection · Industrial robot

## 1 Introduction

Collision detection, as one of most important aspects and preconditions for safe Physical Human-Robot Interaction has been a subject of many researches. Various algorithms have been developed so far for industrial and collaborative robots alike.

While there are some non-model based algorithms for collision detection, most of them rely on the robot model [1–8]. Model based algorithms generate reference values for current or torque which are then compared with corresponding measurements obtained from the robot. Potential collisions are identified as deviations from reference values, and some solutions can even classify them as intentional contacts or real collisions with the environment [1]. Dynamic robot model parameters are mostly identified using experiments performed on the robot [9–18], but there are cases when model is learned from operation [19]. In rare cases, dynamic models can be formed using known parameters of the robot, or are provided by the robot manufacturer.

The work was partially supported by the Serbian Ministry of education, science, and technological development, grant No. TR35003 and H2020 project "A Pan European Network of Robotics DIHs for Agile Production ($DIH^2$)".


K. Berns and D. Görges (Eds.): RAAD 2019, AISC 980, pp. 319–328, 2020.
https://doi.org/10.1007/978-3-030-19648-6_37

Calculation of reference signal values depends on the quality and complexity of the robot model. Models which consider more parameters provide a more realistic reference signal generation, but quality and complexity of the model also affect computational time, influencing the real time aspect of detection.

Robot models often include only parameters related to the robot itself, which means that some other aspects which are not the integral part of the robot, or which are not commonly physically connected to the robot, remain unmodelled. It is highly advisable that robot tools which are connected to the robot most of its operation time should be modelled and considered, such as in [9] or [10]. However, depending on the particular application, the weight and position of centre of mass of objects moved by robot, the force needed to exert or withstand in assembly application are sometimes not taken into consideration. The reason for neglecting might be the complexity of modelling, insufficient knowledge about physical properties of the object or process technology, an unintentional oversight, or any other reason. In some cases, external forces are considered insignificant to the dynamic behaviour of the robot system, and they are intentionally left unmodelled, while some solutions, such as [19], offer means for compensating external influences without explicit model of task dynamics in robots with active torque control. However, when model-based collision detection is considered, all external influences reflect on deviations of measured values from model generated references and may lead to undesired false collision identification. Therefore, external forces which occur with certainty should not be neglected, especially having in mind that robots often perform the same or similar task repeatedly.

The motivation for this paper is to examine the influence of unmodelled external forces on the quality of collision detection on robot performing tasks such as those found in manipulation or assembly applications. External forces in both application types are intentional and occur in each cycle, and they should not be neglected.

The second section describes the general principle by which the experiments were conducted, the procedure and the equipment used. The third section compares measurements from testing the robot experimentally in manipulation tasks with and without actual load. The fourth section performs the experimental analysis of forces and torques which occur during a common types of assembly tasks in which a force needs to be applied to join two parts. Conclusions and discussion of results is presented in the fifth section along with some implications on the quality of collision detection.

## 2 Experiment Description

This paper intends to examine the effect of unmodelled load and external forces on the torques of the robot with experiments performed for different types of manipulation and assembly tasks. The unmodelled influence originating from aforementioned tasks is compared with the effect of collisions on the internal robot torques. To this end, for each task related experiment, three cycles of measurements were performed.

First, a torque profile was recorded from robot executing movement needed to perform the manipulation or assembly task, but without the load or object of assembly itself, so that all recorded torques correspond to signals that would be generated by the optimal robot model which includes only the dynamics of the robot itself.

The second cycle of measurements was recorded from the robot performing the manipulation or assembly task with the load or object which needs to be assembled, therefore also including the external forces related to the load or assembling process.

The third measurement cycle is similar to the first cycle since it was performed on a robot performing its task without the load or assembly object. However, during third measurement cycle, the robot was impacted with different intensities and durations.

With measurements from three cycles, it is possible to better understand the influence of external forces on torques of the robot, and their difference from real impacts.

Experiments were conducted on Denso VP-6242 6-axis industrial robot with 2 kg of payload. The weight of the object which was manipulated is 1.5 kg, which is 75% of the specified maximum load of the robot. For assembly tasks, the forces needed for the assembly process varied from piece to piece. Since VP-6242 does not possess torque sensors, values shown on graphs represent torque values estimated based on current measurements. Measurements range from −100% to 100% of maximum permissible torque for each joint. Sampling time is 0.1 s.

## 3 Manipulation Task Analysis

This section analyzes the influence of unmodelled load-related forces that occur during load manipulation tasks. Related, but different tasks were analyzed in order to identify possible issues related to unmodelled load.

A movement shown on Fig. 1 was designed so that it combines pick and place/palletizing and machine tending tasks into one sequence. The robot picks up the load from one of four positions on the pallet, brings it close to the assumed loading position, changes the orientation of the load, and loads it into the machine.

The pick and place/palletizing aspects of the task need to be observed together because of their similarity. The pick and place part is most interesting from the perspective of the analysis of the repeatability of influence which load picking and releasing has on the torques of the robot when the picking or placing position is the same. The palletizing, or rather de-palletizing aspect of the robot task is interesting to analyze because of different distances requiring different joint configurations.

The first and second pair of picking positions are symmetrical to the robot base which enables examining the effect of load picking to the $2^{nd}$ and the $3^{rd}$ axis of the robot since they contribute the most to the vertical movement. For each pair, from the perspective of those two axes, the movement is the same as if the picking was performed from the same point. However, the distance of second pair of picking points from the robot base is bigger, and it requires different joint configuration. Since the placing point is the same, the effect the load releasing is expected to be the same.

Results shown on Fig. 1 confirm that the influence of unmodelled load is very similar for first two picking points. Shape of deviation from the torque measurement without load is repetitive, as well as the intensities of torque differences. Repeated tests from multiple measurements confirm that the load influence on the $2^{nd}$ axis is predictable and repeatable. Similar conclusion can be made for the second pair of picking points. However, due to increased distance of the second pair of picking points, the difference in torque caused by the load is slightly higher than for the first pair of picking points as on Fig. 1 (middle right). Releasing of the load caused changes which were also repetitive in intensity and shape resembling a step pulse.

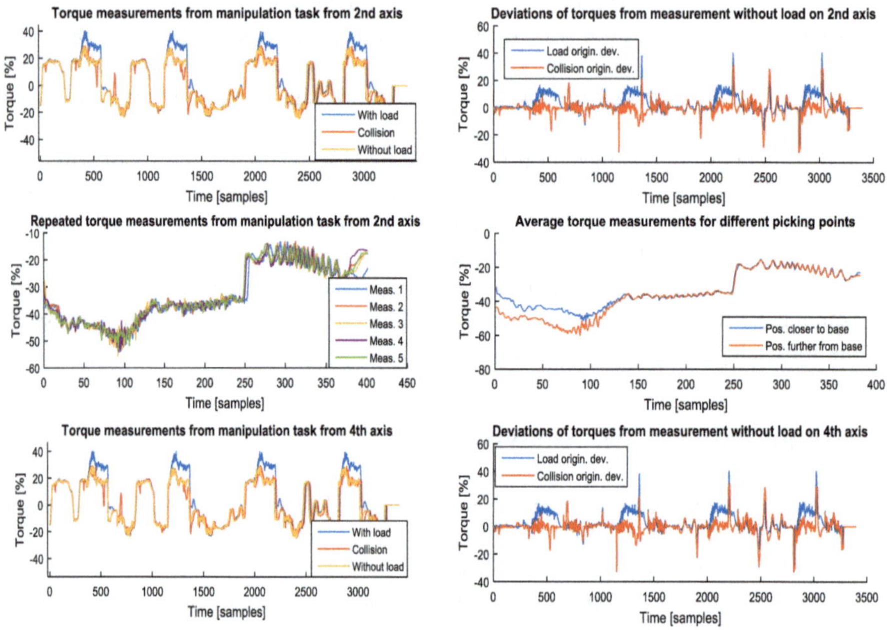


**Fig. 1.** Torque measurements during combined manipulation task: (upper left) Torque measurements with and without load and with collisions. The four segments when the load was picked from different positions are clearly visible as deviations from measurements without load. (upper right) Deviations originating from load and collisions show difference in nature. (middle left) The shown segments of measurements from five consecutive manipulation tasks correspond to periods from when the load was picked from first picking point to the point it was released are repeatable in shape and value. (middle right) Average torque measurements from points closer to the robot base show lower values in sections related to the picking, compared to average torque measurements from those more distant from the base. From the moment the load is close to the loading position, the averaged sections match in values. (lower left) Torque measurements with load are clearly distinguishable from measurements without load from the moment the $4^{th}$ axis starts to move to change loading orientation. (lower right) Shape deviations of torque originating from the load are repetitive and distinguishable from collision induced deviations.

The machine tending aspect of the robot task is intended to use for analysis of the change of flange orientation to the torque measurements. Unlike the pick and place and palletizing experiments, flange is not oriented vertically downwards in this experiment. Orientation of the flange is influenced by motors in last three axes, which generally have lower power, so the load influence will be more noticeable on them.

Results from Fig. 1 (lower left) and Fig. 1 (lower right) show that the influence on the 4th axis of the robot is significant, but only from the moment when the orientation of the flange starts to change. With exception to some occurring peaks, the shape of deviation is related to the change of the 4th axis angle, with highest values being achieved at moments when the loading orientation is reached, and slowly decreasing during the final approach to the loading position due to movement of the 5th axis.

In all parts of the manipulation experiment, the influence of load was predictable and repeatable in shape, making it distinguishable from the impacts in most cases when entire signal is observed as it is visible on Fig. 1 (upper right) and Fig. 1 (lower right). However, the difference between intensities of deviations caused by the load and by collision is not very big, and occasional peaks make them more difficult to tell apart, especially if the distinction needs to be made in matter of samples.

## 4 Assembly Task Analysis

Assembly tasks often require the robot to exert some force in order to join two parts. This section analyses the influence of forces during three different assembly types.

**Snap Fit.** Snap fit is a very common type of assembly process performed by robots. Variations of this process can be found in different fields of industry, depending on the assembly requirements. The experiment was conducted on pieces with annular snap fit, but derived conclusions are applicable on any type of snap assembly.

Results from the testing are shown on Fig. 2. The two peaks on the image originate from two levels of latches and corresponding grooves. It is also noticeable that the second peak has higher value, since the force needed to deform both levels of latches is

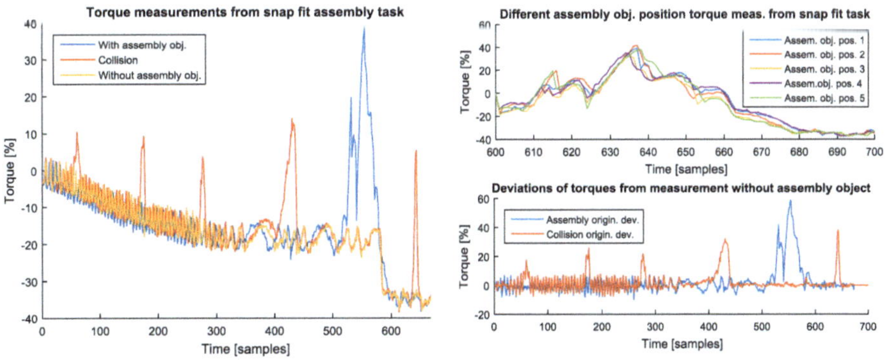


**Fig. 2.** Measurements from snap fit assembly: (left) Torque measurements of robot operation with and without object of assembly as well as with collisions. The double peak originated from two levels of latches and grooves of the object of assembly. (upper right) Five consecutive measurements of assembly task show similar shape, but also differences in intensities in some areas which are result of slight variations of position of the assembly object. (lower right) The shape of the deviation originating from the assembly is similar both in shape and in intensity to the shape of deviations originating from collisions and requires special attention.

higher than the force needed to deform only the first level of latches. When the difference between signals recorded with and without assembly object is compared, it is possible to notice that it resembles a peak belonging to an impact. Repeated tests have shown that while the general shape of the torque is constant, slight variations in the position of the assembly object affect the intensity of peaks.

**Spring Latch.** This type of joining parts is present in applications where it is required to have a good contact, but also allow for easy disassembly. In order to connect the pieces, it is first needed to force the inserted part against a spring, and then move it to a position in which the compressed spring will hold it in place. Assembly and disassembly were performed in a single movement with a two second break between them.

Results on Fig. 3 have shown significant influences on torque, most notably on the 5$^{th}$ axis of the robot. Measurements signals with and without actual assembly piece show great difference, both in assembly and disassembly movement. Although shapes of deviation are roughly symmetrical, tests have shown some differences in intensities of torques exerted on the robot during assembly and disassembly. While collisions with similar shape as parts of assembly induced deviations have been recorded during testing, the overall shape makes them distinguishable from collisions. In particular experiment, the spring induced high torque differences which greatly surpass those of impacts. However, in general case, intensities of assembly and collision induced deviations may happen to be very similar, making them more difficult to tell apart.

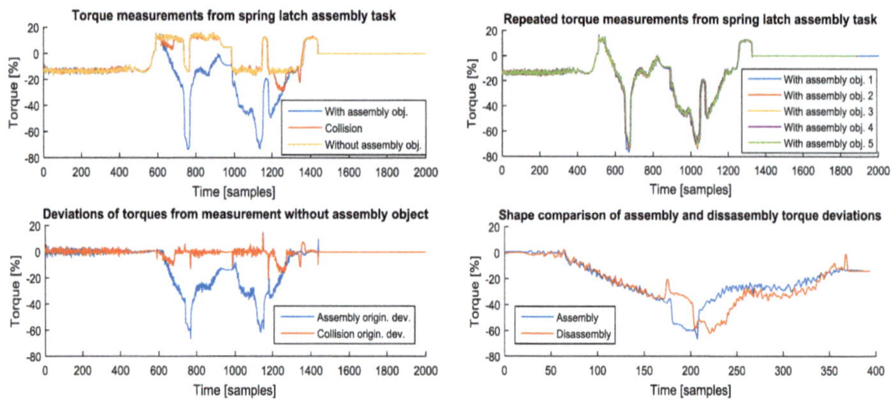


**Fig. 3.** Spring latch task results: (upper left) Torques during assembly and disassembly movement. The deviations originating from the assembly task greatly surpass those induced by collision. (upper right) Torque measurements from 5 consecutive recordings show great similarity and repeatability. (lower left) Deviations induced by assembly task and collisions. Assembly originating deviations show elements of symmetry, but with slightly different intensities. Some collision deviations match sections of the deviations originating from assembly. (lower right) Assembly torque deviation and mirrored disassembly deviation show elements of symmetry.

**Screwing.** Although commonly performed by screwing devices mounted on a robot as a dedicated tool, many robots with unlimited rotation on the 6$^{th}$ axis are used in applications where a part itself is screwed onto another part of the assembly, rather than two parts being fastened by a screw. This is especially the case when a part needs to be rotated less than entire circle in order to screw it in place. In this experiment, both screwing and unscrewing operations were performed and analyzed.

According to expectations, experimental results have shown that the biggest torque difference is on the 6$^{th}$ axis of the robot. The shape of deviation shown on Fig. 4 resembles a ramp signal profile, which corresponds to the nature of the process. In the particular case, the difference can be distinguished from the real collisions due to its duration and intensity, as well as its shape. However, it is possible to imagine that some screwing assemblies might require a higher force. More importantly, collisions need to be identified as soon as possible, which means that decision cannot wait for several samples to make the distinction. The unscrewing action resulted in decrease in intensity of the torque originating from the unscrewing, as it can be seen on Fig. 4 (right). However, the shape of deviation from the measurement without the assembly object has maintained the same shape and nature as with the screwing process.

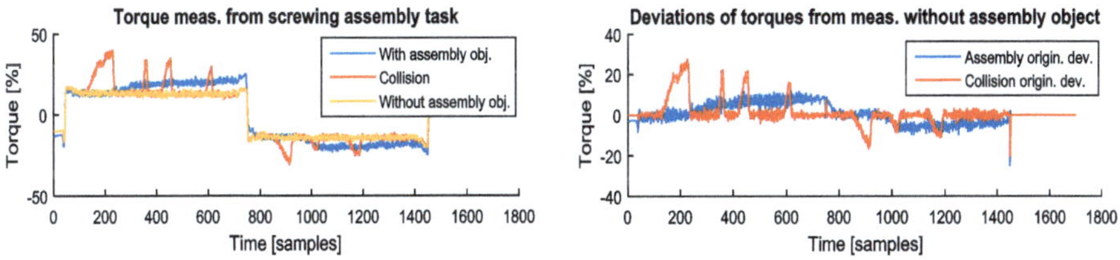


**Fig. 4.** Screwing assembly task: (left) Torque measurements from screwing and unscrewing action resemble ramp signals due to changing forces that act upon the 6$^{th}$ axis. (right) Deviations of torques originating from the task execution and collisions. Though sometimes similar in intensity, the shape of deviations is clearly distinguishable in terms of shape.

## 5 Discussion and Conclusion

The aim of this paper was to examine the influence of unmodelled intentional external forces on the quality of collision detection. To that end, two types of applications which regularly incorporate forces of different profiles and nature were chosen as typical examples.

Experiment with manipulation tasks was performed in a single experiment combining pick and place/palletizing and machine tending tasks. Experiments have shown that the influence of the unmodelled load-related forces has varying effect on the torque measurements depending mainly on the position and orientation of robot flange and consequently the joints affected. It was shown that the disturbances in measurements with and without load can easily be mistaken with collisions with regards to their intensity. However, with exception to variations in intensity, it was shown that the shape of disturbances is repetitive, and that it does not depend on the configuration of robot segments. The aforementioned is important because it means that a disturbance of certain shape can potentially be excluded from being identified as collisions. In [1] it was shown that intentional contact can be identified through use of low and high pass filters with corresponding rules, which implies that influence of load-related disturbances can potentially also be classified as intentional using similar principles with some modifications. The main risk lays in the fact that collisions and unintentional disturbances need to be identified as soon as possible, making it harder to enable processing time needed for avoiding misinterpretation of intentional contacts.

Three types of assembly examples were used for experiments. Snap fit assembly example has shown that profile of torque disturbances it generates has a high potential to be classified as a collision. In particular test case, both the intensity and duration of the peaks in torque originating from connecting two parts closely resemble those of a real collision. Furthermore, it was shown that slight deviations of position of the assembly parts cause variations in intensity of force required to join them. A potential solution to the problem of false identification can be in having higher tolerances for collision detection in a limited time interval within the assembly cycle. Unfortunately, such tolerance modification leads to higher risks of faulty implementation, and potential missed detection of real collision, so this issue deserves special attention.

Forces originating from spring latch assembly example have a high potential to be misclassified, primarily due to their intensity. While shapes of disturbances induced by this assembly type have repeatable profile in particular test, in general case it may depend on positioning accuracy of the assembly parts and repeatability of the spring stiffness. As a general conclusion on this example, the repeatability of the assembly process is the main factor influencing whether or not forces in this type of assembly can reliably be interpreted as intentional, without risking missing real collisions.

Assembly by screwing experiment has shown that the deviations from the measurements from the robot without actual load are repetitive both in shape and in intensity. Although the intensity of the torque deviation was not very significant in the particular performed experiment, in general case the torques needed to apply can easily rise to much higher levels, and be misclassified as collisions. Nevertheless, with adequate positioning accuracy of the assembly parts, a solution can be found to interpret forces during this type of assembly as intentional, as in [1].

As a general conclusion with regards to effect that unmodelled dynamics has on collision detection, the experimental results have shown that influence of unmodelled intentional forces should not be neglected. Failing to do so may lead to false collision detections, while simple increase in detection thresholds may even lead to human injuries. It was shown that some types of forces are predictable and repeatable in shape, making them easier to model and/or include into collision detection algorithm in some other manner. Unfortunately, there are forces which are very difficult to distinguish from the collision because of their shape and/or intensity.

Prompt and timely collision detection poses an additional challenge even for forces which have been shown to be distinguishable from collisions as their correct classification may require additional time for processing.

In a broader sense, varying and unmodelled external force influences have been examined and dealt with in context of impedance control [7, 8], which can be one of directions to elaborate on this topic in future. Nevertheless, the intention of this paper is primarily to point out possible sources of false collision detection, and to raise awareness of importance towards addressing this issue.

As safe physical human-robot interaction is becoming ever more present, results from this paper are shown as a mere reminder on challenges that need more attention in future collision detection related work. Presented examples of different shapes of intentional external forces can be used as guidelines for better understanding their nature and importance they have on the performance of collision detection algorithms.

## References


1. Geravand, M., Flacco, F., De Luca, A.: Human-robot physical interaction and collaboration using an industrial robot with a closed control architecture. In: Proceedings of IEEE International Conference on Robotics and Automation, 2013, pp. 4000–4007 (2013)
2. De Luca, A., Albu-Schäffer, A., Haddadin, S., Hirzinger, G.: Collision detection and safe reaction with the DLR-III lightweight manipulator arm. In: IEEE/RSJ International Conference on Intelligent Robots and Systems (IROS2006), Beijing, China, pp. 1623–1630 (2006)
3. Yamada, Y., Hirasawa, Y., Huand, S., Umetani, Y.: Fail-safe human/robot contact in the safety space. In: IEEE International Workshop on Robot and Human Communication, pp. 59–64 (1996)
4. De Luca, A., Flacco, F.: Integrated control for pHRI: collision avoidance, detection, reaction and collaboration. In: Proceedings of 4th IEEERAS & EMBS Conference on Biomedical Robotics and Biomechatronics, 2012, pp. 288–295 (2012)
5. Yamada, Y., Hirasawa, Y., Huang, S., Umetani, Y., Suita, K.: Human-robot contact in the safeguarding space. IEEE/ASME Trans. Mech. **2**(4), 230–236 (1997)
6. Chen, W., Sun, Y., Huang, Y.: A collision detection system for an assistive robotic. communications in computer and information science, pp. 117–123 (2010)
7. Ferraguti, F., Secchi, C., Fantuzzi, C.: A tank-based approach to impedance control with variable stiffness. In: IEEE International Conference on Robotics and Automation (ICRA), Karlsruhe, pp. 4948–4953 (2013)
8. Schindlbeck, C., Haddadin, S.: Unified passivity-based Cartesian force/impedance control for rigid and flexible joint robots via task-energy tanks. In: IEEE International Conference on Robotics and Automation (ICRA), Seattle, pp. 440–447 (2015)
9. Bahloul, A., Tliba, S., Chitour, Y.: Dynamic parameters identification of an industrial robot with and without payload. In: 18th IFAC Symposium on System Identification, SYSID 2018, IFAC, July 2018, Stockholm, pp. 443–448 < hal-01866381 > (2018)
10. Swevers, J., Verdonck, W., Schutter, J.D.: Dynamic model identification for industrial robots. IEEE Control Syst. Mag. **27**(5), 58–71 (2007)
11. Brunot, M., Janot, A., Carillo, F., Garnier, H.: A pragmatic and systematic statistical analysis for identification of industrial robots. In: Proceedings of the IEEE International Conference on Advanced Intelligent Mechatronics (AIM), 2017, Munich, Germany, pp. 559–564 (2017)
12. Gautier, M.: Numerical calculation of the base inertial parameters of robots. In: Proceedings of the IEEE International Conference on Robotics and Automation (ICRA), 1990, pp. 1020–1025 (1990)
13. Gautier, M., Khalil, W.: On the identification of the inertial parameters of robots. In: Proceedings of the IEEE International on Decision and Control (CDC), 1988, pp. 2264–2269 (1988)

14. Gautier, M., Khalil, W.: Direct calculation of minimum set of inertial parameters of serial robots. IEEE Trans. Robot. Autom. **6**, 368–373 (1990)
15. Janot, A., Vandanjon, P.-O., Gautier, M.: A generic instrumental variable approach for industrial robot identification. IEEE Trans. Control Syst. Technol. **22**(1), 132–145 (2014)
16. Jin, J., Gans, N.: Parameter identification for industrial robots with a fast and robust trajectory design approach. Robot. Comput.-Integr. Manuf. **31**, 21–29 (2015)
17. Khosla, P.K., Kanade, T.: Parameter identification of robot dynamics. In: Proceedings of the IEEE International on Decision and Control (CDC), 1985, pp. 1754–1760 (1985)
18. Walter, E., Pronzato, L.: Identification of parametric models from experimental data. Communications and Control Engineering, 1st edn. Springer, London (1997)
19. Deniša, M., Gams, A., Ude, A., Petrič, T.: Learning compliant movement primitives through demonstration and statistical generalization. IEEE/ASME Trans. Mech. **21**(5), 2581–2594 (2016)

# Use of Bimanual Haptic Teleoperation System for Optimized Task Performance with Adaptive Haptic Tunnel

Jožica Piškur[1(✉)], Sebastjan Šlajpah[1], Bojan Nemec[2], Matjaž Mihelj[1], and Marko Munih[1]

[1] Faculty of Electrical Engineering, University in Ljubljana, Tržaška cesta 25, 1000 Ljubljana, Slovenia
{jozica.piskur, sebastjan.slajpah, matjaz.mihelj, marko.munih}@robo.fe.uni-lj.si
[2] Institut "Jožef Stefan", Jamova cesta 39, 1000 Ljubljana, Slovenia
bojan.nemec@ijs.si

**Abstract.** This paper presents bimanual haptic teleoperation system with implemented haptic tunnel for optimized task performance. The teleoperation system includes two industrial robots as slaves and two haptic robots as masters. The bimanual control is based on impedance control with included adaptive haptic tunnel. Task is performed in relative coordinates where base of the robot is set at end effector of one slave robot while the position of second slave robot is defined relatively to the new base. Haptic tunnel is defined with (a) reference trajectory obtained with initial task execution and described with Gaussian radial basis functions, (b) adaptive radius of the tunnel based on error between reference and traveled trajectory and percentage of the contact with the haptic tunnel. After each task execution the performance is evaluated following by proper adaptation of haptic tunnel. Developed algorithm was tested with a simple assembly task of inserting car bulb into its base.

**Keywords:** Teleoperation · Bimanual control · Haptic tunnel · Impedance control

## 1 Introduction

Industrial and collaborative robots are becoming more user friendly. Their control systems offer variety of tools for improved performance. Control architecture mostly covers point-to-point motion with collision avoidance. By implementing bimanual tasks, an additional bimanual control is needed. Robots are guided accordingly to previously learned trajectories by autonomously adapting the bimanual tasks; two robots can behave as a single robot.

According to Sheridan the teleoperation is human control of remote sensors and actuators, e.g. human-robot interaction [1–3]. The most used type of teleoperation is master-slave system. User operates with master while the slave tracks master's movement. By implementing haptic feedback, operator can feel interaction force from slave through master [4].


K. Berns and D. Görges (Eds.): RAAD 2019, AISC 980, pp. 329–336, 2020.
https://doi.org/10.1007/978-3-030-19648-6_38

In recent years, different methods of adaptive robot learning were introduced [5–7]. Within learning from demonstration showed movement (example) needs to be repeatable in order to provide most optimized data set. This can be achieved by using kinesthetic guidance where the human operator acts as a control system thus ensuring optimization of the system. If demonstration is done via teleoperation, loss of additional information (e.g. depth perception) [8] can be reflected in non-optimized movement including substantial amount of corrections during task performance. This effect can be reduced by increasing telepresence with introduction of haptics.

In this paper bimanual teleoperation system with haptic feedback for optimized task performance is presented. The system was upgraded with an adaptive haptic tunnel to ensure repetitive demonstration as it limits the working area around reference trajectory but still allows minor adjustments for optimizing the movements.

## 2 Methods

The bimanual teleoperation system consists of two slave and two master robots (see Fig. 1). Masters (M1 and M2) are two 7 DOF haptic robots Omega.7, Force Dimension. They are designed as two-hand haptic interface. Slaves (R1 and R2) are two 6 DOF industrial robots Yaskawa Motoman MH5. Each of industrial robots has a JR3 force/torque sensor mounted on their end effectors. Control architecture for Motoman MH5 is based on an open architecture control yControl [9] which enables design of a custom control algorithms via Simulink RealTime. Controller sampling frequency is 4 kHz.

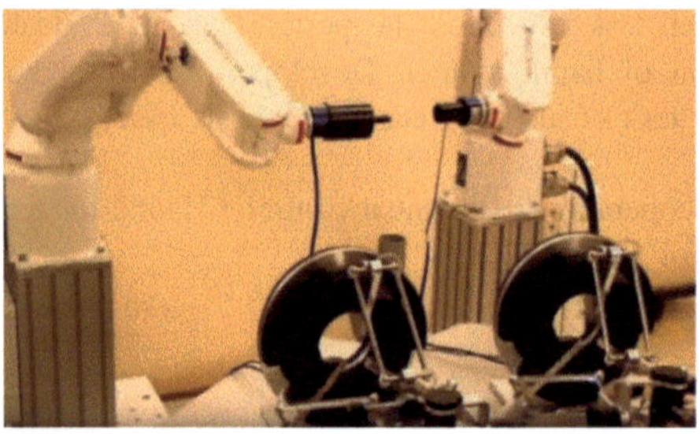

**Fig. 1.** Bimanual haptic teleoperation system consists of four robots (two masters Omega.7 and two slaves Yaskawa Motoman MH5).

### 2.1 Teleoperation Control

The teleoperation control of the slave robots is implemented via virtual proxy. Master is directly linked to a proxy, while adjustable admittance connects slave and proxy as presented in Fig. 2. The forces $\mathbf{F}_{ps}$ acting on a slave via proxy is determined as

$$\mathbf{F}_{\mathrm{ps}} = \mathbf{K}_{\mathrm{p}}\left(\int \zeta \dot{\mathbf{x}}_{\mathrm{m}} dt - \mathbf{x}_{\mathrm{p}}\right) + \zeta \mathbf{K}_{\mathrm{d}} \dot{\mathbf{x}}_{\mathrm{m}} + \mathbf{K}_{\mathrm{ps}}(\mathbf{x}_{\mathrm{m}} - \mathbf{x}_{\mathrm{s}}) \tag{1}$$

where $\mathbf{x}_{\mathrm{m}}$, $\mathbf{x}_{\mathrm{p}}$, and $\mathbf{x}_{\mathrm{s}}$ are master's, proxy's, and slave's position, respectively, $\mathbf{K}_{\mathrm{p}}$ and $\mathbf{K}_{\mathrm{d}}$ proportional and derivative gain, $\mathbf{K}_{\mathrm{ps}}$ gain including position error between master and slave, and $\zeta$ scaling factor between master and slave. The forces $\mathbf{F}_{\mathrm{m}} = -(\mathbf{F}_{\mathrm{ps}} + \mathbf{F}_{\mathrm{e}})$, acting on a master as a feedback, can be presented as a combination of forces $\mathbf{F}_{\mathrm{ps}}$ and external forces $\mathbf{F}_{\mathrm{e}}$ measured by external force/torque sensor. With (2) acceleration of proxy $\ddot{\mathbf{x}}_{\mathrm{p}}$ is determined as a function of mass $m_p$, damping $b_p$, spring $k_p$.

$$\ddot{\mathbf{x}}_{\mathrm{p}} = m_{\mathrm{p}}^{-1}\left(\mathbf{F}_{\mathrm{ps}} - \mathbf{F}_{\mathrm{e}} - b_{\mathrm{p}}\dot{\mathbf{x}}_{\mathrm{p}} - k_{\mathrm{p}}\mathbf{x}_{\mathrm{p}}\right). \tag{2}$$

Admittance parameters determine the behavior of the slave. With integration of proxy's acceleration $\ddot{\mathbf{x}}_{\mathrm{p}}$ desired velocity and position for slave can be obtained. The proxy point can further be connected to a desired reference point in space via variable impedance (spring $k_i$ and damping $b_i$, see Fig. 2). By implementing impedance and feedforwarded force, desired force acting on the robot can be obtained as

$$\mathbf{F}_{\mathrm{s}} = k_{\mathrm{i}}(\mathbf{x}_{\mathrm{p}} - \mathbf{x}_{\mathrm{s}}) + b_{\mathrm{i}}(\dot{\mathbf{x}}_{\mathrm{p}} - \dot{\mathbf{x}}_{\mathrm{s}}) + \mathbf{K}_{\mathrm{f}}\left(k_{\mathrm{i}}(\mathbf{x}_{\mathrm{p}} - \mathbf{x}_{\mathrm{s}}) + b_{\mathrm{i}}(\dot{\mathbf{x}}_{\mathrm{p}} - \dot{\mathbf{x}}_{\mathrm{s}}) - \mathbf{F}_{\mathrm{e}}\right), \tag{3}$$

where $\mathbf{K}_{\mathrm{f}}$ represents force feedforward gain which amplifies the impact of external forces. Forces $\mathbf{F}_{\mathrm{s}}$ are then transformed to desired torques via analytical Jacobian matrix $\tau_{\mathrm{T}} = \mathbf{J}_{\mathrm{A}}^{T}\mathbf{F}_{\mathrm{s}}$

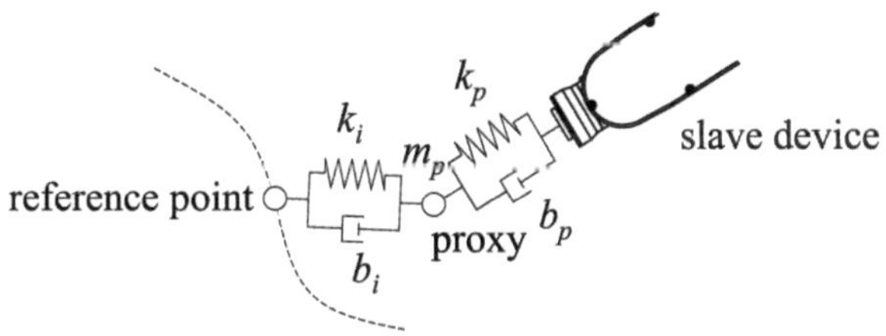


**Fig. 2.** Slave is controlled by a master via proxy; for position tracking impedance control is implemented.

The sum of task's torques $\tau_{\mathrm{T}}$ and torques based on inverse dynamic model $\tau_{\mathrm{IDM}}$ can then be used as a desired torques for the robot's joints, $\tau_{\mathrm{D}} = \tau_{\mathrm{T}} + \tau_{\mathrm{IDM}}$. Mathematical background for attaining inverse dynamic model is presented in [10]. For proper control good estimations of robot's and payload parameters are needed.

## 2.2 Implementation of Bimanual Control

Bimanual robot systems enable to implement human intuitive tasks. Bimanual control represents synchronous movement of two robots as one robot. The R2 symmetrically follows the movement of the R1 and can also perform individual movements.

In this work the bimanual control is implemented through teleoperation. The position of the M1 $\mathbf{x}_{m1}$ affects on both slaves through the forces acting on the slaves $\mathbf{F}_{ps1}$ and $\mathbf{F}_{ps2}$ (see (4) and (5)). Transformation matrix between the M1 and the M2 is identity. Meanwhile the position of the M2 $\mathbf{x}_{m2}$ affects only on the R2. Position $\mathbf{x}_{m2}$ is defined according to the second slave's coordinate system. Other variables used in (4) and (5) are: $\mathbf{x_p}$ and $\mathbf{x_s}$ are proxy's, and slave's position.

$$\mathbf{F_{ps1}} = \mathbf{K_p}\left(\int \zeta \dot{\mathbf{x}}_{m1} dt - \mathbf{x}_{p1}\right) + \zeta \mathbf{K_d} \dot{\mathbf{x}}_{m1} + \mathbf{K_{ps}}(\mathbf{x}_{m1} - \mathbf{x}_{s1}), \tag{4}$$

$$\mathbf{F_{ps2}} = \mathbf{K_p}\left(\int \zeta \dot{\mathbf{x}}_{m2} dt - \mathbf{x}_{p2}\right) + \zeta \mathbf{K_d} \dot{\mathbf{x}}_{m2} + \mathbf{K_{ps}}((\mathbf{x}_{m2} + \mathbf{x}_{m1}) - \mathbf{x}_{s2}). \tag{5}$$

### 2.3 Initial Reference Trajectory

Initial reference trajectory depends on a task to be performed by R1 and R2. In this work the reference trajectory is used as a basic for the haptic tunnel and is defined with radial basis functions (RBF) [11]. RBF is a parametric function whose value depends only on the phase from the initial position between R1 and R2.

Global coordinate system is positioned at a base of the R1 (see Fig. 3). Relative coordinates are introduced when robots R1 and R2 are described as one robot R1R2 where the base of R1R2 is at R1's end effector and the end effector of R1R2 is at R2's end effector. Relative coordinates of the new robot are given with the homogeneous transformations and denoted as $\mathbf{H}_2^{rel}$. Transformation from absolute to relative coordinates is described with

$$\mathbf{H}_2^{\mathrm{rel}} = \mathbf{H}_1^{-1}\mathbf{H}_2^1 = \mathbf{H}_1^{-1}\mathbf{H}_{21}\mathbf{H}_2. \tag{6}$$

The RBF model learns the reference trajectory in relative coordinates $\mathbf{x}_2^{rel}$ with additional approximation parameters as described in [12]. The RBF model describes the point $\mathbf{x}_{\mathrm{RBF}}$ on the reference trajectory as spline of Gaussian functions in dependency of the phase $s$: $\mathbf{x}_{\mathrm{RBF}}(s)$. The phase describes part of the traveled reference trajectory, which begins at $s = 1$ and ends with $s = e^{-2}$.

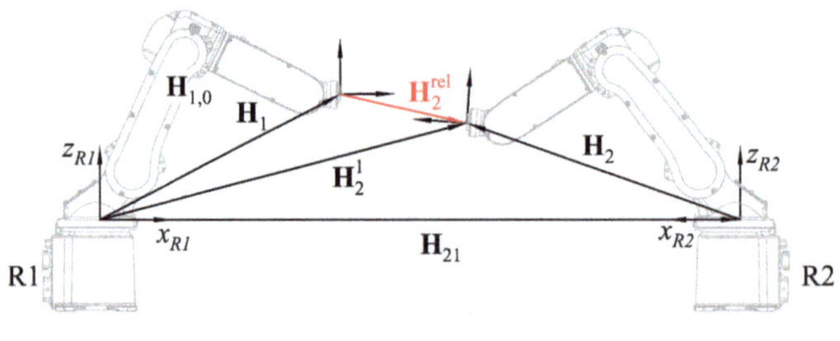


**Fig. 3.** Relations for transformation from absolute to relative coordinates.

## 2.4 Development of the Haptic Tunnel

Haptic tunnel is presented as a tube around reference trajectory, inside which R2 can move freely. Outside rigid walls of the tunnel, the force is applied on the R2 in the direction to the reference trajectory. At first iteration a manually set radius $r_t$ is used to define the haptic tunnel.

The approximation parameters of RBF model are used to calculate the nearest point $\mathbf{x}_{\mathrm{RBF}}(s)$ on the reference trajectory according to the current position of $\mathbf{x}_2^{rel}$ which depends on phase (as described in 2.3). To calculate the nearest phase $s$, numerical method of bisection is used. The phase $s'$ determine previous known point on the reference trajectory $\mathbf{x}_{\mathrm{RBF}}(s')$. At phase $s' - \Delta s$ a next point on the reference trajectory $\mathbf{x}_{\mathrm{RBF}}(s' - \Delta s)$ is defined. Step variable $\Delta s$ is set to 0.1. Between points $\mathbf{x}_{\mathrm{RBF}}(s' - \Delta s)$ and $\mathbf{x}_{\mathrm{RBF}}(s')$ unit vector $\mathbf{t}$ can be defined that is further used for making a projection $u$ of a vector $\mathbf{v}$ (vector connecting current $\mathbf{x}_2^{rel}$ and previous point on the reference trajectory $\mathbf{x}_{\mathrm{RBF}}(s')$; $\mathbf{v} = \mathbf{x}_2^{rel} - \mathbf{x}_{\mathrm{RBF}}(s')$): $u = \mathbf{t} \cdot \mathbf{v}$.

The projection $u$ determines whether the phase $s$ is behind $(u < \varepsilon)$ or after $(u > \varepsilon)$ previous phase. The threshold $\varepsilon$ is set to $\pm 10^{-6} m$. If the phase $s$ is behind the previous phase $s'$, then the new nearest phase it is calculated on interval $[1, s']$ otherwise the $s$ is searched for between $[s', e^{-a_x}]$. After each iteration the phase interval is adapted accordingly. The method of bisection is active until the $\|\mathbf{x}_{\mathrm{RBF}}(s) - \mathbf{x}_{\mathrm{RBF}}(s')\| = u$. For ensuring real time operation, the bisection algorithm was limited to 8 iterations.

**The Interaction Force.** Once the nearest phase $s$ is known then a difference between the position of the R2 $\mathbf{x}_2^{rel}$ and the nearest point $\mathbf{x}_{\mathrm{RBF}}(s)$ is defined as

$$\Delta \mathbf{x}^{rel} = \mathbf{x}_{\mathrm{RBF}}(s) - \mathbf{x}_2^{rel}. \tag{7}$$

The difference is normed $\left\|\Delta \mathbf{x}^{rel}\right\|$ and compared with the radius $r_t$ of the haptic tunnel. If the $\left\|\Delta \mathbf{x}^{rel}\right\| < r_t$, then the R2 is inside the haptic tunnel and there is no interaction force $\mathbf{F}_t = [0 \quad 0 \quad 0]^T$. Opposite if the R2 is outside or on the edge of the haptic tunnel, then the $\mathbf{F}_t$ is applied. This force is also used for haptic feedback to the user via master device. The $\mathbf{F}_t$ is determined by the direction of movement of R2 (see Fig. 4).

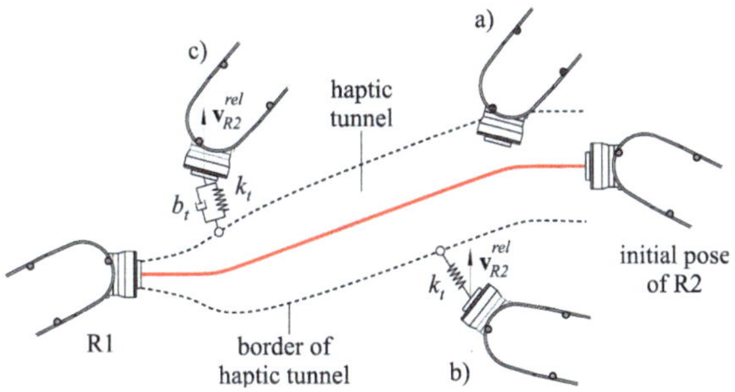


**Fig. 4.** Different scenarios of contact with haptic tunnel: (a) no interaction force, (b) robot moving towards haptic tunnel, (c) robot moving away from the haptic tunnel.

If R2 is moving away from the haptic tunnel, then the interaction force $\mathbf{F}_\mathrm{t}$ is combination of stiffness $\mathbf{F}_\mathrm{s}$ and damping $\mathbf{F}_\mathrm{d}$

$$\mathbf{F}_\mathrm{t} = \mathbf{F}_\mathrm{s} + \mathbf{F}_\mathrm{d} = k_\mathrm{s} \Delta \mathbf{x}_\mathrm{dev}^{rel} + k_\mathrm{d} \Delta \dot{\mathbf{x}}_\mathrm{dev}, \tag{8}$$

where $k_\mathrm{s}$ and $k_\mathrm{d}$ are stiffness and damping parameters, which are set to high values for good haptic feedback. Parameter $\Delta \mathbf{x}_\mathrm{dev}^{rel}$ describes difference between $\mathbf{x}_2^{rel}$ and the nearest point $\mathbf{x}'$ on the border of the haptic tunnel (see (9) and (10)). Velocity $\Delta \dot{\mathbf{x}}_\mathrm{dev} = \mathbf{v}_{R2}^{rel}$ represents the velocity of R2.

$$\mathbf{x}' = \left( \mathbf{x}_\mathrm{RBF}(s) + \frac{\Delta \mathbf{x}^{rel}}{\|\Delta \mathbf{x}^{rel}\|} \cdot r_\mathrm{t} \right), \tag{9}$$

$$\Delta \mathbf{x}_\mathrm{dev}^{rel} = \mathbf{x}' - \mathbf{x}_2^{rel}. \tag{10}$$

If R2 is moving toward the haptic tunnel or is it on its border equal, then the interaction force $\mathbf{F}_\mathrm{t}$ is interpreted as

$$\mathbf{F}_\mathrm{t} = \mathbf{F}_\mathrm{s} = k_\mathrm{s} \Delta \mathbf{x}_\mathrm{dev}^{rel}. \tag{11}$$

### 2.5 Adaptive Radius of the Haptic Tunnel

The radius of the haptic tunnel $r_\mathrm{t}$ can be manually set or can be adapted from previous iterations. The adaptive radius in the haptic tunnel is expected to offer very shallow tracking in the cases of larger trajectory deviations and automatically very precise tracking when number of deviations in the tunnel is smaller. After the initial reference trajectory is described with RBF model (see Fig. 5a), the radius is manually set to a constant value. The first adaptation of radius is then done according to Root-Mean-Square-Error (RMSE) between the reference trajectory and traveled trajectory $b_{rms}$ as a

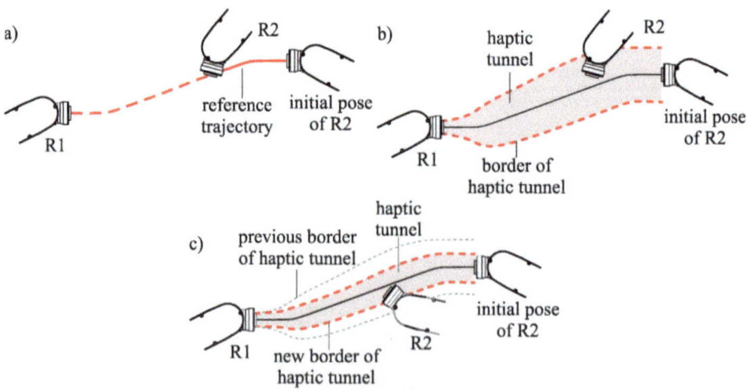


**Fig. 5.** (a) Defining a reference trajectory with RBF, (b) haptic tunnel with preset radius, (c) corrected radius of the haptic tunnel regarding the RMSE between reference and previous traveled trajectory and percentage of contact with the haptic tunnel.

double of RMSE value (see Fig. 5b). In next iterations (see Fig. 5c) of adaptation of the radius of the haptic tunnel, contact of R2 with haptic tunnel border is analyzed.

The radius $r_t$ can also be adapted regarding the number of contacts with the haptic tunnel. If the percentage of contacts are less or equal than 10% of the traveled trajectory then a new radius $r_t$ is adapted as $r_t = r_t' - 1.1\, b_{rms}$, where $r_t'$ is previous radius of the haptic tunnel. If the percentage of contacts are greater than 10% and less or equal to 33% of the traveled trajectory then the radius remains the same: $r_t = r_t'$. If the percentage of contacts are greater than 33% of the traveled trajectory, then the radius $r_t$ is increased as $r_t = r_t' + 0.78\, b_{rms}$.

### 2.6 Experimental Bimanual Task

The designed bimanual teleoperation system with the haptic tunnel was used in a task of car light assembly. R1 holds a platform for a car light which is normally static, so the R1 is in stand-by position, while the R2 is doing the task. The R2 holds, approaches, and inserts a car light bulb into the platform. If the platform is moving, then these movements are considered in the R2 movement (see (5)) as symmetric move. To achieve the task with specific movement, the reference trajectory with the haptic tunnel is created to limit a teleoperation's workspace.

## 3 Results

To evaluate this approach, a RMSE method was used to assess the deviation between reference and traveled trajectory. In addition, the number of contacts with haptic tunnel was also assessed. According to the presented algorithm, in the first iteration the initial reference trajectory was described with the RBF model. The initial radius of the haptic tunnel was set to $r_t = 18$ mm.

In second iteration a haptic tunnel was set. Analysis after task completion showed a $b_{rms} = 5.0$ mm and a percentage of contact with the haptic tunnel border was less than 4%. The radius was corrected to $r_t = 12.5$ mm.

The third iteration showed a RMSE value of $b_{rms} = 3.3$ mm, which was less than in the second iteration, and a percentage of contact with the haptic tunnel border was 11%. In the following iterations the percentage of contact was approximately 12–14%. An additional test with smaller radius of haptic tunnel was done ($r_t = 8$ mm), where the percentage of contact was more than 35% and $b_{rms} = 5.78$ mm.

## 4 Discussions and Conclusions

The radius of haptic tunnel along the reference trajectory was adapted considering the RMSE values between reference and traveled trajectory and the percentage of contacts with the haptic tunnel. For second iteration, a radius of haptic tunnel was set to $r_t = 18$ mm. After task completion, the $b_{rms} = 5.0$ mm with less than 4% of contact. This results in stricter haptic tunnel with radius $r_t = 12.5$ mm. After finished iteration analysis showed $b_{rms} = 3.3$ mm which indicates improved performance due to reduced radius of haptic tunnel. The reduction of haptic tunnel must be done with a precaution as to small radius can lead to stability issues and non-compliant behavior. Experimental tests showed that in our case stability margin is at $r_t = 8\,\mathrm{mm}$.

We designed the bimanual haptic teleoperation system of four robots for learning and performing bimanual tasks such as car light assembly. For optimized task performance an adaptive haptic tunnel was introduced. Trajectory implementation with RBF model enables easy-to-setup haptic tunnels with different areas and adaptive radius (e.g. stop area with a higher damping and smaller radius for more precise task completion). In this way an optimal trajectory can be obtained within few iterations which can be further used as a reference path for teaching by demonstration implementing teleoperation control with haptic modality.

**Acknowledgements.** The work reported in this paper was supported by project No. J2-7360 supported by Slovenian Research Agency (research core funding No. P2-0228).

## References

1. Sheridan, T.B.: Teleoperation, telerobotics and telepresence: a progress report. Control Eng. Pract. **3**(2), 205–214 (1995)
2. Sheridan, T.B.: Space teleoperation through time delay: review and prognosis. IEEE Trans. Robot. Autom. **9**(5), 592–606 (1993)
3. Vertut, J. (ed.): Teleoperation and Robotics: Applications and Technology. Springer, Netherlands (1985)
4. Iida, W., Ohnishi, K.: Reproducibility and operationality in bilateral teleoperation. In: The 8th IEEE International Workshop on Advanced Motion Control, 2004, AMC 2004, pp. 217–222 (2004)
5. Argall, B., Chernova, S., Veloso, M.M., Browning, B.: A survey of robot learning from demonstration. Robot. Auton. Syst. **57**, 469–483 (2009)
6. Calinon, S., Guenter, F., Billard, A.: On Learning, representing, and generalizing a task in a humanoid robot. IEEE Trans. Syst. Man Cybern. Part B (Cybernetics) **37**(2), 286–298 (2007)
7. Billard, A., Calinon, S., Dillmann, R., Schaal, S.: Robot programming by demonstration. In: Siciliano, B., Khatib, O., (eds.) Springer Handbook of Robotics, pp. 1371–1394. Springer Heidelberg (2008)
8. Buss, M., Schmidt, G.: Control problems in multi-modal telepresence systems. In: Advances in Control, London, pp. 65–101 (1999)
9. Mihelj, M., Šlajpah, S., Čepon, P., Munih, M.: yControl - open architecture controller for Yaskawa Motoman MH5 robot. In: 2012 IEEE International Conference on Control Applications, 2012, pp. 1051–1056 (2012)
10. Mihelj, M., et al.: Robot control. In: Robotics, pp. 133–152. Springer, Cham (2019)
11. Nemec, B., Ude, A.: Action sequencing using dynamic movement primitives. Robotica **30**(5), 837–846 (2012)
12. Nemec, B., Žlajpah, L., Ude, A., Šlajpah, S., Piškur, J.: An efficient PbB framework for fast deployment of bi-manual assembly tasks. In: IEEE-RAS 18th International Conference on Humanoid Robots, Beijing, China, vol. 2018, pp. 166–173 (2018)

# Feedforward Control Approaches to Bidirectional Antagonistic Actuators Based on Learning

Nikola Knežević(✉), Branko Lukić, and Kosta Jovanović

School of Electrical Engineering, University of Belgrade, 11120 Belgrade, Serbia
{knezevic, branko, kostaj}@etf.bg.ac.rs

**Abstract.** Safe physical human-robot interaction is a decisive feature in wider adaptation of robots in homes and factories. To that end, a lot of researchers consider new actuation mechanisms and particularly Variable Stiffness Actuators (VSAs) which contribute to robot safety, but also to increase energy efficiency and outperforming rigid actuators in repetitive tasks. However, advantages of VSAs come with their price – issues in design and control of such multivariable non-linear systems. Novel approaches and methods in soft computing methods such as machine learning and neural networks are opening new horizons in VSA control. In this paper, a comparative analysis is carried out between the neural network feedforward control and locally weighted projection regression as a technique for model learning of bidirectional antagonistic VSA – qb move maker pro. Set of measurement is used to create mapping between two motor positions as inputs and measured actuator position and estimated stiffness as outputs. Comparative analysis of the two different approaches for feedforward control observing performances in open loop control, followed by closed loop testing with a simple feedback regulator for fine tuning. Learning techniques result in robust and generalized models that can predict required inputs in ordered to achieve good output tracking.

**Keywords:** Variable Stiffness Actuator · Neural network · Machine learning · Locally Weighted Projection Regression · Model-free feedforward control

## 1 Introduction

For contemporary industrial demands it is no longer enough for a robot to be fast and accurate while executing a task. Increased productivity and flexibility are one of most important demands. To increase productivity and flexibility, a close human-robot collaboration is required. To enable human-robot collaboration, it is very important that robots are compliant in order to ensure safe interaction with the environment. The VSAs are designed to meet these requirements and replace classic rigid actuators.

This paper was partially funded by the Ministry of Education, Science and Technological development of the Republic of Serbia, under contracts TR-35003.


K. Berns and D. Görges (Eds.): RAAD 2019, AISC 980, pp. 337–345, 2020.
https://doi.org/10.1007/978-3-030-19648-6_39

So far, several types of the VSAs had been developed [1]. These VSAs feature elastic nonlinear transmission that can be realized as nonlinear springs or pneumatic muscles. Based on the mechanical construction, the variable stiffness can be achieved by controller, mechanically or by combination of the previous two approaches [1]. The qb move maker pro (from now we will address it as the qb actuator) is a type of the tendon driven bidirectional antagonistic VSA with non-linear spring transmission. The qb actuator, developed by The Natural Motion Initiative, is a low-cost 3D printed bidirectional spring antagonistic actuator. Because of its low price, the qb actuator is an ideal solution for education and research application in the field of robotics and control.

The qb actuator and its structure are shown in Fig. 1. Experiments that demonstrate qb actuator performance and capabilities are described in [2].

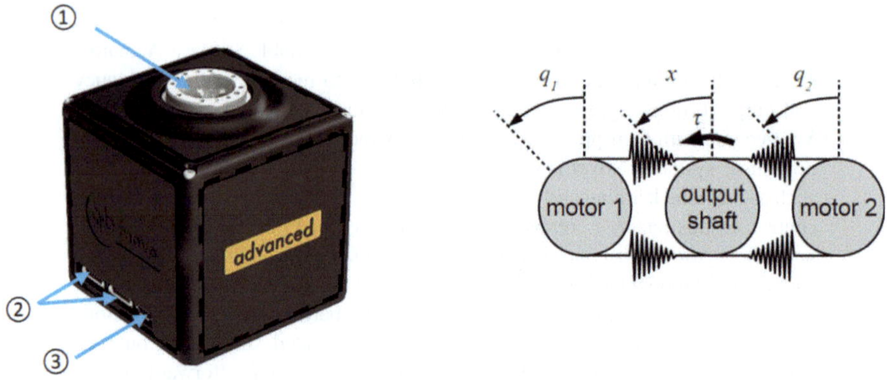


**Fig. 1.** Left: The qb actuator: (1) an output shaft – actuator position; (2) a daisy-chain RS485 connector; (3) a USB port. Right: Functional scheme.

In order to design a high-performance robust controller for VSAs, a good knowledge of the model dynamics is crucial [3–6]. In practice, a mathematical model of the system is not perfect because of the numerical approximation and unmodeled dynamics which are always present. Furthermore, two actuators of the same design do not have the same model due to imperfections in the manufacturing process. In order to avoid development of inaccurate analytical models, learning techniques for mapping static relationships of VSA are applied. Two learning algorithms - Neural networks and Locally Weighted Projection Regression (LWPR) - are used to learn the qb actuator model and map relationships between inputs (motors position) and outputs (actuator position and stiffness).

In Sect. 2, the model of qb actuator bidirectional antagonistic joint is presented. Neural network and LWPR training approaches are present in Sect. 3. Experimental results and comparative analysis of presented learning methods are given in Sect. 4, and final remarks and directions for future work are given in Sect. 5.

## 2 Qb Actuator Static Model

Parameters which should be provided by the manufacturer as VSA specification are suggested in [7]. Datasheet of qb actuator and its specifications is presented in [8]. The qb actuator is a system with two inputs: prime mover motors (antagonistically coupled motor 1 and motor 2) positions ($q_1$ and $q_2$) and two outputs: actuator position and stiffness ($x$ and $S$). The qb actuator has built-in PD controllers for DC motors' position regulation. In general, PID controllers could be used for better motor position reference tracking. The equations that describe actuator equilibrium position ($x_e$) and stiffness of the qb actuator in steady state are provided by the actuator manufacturer and are given by (1) and (2) respectively.

$$x_e = (q_1 + q_2)/2, \tag{1}$$

$$S = a\,k\cosh(a(x - q_1)) + a\,k\cosh(a(x - q_2)), \tag{2}$$

here, cosh is hyperbolic cosine function, while parameters $a = 6.8465$ and $k = 0.0223$ are obtained by system identification described in [7]. The springs in the qb actuator have non-linear force-displacement characteristic, which is very important for controlling the stiffness of the actuator. Some control algorithms developed for the qb actuator were presented in [9, 10].

## 3 Training Methods

Two learning methodologies intended to replace traditional control algorithms and eliminate the need for modeling system dynamics are presented in this paper. Neural network and LWPR are trained to replace the system model in the control scheme. By applying these two methods, inverse mapping of the system is obtained. If the train model is to be considered valid, proper data must be collected. The training data set is formed from real measurements performed on the qb actuator. Prime mover reference signals that result in different output states are used. Both the neural network and LWPR method receive equilibrium position and stiffness of the actuator as inputs and attempt to predict the prime mover position that will result in the desired reference movement. Ideally, the reference signals and movement achieved by the predicted prime mover position should be identical. Accuracy of predicted values is limited by the training set, neural network dimensions, initial LWPR setup and repeatability of the actuator. As noted in [10], five reference position patterns are used for collecting training data. The first pattern assumes a constant difference between positions of prime movers (0, 20, 40, 60 and 70°) which reflects in different stiffness preset, while the position of each prime mover was increased by 5° from −90 to 90°. In this case, constant stiffness periods are expected, but in real scenario this is not exactly the case. In the next four reference patterns, actuator position is increasing from −90 to 90° and difference between position of prime movers was changing increasingly and decreasingly from −60 to 60° by step of 4°. Such prime mover behaviors reflect in saw shape of actuator stiffness. The input and output signals are presented in Fig. 2.

### 3.1 Neural Network Design

Dimensions and design of neural network was determined experimentally. The neural network has an input layer with two inputs, two hidden layers with five neurons and an output layer with two outputs. Sigmoid activation functions in the hidden layers deal with nonlinearity, while the *purelin* activation functions in the output layer provide analogue outputs for controlling qb actuator.

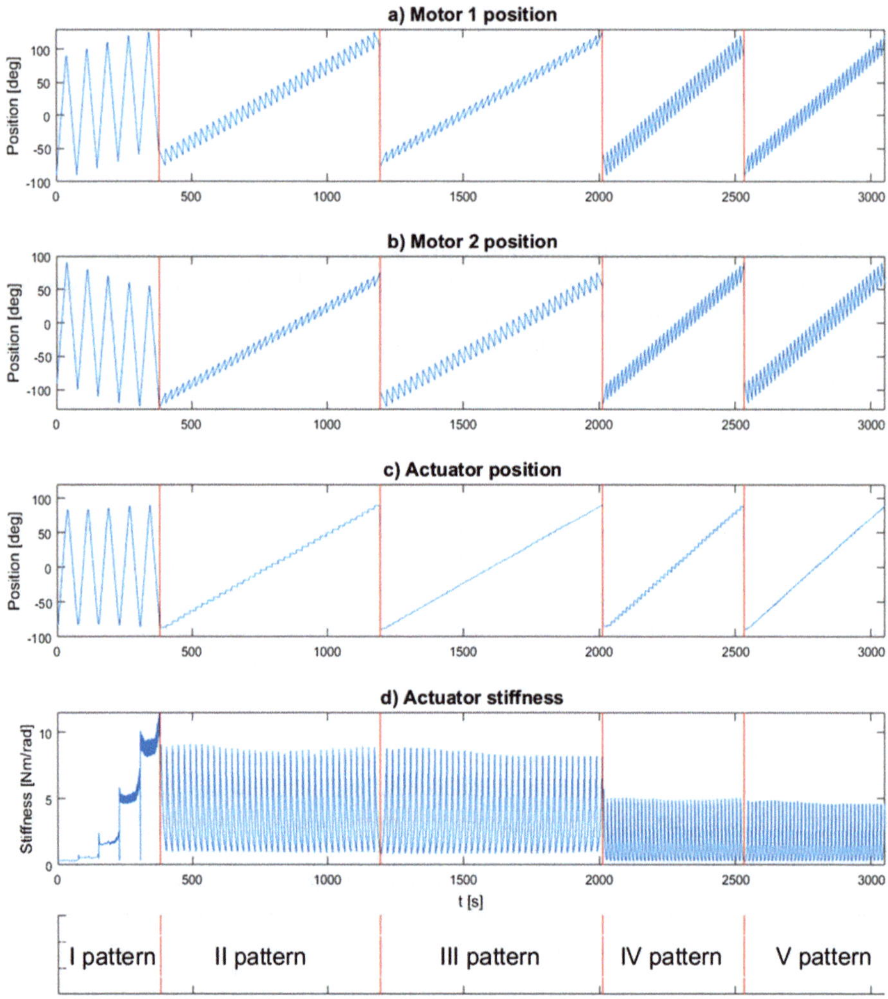


**Fig. 2.** Training data set for five different patterns: (a) Motor 1 position - $q_1$; (b) Motor 2 position - $q_2$; (c) output equilibrium position - $x_e$; and (d) estimated stiffness - $S$.

### 3.2 Locally Weighted Projection Regression

As stated in [11], the LWPR is used to map input/output characteristics of series of elastic actuators. In this paper, this method is used for mapping VSA statically relationships between outputs and inputs. One of the advantages of LWPR is that it can run in real time and incrementally learn model properties [12]. This property is very important when it comes to robotics. In LWPR, the regression function is constructed by blending local linear models, each of which is endowed with a locality kernel that defines the area of its validity. During training, the parameters of models are updated using incremental Partial Least Square. The receptive fields of LWPR are usually modelled by Gaussian kernels, so their activation or response to a given **z** is given by:

$$w_k(z) = \exp\left(-\frac{1}{2}(z - \boldsymbol{c}_k)^T \boldsymbol{D}_k (z - \boldsymbol{c}_k)\right), \tag{3}$$

where $c_k$ is the centre of $k$-th model and $D_k$ is the distance matrix. Treating each output dimension separately, the regression function can be written as:

$$\tilde{f}(z) = \frac{1}{W}\sum_{k=1}^{K} w_k(z)\psi_k(z), \quad W = \sum_{k=1}^{K} w_k(z), \tag{4}$$

$$\psi_k(z) = b_k^0 + \boldsymbol{b}_k^T (z - \boldsymbol{c}_k), \tag{5}$$

where $b_k^0$ and $\boldsymbol{b}_k$ denote offset and slope of the $k$-th model, respectively.

After collecting the data, the model is trained by imposing the condition that the distance matrix needs to be diagonal and contain small values $D = rI$, where $r$ is tunable parameter and $I$ is identity matrix. In the first learning iteration, it is recommended to set $r$ to a small value, somewhere around 0,05. It is then necessary to evaluate the performance of prediction on a test set and repeat the learning procedure with an increasing value of $r$ until the performance is satisfying. A good initial guess of the distance matrix is achieved in the final step.

## 4 Experimental Results

This section presents experimental results in controlling of qb actuator using neural network and LWPR. These two methods are used for representing static equilibrium behavior of the system. They are also used for feedforward control of the actuator. In the previous work [10], the authors pointed out advances of the neural network feedforward control to feedforward control based on analytical modelling and decoupling of such nonlinear and multi-variable system linearized and decoupled using Taylor series approximation. Figure 3 presents performance of neural network and LWPR applied in feedforward control. The qb actuator is an open source, low cost VSA with 3D printed body and coupled springs.

In order to design a robust control algorithm, a different qb actuator, which is not used for training, is used to verify the designed algorithms. One of them is used to collect training data, and the other one is used for testing neural network and LWPR-based control. When feedforward control is used, the response obtained using LWPR exhibits a larger error compared to that obtained using the neural network, as shown in Fig. 3.

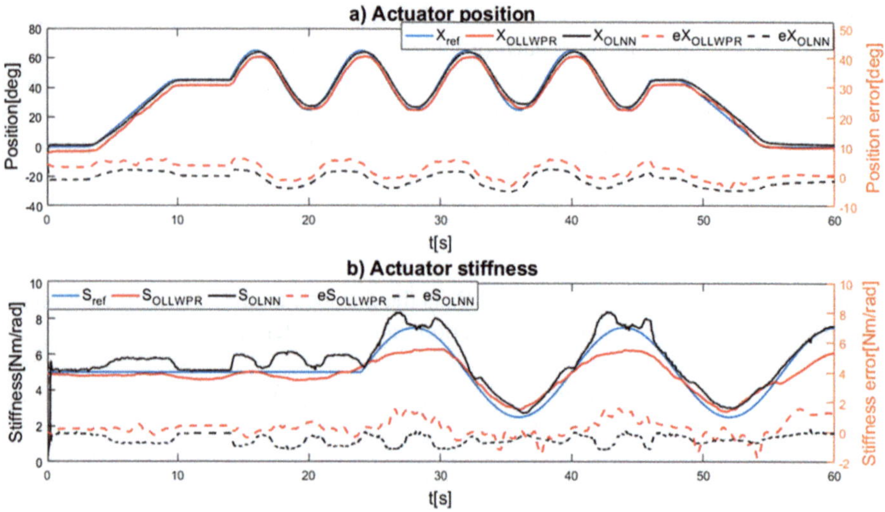


**Fig. 3.** Experimental results of feedforward control based on neural network and LWPR in the open loop: (a) Actuator position tracking; (b) actuator stiffness. Abbreviations OL stands for open loop control, NN for neural network, LWPR for Locally Weighted Projection Regression, eX for error in position tracking and eS for stiffness tracking.

After the feedforward control is validated in the experiment, a feedback component is introduced in order to enhance position and stiffness tracking. Mathematical formulation of the control algorithm is given by (6) and (7), and the Simulink block diagram is shown in Fig. 4.

$$q_{1con} = q_{1ff} + fb_x(e_x) + fb_s(e_s), \tag{6}$$

$$q_{2con} = q_{2ff} + fb_x(e_x) - fb_s(e_s), \tag{7}$$

where $e_x$ and $e_s$ are errors between the desired and measured actuator position and stiffness, respectively. Quantities $q_{1con}$ and $q_{2con}$ are control signals for qb actuator and $q_{1ff}$ and $q_{2ff}$ are feedforward components of control signals. Quantities $fb_x$ and $fb_s$ are feedback components of control signals. These feedback components are obtained using a PD controller for position control (8) and PI controller for stiffness control (9) as derived in [10]

$$PD(s) = 3 + 0.1s/(s+1), \tag{8}$$

$$PI(s) = 3 + 8/s. \tag{9}$$

Experimental results obtained after introducing feedback into the control scheme are given in Fig. 5.

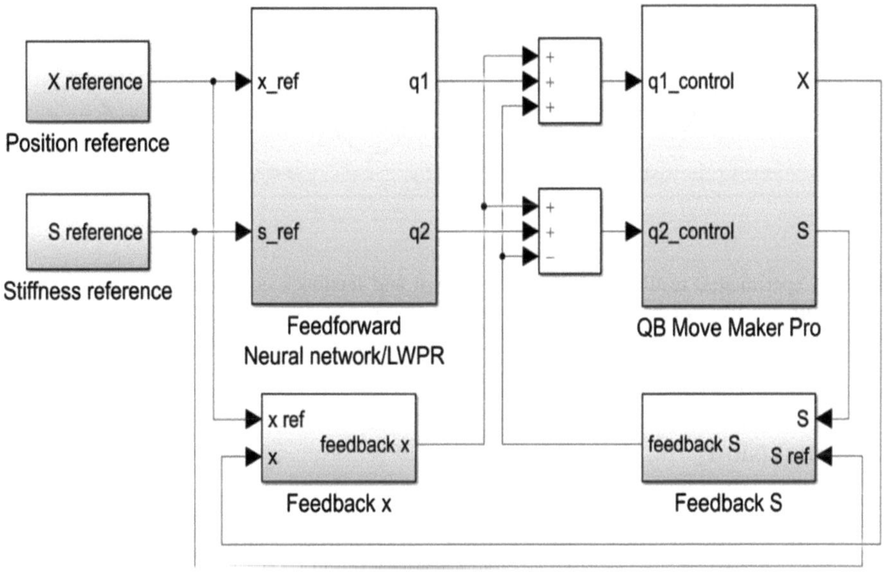


**Fig. 4.** Simulink control scheme for neural network/LWPR feedforward and feedbacks for position and stiffness control.

The results displayed in Fig. 5 demonstrate good position and stiffness tracking when used actuator which did not use for training the neural network and LWPR models. Therefore, good robustness to differences in construction of actuators is demonstrated. These results provide large benefits in practical terms, as it is possible to employ developed control approaches to series of VSA actuators based on qb robotics principle – self-manufacturing of open data, 3D printed and affordable VSA actuators. The algorithms would be applicable to other devices of the same type, regardless of the potential mismatch in their characteristics and therefore inaccuracies in mathematical modeling and model-based control approaches.

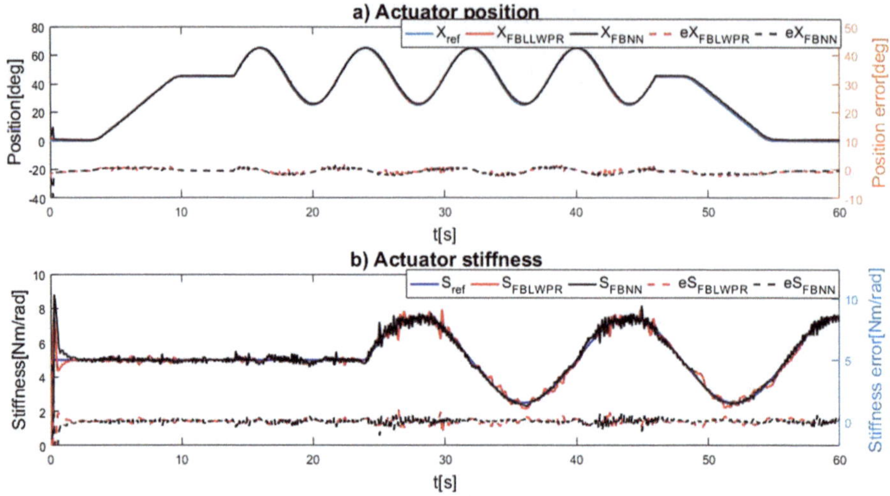


**Fig. 5.** Experimental results based on feedforward and feedback control using neural network and LWPR: (a) Position tracking; (b) stiffness tracking. Abbreviations FB stands for feedback control, NN for neural network, LWPR for Locally Weighted Projection Regression, eX for error in position tracking and eS for stiffness tracking.

## 5 Conclusion

In general, it is very hard to derive VSAs models mathematically. Due to very complex and non-linear transmission mechanisms, the determination of required multi-variable input-output relationships is a challenging task. The nonlinear force-displacement characteristic of the VSA is achieved either by conventional linear springs and non-linear transmission or antagonistically coupled non-linear springs, where symmetry is often doubtful. In this paper, two learning-based methods for control of VSA without exact analytical modeling of the actuator dynamics are presented. It is shown that by applying either of the two methods, neural network and LWPR, the positions of prime movers required for achieving position and stiffness tracking are accurately predicted. Initially, feedforward control is implemented which predicts system inputs (prime mover positions) from the learned model. However, the mismatch between behavior of VSA used for training/learning and the VSA used for testing of the learned feedforward control can cause certain deviations from the desired trajectory. To bridge this gap and make presented control approach based on learning more versatile, the feedback is introduced. In that manner, errors in actuator stiffness and position tracking are reduced, and tracking performance validated even if different actuators are used for learning and testing purposes. Plans for future research include the development of incremental learning algorithms that will works alongside some of model-based algorithms that use feedback linearization, slide control or LQG. These control algorithms should provide online learning possibility and capability to adapt to changes in the system that can occur during exploitation of the actuator.

## References


1. Vanderborght, B., et al.: Variable impedance actuators: a review. Robot. Auton. Syst. **61** (12), 1601–1614 (2013)
2. Melo, K., et al.: Open source VSA-CubeBots for rapid soft robot prototyping. In: Robot Makers - Workshop in conjunction with 2014 Robotics Science and Systems Conference (2014)
3. Hogan, N.: Impedance control: an approach to manipulation. In: Proceedings of the American Control Conference, pp. 304–313 (1984)
4. Palli, G., Melchiorri, C., De Luca, A.: On the feedback linearization of robots with variable joint stiffness. In: 2008 IEEE International Conference on Robotics and Automation, pp. 1753–1759 (2008)
5. Tonietti, G., Schiavi, R., Bicchi, A.: Design and control of a variable stiffness actuator for safe and fast physical human/robot interaction. In: Proceedings of the 2005 IEEE International Conference on Robotics and Automation, pp. 526–531 (2005)
6. Visser, L.C., Carloni, R., Stramigioli, S.: Energy efficient control of robots with variable stiffness actuators. IFAC Proc. Volumes **43**(14), 1199–1204 (2010)
7. Giorgio, G., et al.: Variable stiffness actuators: the user's point of view. I. J. Robot. Res. **34**, 727–743 (2015)
8. Natural Motion Machine Initiative (NMMI) (qbmove - maker pro: agonistic/antagonistic servo-VSA datasheet). https://qbrobotics.com/wp-content/uploads/2016/03/qbmove-Advanced-datasheet.pdf. Accessed 20 Jan 2019
9. Lukić, B., Jovanović, K., Rakić, A.: Realization and comparative analysis of coupled and decoupled control methods for bidirectional antagonistic drives: QBmove maker pro. In: 3rd International Conference on Electrical, Electronic and Computing Engineering (2016)
10. Lukić, B.Z., Jovanović, K.M., Kvaščev, G.S.: Feedforward neural network for controlling qbmove maker pro variable stiffness actuator. In: 2016 13th Symposium on Neural Networks and Applications (NEUREL), pp. 1–4 (2016)
11. Mitrovic, D., et al.: Learning impedance control of antagonistic systems based on stochastic optimization principles. Int. J. Robot. Res. **30**(5), 556–573 (2011)
12. Vijayakumar, S., D'Souza, A., Schaal, S.: Incremental online learning in high dimensions. Neural Comput. **17**(12), 2602–2634 (2005)

# Knowledge Acquisition Through Human Demonstration for Industrial Robotic Assembly

Timotej Gašpar(✉), Miha Deniša, and Aleš Ude

Jožef Stefan Institute, Jamova Cesta 39, 1000 Ljubljana, Slovenia
{timotej.gaspar,miha.denisa,ales.ude}@ijs.si

**Abstract.** With the ambition to introduce robots into assembly lines, not suitable for classical automation, the ease of robot programming is becoming more significant then ever. This paper proposes using various approaches for gaining knowledge from human demonstrations. This knowledge is applied to perform assembly tasks in a industrial robotic cell. Real industrial use case is used for evaluation of proposed approaches. It shows their viability and presents different scenarios which call for different approaches of learning and execution of assembly tasks and its subsets.

**Keywords:** Learning by demonstration · Motion generation · Reconfigurable robotics · Industry 4.0

## 1 Introduction

As more and more large enterprises adopt robots to automate their production lines, classical automation is reaching its saturation point. In order to further increase the productivity of the industrial sector, automation needs to expand into previously untapped areas. Classical automation is challenging to introduce in small to medium enterprises (SMEs). Main hurdles with automation of production lines in SMEs are: small batches of products, variability of products or a family of products, no experts on site, and funding limits for automation investments. Another aspect of robotics, which is being introduced into industrial environments, is collaborative robotics. To reduce the need for experts and make collaborative robots more viable in industrial setting, the ease of (re)programming a robot for a new task is paramount. To ensure the ease of learning assembly tasks, human demonstration can be used to (re)program the robot. In addition, the need for on-site experts is removed, as relatively small amount of expertise is needed for human demonstration.

A popular approach for acquiring new sensorimotor knowledge is programming by demonstration (PbD) [1–3], where a robot observes a human performing a task. These movements can be observed using a marker based system [4,5], vision systems with RGB-D [6] or stereo cameras [7]. Alternatively, the human


K. Berns and D. Görges (Eds.): RAAD 2019, AISC 980, pp. 346–353, 2020.
https://doi.org/10.1007/978-3-030-19648-6_40

can physically move and guide the robot, using the approach called kinesthetic guidance [8,9]. If the movement is captured through kinesthetic guidance, it is already adapted to the robot dynamics and kinematics. The movements, kinesthetically captured in joint space, also preserve the configuration of the robot. Beside classic approaches for point to point motion generation, we also propose using Dynamical Movement Primitives (DMPs) [10–12] to encode a full demonstrated trajectory and store it as a set of parameters.

In the next section we present multiple approaches to extract knowledge from human demonstrations. In addition to kinesthetic guidance, we propose to use an alternative method to ease the process of assembly demonstration. In Sect. 3 we present various robot trajectory generators. In order to transfer gained knowledge to the robot, several motion generation schemes are used. Section 4 tackles evaluation of proposed approaches in a real industrial use case. A reconfigurable robotic work cell, described in our previous work [13], is used as an evaluation environment. The evaluation serves as a proof of concept and shows the feasibility of proposed approaches.

## 2 Data Acquisition

This section presents proposed approaches for data acquisition. While kinesthetic guidance and joystick control can be used for moving the robot, custom button interface is introduced to ease the process of demonstration.

### 2.1 Kinesthetic Guidance

Traditional robot programming approaches require the user to guide and program the robot with its controlling interface. By handling either a touchscreen or a joystick, the user moves the robot to a desired position and defines the type of motion to that configuration. This method turns out to be relatively slow and requires a certain proficiency in robot handling. For these exact reasons modern robots allow for kinesthetic guidance out of the box. Kinesthetic guidance proves to be a considerably more intuitive approach to move the robot during the programming phase.

Our setup consisted of two Universal Robot UR10 arms where kinesthetic guidance is by default implemented with the so called "Gravity Compensation" mode. This kind of control can be achieved by taking the dynamic model of a robot:

$$\boldsymbol{\tau_u} = \widehat{\mathbf{H}}(\mathbf{q})\mathbf{u} + \widehat{\mathbf{C}}(\mathbf{q},\dot{\mathbf{q}})\dot{\mathbf{q}} + \widehat{\mathbf{B}}_f(\dot{\mathbf{q}}) + \widehat{\mathbf{g}}(\mathbf{q}) \tag{1}$$

$\boldsymbol{\tau}_u$ represents the controlled torques in the robot joints, $\mathbf{u}$ is the desired control signal to be regulated, $\mathbf{q}$ is the vector of robot joint angles, while $\widehat{\mathbf{H}}(\mathbf{q})$, $\widehat{\mathbf{C}}(\mathbf{q},\dot{\mathbf{q}})$, $\widehat{\mathbf{B}}_f(\dot{\mathbf{q}})$ and $\widehat{\mathbf{g}}(\mathbf{q})$ represent the estimated inertia, Coriolis, friction and gravity models, respectively. To achieve kinesthetic guidance we set $\mathbf{u}$ to be null. Additionally, because of relatively slow robot motions, we can neglect the $\widehat{\mathbf{C}}(\mathbf{q},\dot{\mathbf{q}})$ term. This results in the following control strategy

$$\boldsymbol{\tau_u} = \widehat{\mathbf{B}}_f(\dot{\mathbf{q}}) + \widehat{\mathbf{g}}(\mathbf{q}) \quad (2)$$

This mode essentially compensates the effects of gravity by estimating the necessary torques in the robot joints with the dynamic model. By applying external torque $\boldsymbol{\tau}_e$ the user can guide the robot kinesthetically without any additional sensing equipment.

### 2.2 Wireless Joystick Control

Due to the UR10 robot lack of sensing equipment in the robot joints, the dynamic model control is based on controlling the torque in the joints by regulating the electrical current to the servo motors. This results in a poor compensation of the friction in the joints which consequently requires significant physical effort to move the robot. It is therefore very challenging to move the robot smoothly and precisely, which is even more notable when kinesthetic teaching of precise trajectory is needed.

To cope with this drawback, we developed an interface that allows the control of the robot in Cartesian space with a wireless joystick. During the development of the joystick control interface we strove to provide an intuitive experience.

The joystick used was a standard Xbox 360 controller, which is widely supported on multiple operating systems (Fig. 1).

Intuitive control was achieved by mapping the controller's analogue stick offsets from its default position to Cartesian space velocities. The user can switch from controlling translational velocities to rotational velocities. The user can move the robot with different velocity settings by pressing a button. This ensures the user to have the option between faster and less precise or slower and more precise motions.

**Fig. 1.** A consumer grade joystick interface that we used to perform precise motions of the robot in Cartesian space.

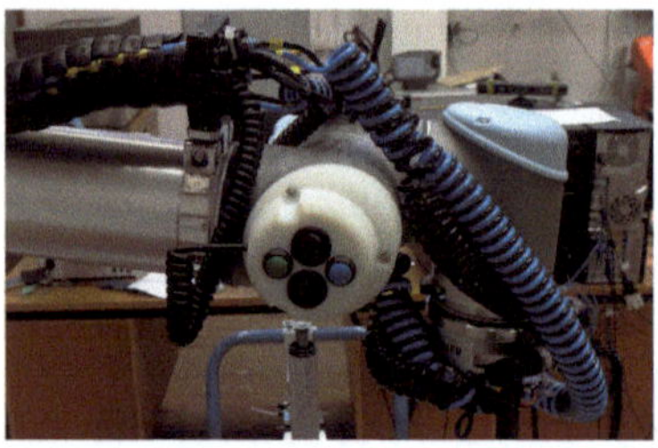

**Fig. 2.** Custom 3-D printer cover serving as an interface for assembly task teaching.

### 2.3 Custom Button Interface

To further increase intuitiveness and accelerate data acquisition we developed an interface with multiple programmable buttons. The interface replaces one of the covers on the robot and it houses 2 buttons and 2 switches with LEDs (Fig. 2). The latter provide a visual reference to the switch's state. The cover was developed to provide buttons that can be programmed for different actions dependent on the needs. In some cases the switches toggle gravity compensation, in others they make an entry of the current joints configuration, pose configuration or whole trajectory into the system database. This allows the user to uninterruptedly record the data needed for the assembly process.

## 3 Robot Trajectory Generation

Following the acquisition of desired poses and trajectories relevant for an assembly process, the next step is to generate appropriate robot movements. We can generate trajectories between two known configurations or replay a whole trajectory previously demonstrated by kinesthetic guidance.

### 3.1 Point to Point Movements

In cases where the motion performed by the robot is relatively simple i.e. the shortest path from one configuration (joint or Cartesian space) to another, we can use simple trajectory generation algorithms. In our experiments we extensively used two such algorithms:

- *Joint space motion with trapezoidal speed profile*
  This trajectory generation algorithm will generate a trajectory in joint space for the robot to reach a desired final configuration from the initial configuration. In this type of movement, the robot reaches the specified desired maximum speed with constant acceleration.
- *Cartesian space straight line & SLERP trajectories following the minimum jerk velocity profile*
  This trajectory generation algorithm will generate a straight line trajectory in Cartesian space with zero initial and final velocities and accelerations in a manner that minimises the jerk throughout the motion. This algorithm is particularly useful when the path from the initial to the final pose of the end effector should follow a straight line. It the context of industrial assembly, we used this algorithm when approaching objects, picking them up, inserting them into a workpiece, rotating them into the workpiece, etc.

In most cases, a sequence of point to point movements can be defined relative to a single point. Instead of demonstrating multiple points that define point to point movements, a single point can be recorded. The other needed points are defined relative to this one.

### 3.2 Free from Movement Representation

With point to point movements we can carry out most steps in an assembly process. However, sometimes it can happen that, due to the robot cell configuration, more point to point movements have to be chained just to avoid possible collisions. This kind of programming requires the user to have some experience in terms of what poses to chose in order to avoid collisions. To offer the user a more intuitive approach for collision free motions we implemented an interface to record whole demonstrated trajectories.

Through the use of kinesthetic guidance the user can guide the robot alongside a complex trajectory while avoiding collisions with either the periphery of the cell or the robot structure itself. In order to minimise and parametrise some aspects of the recorded trajectories we encode them using *Dynamic Movement Primitives* [11]. The representation also provides a method of temporal scaling which makes it possible to increase or decrease the speed of the execution of a previously recorded trajectory.

## 4 Motion Strategy Evaluation Through an Industrial Use Case

The industrial use case, used for evaluation, is a sub set of a glass mounting gripper assembly. It consists of several pick-and-place tasks, (inverse) peg-in-hole actions, positioning of screws, screwing, etc. It was implemented in a modular and reconfigurable work cell [13] consisting of two UR10 robots, tool exchange system, modular beam system, etc. While the robots are controlled by a real-time system using *Matlab Simulink Real-Time (SLRT)* platform [14], the top software layer consists of a *Robot Operating System (ROS)* framework [15], which handles the communication between the modules of the cell. While all the sub-tasks of the studied use case use the proposed approaches, just three of them are presented in this section. They represent different aspects of the whole assembly procedure and cover the full spectrum of all needed tasks.

As described in Sect. 2, kinesthetic guidance and joystick control were used to move the robot during human demonstrations. Kinesthetic guidance was used to record trajectories in joint space, as well as single robot positions in joint or task space. The custom button interface was used to ease the demonstration process, by using its signals as record events. On the other hand, joystick control was used to record just single poses. Recording whole demonstrated trajectories through joystick control is not beneficial, as point to point generated motions exhibit the same behaviour as moving the robot in task space using the proposed joystick control.

The data acquired was stored in a *MongoDB* server [16]. To ensure saved poses and trajectories are available throughout the ROS system we used a ROS implementation of the MongoDB server.

As mentioned before, three sub-tasks are presented in this section. Each of them displays the usefulness of proposed approaches.

**The first sub-task** consists of moving the robot to a so called *initial position* in joint space. Moving to this pose is needed when the other robot is executing a task in a shared workspace. To gain an appropriate pose for each robot's *initial position*, the robot is moved kinesthetically. In order to guide the robot to a desired joint space configuration, joystick control would not be suitable.

When the robot is at an appropriate configuration, the button interface is used to store the joint configuration in the database. When the need for an *initial position* arises in the assembly process, the joint configuration is read from the database and *joint space motion with trapezoidal speed profile* is generated (see Sect. 3.1). Video stills of the robot executing this kind of movement can be seen in Fig. 3.

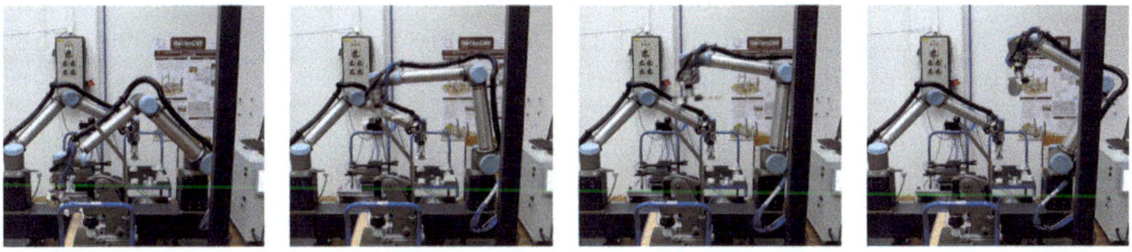

**Fig. 3.** Sub-task of moving the robot to an *initial position.* While kinesthetic guidance was used to record an joint pose, joint space motion with trapezoidal speed profile was executed to move the robot to the stored pose during assembly.

**The second sub-task** consist of a precise placement of an object. The robot must place a metal bushing on a metal shaft, which is already in the press (see Fig. 4). As this sub-task includes a constraint setting, contact with the environment, and precise positioning, kinesthetic guidance would not be suitable. Joystick control, presented in Sect. 2, was used to guide the robot to an appropriate position, where the bushing fit on the base. This demonstrated Cartesian position, stored in the database, was used to define needed relative points. *Cartesian space straight line & SLERP trajectories following the minimum jerk velocity profile* approach was used to generate point to point movements. As it produces a straight line in Cartesian space, appropriate robot trajectories were generated to perform the needed task. Figure 4 shows video stills of the robot successfully executing this sub-task.

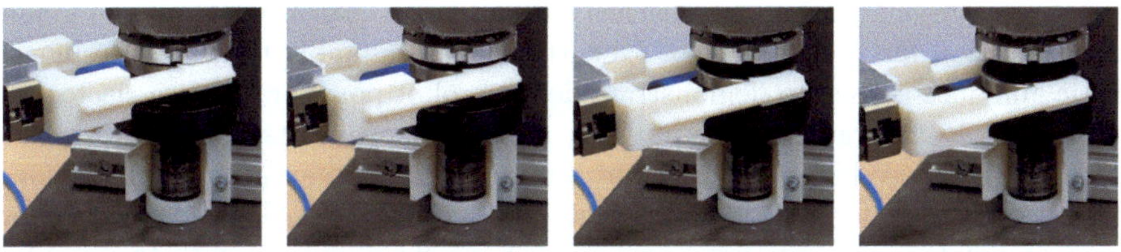

**Fig. 4.** Sub-task of positioning a bushing on a shaft. Due to a constraint setting, contact with the environment, and precise positioning, joystick control was needed for demonstrating appropriate robot position. Cartesian space straight line & SLERP trajectories following the minimum jerk velocity profile were used to successfully move the robot to the demonstrated point.

**The third sub-task** consists of activating the press, which is usually done by a human moving a lever. To automate this sub-task, we used one of the robot arms to move the lever. For this purpose, a human demonstrator moved the robot using kinesthetic guidance. With no expert knowledge of the robot and its surroundings, the human was able to move the robot in such a way to activate the press. As the whole trajectory is important, *Dynamic Movement Primitives* were used to encode it (see Sect. 3.2). Again, the button interface was used to mark the start and end of the desired trajectory, and thus ease the process. The trajectory was then executed during the assembly to effectively use the press. This can be seen on video stills in Fig. 5.

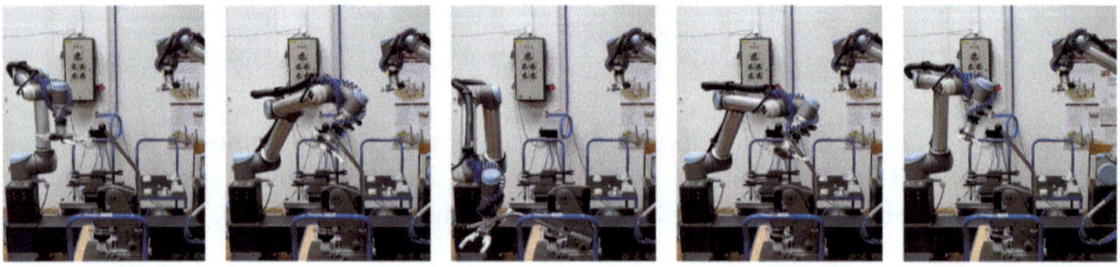

**Fig. 5.** Sub-task of activating a, usually human operated, press. To successfully move the lever, kinesthetic guidance was used to demonstrate the appropriate robot movement. The joint space trajectory was encoded with DMPs. The figure show the execution of the task by the robot.

All these three example sub-tasks represent different aspect of the chosen industrial assembly task. Combinations of proposed approaches were used throughout the whole industrial use-case implementation. They enabled us to successfully program the cell from (non-expert) human demonstration and to execute the presented assembly task.

## 5 Conclusion

This paper presents various approaches for knowledge acquisition through human demonstration for the purpose of executing robot assembly tasks. While kinesthetic guidance is used to demonstrate desired robot poses or whole trajectories, a joystick was integrated for the purpose of precise movements when demonstrating robot poses. While point to point movements were generated from demonstrated poses, free form trajectories were encoded as DMPs. Two approaches for point to point movements were presented: joint space motion with trapezoidal speed profile, and linear Cartesian space motion with minimum jerk velocity profile. Evaluation was presented in a form of a real industrial use case study. It showed that appropriately integrated and executed human demonstration can be used to program a new industrial assembly task with little expertise needed. Evaluation showed that different sub sets of the task can require different approaches of human demonstration and motion execution.

**Acknowledgment.** This work has received funding from the EU's Horizon 2020 IA ReconCell (GA no. 680431) and from GOSTOP programme C3330-16-529000 co-financed by Slovenia and EU under ERDF.

## References

1. Breazeal, C., Scassellati, B.: Robots that imitate humans. Trends Cogn. Sci. **6**(11), 481–487 (2002)
2. Dillmann, R.: Teaching and learning of robot tasks via observation of human performance. Robot. Auton. Syst. **47**(2), 109–116 (2004)
3. Billard, A., Calinon, S., Dillmann, R., Schaal, S.: Robot programming by demonstration. In: Siciliano, B., Khatib, O. (eds.) Handbook of Robotics, Secaucus, NJ, USA, pp. 1371–1394. Springer (2008)
4. Ude, A., Atkeson, C.G., Riley, M.: Programming full-body movements for humanoid robots by observation. Robot. Auton. Syst. **47**(2), 93–108 (2004)
5. Pollard, N.S., Hodgins, J.K., Riley, M.J., Atkeson, C.G.: Adapting human motion for the control of a humanoid robot. In: Proceedings of the IEEE International Conference on Robotics and Automation (ICRA), Washington, DC, USA, pp. 1390–1397 (2002)
6. Shotton, J., Sharp, T., Kipman, A., Fitzgibbon, A., Finocchio, M., Blake, A., Cook, M., Moore, R.: Real-time human pose recognition in parts from single depth images. Commun. ACM **56**(1), 116–124 (2013)
7. Moeslund, T.B., Hilton, A., Krüger, V.: A survey of advances in vision-based human motion capture and analysis. Comput. Vis. Image Und. **104**(2), 90–126 (2006)
8. Hersch, M., Guenter, F., Calinon, S., Billard, A.: Dynamical system modulation for robot learning via kinesthetic demonstrations. IEEE Trans. Robot. **24**(6), 1463–1467 (2008)
9. Deniša, M., Ude, A.: Synthesis of new dynamic movement primitives through search in a hierarchical database of example movements. Int. J. Adv. Robot. Syst. **12**(10), 137 (2015)
10. Ude, A., Nemec, B., Petrič, T., Morimoto, J.: Orientation in cartesian space dynamic movement primitives. In: IEEE International Conference on Robotics and Automation (ICRA), Hong Kong, pp. 2997–3004 (2014)
11. Schaal, S., Mohajerian, P., Ijspeert, A.: Dynamics systems vs. optimal control-a unifying view. Prog. Brain Res. **165**, 425–445 (2007)
12. Ijspeert, A.J., Nakanishi, J., Hoffmann, H., Pastor, P., Schaal, S.: Dynamical movement primitives: learning attractor models for motor behaviors. Neural Comput. **25**(2), 328–373 (2013)
13. Gašpar, T., Ridge, B., Bevec, R., Bem, M., Kovač, I., Ude, A., Gosar, Z.: Rapid hardware and software reconfiguration in a robotic workcell. In: 18th International Conference on Advanced Robotics (ICAR), pp. 229–236 (2017)
14. Simulink Real-Time - Simulink - MATLAB & Simulink. https://www.mathworks.com/products/simulink-real-time.html. Accessed 16 Mar 2017
15. Quigley, M., Conley, K., Gerkey, B., Faust, J., Foote, T., Leibs, J., Wheeler, R., Ng, A.Y.: Ros: an open-source robot operating system. In: ICRA Workshop on Open Source Software, Kobe, Japan (2009)
16. MongoDB. https://www.mongodb.com/. Accessed 4 Feb 2019

# Medical and Rehabilitation Robotics

# Extension of Matlab's PDE Toolbox for Developing Bionic Structural Optimization Methods: Overlapping Region Concept

Yilun Sun(✉) and Tim C. Lueth

Technical University of Munich, Boltzmannstr. 15, 85748 Garching, Germany
yilun.sun@tum.de

**Abstract.** In this paper, we present a novel method called Overlapping Region Concept to extend Matlab's Partial Differential Equation (PDE) Toolbox to apply free-form surface loads during finite element analysis. In our institute, we are developing a toolbox called SG-Library in Matlab to combine multidisciplinary methods to achieve automatic design and manufacturing of medical robots and mechanisms. Recently, we have also integrated the PDE Toolbox into the SG-Library to develop some bionic methods like CAO and SKO to optimize the structure of our robots. During the implementation of these bionic methods, we have encountered the problem of applying free-form surface loads in FEM analysis. The typical workflow of PDE Toolbox separates the surface of a 3D geometry into one or more feature surfaces with sharp boundaries. Boundary conditions can be only applied to such feature surfaces. This is a remarkable drawback for the development of our methods since the surface profile of our model is constantly changing during the shape optimization. Therefore, we need to develop a method to apply free-form surface loads independent from the feature surfaces. In this paper, we use the concept of overlapping region to determine the free-form surface for applying boundary conditions. With this new method, we can achieve robust FEM analysis during the structural optimization process.

**Keywords:** Overlapping Region Concept · Bionic structural optimization · Free-form surface load · PDE Toolbox · FEM analysis

## 1 Introduction

Today with the development of the additive manufacturing technology we are witnessing a big revolution in the mechanical design. Since the additive manufacturing gives us much more design freedom than the traditional manufacturing methods, we are also updating our design logics to explore the potential of this new technology [1]. It means, we can now combine methods from different disciplines to achieve an optimal product design, even if the geometry of the product is too difficult to realize for the traditional manufacturing methods. Therefore, in our institute we are developing a toolbox called SG-Library in Matlab to achieve automatic design of medical robots and mechanisms with multidisciplinary methods [2, 3]. And those robots and mechanisms can be quickly manufactured by different kinds of 3D printers (Fig. 1a and b).


K. Berns and D. Görges (Eds.): RAAD 2019, AISC 980, pp. 357–364, 2020.
https://doi.org/10.1007/978-3-030-19648-6_41

Recently, we have also integrated the finite element method (FEM) into our toolbox by using the PDE Toolbox of Matlab. We are now using several bionic methods like Computer Aided Optimization (CAO) [5, 7] and Soft Kill Option (SKO) [6] to achieve structural optimization of our robots and mechanisms with the help of FEM analysis (Fig. 1c and d). At a later stage, we also plan to use the SG-Library to close the design loop of geometry construction, multibody simulation, structural optimization and additive manufacturing to achieve fully automatic design and manufacturing [4, 8].

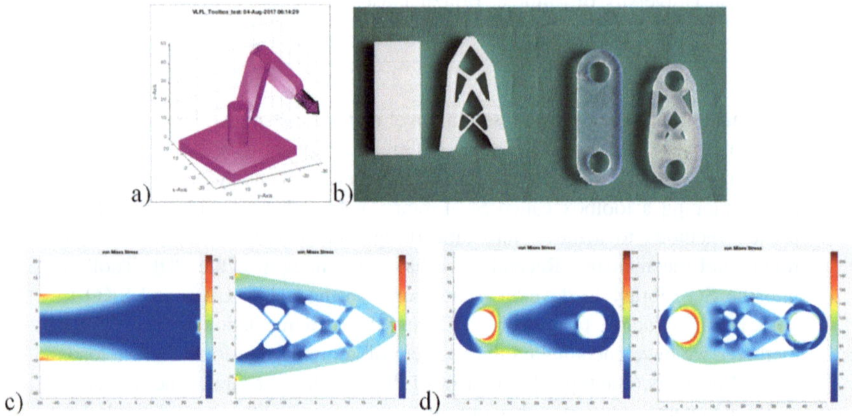

**Fig. 1.** (a) SG-Library for automatic design of medical robots and mechanisms, (b) 3D printed structures which are optimized by bionic methods, (c) FEM analysis of a beam before and after optimization, (d) FEM analysis of a part of mechanism before and after optimization

When performing FEM analysis on a mechanical part, a very important step is applying loads on the model. In the current workflow of PDE Toolbox (see Fig. 2), we can give a STL file or a SG struct (constructed by SG-Library) as the geometry input to create a `PDEmodel` object. PDE Toolbox separates then the entire surface of the model into a list of surface regions which are called feature surfaces [8]. By using the function `pdegplot`, the user can see the indices of these feature surfaces from a plot and then apply the loads on the specific surfaces manually according to the face indices.

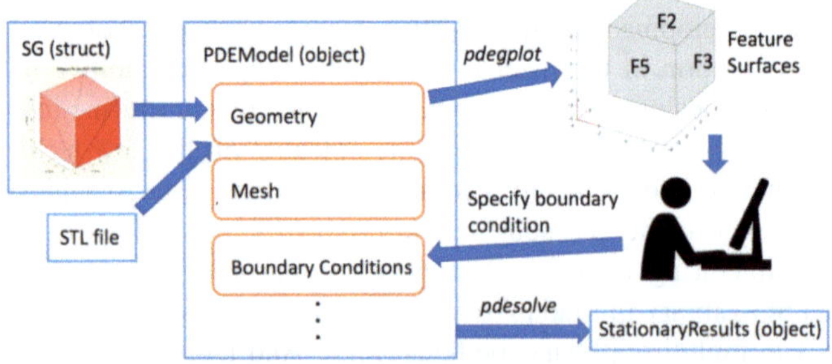


**Fig. 2.** The current workflow of PDE Toolbox for a static FEM analysis

In an early paper [8] we have described the algorithm that PDE Toolbox used to detect different feature surfaces. The basic idea of the algorithm is to look up the facet list of the surface model and find out all the neighboring facets whose dihedral angles are greater than 44°. The edges between those neighboring facets will be detected as the boundary of the feature surfaces.

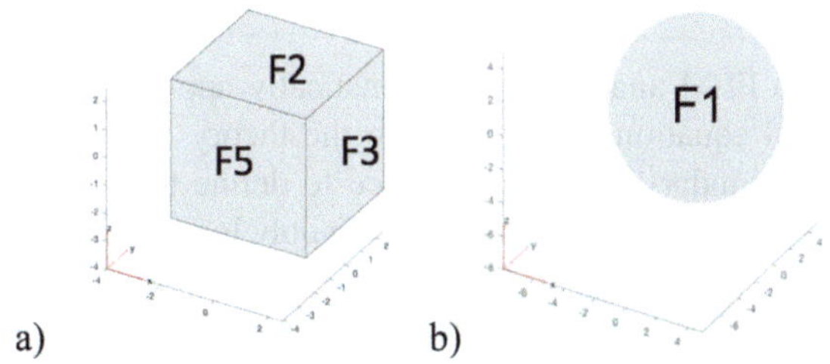


**Fig. 3.** Limit of the feature surface concept: (a) the surface of a box can be divided into 6 feature surfaces, (b) a sphere has only 1 feature surface

However, the functionality of this concept is very limited. Figure 3 shows the feature surfaces of a box and a sphere. The surface of a sphere is much smoother than the box so that only one feature surface is detected. In this case, we have no chance to apply loads on a free-form surface of the sphere if we use the feature surface concept of PDE Toolbox. Generally, if we have a `PDEModel` with a very complicated geometry like the gear in Fig. 4a, we are not able to apply any free-form surface load like the green strip in Fig. 4b since they cannot be detected by PDE Toolbox as single feature surfacc.

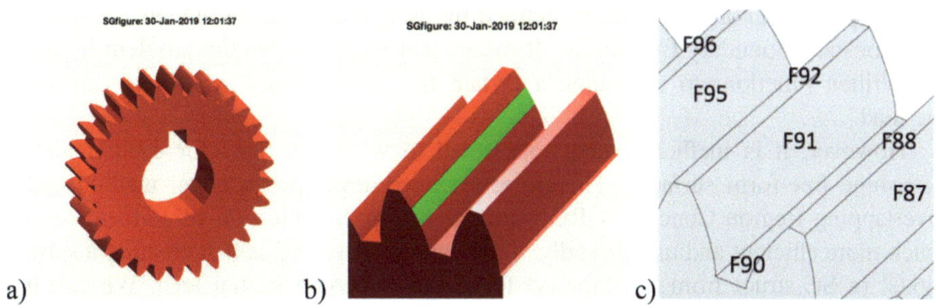


**Fig. 4.** Free-form surface load on a tooth of a gear: (a) a gear which is constructed by SG-Library, (b) a free-form load on a tooth of the gear, (c) feature surfaces of the gear which are calculated by PDE Toolbox

This is a remarkable drawback of PDE Toolbox which may limit the potential of our bionic structural optimization methods since the geometry of our model is always changing during the shape optimization and the indices of the feature surfaces are also inconstant. In this paper, we introduce the concept of overlapping region to determine the area of a free-form surface for applying boundary conditions, which enables the PDE Toolbox to apply any free-form surface loads independent from the feature surfaces.

## 2 Methods and Implementation

Applying a load in a FEM analysis is mathematically applying a boundary condition for partial differential equations. According to the theory of partial differential equations, the Dirichlet Boundary Condition is used to define the fixed part of our model and the Neumann Boundary Condition is used to apply loads [10]. The function in PDE Toolbox to apply boundary conditions for a `PDEModel` is called `applyBoundaryCondition`. In the typical workflow of PDE Toolbox, the syntax of the function for the two boundary conditions in a static problem is as follows [9]:

```
applyBoundaryCondition(model,'dirichlet','face',...
faceID,'u',[0 0 0]);
applyBoundaryCondition(model,'neumann','face',...
faceID,'g',loadvector);
```

From the code above we can see that the variable `faceID` contains the indices of the feature surfaces, on which the boundary conditions are applied. The 'u' in the Dirichlet Boundary Condition is the displacement of each node on the boundary and is set to [0 0 0] since it is fixed. The 'g' in the Neumann Boundary Condition is the load vector. It is clear that we have no chance to apply free-form surface loads or fixation if the fixation and load vector are specified only according to feature surface indices (`faceID`). But from the documentation of PDE Toolbox we know that PDE Toolbox also supports boundary conditions defined by user-written functions which are called Non-constant Boundary Conditions. It means that we can solve the problem by using self-written functions to determine a certain free-form surface where we can apply the load.

However, it is inefficient and impractical to write functions for each model to determine free-form surfaces. Therefore, we have developed a novel method called Overlapping Region Concept (ORC) to make the specification of a free-form surface much more efficient and user-friendly. The basic idea of ORC is first creating a surface model (a SG-struct from SG-Library) to occupy a certain spatial area. We call this surface model a user-defined domain. Then we use the function `SGisInterior` in SG-Library to find out the nodes on the surface of the `PDEModel` which are inside the user-defined domain. Those nodes will be used for applying loads. In this way, we can apply any kind of free-form surface loads. The main function of the ORC method for applying free-form surface loads is as follows:

```
function model = applyNBC_ORC(model,loaddomain,loadvec)
nface = model.Geometry.NumFaces;
if isvector(loaddomain)
    applyBoundaryCondition(model,'neumann','Face',...
    loaddomain,'g',loadvec);
elseif isstruct(loaddomain)
    applyBoundaryCondition(model,'neumann','Face',...
    1:nface,'g',@(region,state)bcMatrix_g(region,...
    state,loaddomain,loadvec),'Vectorized','on');
end
end
```

The variable `loaddomain` in the input can be a vector or a struct. In the case of vector, the `loaddomain` contains the indices of feature surfaces and this function just follows the normal workflow of PDE Toolbox to apply loads according to feature surfaces. In the case of struct, the `loaddomain` is a user-defined domain. Although for the option '`Face`' all the feature faces (`1:nface`) are chosen as loading area, the load vector '`g`' is assigned by the handle function `bcMatrix_g` which is dependent on the region coordinates. We used the function `SGisInterior` in the handle function `bcMatrix_g` to determine the overlapping region. And all nodes inside the overlapping region will be applied with the given load `loadvec`. The handle function `bcMatrix_g` is also written in the same m-file:

```
function g = bcMatrix_g(region,state,loaddomain,loadvec)
% handle function for g matrix (free-form surface loads)
vertex = [region.x' region.y' region.z'];
n = size(vertex,1);
g = zeros(3,n);
% find out the overlapping region
in = SGisInterior(loaddomain,vertex);
if any(in)
    g(:,in) = repmat(loadvec',1,sum(in));
end
end
```

Since the Overlapping Region Concept is also valid for free-form surface fixation, we have also written the function `applyDBC_ORC(model,fixdomain)` to implement the ORC method on the Dirichlet Boundary Condition. Since the principle of this implementation is similar to the case of Neumann Boundary Condition, we will not describe it here again.

# 3 Examples

## 3.1 Free-Form Surface Load for a Ball Bearing

Now we use several FEA applications to show the performance of our new method. The first example is a ball bearing. As is shown in Fig. 3, a sphere only has only one feature surface according to PDE Toolbox, which means with that concept we cannot perform FEM analysis on a ball bearing with a point load on the top. In our application we first construct the ball bearing as a sphere with 5 mm radius (Fig. 5a). With our new method, we use a small cylinder with radius 1 mm on the top of the sphere to define the loading domain (Fig. 5b). The fixation domain is defined by the same cylinder but on the bottom of the sphere. Figure 5c and d shows a plausible result of the FEM analysis of the ball bearing, which means the free-form surface load is successfully applied with our method.

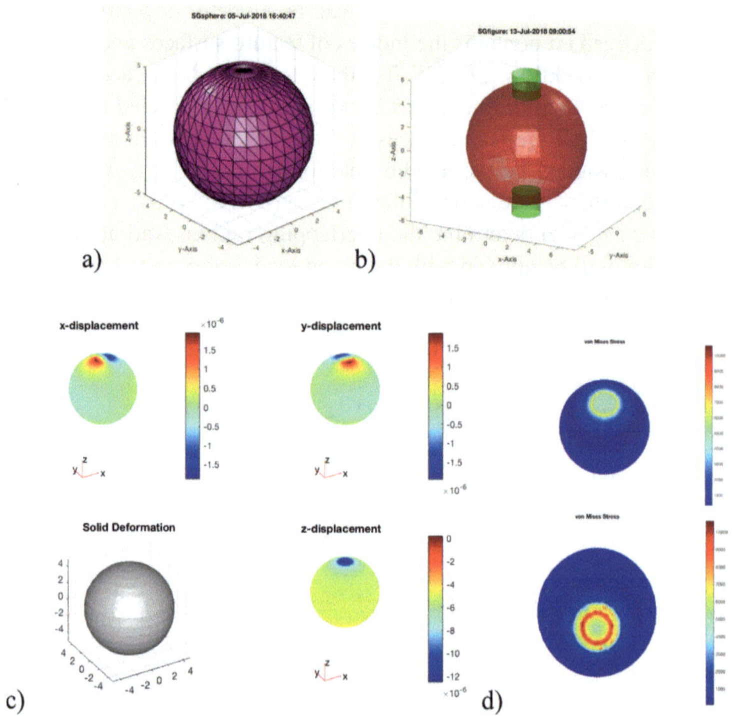


**Fig. 5.** FEM analysis of a ball bearing with the ORC method: (a) the ball bearing, (b) the load domain and fixation domain, (c) displacement of the ball bearing with a point load, (d) Von Mises stress of the ball bearing

### 3.2 Free-Form Surface Load for a Beam During Structural Optimization

As already described, the surface profile of a model is always changing during the bionic shape optimization. Figure 6 shows an example of applying our bionic structural optimization on a beam. At the beginning of the optimization the load is applied on the feature surface F6 in the direction of y axis (Fig. 6a). During the optimization the sharp boundary between loading area and unloaded area has disappeared (Fig. 6b). With our new method a loading domain and a fixation domain are created for applying boundary conditions (Fig. 6c). The result of FEM analysis shows a very equal stress distribution which is reasonable according to the theory of structural optimization [5–7, 11]. Therefore, with the ORC method we can overcome the limits of feature surfaces to make our structural optimization method much more robust.

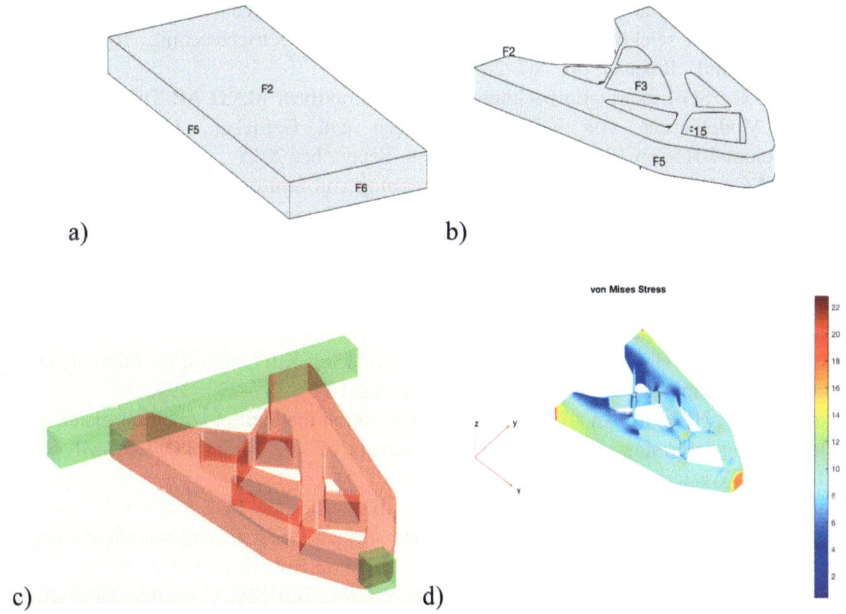


**Fig. 6.** Applying free-form surface load on a beam during the bionic structural optimization process: (a) feature surfaces of the beam before optimization, (b) during the shape optimization the feature surfaces of loading and unloaded area are merged, (c) the load domain and fixation domain, (d) Von Mises stress in the optimized structure

# 4 Conclusion

This paper introduces a new method called Overlapping Region Concept to extend Matlab's PDE Toolbox for applying free-form surface load during FEM analysis. In our institute we have integrated the PDE Toolbox into our self-developed SG-Library in Matlab to implement and develop some bionic methods like CAO and SKO to optimize the structure of our medical robots and mechanisms. The typical workflow of

PDE Toolbox separates the surface of the model into several sub regions with sharp boundaries called feature surfaces for applying boundary conditions. This workflow has limited the development of our optimization method since the surface profile of our model is constantly changing during the shape optimization process. With the ORC method we can apply free-form surface load or fixation during FEM analysis independent from the feature surfaces, which makes our optimization methods more robust.

## References

1. Walzl, A., Buchmayr, B.: Topologieoptimierung – Entwicklungswerkzeug für die additive Fertigung. BHM Berg - Huettenmaenn. Monatsh. **162**(3), 110–116 (2017)
2. Lueth, T.C., Irlinger, F.: Berechnete Erzeugung von dreidimensionalen Oberflächenmodellen im STL-Format aus der Beschreibung planarer Mechanismen für die generative Fertigung durch selektives Lasersintern, in Kolloquium Getriebetechnik: 10. Kolloquium Getriebetechnik, Ilmenau, pp. 267–284 (2013)
3. Lueth, T.C.: SG-Library: Entwicklung einer konstruktiven MATLAB-Toolbox zur räumlichen Modellierung von Körpern, Gelenken und Getrieben, 11. Getriebetechnik-Kolloquium 2015, pp 183–203. TU München, September 2015
4. Lueth, T.C., Irlinger, F: Automatische Generierung von Simscape Multibody Modellen auf der Grundlage von Gelenkgetrieben mit der Matlab-Toolbox SG-Library, 12. Kolloquium Getriebetechnik 2017, Dresden, 20–22 September 2017. TUDpress Verlag der Wissenschaften GmbH
5. Mattheck, C., Burkhardt, S.: A new method of structural shape optimization based on biological growth. Int. J. Fatigue **12**(3), 185–190 (1990)
6. Baumgartner, A., Harzheim, L., Mattheck, C.: SKO (Soft Kill Option): the biological way to find an optimum structure topology. Int. J. Fatigue **14**(6), 387–393 (1992)
7. Umetani, Y., Hirai, S.: An adaptive shape optimization method for structural material using the growing-reforming procedure. In: Proceedings of 1975 Joint ISME-ASME Applied Mechanics Western Conference, pp. 359–365 (1975)
8. Sun, Y., Lueth, T.C.: Extension of the FEM analysis using the PDE-toolbox of matlab with regard to point loads, line loads and freeform surface loads: feature surface concept. In: Proceedings of 2018 IEEE International Conference on Robotics and Biomimetics
9. Partial Differential Toolbox™ User's Guide, Version: R2018b, Copyright 1995–2018 by The Mathworks, Inc
10. Zwillinger, D.: Handbook of Differential Equations, 3rd edn. Academic Press, Boston (1997)
11. Harzheim, L.: Strukturoptimierung, Grundlagen und Anwendungen. Wissenschaftlicher Verlag Harri Deutsch GmbH, Frankfurt am Main (2014)

# Kinematic Design of a Hybrid Mechanism for Bone Reduction Surgery

Terence Essomba(✉) and Sinh Nguyen Phu

National Central University, Zhongli City 32001, Taiwan
tessomba@cc.ncu.edu.tw

**Abstract.** In severe fracture cases, a bone can be completely separated into two fragments. In order to guarantee a re-ossification of the bone, it is mandatory to reposition the bone fragment together. This process requires a delicate surgery called "bone reduction surgery". The most advanced technique relies on the use of a robots to manipulate the bone fragments with higher precision and stability. The present work introduces the kinematic design of a new hybrid mechanical architecture to perform this task. It is made of one 3-$\underline{P}$RP planar mechanism and one 3-R$\underline{P}$S tripod mechanism. Its kinematic model is resolved while taking account the tripod parasitic motion. The workspace of this mechanism is then compared to the standard hexapod mechanism that is widely used in bone reduction surgery. It reveals that the proposed mechanism benefits from a larger workspace.

**Keywords:** Bone reduction surgery · Hybrid mechanism · Planar mechanism · Tripod mechanism · Workspace

## 1 Introduction

In the past few decades, many different techniques and devices have been developed for performing bone reduction surgery such as traction tables, unilateral external fixators, circular external fixators or hybrid mechanisms. If the focus is given on mechanism concept, the literature review of robotic bone reduction surgery can be divided in several categories. Fully embedded architectures are presented as serial or parallel structures that are connected to the patient's anatomy by both their base and end effector. In the case of deported mechanisms, only their end effector is connected to the patient's bone. The first mechanism assisting bone reduction surgery has been by the Dr. Gavril Abramovich Ilizarov. He developed the first circular external fixator with cross wires, named Ilizarov, for repairing complex fractures and non-unions of bones [1]. The Ilizarov external fixator consists in two or more rings that are connected each other by four threaded rods with adjustable nuts. This fixator was patented and accepted for clinical application after his impressive results. A software-based Ortho-SUV Frame (OSF) for treating complex deformities and fractures with single step solution has been developed by Ortho-SUV Ltd. It consists in one base and one mobile rings that are connected to each other by six telescopic struts equipped with scales [2, 3]. The Taylor Spatial Frame (TSF) is a hexapod mechanism hexapod that is now widely used in reduction. This device was used for the first operation in 1995 and patented in 1997 [4].


K. Berns and D. Görges (Eds.): RAAD 2019, AISC 980, pp. 365–373, 2020.
https://doi.org/10.1007/978-3-030-19648-6_42

The standard TSF is composed of two full or partial rings connected by six telescopic struts through multi-axial joints at each end. It enables a 6-DoF manipulation by adjusting six struts so this frame can be used for simultaneous corrections of multidirectional deformities. Compared to Ilizarov fixator and unilateral devices, the TSF fixator offers more accurate results and less complicated modification of the frame during surgery or deformity correction. Parallel mechanical architectures offer higher stiffness, higher accuracy, higher payload and low inertia in comparison to serial architectures. But it provides smaller workspace than serial ones. For this reasons, a hybrid mechanical architecture is proposed for bone reduction surgery. The next Section presents the new hybrid planar-tripod mechanism. Its kinematic model is studied in Sect. 3. The Sect. 4 proposes the workspace analysis of this mechanism. And the Sect. 5 is dedicated to the conclusion.

## 2 Hybrid Planar-Tripod Architecture

In this study, the proposed hybrid mechanism is a combination of two different parallel architectures. It consists of one planar parallel mechanism and one spatial tripod mechanism that are connected in series as illustrated in Fig. 1(a). The planar parallel mechanism is a double triangular architecture that has been introduced by Daniali et al. 1993 in [5]. It has two triangular platforms: one is fixed and considered as the architecture base and one is a movable platform. The mobile platform is connected to the fixed platform by three PRP legs. Each of them has one active prismatic joint attached to the edge of fixed triangular platform and one passive prismatic joint attached on the edge of mobile triangular platform. Both prismatic joints are linked by a passive revolute joint. This planar mechanism provides two linear DoF and one angular DoF. The mechanism fixed on the planar mechanism mobile platform is a tripod architecture. This 3-RPS parallel mechanism was introduced by Hunt in 1983 [6]. It also has two platforms, one is considered as the base and one is a mobile platform that is manipulated with two angular DoF and one linear DoF.

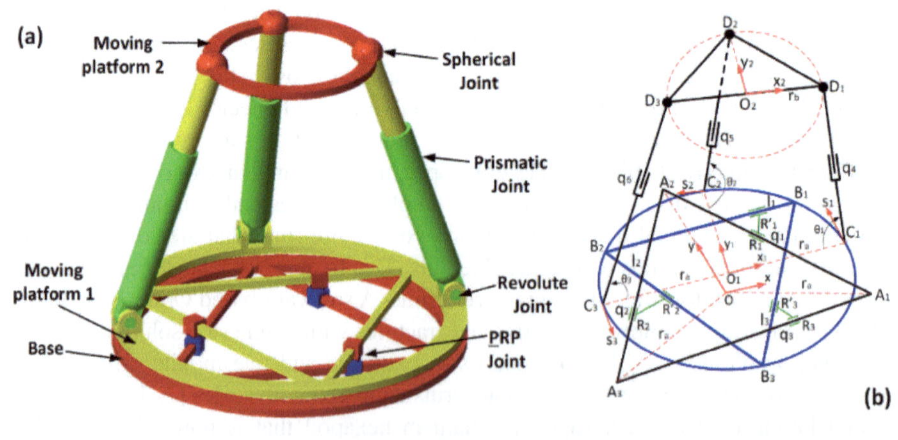


**Fig. 1.** Hybrid planar-tripod mechanism (a). Kinematic drawing of the mechanism (b).

Both platform are connected together by the three R$\underline{P}$S legs. Each leg is connected to the fixed platform by a passive revolute joint and to the mobile platform a passive spherical joint. These two passive joint are linked be an active prismatic joint. The three R$\underline{P}$S leg of the tripod are disposed as an isosceles triangle. This architectural disposition has been chosen in order to suppress the parasitic motion on one axis. By connecting these two 3-DoF mechanisms in series, a hybrid parallel-parallel manipulator is created with a total of six DoF: three linear and three angular. Compared with well-known Stewart platform, this robot has simpler kinematics, control design and a larger workspace. As shown in Fig. 1(b), the original reference frame {**Oxyz**} is attached with the base of the planar mechanism, and the point O is located at the center of the equilateral triangle $A_1A_2A_3$ that forms this base. It is dimensioned by the radius $r_a$ of its circumcircle. Another reference frame {$O_1\mathbf{x_1y_1z_1}$} is attached to the planar mechanism moving platforms, namely the mobile platform 1 (MP1), that is represented by the equilateral triangle $B_1B_2B_3$. This triangle is assumed to have the same size as $A_1A_2A_3$ and is also dimensioned by $r_a$. At the same time, reference frame {$O_1\mathbf{x_1y_1z_1}$} is also attached to the base of the tripod mechanism. It is represented by an isosceles triangle $C_1C_2C_3$ which circumcircle radius is also given by $r_a$. The passive revolute joints of tripod R$\underline{P}$S legs are located at the corners of this triangle. Similarly, the third reference frame {$O_2\mathbf{x_2y_2z_2}$} is attached to the mobile platform of the tripod, namely the mobile platform 2 (MP2) which is the end effector of the entire planar-tripod mechanism. The passive spherical joint of all RPS leg is located at the $D_k$ (k = 1, 2, 3). The point $O_2$ is located at the center the isosceles triangle $D_1D_2D_3$ that represents the MP2. Its dimension is given by the radius $r_b$ of its circumcircle. The planar and tripod mechanisms are both actuated by prismatic joints. On the planar mechanism, the input variables $q_1$, $q_2$ and $q_3$ are therefore defined as the distances between the positions of the active prismatic joints $R_1$, $R_2$ and $R_3$ and their associated corners $A_1$, $A_2$ and $A_3$ respectively. On the tripod mechanism, the variable $\theta_k$ measures the inclination of the R$\underline{P}$S legs by giving the angles ($O_1C_kD_k$), with k = 1, 2, 3. The input variables $q_4$, $q_5$ and $q_6$ generated by the active prismatic joints measure the distance $C_1D_1$, $C_2D_2$ and $C_3D_3$ respectively. The mechanism will be adapted to bone reduction surgery. Each bone fragments will be attached to the manipulator base and its end effector (MP2). By controlling the motion of the MP2, the operator will be able to control the relative motion of one bone fragment from the other.

## 3 Mechanism Kinematic Analysis

The inverse kinematic model of the presented mechanism can be solved in several steps. The first step consists in analyzing separately the kinematic of two mechanism: the 3-$\underline{P}$RP planar mechanism and the 3-R$\underline{P}$S tripod. In the second step, the kinematic of the tripod must be expressed in the reference frame of the entire mechanism base. Another step consist in compensating the parasitic motion of the tripod. First, the inverse kinematics of the triangular planar parallel mechanism is presented. In other words, the position and orientation of the MP1 in the origin reference frame {**Oxyz**} are studied. The double triangular manipulator has 3-DoF: two translations along **x** and **y** axes and one rotation about the **z** axis. The linear position of the MP1 is defined by the vector $\mathbf{o_1} = [x_1\ y_1\ 0]$ expressed in the fixed frame {**Oxyz**} and its angular position is designed by the rotation angle $\gamma_1$ about the **z** axis. It is assumed that there is no distance between the MP1 and the base along the vertical axis. As shown in Fig. 1(b), the position vectors $\mathbf{r_1}$, $\mathbf{r_2}$ and $\mathbf{r_3}$ of the active prismatic joints $R_1$, $R_2$, and $R_3$ are expressed in the fixed reference frame {**Oxyz**} as followed:

$$\mathbf{r}_1 = \begin{bmatrix} \frac{\sqrt{3}}{2} r_a - \frac{1}{2} q_1 \\ -\frac{1}{2} r_a + \frac{\sqrt{3}}{2} q_1 \\ 0 \end{bmatrix}, \mathbf{r}_2 = \begin{bmatrix} -\frac{1}{2} q_2 \\ r_a - \frac{\sqrt{3}}{2} q_2 \\ 0 \end{bmatrix}, \mathbf{r}_3 = \begin{bmatrix} -\frac{\sqrt{3}}{2} r_a + q_3 \\ -\frac{1}{2} r_a \\ 0 \end{bmatrix}, \tag{1}$$

With $r_a$, the radius of the circumcircle of center O passing by $A_1$, $A_2$, and $A_3$. Assuming that $l_k$ ($k$ = 1, 2, 3) is the distance between the corner $B_k$ of the moving equilateral triangle $B_1B_2B_3$ and the closest passive prismatic joints $R'_k$, namely $B_1R'_1$, $B_2R'_2$ and $B_3R'_3$ respectively, then the position vector of the passive prismatic joints $R'_k$ are expressed in the mobile frame {$\mathrm{O_1}\mathbf{x_1y_1z_1}$} as followed:

$${}^1\mathbf{r}'_1 = \begin{bmatrix} \frac{\sqrt{3}}{2} r_a - l_1 \\ \frac{1}{2} r_a \\ 0 \end{bmatrix}, {}^1\mathbf{r}'_2 = \begin{bmatrix} -\frac{\sqrt{3}}{2} r_a + \frac{1}{2} l_2 \\ \frac{1}{2} r_a - \frac{\sqrt{3}}{2} l_2 \\ 0 \end{bmatrix}, {}^1\mathbf{r}'_3 = \begin{bmatrix} \frac{1}{2} l_3 \\ \frac{\sqrt{3}}{2} l_3 - r_a \\ 0 \end{bmatrix}. \tag{2}$$

In the fixed frame {**Oxyz**}, the position of the passive prismatic joints $R'_k$ are calculated as followed:

$$\mathbf{r}'_k = \boldsymbol{O}_1 + \mathbf{R}_z(\gamma_1) \cdot {}^1\mathbf{r}'_k, \tag{3}$$

Where $\mathbf{R_z}(\gamma_1)$ is the rotation matrix of angle $\gamma_1$ about **z** axis. Because the active prismatic joints $R_k$ on the base are attached the passive prismatic joints $R'_k$ by a revolute joint as depicted in Fig. 1(b), the $x$ and $y$ coordinates of two prismatic joints $R_k$ and $R'_k$ in the fixed frame are same that expressed in Eq. (4).

$$\begin{cases} \boldsymbol{r}_{kx} = \boldsymbol{r}'_{kx} \\ \boldsymbol{r}_{ky} = \boldsymbol{r}'_{ky} \end{cases}. \tag{4}$$

After developing the expression given by Eq. (4), the prismatic input coordinates $q_k$ (with $k$ = 1, 2, 3) are isolated and expressed as followed:

$$\begin{cases} (s\gamma_1 + \sqrt{3}c\gamma_1)q_1 = r_a + (\sqrt{3}r_a - 2\mathrm{x}_1)s\gamma_1 + (r_a + 2\mathrm{y}_1)c\gamma_1 \\ (s\gamma_1 + \sqrt{3}c\gamma_1)q_2 = r_a + (\sqrt{3}r_a + \mathrm{x}_1 - \sqrt{3}\mathrm{y}_1)s\gamma_1 + (r_a - \sqrt{3}\mathrm{x}_1 - \mathrm{y}_1)c\gamma_1 \\ (s\gamma_1 + \sqrt{3}c\gamma_1)q_3 = r_a + (\sqrt{3}r_a + \mathrm{x}_1 + \sqrt{3}\mathrm{y}_1)s\gamma_1 + (r_a + \sqrt{3}\mathrm{x}_1 - \mathrm{y}_1)c\gamma_1 \end{cases}, \tag{5}$$

Where $s$(.) and $c$(.) are respectively equal to *sin*(.) and *cos*(.). Next, the position and orientation the MP2 in the moving reference frame {$\mathrm{O_1}\mathbf{x_1y_1z_1}$} attached to the MP1 is presented. The 3-R̲PS parallel mechanism has parasitic motions [7] between the MP2 and the MP1. In order to define these parasitic motions, the vector ${}^{1}\mathbf{o_2} = [x_2 y_2 z_2]^T$ is assumed to be pointing the end effector $O_2$ and is expressed in the reference frame $\{O_1\mathbf{x_1y_1z_1}\}$. Its orientation Yaw-Pitch-Roll are given by the angles $(\gamma_2, \beta_2, \alpha_2)$ respectively rotating about $\mathbf{z_1}$, $\mathbf{y_1}$, and $\mathbf{x_1}$ axis. The position vector of spherical joints $D_1$, $D_2$, $D_3$ are expressed in the coordinate system $\{O_2\mathbf{x_2y_2z_2}\}$ of the MP2 as followed:

$$ {}^{2}\mathbf{d}_1 = \begin{bmatrix} r_b \\ 0 \\ 0 \end{bmatrix}, {}^{2}\mathbf{d}_2 = \begin{bmatrix} 0 \\ r_b \\ 0 \end{bmatrix}, {}^{2}\mathbf{d}_3 = \begin{bmatrix} -r_b \\ 0 \\ 0 \end{bmatrix}. \tag{6} $$

From Fig. 1(b), the position vector of the spherical joints $D_1$, $D_2$, $D_3$ in the frame $\{O_1\mathbf{x_1y_1z_1}\}$ is calculated:

$$ {}^{1}\boldsymbol{d}_k = {}^{1}\boldsymbol{o}_2 + {}^{1}_{2}\mathbf{R} \cdot {}^{2}\boldsymbol{d}_k \tag{7} $$

With ${}^{1}_{2}\mathbf{R}$ the rotation matrix of the MP2 and $k$ = 1, 2, 3. On the other hand, the constraints of the revolute joints $C_k$ ($k$ = 1, 2, 3) impose the following equation:

$$ \begin{cases} {}^{1}\boldsymbol{d}_{1y} = 0 \\ {}^{1}\boldsymbol{d}_{2x} = 0\,. \\ {}^{1}\boldsymbol{d}_{3y} = 0 \end{cases} \tag{8} $$

By solving a systems of equations given by Eq. (8), the parasitic motions can be obtained as:

$$ \begin{cases} \gamma_2 = 0 \\ x_2 = -r_b s\alpha_2 s\beta_2 \end{cases}. \tag{9} $$

According to Eq. (9), the parasitic motion generated by the 3-R̲PS architecture only affects the $x$-component of the end effector position on the tripod reference frame. In other words, the linear position of the end-effector $O_2$ in the horizontal plane of the fixed reference frame will be composed of the linear motion of the MP1 and a uni-directional parasitic motion along the $\mathbf{x_2}$ axis. Finally, the inverse kinematics model of the proposed parallel mechanism will be obtained by coupling the motion of the two parallel mechanism. The parasitic motion of the tripod mechanism relative to the first moving platform will be compensated by the linear motion of the triangular mechanism. The Cartesian coordinates of the end-effector $O_2$ expressed in the reference frame {**Oxyz-**} are given by $\mathbf{o_2} = [x\ y\ z]^T$ and its orientation is presented by Yaw, Pitch, Roll angles $(\gamma, \beta, \alpha)$. The Cartesian coordinate of spherical joints $D_1$, $D_2$, $D_3$ in the reference frame {**Oxyz**}:

$$ \mathbf{d}_k = \mathbf{o}_2 + {}^{0}_{2}\mathbf{R} \cdot {}^{2}\mathbf{d}_k, \tag{10} $$

With ${}^{0}_{2}\mathbf{R}$, the rotation matrix associated with of the MP2 rotations. By developing Eq. (10), the position vector of spherical joints $D_1$, $D_2$, $D_3$ expressed in the reference frame $\{O_1\mathbf{x_1y_1z_1}\}$ of the moving platform 1 is written as:

$$ {}^{1}\mathbf{d}_1 = \begin{bmatrix} r_a - q_4 c\theta_1 \\ 0 \\ q_4 s\theta_1 \end{bmatrix}, {}^{1}\mathbf{d}_2 = \begin{bmatrix} 0 \\ r_a - q_5 c\theta_2 \\ q_5 s\theta_2 \end{bmatrix}, {}^{1}\mathbf{d}_3 = \begin{bmatrix} q_6 c\theta_3 - r_a \\ 0 \\ q_6 s\theta_3 \end{bmatrix}, \quad (11) $$

Where $q_4, q_5, q_6$ are the length of the tripod legs $C_1D_1$, $C_2D_2$, $C_3D_3$ respectively and $\theta_1, \theta_2, \theta_3$ are the rotation angle of the tripod legs. The position vector of spherical joints $D_1$, $D_2$, $D_3$ in the coordinate system $\{\mathbf{Oxyz}\}$ are written:

$$ \mathbf{d}_k = \mathbf{o}_1 + \mathbf{R}_z(\gamma) \cdot {}^{1}\mathbf{d}_k \quad (12) $$

Solving Eq. (10) into Eq. (12), yields:

$$ \begin{cases} x_1 + c\gamma(r_a - q_4 c\theta_1) = x + r_{11} \cdot r_b \\ y_1 + s\gamma(r_a - q_4 c\theta_1) = y + r_{21} \cdot r_b \\ x_1 - s\gamma(r_a - q_5 c\theta_2) = x + r_{12} \cdot r_b \\ y_1 + c\gamma(r_a - q_5 c\theta_2) = y + r_{22} \cdot r_b \\ x_1 + c\gamma(-r_a + q_6 c\theta_3) = x - r_{11} \cdot r_b \\ y_1 + s\gamma(-r_a + q_6 c\theta_3) = y - r_{21} \cdot r_b \end{cases}. \quad (13) $$

Equations given in Eq. (13) are then regrouped into a matrix form as followed:

$$ \mathbf{o}_1 = \begin{bmatrix} x_1 \\ y_1 \\ z_1 \end{bmatrix} = \begin{bmatrix} x + r_b \sin(\alpha)\sin(\beta)\cos(\gamma) \\ y + r_b \sin(\alpha)\sin(\beta)\sin(\gamma) \\ 0 \end{bmatrix} \quad (14) $$

Since the Yaw rotation of the second moving platform is the same as the Yaw angle of the first moving platform. Therefore, the input coordinates $q_k$ ($k$ = 1, 2, 3) can be obtained by substituting Eq. (14) into Eq. (5) and are written as bellow:

$$ \begin{cases} (s\gamma + \sqrt{3}c\gamma)q_1 = r_a + r_a c\gamma + 2yc\gamma - 2xs\gamma + \sqrt{3}r_a s\gamma \\ (s\gamma + \sqrt{3}c\gamma)q_2 = r_a + r_a c\gamma + \sqrt{3}r_a s\gamma - yc\gamma + xs\gamma - \sqrt{3}ys\gamma - \sqrt{3}xc\gamma - \sqrt{3}r_b s\alpha s\beta \\ (s\gamma + \sqrt{3}c\gamma)q_3 = r_a + r_a c\gamma + \sqrt{3}r_a s\gamma - yc\gamma + xs\gamma + \sqrt{3}ys\gamma + \sqrt{3}xc\gamma + \sqrt{3}r_b s\alpha s\beta \end{cases}. \quad (15) $$

The position vector of revolute joints $C_k$ ($k$ = 1, 2, 3) in the coordinate system $\{O_1\mathbf{x_1y_1z_1}\}$ are written:

$$ {}^{1}\mathbf{c}_1 = \begin{bmatrix} r_a \\ 0 \\ 0 \end{bmatrix}, {}^{1}\mathbf{c}_2 = \begin{bmatrix} 0 \\ r_a \\ 0 \end{bmatrix}, {}^{1}\mathbf{c}_3 = \begin{bmatrix} -r_a \\ 0 \\ 0 \end{bmatrix}. \quad (16) $$

The position vector of revolute joints $C_k$ ($k$ = 1, 2, 3) in the coordinate system {**Oxyz**} are written as:

$$\mathbf{c}_k = \mathbf{o}_1 + \mathbf{R}_z(\gamma) \cdot {}^1\mathbf{c}_k. \tag{17}$$

The length of the active prismatic $q_{k+3}(k = 1, 2, 3)$ of tripod R$\underline{\text{P}}$S can then be determined by:

$$q_{k+3} = \|C_k D_k\| = \|\mathbf{d}_k - \mathbf{c}_k\| = \left\|\mathbf{o}_2 + {}^0_2\mathbf{R} \cdot {}^2\mathbf{d}_k - \mathbf{o}_1 - \mathbf{R}_z(\gamma) \cdot {}^1\mathbf{c}_k\right\|, \tag{18}$$

$$\begin{cases} q_4 = \sqrt{(z - r_b s\beta)^2 + (r_a - r_b c\beta + r_b s\alpha s\beta)^2} \\ q_5 = \sqrt{(z + r_b s\alpha c\beta)^2 + (r_a - r_b c\alpha)^2} \\ q_6 = \sqrt{(z + r_b s\beta)^2 + (r_a - r_b c\beta - r_b s\alpha s\beta)^2} \end{cases}. \tag{19}$$

## 4 Mechanism Workspace Analysis

In this section, the workspace of the proposed mechanism is presented. There are various methods for calculating the workspace or boundaries of a mechanism such as using analytical, geometrical or numerical methods. This study uses a numerical method for determining the workspace boundaries of the hybrid planar-tripod mechanism manipulator. As previously mentioned, this hybrid mechanism is made of one planar and one tripod mechanism, so the workspace analysis of the entire mechanism can be separated into two different studies. Therefore, the workspace of the mechanism is shown in two different coordinate systems: ($x$, $y$, $\gamma$) and ($\alpha$, $\beta$, $z$) that are respectively associated with the planar and tripod mechanisms. A range of translational motion ($x$, $y$, $z$) and the orientation angles ($\alpha$, $\beta$, $\gamma$) are discretized into sampling points with a desired resolution. For each point, the corresponding mechanism input variables ($q_1$, …, $q_6$) are calculated using the inverse kinematics model given by Eqs. (15) and (19). Then, the $q_i$ values are checked to make sure they are within the stroke range limits given by [$q_{i\min}$; $q_{i\max}$]. In order to study the suitability of the proposed hybrid planar-tripod mechanism with bone reduction surgery, its workspace is compared to the Stewart platform which is the most commonly used architecture in the medical application. To insure a fair comparison, the Stewart platform and the hybrid mechanism are set with the same general dimensions as illustrated in Fig. 2. Their input variables range of motion are also adjusted similarly. Their upper and lower rings are set to 380 mm diameter and the range of motion of their input prismatic joints from 175 to 250 mm. For each mechanism, the ($x$, $y$, $\gamma$) and ($\alpha$, $\beta$, $z$) coordinate system workspaces are respectively displayed in Fig. 3(a) and (b).

It reveals that while the maximum ($xy$) planar motion range is at $\gamma = 0$ for the Steward platform, it is around $\gamma = 23°$ for the hybrid mechanism. In this configuration, the motion range of the hybrid mechanism is 17% larger on **x** axis and 13% larger on

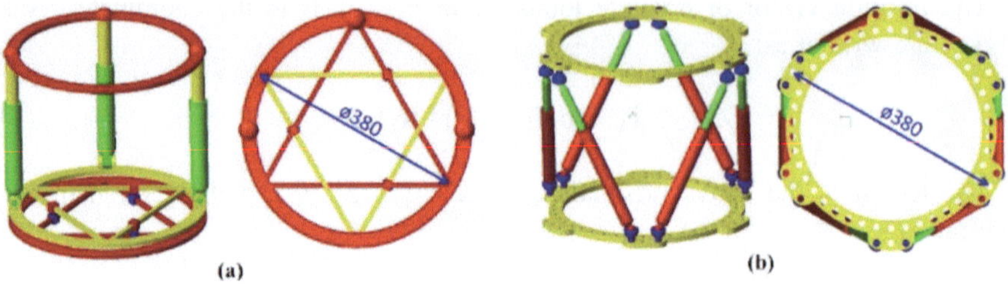


Fig. 2. (a) The proposed hybrid robot. (b) Stewart platform.

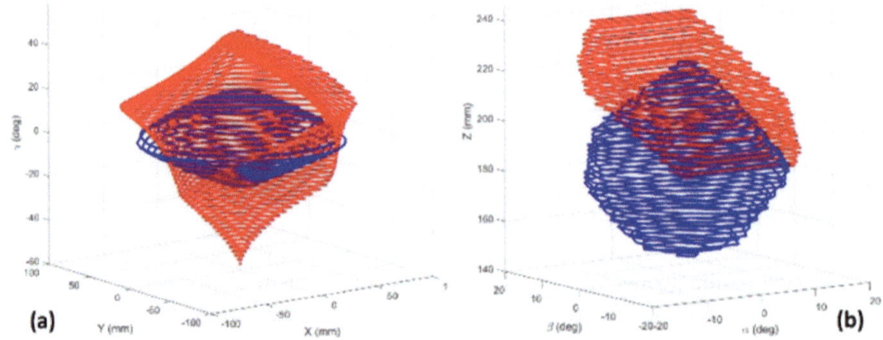


Fig. 3. Workspace representation of the hybrid mechanism (red) and the Stewart platform (blue) in the $(x, y, \gamma)$ coordinate system (a) and in the $(\alpha, \beta, z)$ coordinate system (b).

the **y** axis, while the Steward platform has vertical motion range 3% larger. Although the Steward platform angular motion is 21% large on angle $\beta$, the hybrid mechanism has an angular motion 16% larger on angle $\alpha$ and 162% larger on angle $\gamma$.

## 5 Conclusion

In the present study, the kinematic design of a new hybrid architecture is proposed for bone reduction surgery. It is based on a 3-PRP planar mechanism attached with a 3-RPS tripod mechanism. The kinematic model of the entire mechanism is provided while considering the parasitic motion issued from the tripod mechanism end effector rotation. Due to its specific architectural arrangement, this parasitic motion is generated on only one axis of the tripod base. Also, the planar mechanism kinematic model takes into account this parasitic motion and it can totally compensate this motion error. When compared to the classical Taylor mechanism that is mostly used in this application, the presented hybrid planar-tripod mechanism has a larger workspace, making the proposed mechanism a good candidate for bone reduction surgery.

## References


1. Ilizarov, G.A., Ledyaev, V.I.: The replacement of long tubular bone defects by lengthening distraction osteotomy of one of the fragments. Clin. Orthop. Relat. Res. **280**, 7–10 (1992)
2. Solomin, L.N., Paley, D., Shchepkina, E.A., Vilensky, V.A., Skomoroshko, P.V.: A comparative study of the correction of femoral deformity between the Ilizarov apparatus and Ortho-SUV frame. Int. Orthop. **38**, 865–872 (2013)
3. Skomoroshko, P.V., Vilensky, V.A., Hammouda, A.I., Fletcher, M.D.A., Solomin, L.N.: Mechanical rigidity of the Ortho-SUV frame compared to the Ilizarov frame in the correction of femoral deformity. Strat. Trauma Limb Reconstr. **10**(1), 5–11 (2015)
4. Keshet, D., Eidelman, M.: Clinical utility of the Taylor spatial frame for limb deformities. Orthop. Res. Rev. **9**, 51–61 (2017)
5. Daniali, M., Zsombor-Murray, H.P., Angeles, P.J.: The kinematics of 3-DoF planar and spherical double-triangular parallel manipulators. Comput. Kinemat. Solid Mech. Appl. **28**, 153–164 (1993)
6. Hunt, K.H.: Structural kinematics of in-parallel-actuated robot-arms. J. Mech. Transm. Autom. Des. **105**(4), 705–712 (1983)
7. Li, Q., Chen, Z., Chen, Q., Wu, C., Hu, X.: Parasitic Motion Comparison of 3-PRS Parallel Mechanism with Different Limb Arrangements. Robot. Comput. Integr. Manuf. **27**, 389–396 (2011)

# Pneumo-tronic Perturbator for the Study of Human Postural Responses

Daniela Maffiodo[1], Walter Franco[1], Carlo De Benedictis[1], Maria Paterna[1], Giovanni Gerardo Muscolo[1], Silvestro Roatta[2], Carlo Ferraresi[1(✉)], and Zeevi Dvir[3]

[1] Department of Mechanical and Aerospace Engineering, Politecnico di Torino, Torino, Italy
carlo.ferraresi@polito.it
[2] Department of Neuroscience, University of Torino, Torino, Italy
[3] Department of Physical Therapy, Tel Aviv University, Tel Aviv, Israel

**Abstract.** This paper describes the design and operational principles of a device that imparts a well-controlled mechanical force or impulse, a so-called perturbation, to a pre-selected point on the surface of the human body. This perturbator will be integrated within a system aimed at measuring and evaluating human postural reaction in a clinically meaningful way. The ease of use and versatility of the device renders it suitable for manual operation but it can also be integrated in a robotized system. The hardware, control law and characterization of the perturbator are presented. Preliminary results indicate that the device is able to generate repeatable perturbations with characteristics appropriate to the intended application. Further improvements are discussed and proposed.

**Keywords:** Postural control analysis · Pneumatic actuation · Human response · Perturbation · Human balance

## 1 Introduction

Maintenance of body balance, using the so-called postural reactions, is a very fundamental neuro-motor skill necessary for the performance of everyday activities. Compromise to the postural control system may have significant consequences including an increased risk of falling, which might lead to subsequent injuries, and reduced quality of life. Therefore, a quantitative analysis of postural stability is important for assessing the various associated impairments as well as for evaluating the efficacy of interventions such as rehabilitation or pharmacotherapy [1].

Postural control involves the interaction of three sensory systems: the somatosensory system, the vestibular system [2, 3] and gaze control [4, 5]. Postural deficits may be better understood by examining the postural control response to an external perturbation (e.g. a mechanical, visual or auditory). A possible scheme of the human balance control system subject to external perturbations [6] is shown in Fig. 1. The *global sensory system* measures the current postural status (human response) whereas the *controller* is designed to reduce the error between desired and perceived posture, the latter being estimated by the *sensory integration system* that processes the


K. Berns and D. Görges (Eds.): RAAD 2019, AISC 980, pp. 374–383, 2020.
https://doi.org/10.1007/978-3-030-19648-6_43

information provided by the sensors. Among the several kinds of external perturbations, we focused on the mechanical ones. In these cases, the perturbation often consists of a movement of the support surface [7, 8], but it can also be applied directly to upper body segments, e.g. by pulling [9–12] or pushing [13] the trunk, shoulder or pelvis. Past approaches proposed to use an external perturbation applying bulky and rather expensive devices which limit the mobility of these systems and the ecological validity and feasibility of the clinical trials. In these applications, the point of application of the perturbation is often predictable, thus affecting the response of the patients [14]. In most of the cases, the devices described in previous studies did not allow control and/or adjustment of the force as well as the impulse (time-integral of the contact force) magnitude that were applied to the subject, making the analysis and comparison of the results difficult.

Furthermore, standardization of the measurement of postural reactions was never achieved, reflecting among other things, the diversity of technological approaches to the problem. Some solutions were based on the use of force platforms to track the center of pressure (COP) of the ground-foot interface [15, 16] whereas others applied accelerometers and inertial units [17–20], surface electromyography [21] or motion analysis [22].

This paper describes the design and function of a new mechanical perturbating device capable of imparting a targeted, force-controlled impulse to the human body. This perturbator may be carried by the examiner/clinician or secured to a stable base of support. The former option is enabled by the specific features of the device namely it is simple, lightweight, transportable and adaptable to the patient's characteristics and the clinical task at hand. The design is based on a commercial pneumatic cylinder controlled by means of flow proportional control valves. Pneumatic actuators are usually inexpensive and are suitable for the use in clean environments. Although there are several nonlinearities involved in pneumatic actuation systems, which encumber their control, a number of previously reported control algorithms [23–26] enable a high level of performance.

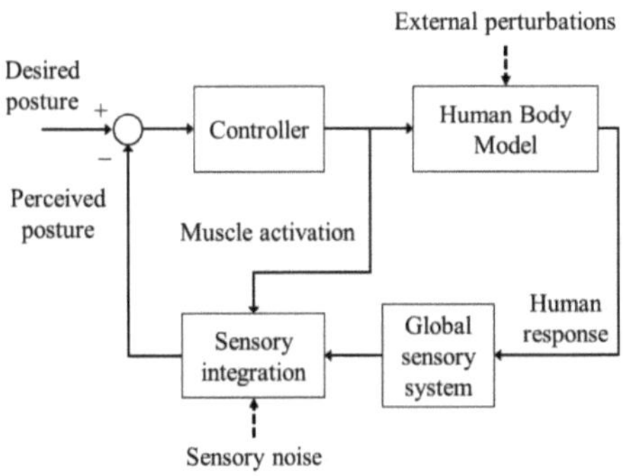


**Fig. 1.** Scheme of the human balance control system subject to external perturbations.

## 2 Materials and Methods

This study focuses on the design of an automatic perturbator (AP) used for studying dynamic balance in patients. The 'business end' of the AP which is momentarily interfaced with the human body, termed the striker, is intended to generate a controlled perturbation to the body. The main objective in both healthy subjects and patients is to analyze comparatively the response pattern to the perturbation and in the latter, to identify relevant correlates to the severity of the disorder and its variations due to intervention. In this study, the contact force between the striker and the subject body is considered to be the only controllable external perturbation while the other kinds of stimuli such as visual or auditory, are considered noise and should be limited to avoid undesired effects on the body response evoked by the perturbation.

Among the several types of mechanical perturbations proposed in the literature, very limited attention has been given to the application of a push force on the subject. Despite the simple conception of such a stimulus, the perturbation given by a push is not simple to define and to control. The perturbation generated by the AP is tailored individually to the subject in terms of timing, amplitude and direction of the stimulus. The entity of the perturbation can be defined through several physical quantities: in addition to the waveform of the contact force between the striker and the stricken body, the peak force as well as the impulse may be considered. Unlike the traditional approaches (moving base platform or cable-based systems), the AP is not limited in terms of the direction as well as the point of application of the perturbation. Two separate working configurations of the AP have been considered and analyzed:

- fixed: in which the AP is connected to a fixed-frame (e.g. a pole, or the end-effector of a manipulator) thus providing a rigid connection to the environment;
- hand-held: in which the AP may be handled by the clinician, who acts as a viscoelastic constraint towards the environment.

It should be emphasized that the second configuration represents undoubtedly an easier-to-use and compact solution compared to its fixed-frame counterpart, albeit introducing additional uncertainties to the characterization of the device, which could affect negatively the performance of the control system and the accuracy of the analysis. The proposed device and its architecture are shown in Fig. 2, constituted by the following elements:

1. a linear double acting pneumatic actuator (MetalWork, mini-cylinder Series "ISO 6432", bore diameter 25 mm, stroke 120 mm);
2. two flow proportional control valves (CKD, 3AF2, 0–10 V of analog input, working pressure range 0–0.97 MPa) and dedicated control drivers (CKD APC-23);
3. a uniaxial load cell (Dacell UMM, rated capacity about 500 N, nonlinearity, hysteresis and repeatability 0.1% of the rated capacity) and signal conditioner (DEWETRON, DEWE-RACK-4);
4. an end striker, integral with the load cell, covered by a thick layer (10 mm) of a synthetic foam material working as a deformable buffer;
5. two handles (in aluminum) used to maneuver the device;
6. a trigger button to enable the perturbation.

The complete perturbation system includes an air compressor and pressure regulator (gauge supply pressure set to 3.5 bar); a real-time system (dSPACE), Multi-Channel A/D Board DS2002 (16 bit, 32 channels) for signal acquisition and D/A board DS2101 (12 bit, 5 channels) for control generation; a Personal Computer with MATLAB®-Simulink® Real-Time (The MathWorks) and ControlDesk® (dSPACE) software.

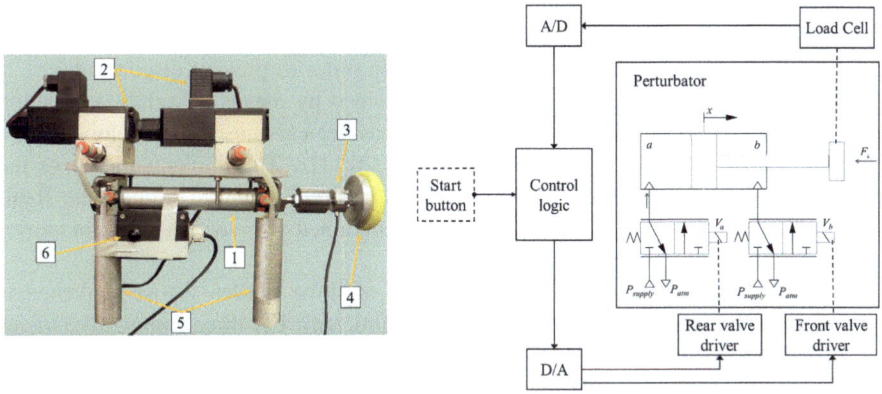


**Fig. 2.** Photo (left) and a scheme (right) of the perturbation system. The components highlighted are: 1 – actuator, 2 – valves, 3 – load cell, 4 – buffer, 5 – handles, 6 – trigger button.

The stroke of the pneumatic cylinder has been considered appropriate for the application, since the clinician is expected to stand near the subject, placing the end of the perturbator at short distance from the body (about 100 mm). This actuator was chosen for its large availability and low cost and it was tested to evaluate its performance in a demanding application as the one proposed in this paper. However, more expensive solutions with significantly higher dynamic performance are available (e.g. low friction metal seal cylinders), which may be tested in the future. With reference to Fig. 2, the two three-way valves have been connected respectively to the rear (a) and front (b) chambers of the cylinder, in order to control the input-output mass flow rate from both chambers. In this work, the system has been designed in order to perform the tracking of a reference contact force profile with short duration (typically in the range 250–500 ms) and amplitude in the range 20 to 100 N. Given the limited size of the actuator, these high-flow rate valves were considered appropriate for the purpose and able to fulfill the dynamic requirements of the stimulus. However, the presence of significant friction in the pneumatic actuator chosen could limit the performance and accuracy of the force tracking, mainly due to the stick-slip effect.

The valves have been placed on a base integral with the actuator's body, as shown in Fig. 2, in order to shorten as much as possible the pipes towards the chambers, thus optimizing the dynamics of the pneumatic circuit. A metal cylindrical support has been fixed at the end of the piston rod and used to carry the uniaxial load cell. The range of the load cell, considerably higher than the amplitude of the contact force, has been chosen to avoid unsafe conditions for the transducer in case of unexpected instabilities in the control logic and for characterization purposes. The sensor has been preliminary calibrated and zeroed through the conditioning software before each set of trials. The presence of a uniaxial force transducer rather than a multi-axis sensor significantly simplifies the acquisition and processing of the impact data. However, it requires appropriate care and training of the operator to avoid bending of the load cell rod during the application of the perturbation. The free end of the rod is rigidly connected to an aluminum cylindrical component (about 50 mm of diameter) covered by a deformable synthetic foam, the latter acting as the soft interface between the actuator and the body of the subject. This component distributes the contact pressure on a wider area than the sectional area of the load cell rod in order to prevent discomfort.

The acquisition of the sensor output is performed by means of a data acquisition and real-time system by dSPACE. Such system has been also used to develop and test the control logic. The sampling rate was set to 200 Hz. The implementation of the control algorithm has been enabled using MATLAB®-Simulink® through the Real-Time™ Toolbox, while the user interface for the parameter's selection and data viewer has been developed in ControlDesk®.

The working condition of the AP consists of three phases whose sequence is initiated by an external trigger (activated by the clinician). The three consecutive stages are the approach phase, the strike phase and the return phase (see Fig. 3). In the absence of an external command, the system remains in an idle condition, which represents the fourth stage of operation. For these reasons, the control logic has been conceived as a finite-state machine and implemented directly through the Stateflow® environment in Simulink®. The approach phase is triggered directly by the occurrence of the start command and consists of the extension of the piston. When the contact force exceeds a defined threshold (selected as the 10% of the reference force magnitude), the system moves to the strike stage, tracking the desired force profile. The transition to the return phase is triggered by the fall of the reference force signal, then the idle condition is reached after a time-out period of 4 s since the start command.

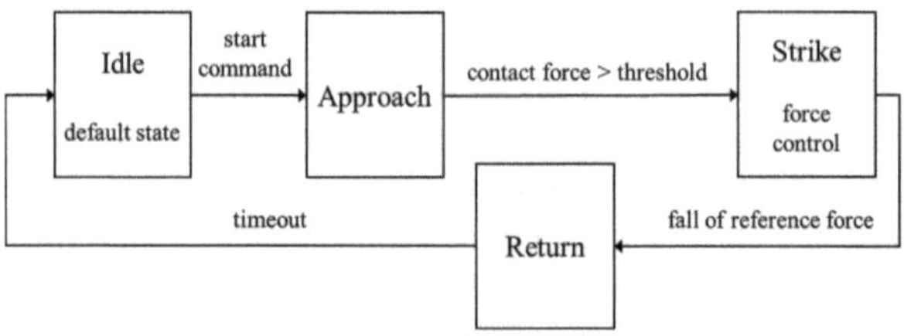


**Fig. 3.** General scheme of the control logic sequence.

During the approach and return phases, the system has been set to drive the control valves with constant voltage signals (open-loop) due to the inability of the controller to continuously monitor the piston position or velocity due to lack of specific transducers. Such values were selected by preliminary trials, verifying that the striker would not impact the subject's body (or a fixed target) with high speed, in order to avoid the risk of a significant overshoot in the contact force produced by the passive response of the stricken body. The resulting moving speed of the piston rod was considered sufficient for the application, always producing extension and retraction phases lasting less than 1.5 s with good repeatability. Since any overshoot in the contact force may harm the subject, it is possible to drive the piston at a very high speed only during the return stage. To perform a better control of the impact, it may be necessary to monitor the actual velocity of the rod with a transducer, thus bringing to a more complex control law. The strike phase, in which the interaction between the striker and the target has to be kept under control, has been designed by implementation of a PI controller, evaluating the error between the required force profile (in this case represented by a rectangular function defined by a duration and an amplitude) and the actual contact force measured by the load cell and filtered by a Butterworth low pass filter ($4^{th}$ order, cut frequency 10 Hz), designed in MATLAB® (Signal Processing Toolbox™) and necessary to remove the effect of the powerline noise. The control system had to ensure the correct synchronization between the reference signal and the measured one. The tuning of the PI controller was performed by a semi-empirical iterative approach (i.e. Ziegler-Nichols) for different durations and amplitudes of the stimulation. This controller has been chosen for its simple formulation and implementation, thus not taking directly into account the nonlinearities within the system, that may require a more advanced design (e.g. sliding mode control).

The device has been characterized by performing several trials in laboratory: the dynamic performance of the system has been assessed with the actuator fixed on the work bench (configuration 1), testing contact force profiles lasting 250 and 500 ms and considering different initial distances between the striker and a fixed target (55 and 80 mm). Then, the AP was tested in a hand-held condition (configuration 2) by a skilled operator. The operator applied the perturbations while trying to keep the same initial distance between the striker and the fixed target.

## 3 Results of the Experimentation and Discussion

Figure 4 shows the results of the first characterization trials performed with the AP in fixed (left) or hand-held (right) configuration. In both cases the tracking of the force profile seems to be insufficiently accurate, demonstrating slow transients in the initial and final phases of the strike. The overall duration of the stimulation reaches about 350–400 ms, while the amplitude of the contact force remains closer to the reference level. In both cases, however, the system produced quite repeatable results, showing a better performance in the fixed condition. This result was expected since the performance of the control system was affected by the mechanical impedance of the operator in the hand-held configuration as well as by the repeatability of targeting the specific point on the subject's body, an inherent attribute of the operator. The size of the actuator could also affect the dynamic performance of the device, thus a more compact cylinder may be necessary to get a fast-responsive system. To assess the repeatability of the trials, the time integral of each perturbation was calculated and compared with the impulse of the reference signal. For the 50 N set-level trials (reference impulse equal to 12.50 Ns, Fig. 4, bottom), in the fixed-frame and hand-held configurations, the impulse was $13.88 \pm 0.05$ and $12.49 \pm 0.24$ Ns, respectively. This outcome indicates good repeatability among the several trials. The force tracking shows acceptable dynamics, considering the step variation of the reference signal. This result was reconfirmed for the trials performed with a reference signal of 75 N amplitude (Fig. 4, bottom), as well as for different durations of the perturbation (Fig. 5).

Since in the hand-held set-up the operator cannot easily keep a constant initial distance between himself and the body target, we tested the performance of the system for different values of the initial distance (80 and 55 mm) between the (fixed) actuator and the target. The results of this analysis are shown in Fig. 6: the tracking of the force profile seems to be slightly affected by the initial distance, showing an overshoot of the contact force at 55 mm but not at 80 mm.

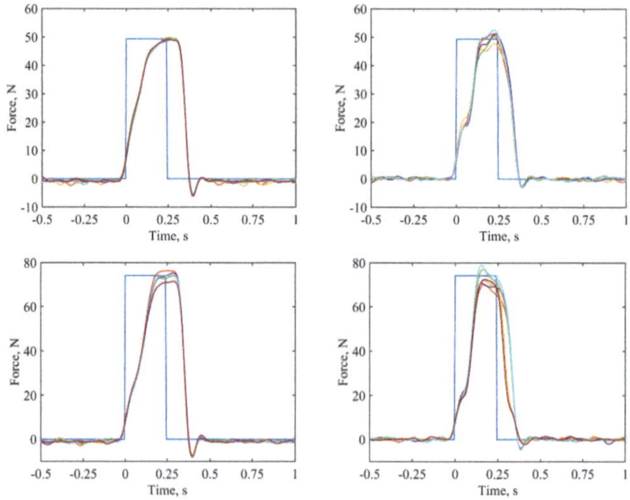


**Fig. 4.** Force tracking in fixed (left, top and bottom) or hand-held (right, top and bottom) configuration. The figures on top refer to a force reference of 50 N, the bottom ones to 75 N. In all trials, the duration of the stimuli was 250 ms.

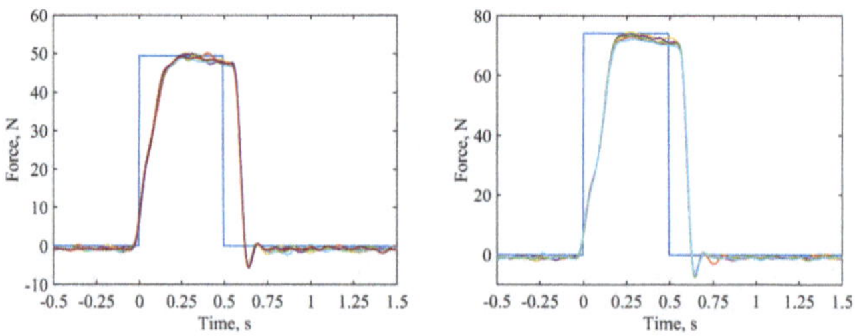


**Fig. 5.** Force tracking in fixed condition for 50 N (left) and 75 N (right) perturbations lasting 500 ms.

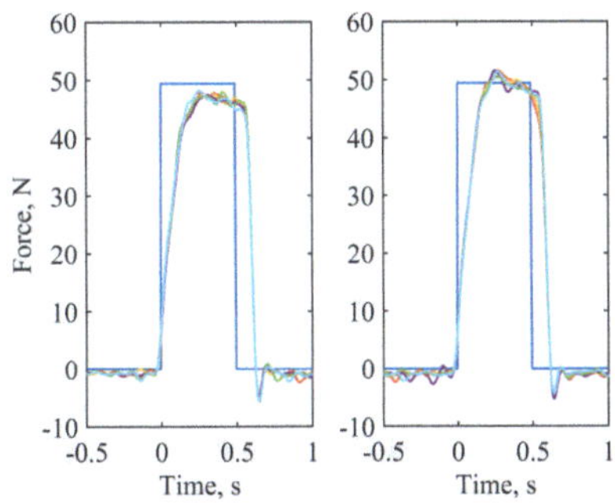


**Fig. 6.** Force tracking in fixed condition for different values of the initial distance between the perturbator and a fixed target: 80 mm (left) and 55 mm (right).

Figure 6 refers to 500 ms perturbations, almost coincident results were produced by 250 ms perturbations. This outcome may suggest an improvement in the design of the controller, e.g. by performing a closed-loop control of the speed of the piston rod during the approach phase, leading to a more robust control law. Preliminary tests of the AP, using a small group of healthy volunteers, has already indicated that the perturbations imparted by this apparatus are well tolerated and resulting in no feeling of discomfort.

# 4 Conclusion

Past approaches to imparting perturbations to human subjects during clinical postural control tests involved non-scalable systems, had a very limited number of points used for applying the perturbation, or did not allow a real-time control of the stimulus. This work presents a novel pneumatic perturbation system capable of imparting customizable force perturbations directly to the body of a subject.

The system has been conceived to control the intensity of the perturbation not only in terms of its peak force but also in those relating to more significant parameters such as the value of the force impulse. Some experimentations with the system fixed on a structure or hand-held by an operator are presented. A comparison of the results highlights the robustness of the device, given that only slight differences are noticeable by comparison of the fixed and hand-held configurations.

Future steps will involve a more robust control law, e.g. based on the monitoring of the actuator stroke and velocity, and the development of a model of the human stance control system aimed at interpretation of the body response under external perturbations. Different architectures, e.g. based on more compact pneumatic cylinders with low friction, as well as on electromechanical actuators, shall be tested to improve the tracking of the force profile.

## References


1. Visser, J.E., et al.: The clinical utility of posturography. Clin. Neurophysiol. **119**, 2424–2436 (2008)
2. Horak, F.B., Macpherson, J.M.: Postural orientation and equilibrium. In: Handbook of Physiology, pp. 253–292 (1996)
3. Horak, F.B., Nashner, L.M., Diener, H.C.: Postural strategies associated with somatosensory and vestibular loss. Exp. Brain Res. **82**, 167–177 (1990)
4. Pankanin, E., Dobosiewicz, A.M., Mietkowska, P.: Static posturography as an instrument to assess the balance among athletes. J. Educ. Health Sport. **8**(5), 216–225 (2018)
5. Gandelman-Marton, R., Arlazoroff, A., Dvir, Z.: Ocular dominance and balance performance in healthy adults. Gait Posture **31**, 394–396 (2010)
6. van der Kooij, H., et al.: An adaptive model of sensory integration in a dynamic environment applied to human stance control. Biol. Cybern. **84**, 103–115 (2001)
7. Farahpour, N., et al.: Electromyographic responses of erector spinae and lower limb's muscles to dynamic postural perturbations in patients with adolescent idiopathic scoliosis. J. Electromyogr. Kinesiol. **24**(5), 645–651 (2014)
8. Grassi, L., et al.: Quantification of postural stability in minimally disabled multiple sclerosis patients by means of dynamic posturography: an observational study. J. Neuro-Eng. Rehabil. **14**(1), 4 (2017)
9. Colebatch, J.G., Govender, S., Dennis, D.L.: Postural responses to anterior and posterior perturbation applied to the upper trunk of standing human subjects. Exp. Brain Res. **234**, 367–376 (2016)
10. Martinelli, A.R., et al.: Light touch modulates balance recovery following perturbation: from fast response to stance restabilization. Exp. Brain Res. **233**(5), 1399–1408 (2015)
11. Mansfield, A., et al.: Effect of a perturbation-based balance training program on compensatory stepping and grasping reactions in older adults: a randomized controlled trial. Phys. Ther. **90**(4), 476–491 (2010)
12. Mansfield, A., Maki, B.E.: Are age-related impairments in change in support balance reactions dependent on the method of balance perturbation? J. Biomech. **42**, 1023–1031 (2009)
13. Chen, B., Lee, Y.J., Aruin, A.S.: Role of point of application of perturbation in control of vertical posture. Exp. Brain Res. **235**(11), 3449–3457 (2017)
14. He, J., Tian, C.X.: Adaptive postural control for a repeated perturbation. In: 18th International Annual Conference IEEE Transactions Automatic Control, vol. 36, pp. 322–332 (1996)
15. Prieto, T., et al.: Measures of postural steadiness: differences between healthy young and elderly adults. IEEE Trans. Biomed. Eng. **43**(9), 956–966 (1996)
16. Ruhe, A., Fejer, R., Walker, B.: The test-retest reliability of centre of pressure measures in bipedal static task conditions - a systematic review of the literature. Gait Posture **32**(4), 436–445 (2010)
17. Moe-Nelssen, R., Helbostad, J.L.: Trunk accelerometry as a measure of balance control during quiet standing. Gait Posture **16**, 60–68 (2002)
18. Alberts, J.L., et al.: Using accelerometer and gyroscopic measures to quantify postural stability. J. Athl. Train. **50**(6), 578–588 (2015)
19. Hubble, R.P., Naughton, G.A., Silburn, P.A., Cole, M.H.: Wearable sensor use for assessing standing balance and walking stability in people with Parkinson's Diseases: a systematic review. PLoS One **10**(4), e0123705 (2015)

20. Heebner, N.R., et al.: Reliability and validity of an accelerometry based measure of static and dynamic postural stability in healthy and active individuals. Gait Posture **41**, 535–539 (2015)
21. Henry, S.M., Fung, J., Horak, F.B.: EMG Responses to maintain stance during multidirectional surface translations. J. Neurophysiol. **80**, 1939–1950 (1998)
22. Rossi, S., et al.: Compensation to whole body active rotation perturbation. Gait Posture **39**, 621–624 (2014)
23. Faudzi, A., Mustafa, N., Osman, K.: Force control for a pneumatic cylinder using generalized predictive controller approach. Math. Probl. Eng. **2014**, Article ID 261829 (2014)
24. Driver, T., Shen, X.: Pressure estimation-based robust force control of pneumatic actuators. Int. J. Fluid Power **14**, 37–45 (2013)
25. Chiaverini, S., Sciavicco, L.: The parallel approach to force/position control of robotic manipulators. IEEE Trans. Rob. **9**(4), 361–373 (1993)
26. Stanisic, R.Z., Fernandez, A.V.: Simultaneous velocity, impact and force control. Robotica **27**(7), 1039–1048 (2009)

# Tilt-Twist Method Using Inertial Sensors to Assess Spinal Posture During Gait

Elisa Digo[1(✉)], Giuseppina Pierro[1], Stefano Pastorelli[1], and Laura Gastaldi[2]

[1] Department of Mechanical and Aerospace Engineering, Politecnico di Torino, Turin, Italy
{elisa.digo, stefano.pastorelli}@polito.it, giuseppina.pierro@studenti.polito.it

[2] Department of Mathematica Sciences "G.L. Lagrange", Politecnico di Torino, Turin, Italy
laura.gastaldi@polito.it

**Abstract.** In the clinical context, the need to estimate spinal posture during gait is constantly growing. The most functional way to achieve this goal is first to model the rachis as a multibody structure with rigid segments and second to apply the tilt-twist method. Inertial Measurement Units (IMUs) are the suitable instrumentation to do this because they are portable, low cost, not invasive and free from laboratory constraints. The aim of this pilot study was the assessment of spinal angles by applying the tilt-twist method to IMUs data. A marker stereo-photogrammetric system (Optitrack) was adopted as gold standard. Three IMUs (MTx Xsens) were positioned on C7, T12 and S1 vertebral levels. A young healthy subject performed a gait trial at a self-selected speed. Data analysis focused on rotation matrices obtained simultaneously from both the instrumentations. Post-processing algorithms identified movement values of flexion-extension and lateral bending from both IMUs and stereo-photogrammetric system. Comparison graph with the obtained angular patterns showed very similar trends for the three spinal segments. Inertial sensors are suitable to be used to assess spinal posture during gait.

**Keywords:** Tilt-twist method · IMUs · Multibody · Spinal angular motion · Gait

## 1 Introduction

Many pathologies involve posture disorders that negatively affect patients' quality of life. Among these disturbs, one of the most common is Parkinson's disease (PD), a neurodegenerative pathology that provokes bradykinesia, rigidity and tremor [1]. Patients with PD often develop an abnormal posture, usually consisting of a shoulders rounding, that can progress in more severe conditions such as scoliosis or Pisa syndrome [2]. For all these reasons, rachis motion analysis during daily movements is becoming increasingly important. To assess its motion, first the rachis has to be modeled as a multibody structure with rigid links consisting of vertebral segments.


K. Berns and D. Görges (Eds.): RAAD 2019, AISC 980, pp. 384–392, 2020.
https://doi.org/10.1007/978-3-030-19648-6_44

According to this model, relative angles between two adjacent segments during movement can be calculated, borrowing from robotics essential knowledge [3, 4].

Several motion capture systems have already investigated locomotion in different conditions: neurological diseases [5], orthopedic disturbs [6, 7], aging process [8] and obesity [9, 10]. Among different instrumentations, the use of Inertial Measurement Units (IMUs) constantly increases because they are portable, low cost and not invasive. Compared to other instruments, IMUs can be affected by the drift problem, which occurs by integrating the signals to obtain speeds and displacements. Despite this, IMUs give the possibility to overcome the limits of traditional motion capture systems such as instrumented mats or multi-camera systems [11]. Indeed, IMUs allow to record a movement in its natural environment and not only in a lab. This advantage makes them suitable for the clinical context and for the assessment of spinal posture.

The most effective technique to estimate spine angles by using IMUs is the tilt-twist method, which calculates the planar movements of a vertebra with respect to the lower one [12]. Geometrical interpretations and advantages of this method with respect to Euler angles are described in detail in [13]. In previous works tilt-twist method was applied to data extracted from IMUs to estimate spine angles during elementary movements performed standing still [14, 15]. Other works analyzed the spinal angular motion during gait by using optoelectronic systems [16–20].

The focus of the present pilot study was the assessment of spinal angles during gait by applying tilt-twist method with inertial sensors. More specifically, a stereo photogrammetric system was used as reference gold-standard. Three IMUs were positioned on three vertebral levels: C7, T12 and S1. One young healthy subject was tested during gait at a self-selected speed. Results from IMUs were compared with those obtained from the optical motion capture system, with the attempt to validate the tilt-twist method applied to different axes.

# 2 Tilt-Twist Method

## 2.1 Theoretical Aspects

According to the tilt-twist method, the two adjacent halves of a joint can be modeled as two cylinders. For the vertebral column, these two halves are two vertebrae. Based on the movement of a vertebra relative to the other, three angles are generated: tilt, tilt-azimuth and twist angles. The tilt angle $\varphi$, developed when the upper vertebra leans on the lower one, is always positive in a range from 0° to 180°. The direction of the bending identified by the tilt angle is described by the tilt-azimuth angle $\theta$, which ranges from −180° to 180°. Since these two angles are not sufficient to define a complete angular joint movement, also the twist angle $\tau$ is necessary. It identifies the upper vertebra "twist" movement around its axis and ranges from −180° to 180° [12].

Tilt, tilt-azimuth and twist angles can be calculated considering the vertebra movement from an initial to a final orientation. Two orthogonal reference frames, with the origins coincident with the center of the vertebral body, are considered: $X_S Y_S Z_S$ (vector units $\mathbf{i}$, $\mathbf{j}$, $\mathbf{k}$) and $X'_S Y'_S Z'_S$ (vector units $\mathbf{i}'$, $\mathbf{j}'$, $\mathbf{k}'$), that identify vertebra's initial and final condition respectively. Considering the projections of the final vectors on the planes formed by the initial reference frame, it is possible to evaluate the three angles (Fig. 1).

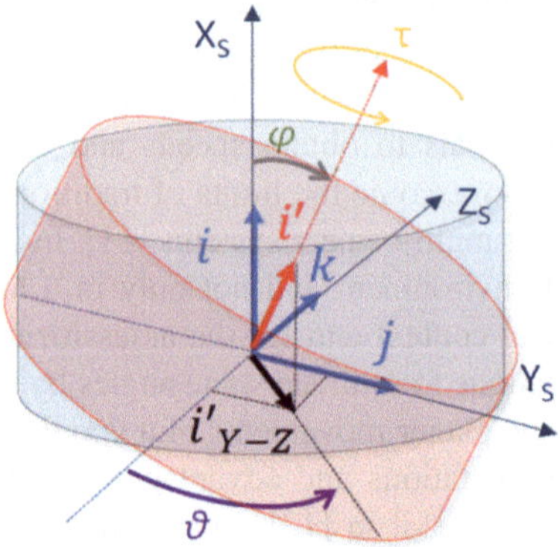


**Fig. 1.** Tilt, tilt-azimuth and twist angles.

Their mathematical formulas are:

$$\theta = atan2(i'_y, -i'_z) \tag{1}$$

$$\varphi = atan2(i'_y \sin\theta - i'_z \cos\theta, i'_x) \tag{2}$$

$$\tau = atan2(k'_x \sin(\varphi/2) \sin\theta - k'_y \cos(\varphi/2),\ k'_z \cos(\varphi/2) + k'_x \sin(\varphi/2) \cos\theta) \tag{3}$$

From these angles, it is possible to extract the values of flexion-extension, lateral bending and axial rotation of the vertebra. The flexion-extension (FE) and the lateral bending (LB) are both defined in terms of tilt and tilt-azimuth angles in Eqs. (4) and (5). The axial rotation, instead, coincides with the twist angle value previously estimated. Considering Eq. (4), flexion is positive and extension is negative. According to Eq. (5), LB is positive to the right side and negative to the left one.

$$FE = \varphi \cos\theta \tag{4}$$

$$LB = -\varphi \sin\theta \tag{5}$$

## 2.2 Practical Aspects

Human spine can be modeled as a multibody system with rigid links. Considering one IMU integral with each segment, the tilt-twist method described above can be applied to rotation matrices, in order to evaluate spine segments relative planar movements. IMUs adopted in this study are the Xsens MTx, for which the output is a matrix with sixteen columns (Eq. (6)). The first element of every row is time count; the second, the third and the fourth ones are accelerations along the three axes; the fifth, the sixth and the seventh elements are angular velocity values around the three axes. The last nine elements express the sensor orientation in the global coordinate frame (Eq. (7)).

$$row = \left[t\ acc_x\ acc_y\ acc_z\ gyr_x\ gyr_y\ gyr_z\ R_{11}\ R_{21}\ R_{31}\ R_{12}\ R_{22}\ R_{32}\ R_{13}\ R_{23}\ R_{33}\right] \quad (6)$$

$$^{G}R_S = [R_{11}\ R_{21}\ R_{31}\ R_{12}\ R_{22}\ R_{32}\ R_{13}\ R_{23}\ R_{33}] \quad (7)$$

Each vector $^{G}R_S$ can be rearranged in a 3x3 matrix (Eq. (8)):

$$^{G}R_S = \begin{bmatrix} R_{11} & R_{12} & R_{13} \\ R_{21} & R_{22} & R_{23} \\ R_{31} & R_{32} & R_{33} \end{bmatrix} \quad (8)$$

The spine neutral position is obtained asking the subject to stand still for 3 s. Orthonormal matrices measured by IMUs during this interval are used to refer sensors orientations to the neutral position of the subject ($^{N}R_S$). In order to obtain the angular motion of a vertebral segment with respect to the lower one, a sensor orientation is referred to the one positioned lower. In Eq. (9) the result $^{S\_inf}R_{S\_sup}$ is the orientation of the superior sensor with respect to the inferior one. The two terms after the equal sign express respectively the inverse of the inferior sensor orientation ($^{S\_inf}R_N$) and the superior sensor orientation ($^{N}R_{S\_sup}$) both in the neutral position.

$$^{S\_inf}R_{S\_sup} = {}^{S\_inf}R_N * {}^{N}R_{S\_sup} \quad (9)$$

Tilt, tilt-azimuth and twist angles and consequently planar movements of FE and LB can be evaluated with respect to the neutral position.

## 3 Materials and Methods

A young healthy woman with no musculoskeletal or neurological disease (age: 26 years, BMI: 19.83 kg/m$^2$) participated in the study giving her informed consent. The rachis was ideally divided into three segments, each of them integral with an IMU and a marker triad: S1, T12-S1 and C7-T12.

### 3.1 Instruments

Two different motion capture instruments were adopted: an inertial sensors system (MTx Xsens) and an optoelectronic system (Optitrack V120:Trio) as reference.

**Xsens.** The inertial sensor system was composed of three MTx Xsens. Each of them contained tri-axial accelerometer, gyroscope and magnetometer. A sampling frequency of 50 Hz was selected for all the IMUs. Accelerometers and gyroscopes measurement ranges were set respectively to ± 5G and ±1200 dps. The first IMU was fixed at C7 vertebra level with double-sided adhesive tape. The second and the third ones were placed respectively at T12 and S1 vertebrae levels through elastic bands. All IMUs axes were oriented with x-axes pointing upward, z-axes opposite to the walking direction and y-axes creating a Cartesian frame with the other two (Fig. 2A). A chain of three MTx was created by using cables. The S1-IMU was also connected to the Xbus Master, the control unit able to send synchronous, sampled and digital data from IMUs to PC. Recordings were made with the software MT Manager.

**Optitrack.** The optoelectronic system was composed of two self-contained and pre-calibrated Optitrack V120:Trio bars and ten passive reflective markers. Each bar had 3 cameras able to detect the infrared light of markers and presented its local reference system in the middle of the central camera. A sampling frequency of 120 Hz was selected for both bars. Three markers (6 mm of diameter) were positioned on each IMU (Fig. 2B). The first marker was placed at the center of the IMU's local reference frame. The second marker was positioned 5 cm above the first one with a vertical alignment along the IMU's x-axis. The third marker was positioned halfway, moving it 4 cm to the left side. A tenth additional marker on the right heel was positioned to synchronize the two bars. Recordings were made with the software Motive.

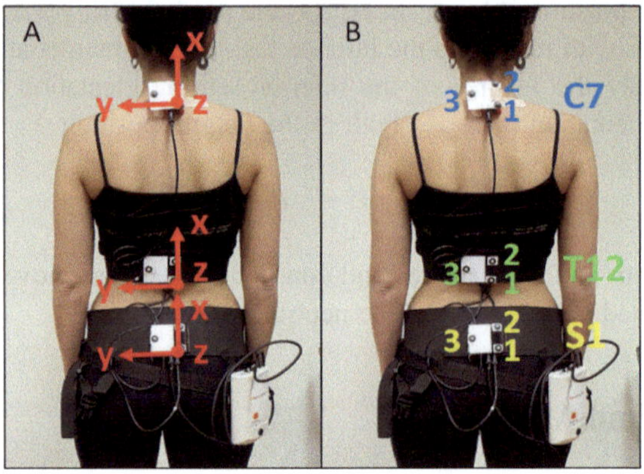


**Fig. 2.** (A) IMUs positions and reference systems. (B) Markers positions.

## 3.2 Protocol

The test was conducted indoor. Spatial synchronization of the two Optitrack bars was obtained by positioning three markers on the floor along the path. A static recording was made before and after the test in order to obtain the coordinates of these fixed markers. Temporal synchronization of the two motion capture systems was achieved by asking the subject to impact her right heel on the floor before starting to walk. In this way, the zero instant for the two bars was the frame corresponding to the minimum of the heel marker trajectory. The zero frame for the three IMUs was the one corresponding to the maximum peak of vertical acceleration from the T12 level IMU. The subject was first asked to stand still for a few seconds and then to walk along a 6-m path always in the same direction, at a self-selected speed and for five times. Data were acquired with the two motion capture systems at the same time.

### 3.3 Signal Processing and Data Analysis

Signal processing and data analysis were conducted by using Matlab® routines.

**Xsens.** IMU's accelerations and angular velocities were extracted. Gait events and the gait cycle were defined according to [21]. For every walk a complete gait cycle between two consecutive heel-strikes of the same foot was analyzed. Correspondingly, the tilt-twist method was applied to calculate FE and LB values.

**Optitrack.** From the static recording of the fixed markers on the floor a Global Reference System (GCS) was created. The Z-axis was aligned and opposite to the walking direction; the X-axis was vertical with respect to the floor; the Y-axis formed a Cartesian triad with X and Z. Markers coordinates obtained from the two bars during gait were expressed into this GCS by using rotational matrices. Subsequently, they were sub-sampled at 50 Hz in order to be easily compared with IMUs data. The markers triad on each sensor was used to create a local reference frame coincident with the IMU's one. Planar motions obtained with the tilt-twist method were smoothed by using the Matlab® Smoothing Spline.

## 4 Results and Discussion

Angular orientation of the spine segments obtained with both motion capture systems were averaged among the five walks (Fig. 3). Segments Ranges of Motion (ROM) obtained with Xsens and Optitrack are reported in Table 1.

**Table 1.** Optitrack (Op) and Xsens (Xs) ROM (°) for FE and LB with differences (°).

| | S1 | T12-S1 | C7-T12 |
|---|---|---|---|
| FE | \|4.8° Op – 5.3° Xs\| = 0.5° | \|6.8° Op – 5.7° Xs\| = 1.1° | \|2.8° Op – 3.2° Xs\| = 0.4° |
| LB | \|11.5° Op – 11.7° Xs\| = 0.2° | \|12.2° Op – 13.3° Xs\| = 1.1° | \|6.8° Op – 6.6° Xs\| = 0.2° |

The focus of the present study was the application of tilt-twist method to IMUs data to evaluate spinal angular motion. In other papers [14, 15] the tilt-twist method was already applied to IMUs in order to evaluate spinal posture in elementary movements performed standing still. In this work the same tilt-twist method was developed, but it was applied during gait instead of elementary movements. Other works already estimated spine angles during gait, but only with optoelectronic systems [16–20]. The patterns obtained with the stereo-photogrammetric system and inertial sensors show a very good correspondence (Fig. 3) for both FE and LB and for all the segments. ROM of LB (Table 1) are greater than those of FE for both Optitrack and Xsens. Xsens ROM differ from Optitrack ROM by a value between 0.2° and 1.1°. These differences can be considered acceptable according to similar results in literature [15].

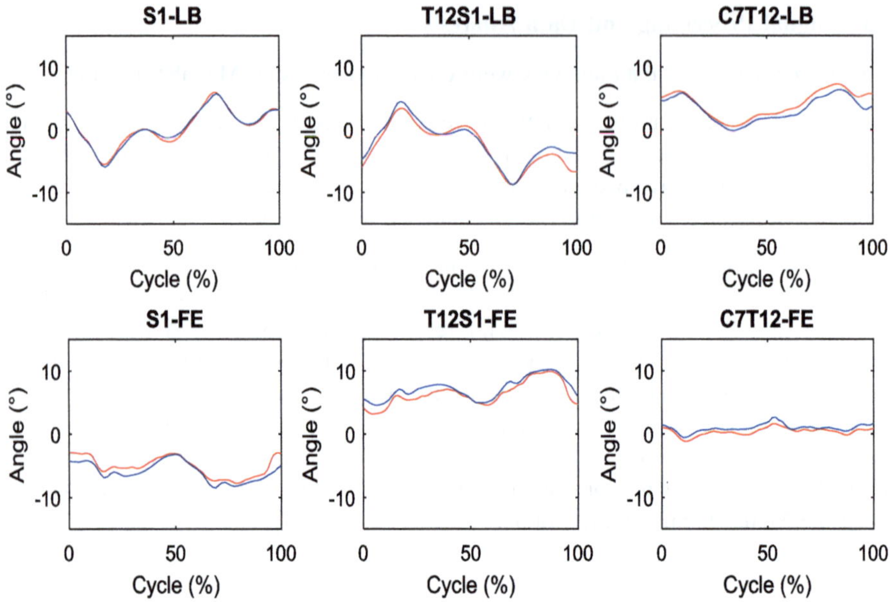


**Fig. 3.** Comparison between planar angles values obtained with Optitrack (red line) and Xsens (blue line). LB = lateral bending; FE = flexion-extension.

The application of this method to IMUs data represents a useful tool especially in a clinical context, due to the need to evaluate the effectiveness of a rehabilitation program over time. Particularly, the possibility to obtain vertebral segments ROM during gait in a simple and fast way takes on great importance. Furthermore, despite the less dense segmentation obtained with IMUs compared to markers, the use of inertial sensors without the laboratory constraints typical of optical motion capture systems allows to analyse patients' posture also in a domestic environment and consequently opens the door to tele-rehabilitation.

## 5 Conclusion

Results obtained by comparing IMUs data with a gold standard methodology show that this new technology can be used to assess human spine posture during gait. The final aim of this pilot study is to apply the presented methodology to patients with Parkinson's disease or Pisa syndrome, to evaluate the effectiveness of their rehabilitation treatments on trunk postural control.

**Acknowledgments.** The present research has been partially supported by MIUR grant Dipartimenti di Eccellenza 2018–2022 (E11G18000350001).

## References


1. Tysnes, O.B., Storstein, A.: Epidemiology of Parkinson's disease. J. Neural Transm. **124**, 901–905 (2017). https://doi.org/10.1007/s00702-017-1686-y
2. Doherty, K.M., van de Warrenburg, B.P., Peralta, M.C., Silveira-Moriyama, L., Azulay, J.P., Gershanik, O.S., Bloem, B.R.: Postural deformities in Parkinson's disease. Lancet Neurol. **10**, 538–549 (2011). https://doi.org/10.1016/S1474-4422(11)70067-9
3. Esat, I.I., Ozada, N.: Articular human joint modelling. Robotica **28**, 321–339 (2010). https://doi.org/10.1017/S0263574709990592
4. Gastaldi, L., Lisco, G., Pastorelli, S.: Evaluation of functional methods for human movement modelling. Acta Bioeng. Biomech. **17**, 31–38 (2015). https://doi.org/10.5277/ABB-00151-2014-03
5. Sofuwa, O., Nieuwboer, A., Desloovere, K., Willems, A.M., Chavret, F., Jonkers, I.: Quantitative gait analysis in Parkinson's disease: comparison with a healthy control group. Arch. Phys. Med. Rehabil. **86**, 1007–1013 (2005). https://doi.org/10.1016/j.apmr.2004.08.012
6. Weidow, J., Tranberg, R., Saari, T., Kärrholm, J.: Hip and knee joint rotations differ between patients with medial and lateral knee osteoarthritis: gait analysis of 30 patients and 15 controls. J. Orthop. Res. 1–9 (2006). https://doi.org/10.1002/jor.20194
7. Kay, R.M., Dennis, S., Rethlefsen, S., Reynolds, R.A.K., Skaggs, D.L., Tolo, V.T.: The effect of preoperative gait analysis on orthopaedic decision making. Clin. Orthop. Relat. Res. **372**, 217–222 (2003). https://doi.org/10.1097/00003086-200003000-00023
8. Hartmann, A., Luzi, S., Murer, K., de Bie, R.A., de Bruin, E.D.: Concurrent validity of a trunk tri-axial accelerometer system for gait analysis in older adults. Gait Posture **29**, 444–448 (2009). https://doi.org/10.1016/j.gaitpost.2008.11.003
9. Agostini, V., Gastaldi, L., Rosso, V., Knaflitz, M., Tadano, S.: A wearable magneto-inertial system for gait analysis (H-Gait): validation on normalweight and overweight/obese young healthy adults. Sensors (Switzerland) **17**, 2406 (2017). https://doi.org/10.3390/s17102406
10. Lai, P.P.K., Leung, A.K.L., Li, A.N.M., Zhang, M.: Three-dimensional gait analysis of obese adults. Clin. Biomech. **23**, 2–6 (2008). https://doi.org/10.1016/j.clinbiomech.2008.02.004
11. Muro-de-la-Herran, A., García-Zapirain, B., Méndez-Zorrilla, A.: Gait analysis methods: an overview of wearable and non-wearable systems, highlighting clinical applications. Sensors (Switzerland) **14**, 3362–3394 (2014). https://doi.org/10.3390/s140203362
12. Crawford, N.R., Yamaguchi, G.T., Dickman, C.A.: A new technique for determining 3-D joint angles: the tilt/twist method. Clin. Biomech. **14**, 153–165 (1999)
13. Bonev, I., Zlatanov, D., Gosselin, C..: Advantages of the modified Euler angles in the design and control of PKMs. In: Proceeding Parallel Kinematic Machines International Conference, pp. 171–188 (2002)
14. Goodvin, C., Park, E.J., Huang, K., Sakaki, K.: Development of a real-time three-dimensional spinal motion measurement system for clinical practice, pp. 1061–1075 (2006). https://doi.org/10.1007/s11517-006-0132-3
15. Bauer, C.M., Rast, F.M., Ernst, M.J., Kool, J., Oetiker, S., Rissanen, S.M., Suni, J.H., Kankaanpää, M.: Concurrent validity and reliability of a novel wireless inertial measurement system to assess trunk movement. J. Electromyogr. Kinesiol. **25**, 782–790 (2015). https://doi.org/10.1016/j.jelekin.2015.06.001
16. Leardini, A., Biagi, F., Merlo, A., Belvedere, C., Benedetti, M.G.: Multi-segment trunk kinematics during locomotion and elementary exercises. Clin. Biomech. **26**, 562–571 (2011). https://doi.org/10.1016/j.clinbiomech.2011.01.015

17. Leardini, A., Berti, L., Begon, M., Allard, P.: Effect of trunk sagittal attitude on shoulder, thorax and pelvis three-dimensional kinematics in able-bodied subjects during gait. PLoS One **8**, 1–7 (2013). https://doi.org/10.1371/journal.pone.0077168
18. Needham, R., Naemi, R., Healy, A.: Multi-segment kinematic model to assess three-dimensional movement of the spine and back during gait, pp. 1–12 (2015). https://doi.org/10.1177/0309364615579319
19. Ceccato, J.C., de Sèze, M., Azevedo, C., Cazalets, J.R.: Comparison of trunk activity during gait initiation and walking in humans. PLoS One **4**, e8193 (2009). https://doi.org/10.1371/journal.pone.0008193
20. Crosbie, J., Vachalathiti, R., Smith, R.: Patterns of spinal motion during walking. Gait Posture **5**, 6–12 (1997). https://doi.org/10.1016/S0966-6362(96)01066-1
21. Panero, E., Digo, E., Agostini, V., Gastaldi, L.: Comparison of different motion capture setups for gait analysis : validation of spatio-temporal parameters estimation. In: MeMeA 2018 - 2018 IEEE International Symposium on Medical Measurements and Applications. Proceedings, pp. 1–6 (2018). https://doi.org/10.1109/memea.2018.8438653

# Wrist Rehabilitation Equipment Based on the Fin-Ray® Effect

Andrea Deaconescu(✉) and Tudor Deaconescu

Transilvania University, Brasov, Romania
{deacon, tdeacon}@unitbv.ro

**Abstract.** A swift post-traumatic recovery of upper limbs can be achieved best by means of dedicated rehabilitation equipment. A speedy recovery process ensures the early reintegration of patients into society. The rehabilitation equipment proposed in this paper is conceived for the simultaneous passive mobilization of the radiocarpal, metacarpophalangeal and interphalangeal joints. The paper presents and discusses the construction and actuation system of the equipment. The elements of novelty put forward by this equipment refer to the Fin-Ray® effect underlying the design of the hand support and to its operation by means of a pneumatic muscle – an actuator with inherently compliant behavior. The discussion includes the occurring of hysteresis, and concludes that it does not affect the efficiency of the rehabilitation exercises.

**Keywords:** Rehabilitation equipment · Fin-Ray® effect · Pneumatic muscle

## 1 Introduction

The dysfunctionalities caused by trauma of the extremities and particularly of the upper limbs have a high rate of functional recovery as well as real chances of swift reintegration into everyday work or home activity. The development of efficient therapeutic methods has improved patient recovery periods and reduced the failure rate of the treatment of such pathologies [1]. A significant part of the general approach to ensuring swift patient recovery is the use of rehabilitation equipment for the joints of the upper limb.

The hand is the most complicated osteo-articular segment of the body that has evolved into a complex structure and is able of performing functions perfectly adapted to the wide range of human activities. The hand, wrist included, has functions that include prehension (gripping) and discriminatory sensitivity and is an organ that stands for the individual's personality, expressivity and professionalism. The remainder of the upper limb, including shoulder, arm, elbow and forearm merely fulfil the task of positioning the hand for the performance of certain functions.

While its integrity is paramount to daily functioning, the hand's joint – the wrist – is exposed to the risk of several affections and trauma in particular. Wrist trauma is followed, initially, by immobilization in a cast. It is known that prolonged immobilization causes muscle hypertrophy, a diminished metabolism, bone demineralization and dysfunctions of the circulator respiratory system.


K. Berns and D. Görges (Eds.): RAAD 2019, AISC 980, pp. 393–401, 2020.
https://doi.org/10.1007/978-3-030-19648-6_45

Muscle hypertrophy entails the diminishing of the muscle microfiber diameter, a condition that in severe cases may even lead to muscle atrophy. For the recovery of muscle tone passive kinetic therapy is recommended, or, in certain cases, excitomotor therapy, like in the case of muscle atrophy following denervation. Lack of motion caused by muscle shrinkage can lead to muscle-tendon retractions that affect the system of capsular ligaments. In order to avoid joint blockage, the physical therapist has to mobilize and strain the joints at maximum capacity.

In the case of wrists, the applicability and success of motion-based rehabilitation depend on the location of the fracture (intra- or extra-articular) and on the type of received treatment (surgical or non-surgical). The rehabilitation protocol is aimed at restoring the functionality of the affected area as well as of the entire limb while ensuring pain control. Initially, in order to avoid excessive local straining, light exercising is recommended consisting of compression/elevation and active and passive finger motion [2].

Rehabilitation procedures must address the motor status (muscle tone), the sensorial status (proprioception) and last but not least the patient's psycho-social and occupational status.

One of the applicable procedures for the rehabilitation of the joints of the hand is based on continuous passive motion (CPM). In this case the injured joints are mobilized by means of mechanized devices instead of deploying the patient's muscles. The role of CPM is to prevent the generation of fibrous tissue and to reduce joint stiffness.

The structure of CPM rehabilitation equipment allows for applying the optimum rehabilitation motions to the affected segment. Passive joint mobilization entails controlling the applied forces, the amplitudes of motion, the velocities, accelerations, and the duration and frequency of the rehabilitation exercises. The possibility of modifying between certain limits of all these quantities allows the adjustment of rehabilitation exercises to the clinical state of the patients and to their pain threshold.

The market of rehabilitation equipment for the joints of the hand is abundant and dominated by electrically actuated systems. Equipment worth mentioning in this respect is Kinetec Maestra Hand & Wrist CPM [3], W2 Wrist CPM – provided by QAL Medical [4] or WaveFlex Hand from Remington Medical [5]. All these devices allow the performance of the rehabilitation motions within the limits imposed by the biomechanics of the joints of the hands.

An extremely important requirement to be met by rehabilitation equipment is to be capable of compliant behavior, which is the ability of the system to adapt swiftly to any given concrete situation defined by the state of the patient, possibly different from the one that had been envisaged initially. Electrically actuated equipment like the systems enumerated above meet this requirement to a lesser degree, as a compliant behavior would entail a drastic increase of the sensorization and adequate design of the control diagrams [6].

A solution for avoiding the excessive sensorization of rehabilitation equipment is to deploy adjustable compliance actuators (ACAs). The main quality of such actuators is their safe interaction with human operators şi their ability of storing and releasing energy into passive elastic elements [7]. Such an actuator is the pneumatic muscle that has an inherently compliant behavior due to air compressibility.

At present few pneumatic muscle actuated rehabilitation systems of the joints of the hand are available to patients, as most are still prototypes (patents). Some such achievements should be mentioned, like the Hand Mentor Pro of Motus Nova [8] or the EXOWRIST model described in [9].

Starting from the current state of pneumatic muscle actuated equipment this paper proposes a novel constructive solution of such a system, capable of mobilizing simultaneously the wrist and finger joints. The source for the motions is a pneumatic muscle and the palm support is a new bio-inspired concept based on the Fin-Ray® effect specific to fish fins.

## 2 Design Requirements

The functional requirements such rehabilitation equipment has to meet are underlain by a thorough study of the biomechanics of the joints of the hand. Based on the study the limits of the motions carried out by each component of the wrist were determined, as well as the forces and torques necessary for such motions.

The human hand includes 27 bones grouped into 8 carpal bones, 5 metacarpal bones and 14 phalanges. This entire osteo-articular assembly makes possible a wide range of motions like flexion, extension, circumduction, radial and ulnar deviation. Figure 1 presents the motions of interest that have to be induced by the proposed rehabilitation equipment. These are flexion/extension of the wrist and of the joints of the fingers.

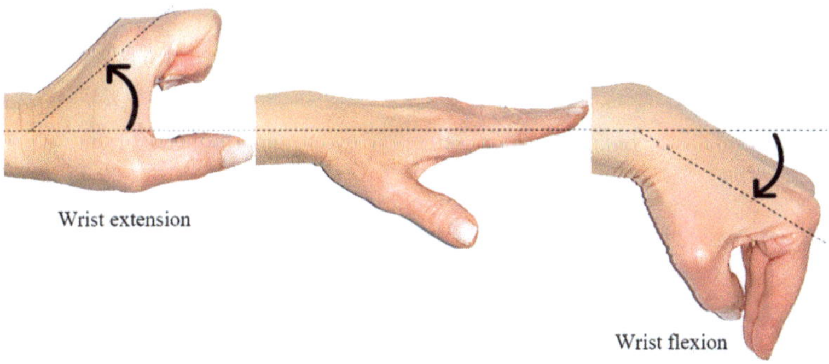


**Fig. 1.** Flexion/extension of wrist and fingers.

Palmar flexion is obtained by rotating the palm in volar direction (downwards), and zero position is reached when the forearm is in flexion at 90° and in pronation. The extension of the fist or dorsiflexion is obtained by rotating the palm in dorsal direction. The zero position of the fingers is considered to correspond to extended fist and fingers. Downward rotation entails conducting flexion motions, while the upward motion entails the extension of the fingers and large rotation angles correspond to hyperextension.

Table 1 features the limits of the motions allowed by these joints, and also the maximum values of the displacement speeds that can be achieved by rehabilitation equipment currently available on the marketplace [10].

**Table 1.** Limits of wrist and finger motions.

| Joint | | Motion | Rotation angle [°] | Rotation speed [°/min] |
|---|---|---|---|---|
| Wrist | | Flexion | 0 to 80 | 30 to 180 |
| | | Extension | 0 to 70 | |
| Joints of the fingers | Metacarpophalangeal | Flexion | 0 to 90 | 55 to 440 |
| | | Extension | 0 to 90 | |
| | Proximal | Flexion | 0 to 120 | |
| | Distal | Flexion | 0 to 90 | |
| | | Extension | 0 to 90 | |

## 3 Constructive Principle of the Rehabilitation Equipment

Figure 1 suggests that a maximum flexion of the fist can be obtained by extending the fingers at the same time, while a maximum extension of the fist is reached when the fingers are flected. Starting from this observation and identifying a certain similarity to the motions of the fins of some fish (stingray), a novel system for the palm support used by patients could be conceived based on the Fin-Ray® effect (Fig. 2).

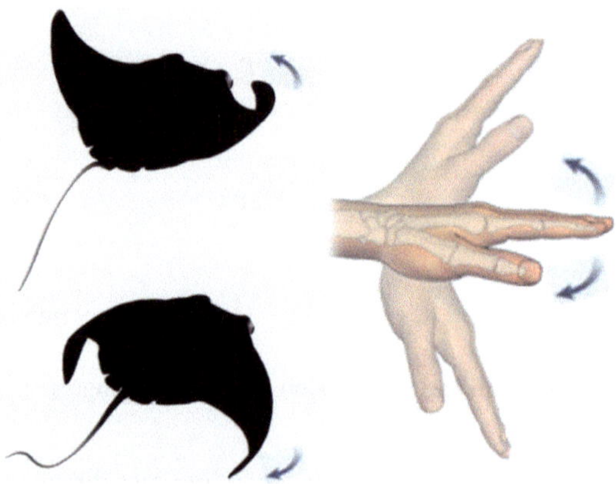

**Fig. 2.** Motions of stingray fins versus motions of the human palm.

The Fin-Ray® effect was a discovered by Leif Kniese in 1997 şi and first used for a practical application by EvoLogics GmbH Berlin [11]. A construction based on the Fin-Ray® effect is obtained by series-connecting several bar mechanisms displayed in a pyramid structure, consisting of two longitudinal fibers placed at a certain angle one to another and linked together by further transversal fibers that keep the entire assembly together. All fibers are inter-connected by flexible ties that allow their relative mobility.

The similarity of the motions of fish fins to those of the hand consists in the fact that both motions are generated by antagonist muscles located on each side of the respective skeleton. However, unlike fish fin motions that are symmetrical in relation to the rest (relaxed) position of the fin, the motions of the hand and fingers are asymmetrical, as the angles of flexion and extension, respectively, have different values.

These observations have led to the design and development of a constructive solution based on the Fin-Ray® effect for supporting the patient's hand on the rehabilitation equipment (Fig. 3) [12].

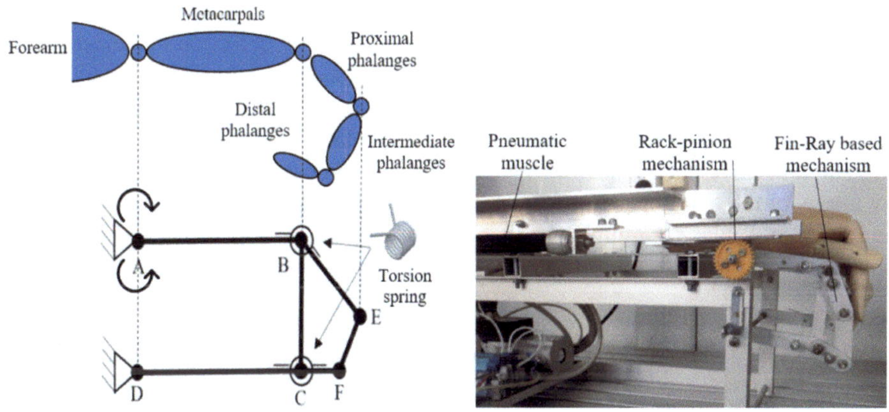


**Fig. 3.** Construction of Fin-Ray® effect based hand support.

The proposed equipment for the mobilization of the hand joints consists of two rigidly parallel-connected Fin-Ray® type mechanisms, placed at an adequate distance such as to accommodate the human palm. Each of these two mechanisms includes two series-linked bar mechanisms, ABCD and BCFE, using joint A as the entry point for the motion of the entire assembly. The two bar mechanisms are linked by joints B and C that include each a torsion spring with a 270° twist angle that have the role of cancelling one of the two degrees of mobility of the entire assembly. Thus, the hand support has a remaining degree of mobility equal to one, the motion of element BE being dependent on the motion of lever AB.

The rotation of joint A is generated by a rack and pinion mechanism. Feeding the pneumatic muscle compressed air at various pressures allows adjusting the displacement of the rack. The pinion in joint A rotates the hand support and thus generates the passive mobilization of the patient's hand. Figure 4 illustrates several sequences of the motions carried out by this mechanism.

The rotation of the radiocarpal joint is ensured by quadrangle ABCD, while quadrangle BCFE is responsible for the mobilization of the finger joints. When lever AB is lifted it causes the extension of the palm, while at the same time the flexion of the fingers is achieved by means of quadrangle BCFE. When lever AB rotates in the

**Fig. 4.** Motions of the hand support.

opposite direction (clockwise) the palm carries out a motion of flexion the fingers being in extension. It is important to point out that this system allows the simultaneous mobilization of the wrist and the finger joints, an exercise not permitted by rehabilitation equipment currently available on the marketplace.

The kinematic analysis of the rehabilitation equipment was conducted graphically by means of the 'Mechanism' module available in Pro/Engineer software. The results, as shown in Table 2, highlight the theoretical limits of the motions reachable by the device.

**Table 2.** Limits of the motions carried out by the rehabilitation equipment.

| | Wrist | | Metacarpophalangeal joints | |
|---|---|---|---|---|
| | Flexion | Extension | Flexion | Extension |
| Indicated in literature | 80° | 70° | 90°–100° | 0°–90° |
| Attained by the equipment | 80° | 70° | 90° | 9,67° |

The obtained values confirm that the rehabilitation equipment ensures the mobilization of the joints over their entire natural mobility range as indicated in literature.

Various angular positions of the hand support are attained by feeding the pneumatic muscle compressed air at different pressures. A DMSP-10-300N pneumatic muscle made by Festo, Germany was selected for the discussed rehabilitation equipment. For a 6 bar compressed air feeding pressure the muscle contracts axially by 60 mm and causes a displacement of the rack of the same magnitude. The minimum force generated in this case is of 35 N, sufficient for the mobilization of the hand. Figure 5 shows the pneumatic actuation diagram of the rehabilitation equipment.

The axial contraction of the pneumatic muscle is proportional to the feed pressure. A higher feed pressure will cause the muscle to contract more, and consequently the mechanism conceived for the mobilization of the joints of the hand will rotate towards its maximum limits. The pneumatic muscle is fed compressed air by means of a proportional pressure regulator that is controlled by either the physical therapist or the

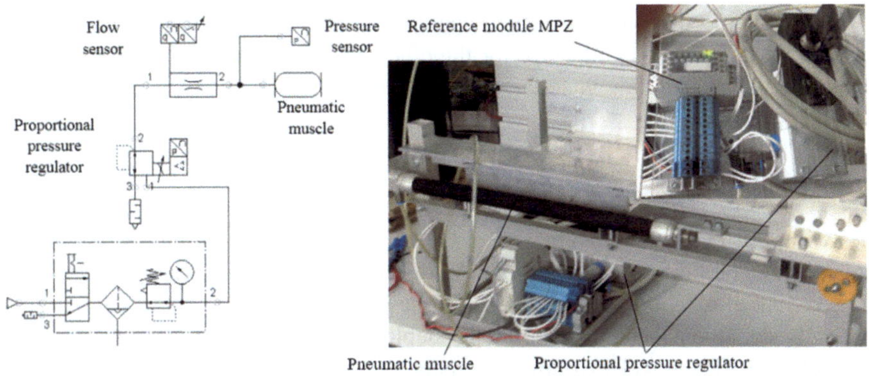


**Fig. 5.** Control of the rehabilitation equipment.

patient herself according to the requirements of the rehabilitation exercise and the respective pain threshold.

A disadvantage related to the utilization of pneumatic muscles is the occurrence of hysteresis. This phenomenon was studied on the proposed equipment by feeding the pneumatic muscle with compressed air at a pressure increasing from 0 to 6 bar, and decreasing from 6 to 0 bar, respectively. The flexion/extension angles of the radio-carpal joint (the wrist angles) for various pressures were measured and plotted into the graph shown in Fig. 6.

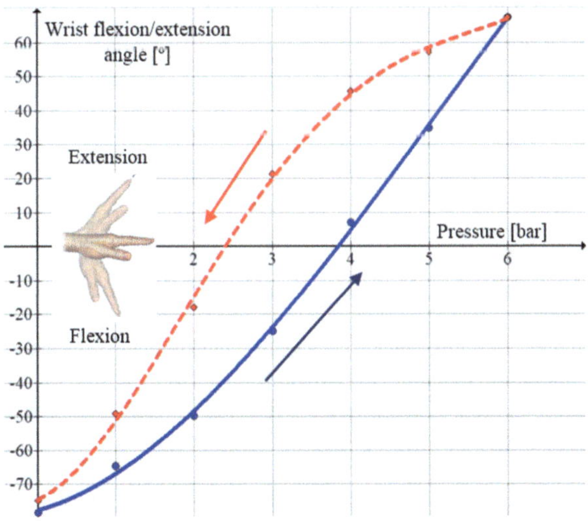


**Fig. 6.** Variation of the flexion/extension angle versus air pressure.

The pronounced hysteresis in the behavior of the rehabilitation equipment holds, however, of no special significance for the efficiency of the rehabilitation exercise, as the accuracy of positioning is not an essential requirement in this type of applications.

## 4 Conclusion

The swift post-traumatic rehabilitation of the upper limbs is a requirement for ensuring patient reintegration into everyday life. Within this context the paper puts forward a novel variant of rehabilitation equipment of the joints of the hand that benefits from an inherently compliant behavior due to the deployment of a pneumatic muscle as the motor element. The novelty of this equipment consists in the proposed solution for the hand support, based on the Fin-Ray® effect.

The paper presents and discusses the current stage of the authors' research related to the rehabilitation of the hand joints, namely construction and actuation of the equipment. The next stage of research envisages extending tests - that so far have been conducted on healthy persons - to clinical tests on informed and consenting patients. The completed experiments have revealed the limits of the motions allowed by the equipment, that cover the motion range of a healthy human hand.

The occurrence of hysteresis is also discussed, a phenomenon that does not affect the expected performance of the equipment.

Compared to devices currently available, the equipment proposed in this paper ensures the simultaneous mobilization of the wrist and the finger joints.

The overall conclusion of the paper is that the proposed rehabilitation equipment represents an innovation compared to existing devices and as such an efficient and desirable alternative to these.

## References


1. Bulbul, M.K.: Study on Carpal Tunnel syndrome incidence, a tardy complication of distal meta- and physeal fractures of the forearm in correlation with their treatment. Ph.D. thesis, Oradea University (2015). (in Romanian)
2. Skirven, T.M.: Rehabilitation of the Hand and Upper Extremity, vol. 1. Elsevier Mosby, Philadelphia (2011)
3. Maestra. Kinetec. http://kinetec.fr/en/kinetec-selection/cpm-continuous-passive-motion/attelle-kinetec-maestra-detail.html. Accessed 15 Aug 2017
4. W2 Wrist CPM. http://qalmedical.com/w2-wrist-cpm-device/. Accessed 10 Sept 2018
5. WaveFlex Hand. http://www.remingtonmedical.com/product/detail/A1. Accessed 20 Sept 2018
6. Deaconescu, T., Deaconescu, A.: Pneumatic muscle-actuated adjustable compliant gripper system for assembly operations. Strojniški vestnik J. Mech. Eng. **63**(4), 225–234 (2017)
7. Van Ham, R., Sugar, T.G., Vanderborght, B., Hollander, K.W., Lefeber, D.: Compliant actuator designs. IEEE Robot. Autom. Mag. **16**, 81–94 (2009)
8. The Hand Mentor Pro. https://motusnova.com/products/hand-mentor-pro/. Accessed 22 Nov 2018

9. Andrikopoulos, G., Nikolakopoulos, G., Manesis, S.: Motion control of a novel robotic wrist exoskeleton via pneumatic muscle actuators. In: 2015 IEEE 20th Conference on Emerging Technologies & Factory Automation (ETFA), pp. 1–8 (2015)
10. Petre, I., Deaconescu, A., Sârbu, F., Deaconescu, T.: Pneumatic muscle actuated wrist rehabilitation equipment based on the fin ray principle. Strojniški vestnik J. Mech. Eng. **64** (6), 383–392 (2018)
11. Kniese, L.: Load carrying element with flexible outer skin. EP1040999A2. Patent (1999)
12. Filip, O., Deaconescu, T.: Mathematical modelling of a Fin Ray type mechanism, used in the case of the wrist rehabilitation equipment. In: 4th International Conference on Computing and Solutions in Manufacturing Engineering, Brasov, vol. 94 (2017). https://doi.org/10.1051/matecconf/20179401006

# A Two-Degree of Freedom Mobile Ankle Rehabilitation Unit (MARU) to Improve and Track Joint Mobility

Samuel Detzel[1(✉)], Alexandra Mercader[1], Christian Dietz[1], Go Nakamura[2], and Tim C. Lueth[1]

[1] Technical University of Munich, 85748 Garching b. München, Germany
samuel.detzel@tum.de

[2] Robot Rehabilitation Center in the Hyogo Institute of Assistive Technology, 1070 Akebonocho, Nishiku, Kobe, Japan

**Abstract.** Rehabilitation for conditions affecting the knee and ankle involve regular mobilization exercises. In order to facilitate the adherence to a rehabilitation plan, we present a mobile and simple to use device consisting of a base and an actuated foot plate. The device can be used for both mobilization and strengthening exercises. Apart from rotation around the ankle, the movement of the foot in the sagittal plane is controlled to imitate walking using a four-bar linkage, potentially expanding the benefits of the exercises to the knee. The trajectory can be fitted to the patient's individual needs by adjusting the linkage dimensions. Using torque sensors, the exerted forces are monitored and the actuation can be adjusted accordingly. The device can be easily controlled by the patient through a smartphone app, which also supplies additional features such as progress monitoring, with a separate interface for the treating physician.

**Keywords:** Rehabilitation · Human-Robot interfaces · Personal robots

## 1 Introduction

### 1.1 The Need for Mobile Rehabilitation

The large number of people suffering from a stroke that subsequently require rehabilitation in the form of mobilization exercises complicates the task for rehabilitation centers, where the capacity is limited because of a shortage of care workers and high costs for health insurances [1]. In addition, joint pain due to exercise activities or aging are common in today's society [2]. Suffering patients need early rehabilitation, which could be complemented by training at home between sessions or after rehabilitation time [1].

### 1.2 State of the Art

The existing mechanisms for ankle rehabilitation belong to two different categories: wearable [3–5] and non-wearable devices [6–10]. Another differentiating factor is the number of degrees of freedom (DOF). In [8], six motors are built into a platform to actuate a parallel four-bar mechanism. The idea behind this is to allow the ankle to


K. Berns and D. Görges (Eds.): RAAD 2019, AISC 980, pp. 402–409, 2020.
https://doi.org/10.1007/978-3-030-19648-6_46

perform its six possible movements in the coronal, sagittal and horizontal axis. In [9], a two DOF mechanism is presented, performing inversion/eversion and flexion/dorsiflexion. The rehabilitation of these last two movements influence walking and other basic tasks of the ankle. Wearable devices are mostly one DOF to help patients climb staircases [5] or improve post-stroke mobility [4]. In [3], a four DOF mechanism with artificial muscles was developed to mimic natural ankle movements, however, some post-stroke patients do not have the necessary mobility to wear these devices.

### 1.3 Project Definition

In the rehabilitation process, wearable devices are mostly used to improve the gait of the patient and to guide the ankle in a walking-like motion. These devices are limited to patients who are already able to stand up [4]. For patients who have to stay in a sitting position, the known platforms rotate in various orientations which only stimulates the ankle joint. For most patients, the ankle joint is not the only cause of pain [11].

The mechanism presented in this article has two DOF, limiting the number of actuators and thus facilitating the transportation and use of the system at home. In addition, this system supports inversion and eversion movements as well as plantar flexion and dorsiflexion. The choice of a four-bar mechanism adds two additional features, which are up and down as well as forward and backward movements. Comparable devices do not carry out these movements, which are similar to walking [12]. In Fig. 1, the various available systems are classified into two categories, with some common features. In order to combine the benefits of both categories, the suggested platform provides a framework for early rehabilitation while imitating the walking movement.

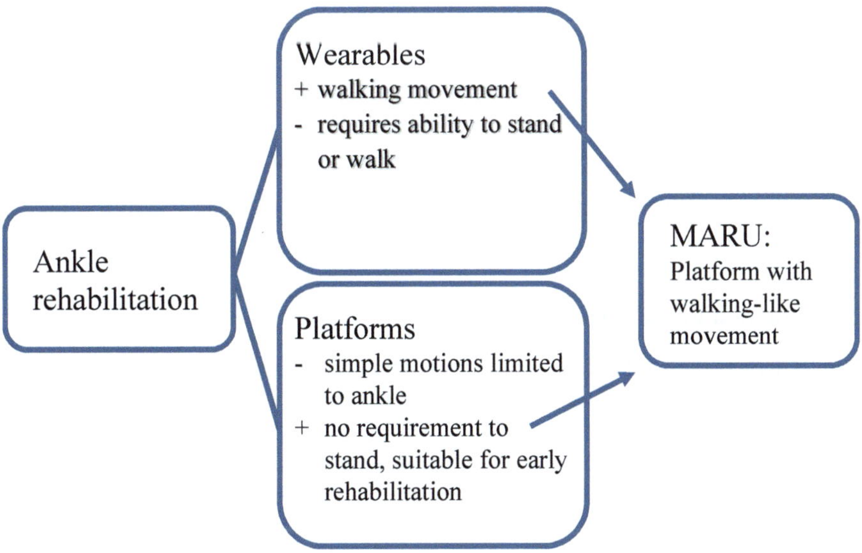


**Fig. 1.** Ankle rehabilitation systems with their advantages and limitations

## 2 System Description

### 2.1 Structural Design

The central feature of the mobile ankle rehabilitation unit (MARU) is a footplate, on which the patient places their affected foot. With the help of straps, the foot is fixed to the plate. Using series elastic actuators (SEA), the movement of the foot plate is controlled with two independent DOF. The first controls the inversion/eversion of the foot, i.e. the rotation around the toe-heel axis. A second DOF is used in combination with a four-bar linkage to control all movement in the sagittal plane. These movements (lifting/lowering, moving front/back and rotating the foot around the ankle to lift/lower the toes) are prescribed by the linkage in a way that resembles the natural gait pattern, see Fig. 2.

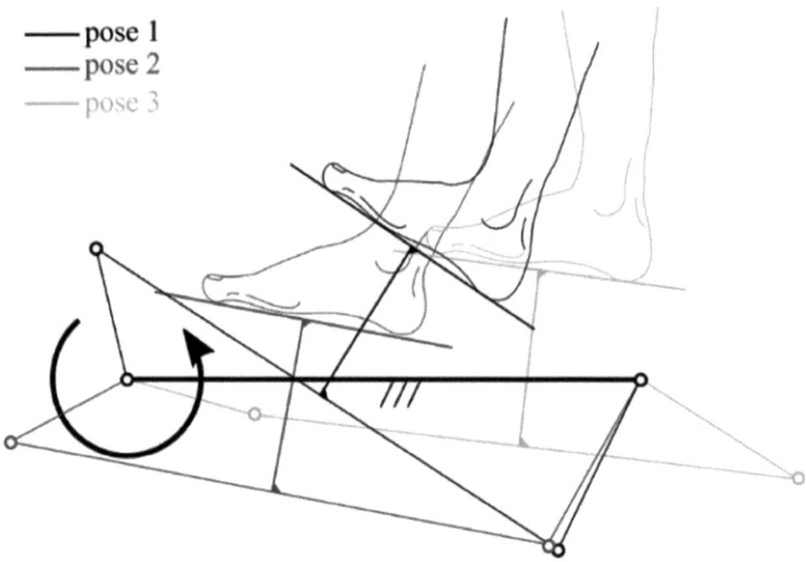


**Fig. 2.** Kinematic sketch with three exemplary poses of the foot and linkage. Pose 1: lowering foot; pose 2: moving foot backwards; pose 3: lifting foot

### 2.2 Actuation and Control

Figure 3 depicts the hardware setup of the MARU system consisting of actuators (1), a central control platform (2), a tablet computer running a web application used by the physician to access historical training data (3) as well as a smartphone for the patient to control the current training session (4). In our prototype, a notebook computer was used as the control platform. However, any platform capable of running Matlab including small single board computers could be used. This makes it possible to integrate the control platform into the MARU base box. Three Hebi X8-9 [13] modules are used as actuators and sensors: There is one module to control eversion and two modules to move the four-bar linkage. The modules are able to sense the motor position as well as the torque applied to the motors. Cat5 Ethernet cables are used to connect the actuators to the control platform. As each HEBI Module incorporates a three-port Ethernet switch, a line topology can be used. There are two user interfaces: A web application for physicians designed to run on a tablet or desktop computer and a smartphone application for patients. The web application for physicians allows to access and display historical training data such as how many rounds have been completed each day. The smartphone application is used by the patient to start/stop the current training session and to adjust training parameters such as motor speed. Both user interfaces are connected to the control platform via Wi-Fi.

**Fig. 3.** MARU hardware

The software architecture of the MARU system is shown in Fig. 4. Flow of information is depicted by arrows. Solid arrows represent direct function calls and dotted arrows represent information passed in a serialized manner via network protocols. Note that only the components marked by (*) were implemented by us. The HEBI Matlab library (1) to control the HEBI modules was provided by the manufacturer of the actuators. The central part of the MARU control platform is a Matlab instance running a control loop (2): It interfaces with the vendor-specific HEBI Matlab library (1) to query sensor feedback from the actuators and to send control data back to the actuators. The vendor-provided firmware of the internal electronics of each actuator implements a PID-controller which ensures that target positions and velocities are reached.

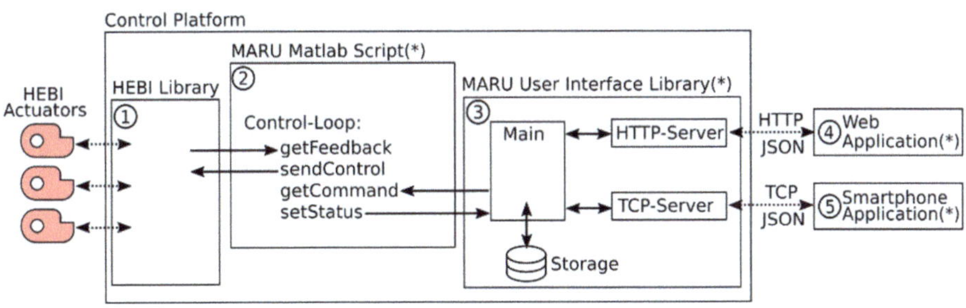


**Fig. 4.** MARU software architecture

The control loop also interfaces with a so-called MARU User Interface Library (3) written in Java to allow user interaction. It queries the User Interface Library for new user commands to be executed (e.g. 'start training' or 'stop training') and reports the current status back to the user interface. This status carries information such as the current training speed, the current training mode (active, passive) or the number of rounds that have been completed.

The User Interface Library (3) provides a persistent storage method to save historical training data. When the User Interface Library is initialized, it spawns a HTTP-Server for the physician's interface as well as a TCP-Server for the patient's interface. The HTTP-Server serves the web-application for the physician written in HTML and JavaScript. Once the main HTML webpage is served to the web browser, the embedded JavaScript code requests new data via periodic HTTP requests from the HTTP server. The server replies with up-to-date data encoded as JSON which is then rendered in the web browser. The smartphone application directly connects to the TCP-Server. Data exchanged between the smartphone application and the User Interface Library's TCP-Server is also encoded as JSON.

The motors are coordinated by a Matlab script (2) that controls two modes: the passive and the active mode. The passive mode is relevant in a first step for early rehabilitation. The motors run at a prescribed speed, allowing the patient to follow the movement of the platform without effort. The four-bar mechanism creates a continuous periodic movement. The foot platform controlled by a third motor performs a constant sinusoidal movement. The minimum and maximum angles of the platform may be adjusted, which ensures effective control of physical rehabilitation. The active mode, in a second step, is suitable for patients who wish to increase the difficulty and pursue their rehabilitation. This mode activates the muscles by only moving when the force applied to the motors is greater than a threshold value. This value can be adjusted as well, according to the desired difficulty of the training.

### 2.3 Automated Design Process

In order to optimally fit the four-bar linkage to the patient's needs, the desired movement is provided by the physician or automatically suggested among the available basic movements, which have been designed taking into account the range of motion that a seated person can achieve and the angle of the foot that is necessary for the mechanism to work properly in the active mode. Once a set of link lengths has been chosen, the functionality of the resulting four-bar linkage is checked via simulation. If the simulation succeeds, the manufacturing files for the resulting mechanism are automatically generated by the software to allow quick production with a laser cutter, see Fig. 5.

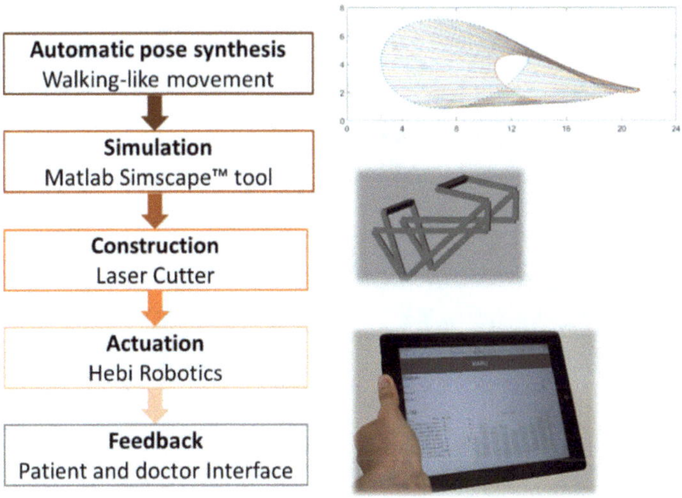


**Fig. 5.** Design and usage process, from patient-individual design of the mechanism to the analysis of recorded training data.

## 2.4 Realization

A working prototype was manufactured and assembled, see Fig. 6. In order to minimize the danger of pinching, the linkages (1) are placed within a base box (2), thereby protecting the user. The base box, linkages and supporting footplate (3) were manufactured from acrylic glass cut into shape using a laser cutter and glued together using an acrylic glue. Linkages of different length can therefore be manufactured in a short time frame. This offers the possibility to adjust the kinematic path individually for each patient. The upper part of the footplate consists of an array of spring-supported square elements manufactured using selective laser sintering of PA2200 [14]. The stiffness of the springs as well as the height of the individual elements were modeled to ergonomically fit a human foot and can be automatically generated to fit an individual patient. Due to constraints in space and to reduce the complexity of the prototype, plain bearings were used in the joints of the linkage. The three series-elastic actuators (4) control the two DOF. With a peak torque of 20 Nm [13] each, there is no need for a transmission and the actuators directly drive the linkage and foot plate. For reasons of stability and to prevent torsion within the linkage, two actuators are used to control the single DOF of the four-bar linkage. They are programmatically linked so that they always mirror each other's movement. The actuators are equipped with torque sensors and can therefore deliver real-time measurement of the patient's effort, a prerequisite for the different control strategies and training modes.

To begin the training cycle, the patient sits on a chair, with MARU on the floor in front of them. The foot that should be treated is placed on the footplate and secured with the Velcro straps. To start the training, the patient takes their personal smartphone, opens the MARU application and selects the desired training mode. The smartphone

**Fig. 6.** Prototype of the mobile ankle rehabilitation unit. (1) four-bar linkage, (2) base box, (3) foot plate, (4) actuators

application can also be used to adjust the training speed. While the training is performed, the number of completed rounds is counted and displayed to the patient.

The patient can stop training by clicking a button in the MARU smartphone application. At any time, the physician can connect to the system via Wi-Fi and display the history of progression of the training via a web application either on a tablet or on a computer. A wireless control concept via smartphones and tablets was chosen because the training device is placed on the floor and buttons on the training device itself would not be easily accessible to the patient.

## 3 Future Work

While the prototype is functional and preliminary tests gave promising results, additional work needs to be done. This includes evaluating the benefit to the rehabilitation process as well as patient's and physician's acceptance.

There is also room for improvement on the technical side: The prototype presented in this work used a notebook computer to run the control software. In the future, this software should be run on a small single-board-computer that can be integrated into the MARU system.

## References


1. Mokrusch, T.: Frühreha von Schlaganfallpatienten. Gesetzesreform verschlechtert Therapiechancen. f&w - Reha-Rubrik 8/2018 (8), 752–753 (2018)
2. Robert-Koch-Institut: Journal of Health Monitoring. Prävalenz von Gelenkschmerzen in Deutschland. RKI-Bib1 (Robert Koch-Institut) (2017)
3. Park, Y.-L., Chen, B.-r., Pérez-Arancibia, N.O., Young, D., Stirling, L., Wood, R.J., Goldfield, E.C., Nagpal, R.: Design and control of a bio-inspired soft wearable robotic device for ankle-foot rehabilitation. Bioinspir. Biomim. **9**(1), 016007 (2014)
4. Takahashi, K.Z., Lewek, M.D., Sawicki, G.S.: A neuromechanics-based powered ankle exoskeleton to assist walking post-stroke: a feasibility study. J. NeuroEng. Rehabil. **12**(1), 1–13 (2015)
5. Yeung, L.-F., Ockenfeld, C., Pang, M.-K., Wai, H.-W., Soo, O.-Y., Li, S.-W., Tong, K.-Y.: Design of an exoskeleton ankle robot for robot-assisted gait training of stroke patients. In: IEEE International Conference on Rehabilitation Robotics, pp. 211–215 (2017)
6. Cioi, D., Kale, A., Burdea, G., Engsberg, J., Janes, W., Ross, S.: Ankle control and strength training for children with cerebral palsy using the Rutgers Ankle CP: a case study. In: IEEE International Conference on Rehabilitation Robotics (2011)
7. Hau, C.T., Gouwanda, D., Gopalai, A.A., Yee, L.C., Akhtar Binti Hanapiah, F.: Design and development of platform ankle rehabilitation robot with Shape Memory Alloy based actuator. In: Annual International Conference of the IEEE Engineering in Medicine and Biology Society, pp. 946–949 (2017)
8. Li, W., Huang, J., Wang, C., Duan, L., Liu, Q., Yang, S., Shang, W., Shen, Y., Lin, Z., Lu, Z., Chen, X., Xia, J., Wu, Z.: Design of 6-DOF parallel ankle rehabilitation robot. In: 2018 IEEE International Conference on Cyborg and Bionic Systems (CBS), pp. 532–536 (2018)
9. Racu, C.: Preliminary ideas on the development of a new ankle rehabilitation device. In: The 14th IFToMM World Congress (2015)
10. Wang, C., Wang, L., Qin, J.: Development of an ankle rehabilitation robot for ankle training. In: In conjunction with 2015 IEEE International Conference on Automation and Logistics. IEEE International Conference on Information and Automation (ICIA), The Old Town of Lijiang, Yunnan, China, 8–10 August 2015 (2015)
11. Kurth, B.-M.: Gesundheit in Deutschland. Gesundheitsberichterstattung des Bundes: gemeinsam getragen von RKI und DESTATIS. Beiträge zur Gesundheitsberichterstattung des Bundes. Robert Koch-Institut, Berlin (2015)
12. Ngamsuriyaroj, S., Chira-Adisai, W., Somnuk, S., Leksunthorn, C., Saiphim, K.: Walking gait measurement and analysis via knee angle movement and foot plantar pressures. In: 2018 15th International Joint Conference on Computer Science and Software Engineering (JCSSE), pp. 1–6 (2018)
13. HEBI Robotics: X-SERIES Actuator ®. Technical Specifications. www.hebirobotics.com. Accessed 31 Jan 2019
14. EOS GmbH: PA2200 Performance 1.0. PA12 (2018). www.materialdatacenter.com. Accessed 31 Jan 2019

# Model Based Analysis of Trunk Exoskeleton for Human Efforts Reduction

Elisa Panero[1(✉)], Giovanni Gerardo Muscolo[1], Stefano Pastorelli[1], and Laura Gastaldi[2]

[1] DIMEAS-Department of Mechanical and Aerospace Engineering, Politecnico di Torino, Turin, Italy
elisa.panero@polito.it

[2] DISMA-Department of Mathematical Sciences, Politecnico di Torino, Turin, Italy

**Abstract.** Recent studies highlighted the importance of assisting workers for human efforts reduction in manual handling and lifting tasks by using wearable exoskeletons. In this paper, several configurations of a trunk exoskeleton in terms of hinge joint positions are investigated with the attempt to identify the best ones for human efforts reduction. Both human joints loads and interface forces are considered and compared through simulations. The proposed computational approach may be the starting point for the analysis of design and development of effective human assistance devices.

**Keywords:** Trunk exoskeleton · Industrial handling and lifting strategies · Human efforts reduction · Wearable robotics

## 1 Introduction

Work Related Musculoskeletal Disorders (WMSDs) represent a fundamental health problem for work population [1]. The human spine results one of the most affected body parts, because entailed in numerous working tasks. Recent studies highlighted that more than 50% of workers suffer from low back pain, most of them are involved in industrial occupations [2]. All these considerations stress the importance of assisting human workload activity in industry. Safety and health organizations such as OSHA and NIOSH have provided detailed guidelines for WMSDs prevention [3, 4]. At the same time, several researchers have been working on risk-assessment methods consisting in self-report analysis, expert observation and direct measurement of risks [5].

Exoskeletons are wearable devices that can assist users to replace muscle functionality or to offer normal/amplified human capability. Exoskeletons seem to be a promising solution to reduce body physical efforts for workers, by providing support and assistance during selected tasks. Recently, back support devices had been developed, both passive as Laevo and PLAD [6], and powered, as the Hyundai H-WEX and Cyberdyne HAL [7]. Evaluation of industrial exoskeletons functionality was assessed with experimental functional tests [8, 9], EMG signals [10] and qualitative questionnaires [8, 11]. However, numerous challenges are still open.


K. Berns and D. Görges (Eds.): RAAD 2019, AISC 980, pp. 410–418, 2020.
https://doi.org/10.1007/978-3-030-19648-6_47

Most of the studies concerning the development of an industrial trunk-support exoskeleton rely on simple two-dimensional models of human body [12]. These models consider the trunk as one rigid segment (neglecting the spine degrees of freedom) articulated to the lower limbs by the hip joints. Only the recent approach proposed by Millard [13] investigates the contribution of two joints (waist and hip) in trunk flexion-extension. Moreover, all the proposed solutions considered the alignment of exoskeleton joints with the human hips, without inspecting other possible device configurations.

The main object of this work is to investigate different exoskeleton joint positions in order to evaluate their influence on the requested human efforts during manual handling and lifting tasks. The simulation approach through a 3D multibody model allows considering different lifting strategies and exoskeleton joints positions. Through the evaluation of the net moments at human joints and the interface forces exchanged with the device, it is possible to identify joints positions that allow the reduction of required human efforts. At the same time, the analysis stresses how some exoskeleton configurations might negatively affect the human body and the importance of the correct development of the interaction surfaces.

# 2 3D Multibody Model

## 2.1 Human Model

A 3D human model described by 15 rigid links, 14 joints and a total of 34 degrees of freedom was implemented in Matlab-Simscape software. The joints range of motion and local reference systems are defined according to ISB standards [14]. The model reproduces a 50 percentile Italian man (mass 75 kg, height 1.71 m), inertial and mass parameters are defined according to [15]. Standing posture is assumed as neutral reference position, while positive angles refer to flexion, negative to extension.

In biomedical applications, numerous musculoskeletal models and biomechanical software are available, with the attempt to represent the complexity of human body and to understand the different joints and muscles contributions during motion. Nevertheless, the majority of previous human biomechanical models developed for estimating loads at lower back during flexion-extension with assisting device [12] depicts the human trunk as one single rigid segment. This solution considers trunk and legs connected by hip joints and it simplifies the analysis, but it neglects the difference between trunk flexion and hip flexion. In the present model, the human trunk is not considered as a single segment, but it is divided in two different links: pelvis and torso. A three rotational DOFs joint, named waist joint, connects the torso and pelvis links. This joint is located in correspondence of the last lumbar vertebra, which results the most loaded point of human spine during flexion-extension and, consequently, the most affected by low back pain. Moreover, the introduction of the waist joint allows separating the trunk and hip flexion lifting strategies and analyzing the different joints contribution. Figure 1A depicts a graphical representation of human trunk separation in torso and pelvis segments, with highlighted position of waist and hip joints. The present 3D model has been previously validated [16] comparing the waist torque obtained from simulated results with the experimental data from literature, both during stoop (legs extended) and squat (knees flexed). Different time intervals were considered, to depict human motion at several velocities. Validation results pointed out a good reliability of the model.

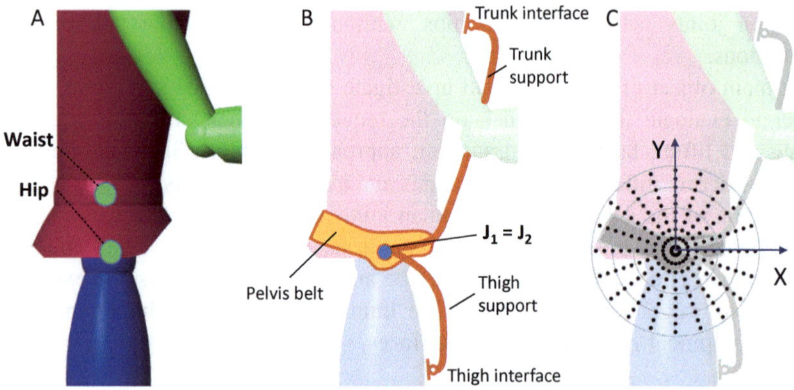


**Fig. 1.** (A) Human body model, separating waist and hip joints. (B) Human body model wearing the exoskeleton in the nominal configuration. (C) Mapping of exoskeleton joints configurations.

### 2.2 Exoskeleton Model

An exoskeleton structure with a total mass of 2.5 kg is considered. It consists of four rigid elements as reported in Fig. 1B: trunk support, pelvis belt and two thigh supports. Left and right pairs of joints connect the rigid bodies: hinge joints $J_1$ connect pelvis belt and thigh supports, while hinge joints $J_2$ connect each thigh support with the trunk support. The nominal configuration of exoskeleton considers joints $J_1$ and $J_2$ coaxial with human hip joint (Fig. 1B). The human-device interfaces are modeled as kinematic connections at trunk and thighs contact points, while the pelvis belt is fixed to human pelvis. Contacts allow translation along vertical direction and rotation in the sagittal plane, so the forces transmitted by the exoskeleton are always perpendicular to the human body. The assistance of the exoskeleton is modeled as torques applied in joints $J_2$. In particular, in this specific analysis, the torques are assumed constant and equal to 20 Nm for each side, for a total assistance of 40 Nm. The value represent the maximal assistive torque provided by exoskeleton prototypes [17] and it was assumed constant in order to avoid possible effects between simulations due to assistance.

## 3 Simulation

### 3.1 Human Joint Kinematics in Lifting Movements

Starting from the reference standing posture, three main phases can be highlighted for stoop and squat motion strategies. The first step is the flexion phase, where the human body flexes in order to reach the object. The second step is the grasping phase, where the human body maintains the flexed posture and lifts an external load, which in the present study is modeled as a 15 kg external mass. Finally, during the extension phase, the human body returns at the starting point, lifting the external load. The movement takes place only in the sagittal plane, being symmetric between left and right side [16].

With the attempt to simulate the two lifting strategies, different kinematics of human joints must be described. Starting from the reference standing posture, Table 1 shows the maximum angle for waist and lower limb joints, as registered in experimental tests and reported in [18]. Positive values refer to flexion angles, while negative to extension ones. During stoop motion, the human body flexes the trunk with extended legs. The squat motion requires lower flexion at the waist joint, but involves the legs joints, with large flexion both at hip and knee joints. Concerning the upper body, the kinematic inputs is the same for both motions (from 0° to 80° flexed shoulders during grasping phase and from 0° to 60° flexed elbows during extension phase).

**Table 1.** Maximum kinematic values of human joints during stoop and squat. Positive values refer to flexion angles, negative to extension angles between segments.

| Human joint kinematics | | | | | |
|---|---|---|---|---|---|
| **Stoop Motion** | Waist | +100° | **Squat Motion** | Waist | +55° |
|  | Hip | −5° |  | Hip | +80° |
| | Knee | 0° | | Knee | +110° |
| | Ankle | −10° | | Ankle | +40° |

### 3.2 Exoskeleton Joints Configuration

Considering the model approach, most of previous studies positioned exoskeleton joints aligned with the human hip joints. In addition, developed and commercial devices indicate to align the exoskeleton assisting joints with the human hip joints during wearing. In this study, the coaxial configuration of exoskeleton joints and human hip joint is the nominal starting point (Fig. 1B). In order to investigate the effects of the position of the exoskeleton hinges on human loads, the effect of their position in the sagittal plane was analyzed. In a previous study [16], only the hinge joint $J_2$ was moved inside a circle space (0.2 m radius) obtaining 100 random configurations, while the hinge $J_1$ axis was maintained aligned with the hip joint. Moving only the assistance hinge exoskeleton may require more complex mechanical structure arrangements. The introduction of transmission elements might be necessary to connect exoskeleton parts, causing an increasing of the device weight and space encumbrance. For all these reasons, in the present study both $J_1$ and $J_2$ are shifted, maintaining their coaxial relation.

A parameter set of 200 position configurations is defined. The translation is circumscribed in a circle (radius 0.2 m) around the starting position. Figure 1C depicts the map of exoskeleton joints translations. Considering the sample set configurations, the simulation starts from the point in the first circle near the nominal configuration (X = 0, Y = 0), with coordinates: X = 0.01 m, Y = 0 m and continues in the same circle with a counterclockwise motion. Twenty points are in the same circle. The 21st point has the same y coordinate as the 1st point but translated in the second circle (X = 0.03, Y = 0), and so on until the last external circle (X = 0.19, Y = 0). This strategy, compared to previous random samples definition [16], allow mapping the circle space with a specific order and homogenous distribution.

### 3.3 Simulation Description

The 3D model was dynamically simulated with a total time interval of 6 s, two seconds for each motion phase (flexion, grasping, and extension). Kinematics of human body during stoop and squat (Table 1) describes the inputs. The exoskeleton joint position was selected as the varying element between simulations. The exoskeleton contribution has been maintained constant through and between simulations in order to highlight the possible effects only due to exoskeleton joint position. For results comparison, also human body without the exoskeleton and with the exoskeleton in the nominal configuration (exoskeleton joints coaxial with human hip joint) were considered. The dynamic results as the net joints moments at waist and hip joints and the interface forces at trunk and thigh contact points identified the outputs of interest. Force impulses (Ns) at interface points (trunk and thigh) and torque impulses (Nms) at human joints (waist and hip) have been considered as an overall performance index for comparison among configurations. Impulses are calculated as the integral of forces and torques over the simulation interval (6 s).

## 4 Results and Discussion

Figure 2A depicts the total impulses of the output variables related to exoskeleton joints Y positions during stoop. Results point out several aspects. First, different configurations of joints affect interface forces and human joint torques, both in stoop and squat strategies. The configurations with a large translation from the origin point seems to produce higher impulses, especially for the interface forces (red and green dots). This problem can be partially reduced with a proper design of point of contacts, considering a larger area and distribution of forces. As alternative, the developer can identify a force threshold not to be exceed. Considering the physiological pressure limits of 32 mmHg (4.5 kPa) for skin capillarity flow [19], hypothesizing a contact area of 250 $cm^2$ and taking into account the rigid connection without modelling soft tissue dissipation of force, all the configurations in stoop or squat exceeding the threshold of 250 N were discarded. Figure 2B stresses the discarded critical conditions (purple dots).

Another important analysis consists in the evaluation of human joint loads. Figure 3 show net moments results at waist human joints during stoop (A) and squat (B). Positive values depict flexion net joint moments, while negative the extension ones. The grey area depict the range of results obtained with different exoskeleton hinges positions. The graphical representations highlight one important aspect: the nominal exoskeleton joint position aligned with human hip (black dash-dotted line)

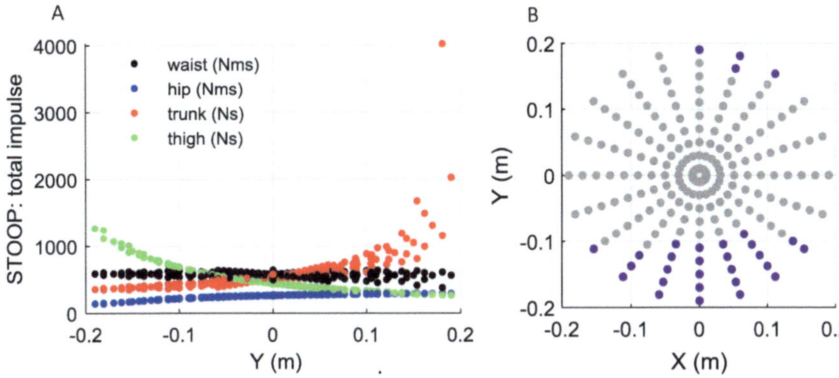


**Fig. 2.** (A) Force impulses at interfaces (trunk/thigh) and torque impulses at human joints (waist/hip) during stoop, related to the Y translation. (B) Exoskeleton joints translations with highlighted critical conditions (purple dots).

cannot be considered as the best configuration for reducing waist loads. Moreover, all the configurations allow reducing the human waist loads compared to results obtained without the exoskeleton (red line). Analog considerations can be pointed out for the hip loads. Both in stoop and squat, all the exoskeleton configurations entail a negative torque during the first flexion angles due to the constant contribution of the support. It may cause not only perceived discomfort by the worker, but also an increase of human muscle activity to drive the exoskeleton during the flexion phase. The drawback can be overcome with the development of a suitable powered device with no assistance in the previous flexion angles. This aspect reveals the necessity of identifying an appropriate law of exoskeleton assistance during the exoskeleton design and characterization. After the identification of a proper exoskeleton joints position, it might be the next step of the current research.

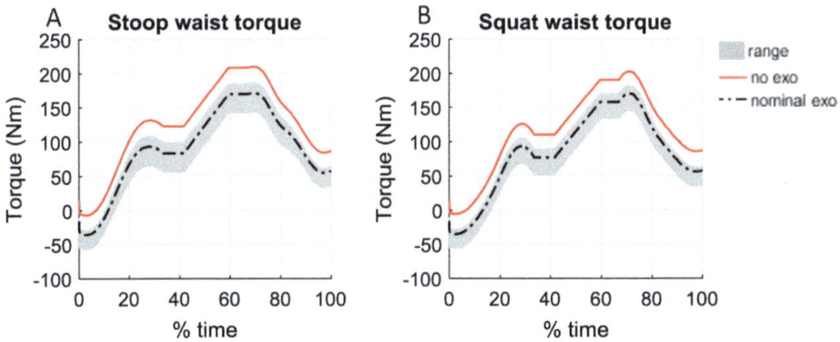


**Fig. 3.** Net joint moments at waist human joints in stoop (A) and squat (B) lifting motion.

Figure 4 shows impulses of waist and hip torques (Nms) and trunk and thigh interface forces (Ns) both in stoop (A) and squat (B) after the elimination of critical conditions. The developer can define the position of exoskeleton joints choosing the desired effort reduction and, at the same time, limiting the interface forces. During squat, due to the flexion-extension of knees and ankles, impulses at the thigh interface are higher than in stoop. Based on motions, the human body is differently affected by the device. It is fundamental to identify and correctly characterize the industrial task of interest.

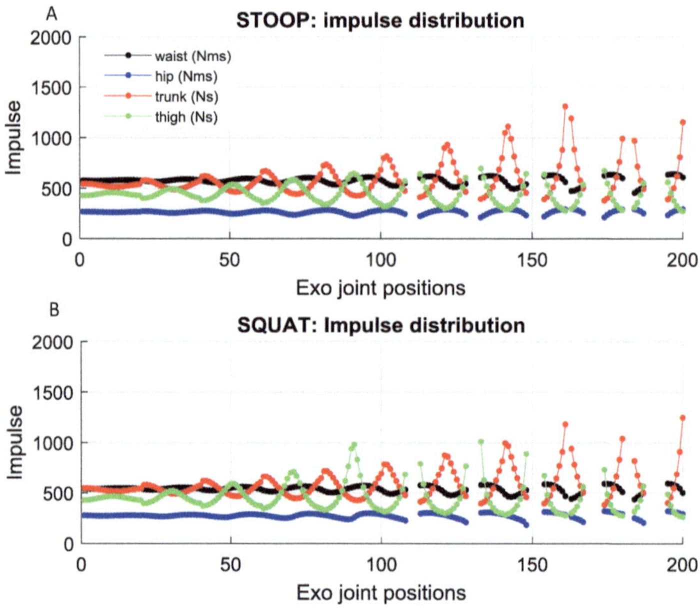


**Fig. 4.** Impulses results of waist and hip torques, and trunk and thigh interface forces through simulations, both in stoop (A) and squat (B) motion.

## 5 Conclusion

Human efforts during stoop and squat lifting strategies may be reduced if a wearable exoskeleton is appropriately designed. The presented analysis underlines the importance to have a detailed human body model for helping in human wearable systems characterization. The output of this research is useful for designing an exoskeleton, which helps to reduce human efforts. The proposed method might be an important starting point for the investigation of different mechanical design and assistance of the device as for the influence estimation on the body loads during specific tasks. Comparing different arrangements, it might possible to select the most suitable structure in terms of human efforts reduction, wearability and limited discomfort. The computational approach reveals to be a promising strategy for analysis of human-exoskeleton interaction.

**Acknowledgement.** The present research has been partially supported by MIUR grant Dipartimenti di Eccellenza 2018–2022 (E11G18000350001).

# References

1. Silva, C., Barros, C., Cunha, L., Carnide, F., Santos, M.: Prevalence of back pain problems in relation to occupational group. Int. J. Ind. Ergon. **52**, 52–58 (2016). https://doi.org/10.1016/J.ERGON.2015.08.005
2. De Looze, M.P., Bosch, T., Krause, F., Stadler, K.S., O'Sullivan, L.W.: Exoskeletons for industrial application and their potential effects on physical work load. Ergonomics **59**, 671–681 (2016). https://doi.org/10.1080/00140139.2015.1081988
3. Mirka, G.A.: Development of an ergonomics guideline for the furniture manufacturing industry. Appl. Ergon. **36**, 241–247 (2005). https://doi.org/10.1016/J.APERGO.2004.10.003
4. Bernard, B.P., Putz-Anderson, V.: Musculoskeletal disorders and workplace factors; a critical review of epidemiologic evidence for work-related musculoskeletal disorders of the neck, upper extremity, and low back (1997)
5. Wang, D., Dai, F., Ning, X.: Risk assessment of work-related musculoskeletal disorders in construction: state-of-the-art review. J. Constr. Eng. Manag. **141**, 04015008 (2015). https://doi.org/10.1061/(ASCE)CO.1943-7862.0000979
6. http://en.laevo.nl/. Accessed 21 Nov 2018
7. Ko, H.K., Lee, S.W., Koo, D.H., Lee, I., Hyun, D.J.: Waist-assistive exoskeleton powered by a singular actuation mechanism for prevention of back-injury. Rob. Auton. Syst. **107**, 1–9 (2018). https://doi.org/10.1016/j.robot.2018.05.008
8. Spada, S., Ghibaudo, L., Gilotta, S., Gastaldi, L., Cavatorta, M.P.: Analysis of Exoskeleton introduction in industrial reality: main issues and EAWS risk assessment. In: International Conference on Applied Human Factors and Ergonomics, pp. 236–244. Springer, Cham (2018)
9. Spada, S., Ghibaudo, L., Carnazzo, C., Di Pardo, M., Chander, D.S., Gastaldi, L., Cavatorta, M.P.: Physical and virtual assessment of a passive exoskeleton. Adv. Intell. Syst. Comput. 247–257 (2019). https://doi.org/10.1007/978-3-319-96068-5_28
10. Bosch, T., van Eck, J., Knitel, K., de Looze, M.: The effects of a passive exoskeleton on muscle activity, discomfort and endurance time in forward bending work. Appl. Ergon. **54**, 212–217 (2016). https://doi.org/10.1016/J.APERGO.2015.12.003
11. Spada, S., Ghibaudo, L., Carnazzo, C., Gastaldi, L., Cavatorta, M.P.: Passive upper limb exoskeletons: an experimental campaign with workers. Adv. Intell. Syst. Comput. 230–239 (2019). https://doi.org/10.1007/978-3-319-96068-5_26
12. Toxiri, S., et al.: A wearable device for reducing spinal loads during lifting tasks: biomechanics and design concepts. In: 2015 IEEE International Conference on Robotics and Biomimetics, IEEE-ROBIO 2015, pp. 2295–2300 (2015). https://doi.org/10.1109/robio.2015.7419116
13. Millard, M., Sreenivasa, M., Mombaur, K.: Predicting the motions and forces of wearable robotic systems using optimal control. In: Frontiers in Robotics and AI, pp. 1–12 (2017)
14. Wu, G., et al.: ISB recommendation on definitions of joint coordinate system of various joints for the reporting of human joint motion—part I: ankle, hip, and spine. J. Biomech. **35**, 543–548 (2002). https://doi.org/10.1016/S0021-9290(01)00222-6
15. de Leva, P.: Adjustments to Zatsiorsky-Seluyanov's segment inertia parameters. J. Biomech. **29**, 1223–1230 (1996). https://doi.org/10.1016/0021-9290(95)00178-6

16. Panero, E., Muscolo, G.G., Pastorelli, S., Gastaldi, L.: Influence of hinge positioning on human joint torque in industrial trunk exoskeleton. In: 15h IFToMM World Congress (2019)
17. Toxiri, S., et al.: Towards standard specifications for back-support exoskeletons. In: Biosystems and Biorobotics (2019)
18. Hwang, S., Kim, Y., Kim, Y.: Lower extremity joint kinetics and lumbar curvature during squat and stoop lifting. BMC Musculoskelet. Disord. **10**, 15 (2009). https://doi.org/10.1186/1471-2474-10-15
19. Sposito, M., et al.: Towards design guidelines for physical interfaces on industrial exoskeletons: overview on evaluation metrics. In: International Symposium on Wearable Robotics, pp. 170–174. Springer, Cham (2019)

# Safety Related Devices and Applications

# Validation of an Ergonomic Contactless Master Device for a Teleoperation Platform

Juan Sandoval(✉), Med Amine Laribi, and Saïd Zeghloul

Department of GMSC, Pprime Institute CNRS, ENSMA,
University of Poitiers, UPR 3346, Poitiers, France
{juan.sebastian.sandoval.arevalo,med.amine.laribi,
said.zeghloul}@univ-poitiers.fr

**Abstract.** In this paper we evaluate the accuracy of a contactless 6-DoF master device for a teleoperation platform. The master device is an optical tracking system, i.e. Leap Motion, capable of recognize and track the hand movements. A method to evaluate its accuracy by using a Qualisys motion capture system composed of 8 high-resolution cameras is proposed. In addition, the teleoperation control architecture is presented, where a torque-controlled robot is employed as the slave device, i.e. a 7-DoF Franka Emika robot. The results presented in this paper allow to validate the use of the Leap Motion as an accurate master device in a teleoperation chain while also to evaluate its operational limitations.

**Keywords:** Teleoperation platform · Optical tracking system

## 1 Introduction

Teleoperation platforms classically use contact-based master devices, especially conceived to give users the impression of physically operate the instrument held by the slave robot. Moreover, in most of the cases, these master devices are provided with haptic features, allowing users to feel the exchanged forces between the slave device and the environment [1]. Several contact-based haptic devices have been developed in past years for industrial procedures [2], gaming [3] and medical applications [4, 5].

Recent technological advances have made possible the development of new innovative master devices for teleoperation platforms, such as the case of contactless solutions, generally based on cameras systems. These new systems have enlarged the range of applications for teleoperation platforms to those cases where the hand contact is not suitable, e.g. for aseptic reasons in medical applications, for restricted space constraints or for low costs requirements. For instance, a gaze tracking system have been used [6] to control a robotic endoscope in the context of minimally invasive surgery (MIS). Camera systems allowing the 3D recognition have also been used for teleoperation applications. For example, Sripada et al. identified the skeleton movements through a Microsoft's Kinect ® camera to tele-operate a humanoid robot [7].


K. Berns and D. Görges (Eds.): RAAD 2019, AISC 980, pp. 421–429, 2020.
https://doi.org/10.1007/978-3-030-19648-6_48

Leap motion[1] is an optical sensor allowing the online hand gestures recognition, based on two cameras and three IR sensors. Besides its low cost, this system has various benefits: portability, compactness, ideal for ergonomic solutions, ease-to-use and so on. Although the main sector of application is the virtual reality video games, it has also been explored for medical applications. For instance, Zhou et al. proposed to perform open surgery tasks by tele-operating the Taurus robot with the Leap Motion sensor [8]. Few studies also propose the use of the Leap Motion sensor for rehabilitation tasks [9].

In order to validate the use of the Leap Motion sensor on medical and industrial applications, it is essential to verify its accuracy. On this way, Weichert et al. employed an industrial robot of accuracy = 0.2 mm to validate the repeatability and accuracy of the system [10]. On the same way, Guna et al. proposed to corroborate the accuracy by using a motion capture system composed of 8 high-resolution camera system [11]. Nevertheless, the study was limited to static measurements.

We propose a new protocol to verify the accuracy of the system by using a high-quality motion capture (MoCap) system composed of 8 high-resolution cameras, i.e. Qualisys Miqus[2] system, by focusing our study on the dynamic hand's orientation recognition (3-DoF), where a precise limit range definition may be suitable for certain applications.

The paper is organized as follows. In Sect. 2 we briefly present the features of the sensor, the proposed protocol to evaluate its accuracy and the obtained results. We describe in Sect. 3 the teleoperation platform we develop at the Pprime Institute. Conclusions about the obtained results are provided in the last section.

## 2 Accuracy Validation of the Contactless Master Device

In this section we evaluate the accuracy of the Leap Motion sensor on position and orientation, through the Qualisys Miqus MoCap system, using a protocol explained below.

### 2.1 Leap Motion Sensor

The Leap Motion sensor, developed by Leap Motion Inc., is a non-contact finger/hand tracking sensor. Some preliminary studies have been carried out for robot manipulation. For instance, Zubrycki et al. use a Leap Motion sensor to control a 3-finger gripper [12], Guerrero Rincon et al. develop an interface to control a robotic arm [13], Marin et al. report the first attempt to detect gestures from the data combination of Leap Motion and Kinect [14, 15].

The use of Leap Motion for hand tracking and gesture recognition is well reported in the literature; however, due to the occlusion problem between fingers, an ideal palm orientation is requested. Moreover, although its workspace in position (3-DoF) is well

[1] https://www.leapmotion.com.

[2] https://www.qualisys.com.

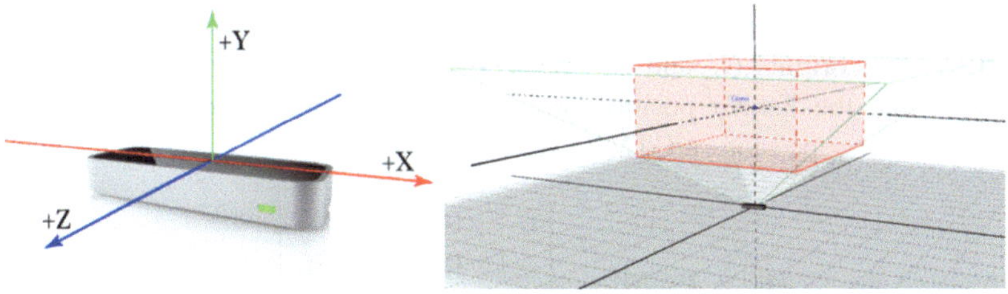


**Fig. 1.** Leap Motion sensor and its workspace.

determined (see Fig. 1), Leap Motion sensor has only two cameras to detect the hand gestures, which may cause loss of accuracy when singularity postures are adopted. Thus, it is essential to study and identify an ideal operational range in orientation, in order to consider it for the control of the teleoperation system.

## 2.2 Experimental Accuracy Evaluation

The protocol to experimentally evaluate the accuracy of the Leap Motion hand tracking system is based on the use of the motion capture system Qualisys. The experimental setup is shown on Fig. 2-a. The Mocap system is composed of eight high-resolution cameras that can detect a desired number of reflective markers. All the cameras send information to the Host PC running Qualisys Track Manager (QTM) for the data processing. This specific software allows to identify and visualize the detected markers in a 3D environment.

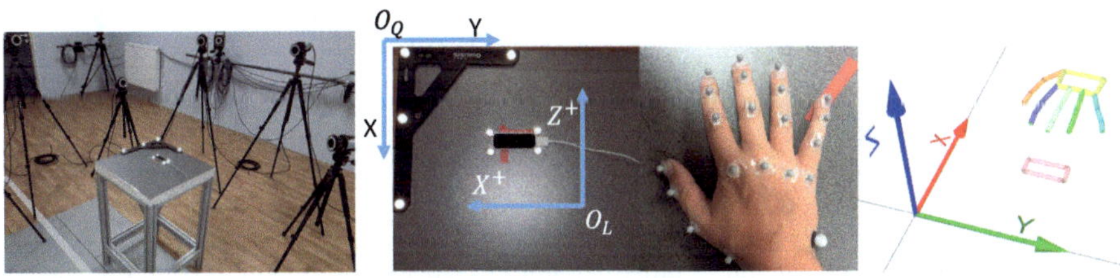


(a) (b)

**Fig. 2.** Experimental setup with Leap Motion sensor and Qualisys motion capture system.

Twenty-one markers have been used on a hand (3 markers by finger and 6 on the palm) as shown on Fig. 2. Marker trajectories have been recorded and processed during real hand motions. The frequency of motion acquisition was fixed to 120 images per second. Using QTM software, we have been able to reconstruct each motion by regrouping markers into "segments" (see Fig. 2). Each segment is defined as a set of markers belonging to the same solid. Six segments have been considered on the hand: thumb finger, index finger, middle finger, ring finger, little finger and palm.

Figure 2-b shows the coordinates systems for the Leap Motion and Qualisys, defined by their origins $O_L$ and $O_Q$, respectively. The manipulation coordinates are merged so that the hand's position and gestures registered into the demonstration workspace are mapped into a QTM's workspace, by using:

$$\mathbf{M}^Q = \mathbf{T}^{QL}\mathbf{M}^L = \begin{bmatrix} 0 & 0 & -1 & \\ -1 & 0 & 0 & QL \\ 0 & 1 & 0 & \\ 0 & 0 & 0 & 1 \end{bmatrix} \mathbf{M}^L \quad (1)$$

An example of recorded gestures is the simple translational motion shown in Fig. 3-a, where the curves are drawn using the information provided by the Leap Motion and Qualisys, respectively. The workspace dimension for each axis is 235 mm along $\mathbf{X}^+$, 235 mm along $\mathbf{Y}^+$ and 147 mm along $\mathbf{Z}^+$, according to the Producer company.

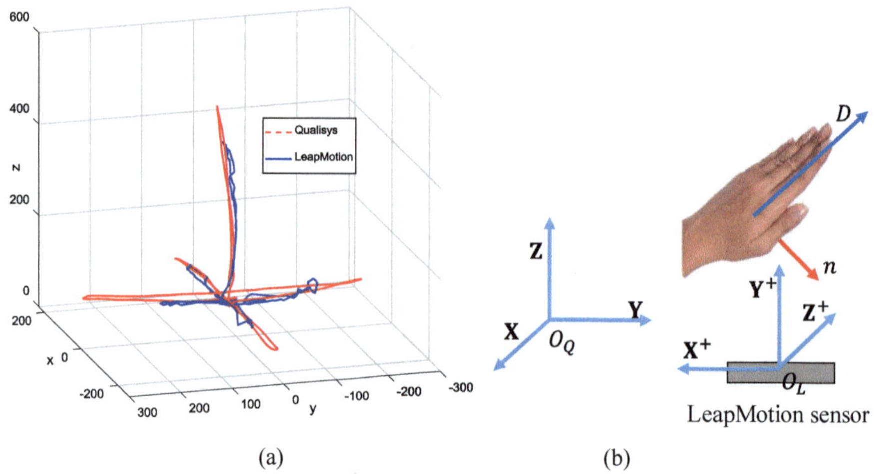


**Fig. 3.** (a) Example of translation recorded motion. (b) Vector description of palm orientation.

In order to evaluate the orientation accuracy of the Leap Motion, we conducted simple successive rotations. For the recorded motion, the orientation error between Leap Motion and Qualisys output are computed. The error is computed as follows:

$$e = \frac{1}{N}\sum\nolimits_{i=1}^{N} (L_i - Q_i)^2 \quad (2)$$

With $L_i - Q_i$ the vector computing the orientation differences and $N$ the number of considered positions.

The palm orientation is defined through the coordinates of the normal vector $\mathbf{n}$, as shown in Fig. 3-b. The rotation angle of the palm, $\beta$, is defined as the angle between the normal vector $\mathbf{n}$ and the $-\mathbf{Y}^+$ axis. The initial orientation, i.e. $\beta = 0$, is defined as the palm flat to the sensor. The coordinates of the direction vector **D** are used to compute the hand orientation angle $\alpha$ with respect to the $\mathbf{Z}^+$ axis.

An example of recorded rotation motion is shown in Figs. 4 and 5. The computed orientation angles $\beta$ and $\alpha$ are depicted in the Fig. 5. Two 3-D trajectories can be distinguished, the first one provided by the normal vector tip variation (Fig. 4-a) and the second one by the direction vector tip variation (Fig. 4-b).

The Leap Motion is then able to easily detect the palm orientations by delivering the coordinates of the normal $\boldsymbol{n}$ and direction $\boldsymbol{D}$ vectors, respectively. It is possible to observe that the normal vector is wrongly identified when the palm reaches a vertical posture, which represents a limitation of this contactless sensor. This issue is evidenced on the evolution of the computed orientation angle $\beta$, as shown in Fig. 5-a. A lag between the two curves is noticed each time a vertical palm configuration is reached. Otherwise, the direction tracking is well done and the corresponding angle, $\alpha$, is correctly computed, as verified in Fig. 5-b.

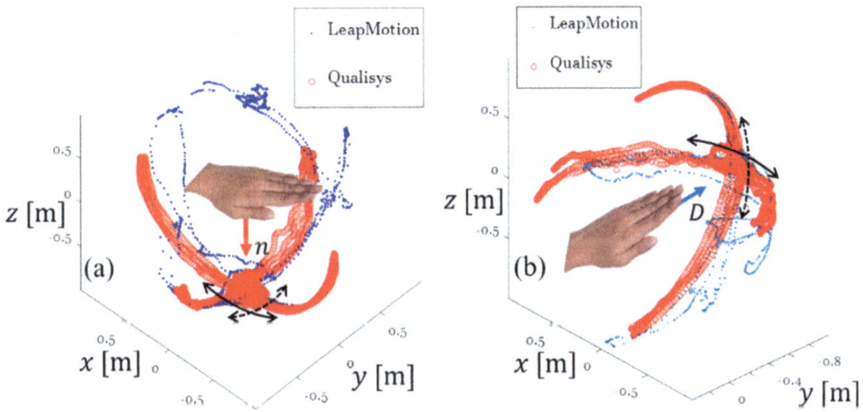


**Fig. 4.** Hand direction and normal vectors.

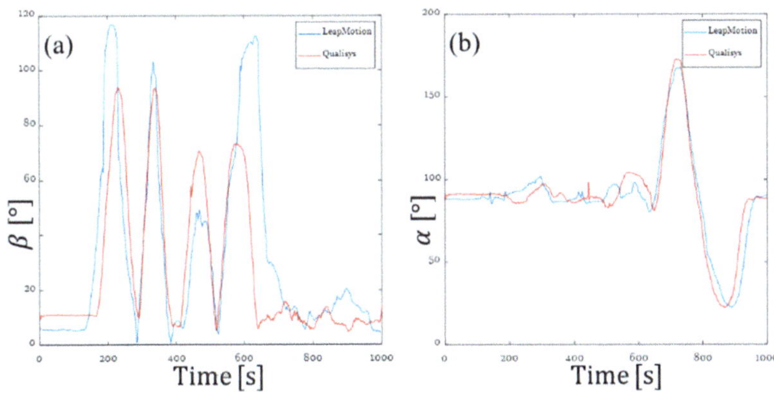


**Fig. 5.** Palm orientation angles.

The problem discussed below is due to the occlusion between the fingers. Actually, when the rotation angle of the palm is close to 90°, fingers might be occluded between them, i.e. singularity problem. This explains the gap observed on the curve of the normal vector tip variation for the Leap Motion. Out of occlusion, Leap Motion sensor effectively follows the hand movements. According to our observations, we confirm the recommendations made by the developer, defining an optimal behavior of the sensor when the palm rotates less than 60°. Out of this limitation, the orientation tracking performance using the Leap Motion is acceptable, according to the obtained error given in Table 1.

**Table 1.** Orientation errors measured until a maximum palm rotation of 60°.

| Orientation angle | $\beta$ | $\alpha$ |
|---|---|---|
| Error in [°] | 0.0607 | 0.0141 |

## 3 Teleoperation Platform

The robotics team of Pprime Institute develops teleoperation platforms based on collaborative robots. These platforms are mainly developed for medical applications, such as Doppler sonography or minimally invasive surgery applications. On this context, we propose a new teleoperation platform using the Leap Motion sensor and a 7-DoF Franka Emika robot. We use ROS to establish the data exchanges between the Sensor and the robot. Figure 6 shows a bloc diagram of the proposed platform.

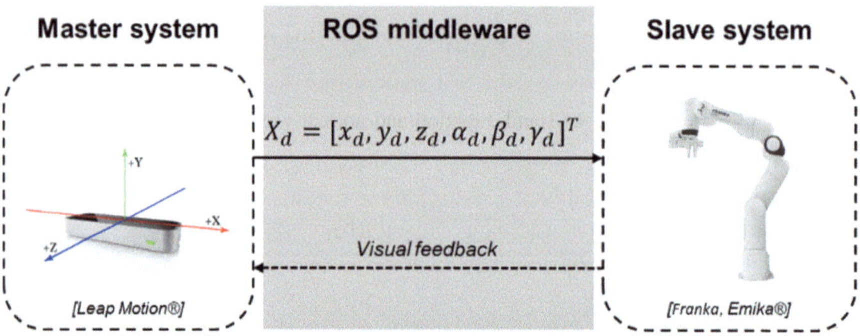


**Fig. 6.** Teleoperation platform using a contactless master device and a collaborative robot.

Franka Emika is a 7-DoF ($n = 7$) torque-controlled robot useful to coexist with human in a shared common workspace. We can then implement a compliant strategy to guarantee a safe physical human-robot interaction. Thus, the torque control input $T_i$ is defined as follows:

$$T_i = J^T\left[K_{p_x}(X_d - X_c) - K_{d_x}\dot{X}_c\right] - N(q_c)\xi + \hat{C}(q_c, \dot{q}_c) + \hat{g}(q_c) \tag{3}$$

The Leap Motion sensor provides the desired pose vector $X_d \in \Re^m$ of dimension $m = 6$. The current pose is represented by $X_c \in \Re^m$. The torque input compensates the inertial and dynamic effects by including an estimation of the Coriolis $\hat{C}(q_c, \dot{q}_c) \in \Re^n$ and gravity $\hat{g}(q_c) \in \Re^n$ torques, calculated according to the current joint position vector $q_c \in \Re^n$. Moreover, redundancy of the robot can be exploited by performing additional tasks through the vector $\xi \in \Re^n$ projected to the null-space by a projector $N(q_c)$.

Compliant motion intensity is regulated by the choice of the cartesian stiffness constant $K_{p_x}$ and the cartesian damping constant $K_{d_x}$. Figure 7 shows a translational trajectory example performed by the teleoperation platform, where the delay between the two curves is explained by the compliant strategy, i.e. there is an inversely proportional relationship between the delay and $K_{p_x}$. The constant values were fixed at $K_{p_x} = [200\,\mathrm{N/m}, 200\,\mathrm{N/m}, 200\,\mathrm{N/m}, 10\,\mathrm{N}\cdot\mathrm{m/rad}, 10\,\mathrm{N}\cdot\mathrm{m/rad}, 10\,\mathrm{N}\cdot\mathrm{m/rad}]$ and $K_{d_x} = 2\sqrt{K_{p_x}}$.

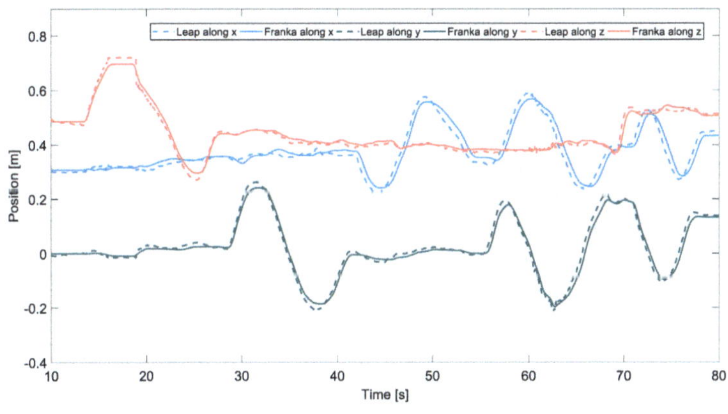


**Fig. 7.** Experimental trajectory performed by the teleoperation platform.

## 4 Conclusion

A contactless master device, i.e. Leap Motion sensor, for a teleoperation platform has been evaluated in this paper. A new protocol has been used to evaluate the orientation accuracy of the sensor, evidencing a loss of accuracy when the palm orientation is close to the vertical posture. This problem is explained by the stereo camera principle used by the leap, where only two cameras are used to make a 3D reconstruction, leading to the singularity postures. An orientation limit should be considered for any application using the Leap Motion sensor. Other possible solutions will be explored to cope with the orientation issue in our future research work, concerning the combination of two Leap Motion sensors. The redundancy will allow to avoid the singularity problem and provide a full orientation range to the contactless master device.

**Acknowledgements.** This research was funded by the region "Nouvelle-Aquitaine" (program HABISAN 2015–2020) with the financial participation of the European Union (FEDER/ERDF, European Regional Development Fund). This work was also sponsored by the French government research program *Investissements d'avenir* through the Robotex Equipment of Excellence (ANR-10-EQPX-44).

## References

1. Burdea, G., et al.: Dextrous telerobotics with force feedback-an overview. Robotica **9**, 291–298 (1991). Part 3
2. Thurfjell, L., McLaughlin, J., Mattsson, J., Lammertse, P.: Haptic interaction with virtual objects: the technology and some applications. Ind. Robot **29**(3), 210–215 (2002)
3. Basori, A.H., Daman, D., Sunar, M.S., Bade, A.: The potential of human haptic emotion as technique for virtual human characters movement to augment interactivity in virtual reality game. Int. J. Virtual Reality (IJVR) **7**, 27–32 (2008)
4. Yamamoto, T., Abolhassani, N., Jung, S., Okamura, A.M., Judkins, T.N.: Augmented reality and haptic interfaces for robot-assisted surgery. Int. J. Med. Robotics Comput. Assist. Surg. **8**, 45–56 (2012)
5. Saafi, H., Laribi, M.A., Zeghloul, S.: Optimal torque distribution for a redundant 3-RRR spherical parallel manipulator used as a haptic medical device. Robot. Auton. Syst. **89**, 40–50 (2017)
6. Fujii, K., Salerno, A., Sriskandarajah, K., Kwok, K., Shetty, K., Yang, G.: Gaze contingent cartesian control of a robotic arm for laparoscopic surgery. In: 2013 IEEE/RSJ International Conference on Intelligent Robots and Systems, Tokyo, pp. 3582–3589 (2013)
7. Sripada, A., et al.: Teleoperation of a humanoid robot with motion imitation and legged locomotion. In: 2018 3rd International Conference on Advanced Robotics and Mechatronics (ICARM), Singapore City, Singapore, pp. 375–379 (2018)
8. Zhou, T., Cabrera, M.E., Wachs, J.P.: A comparative study for touchless telerobotic surgery. In: Ritacco, L., Milano, F., Chao, E. (eds.) Computer-Assisted Musculoskeletal Surgery. Springer, Cham (2016)
9. Webster, A., Poyade, M., Rea, P., Paul, L.: The co-design of hand rehabilitation exercises for multiple sclerosis using hand tracking system. In: Rea, P. (ed.) Biomedical Visualisation, Volume 1. Series: Advances in Experimental Medicine and Biology, 1. Springer Nature (2019, in Press)
10. Weichert, F., Bachmann, D., Rudak, B., Fisseler, D.: Analysis of the accuracy and robustness of the leap motion controller. Sensors **13**(5), 6380–6393 (2013)
11. Guna, J., Jakus, G., Pogačnik, M., Tomažič, S., Sodnik, J.: An analysis of the precision and reliability of the leap motion sensor and its suitability for static and dynamic tracking. Sensors **14**(2), 3702–3720 (2014)
12. Zubrycki, I., Granosik, G.: Using integrated vision systems: three gears and leap motion, to control a 3-finger dexterous gripper. In: Recent Advances in Automation, Robotics and Measuring Techniques, pp. 553–564. Springer (2014)

13. Guerrero-Rincon, C., Uribe-Quevedo, A., Leon-Rodriguez, H., Park, J.-O.: Hand-based tracking animatronics interaction. In: 2013 44th International Symposium on Robotics (ISR), pp. 1–3. IEEE (2013)
14. Marin, G., Dominio, F., Zanuttigh, P.: Hand gesture recognition with leap motion and kinect devices. In: 2014 IEEE International Conference on Image Processing (ICIP), pp. 1565–1569. IEEE (2014)
15. Marin, G., Dominio, F., Zanuttigh, P.: Hand gesture recognition with jointly calibrated leap motion and depth sensor multimedia tools and applications. Multimed. Tools Appl. **75**(22), 14991–15015 (2015)

# Collaborative Robotics Safety Control Application Using Dynamic Safety Zones Based on the ISO/TS 15066:2016

Vincenzo Di Cosmo[1,2(✉)], Andrea Giusti[1], Renato Vidoni[2], Michael Riedl[1], and Dominik Tobias Matt[1,2]

[1] Fraunhofer Italia, Via A. Volta 13, 39100 Bolzano, Italy
v.dicosmo94@gmail.com, {andrea.giusti,michael.riedl, dominik.matt}@fraunhofer.it
[2] Free University of Bolzano, Piazza Università 1, 39100 Bolzano, Italy
renato.vidoni@unibz.it

**Abstract.** In this work we propose a safety control approach for industrial applications of collaborative robotics based on the requirements of the latest safety standards. The safety requirements are verified by performing intersection tests between bounding volumes whose dimensions are computed online, based on the speed of the manipulator along its stopping trajectory. This approach aims at tackling the limitations that impair the contribution of the robot to the collaborative task due to the enforcement of excessively conservative safety constraints. It does so neither undermining the robustness of the algorithm nor increasing excessively its complexity. To this end, implementation and computational simplicity have been preferred by (1) evaluating online only the largest contributions to the protective separation distance required by the technical specification, and (2) using a conservative approach for less predictable data. The proposed method has been tested by means of an experiment for collaborative pick-and-place operations.

**Keywords:** Collaborative robotics · Speed and separation monitoring · Bounding volumes · ISO/TS 15066:2016 · Dynamic safety zones · Industry 4.0

## 1 Introduction

The manufacturing industry is currently undergoing significant changes as the fourth industrial revolution arises. This trend is also known as Industry 4.0 [1]. In this regard, collaborative robotics can be seen as one of the most promising technologies supporting this trend [2]. In fact, this technology can enable the combination of force, strength and repeatability of machines with the dexterity and cognitive capacities of humans allowing them to work together sharing a collaborative workspace.

The Human-Robot Collaboration (HRC) paradigm introduces novel risks and thus specific safety requirements with respect to traditional industrial robots [2, 3]. To address this issue, the International Organisation for Standardisation (ISO) introduced four methods of operation for HRC applications with the ISO standards [4] and [5]: i.e. hand guiding, safety-rated monitored stop, speed and separation monitoring, and power


K. Berns and D. Görges (Eds.): RAAD 2019, AISC 980, pp. 430–437, 2020.
https://doi.org/10.1007/978-3-030-19648-6_49

and force limiting. These were further outlined in the successive technical specification [6]. Focus of this work is the speed and separation monitoring paradigm which supports high flexibility and applicability due to its task independent implementation. This method of operation allows the robot and a human operator to work concurrently within the collaborative workspace as long as a protective separation distance (PSD) requirement between them is respected. In this work, we present an effective way to verify the PSD requirement based on velocity dependent safety zones (defined in Sect. 2), hereby implemented as bounding volumes (BV) [7].

As the literature review [3] reports, there is a large body of research on HRC safety and most of the reported works are focused on injury minimization by collisions and collision avoidance strategies. Principles of both research areas are often combined. Established real-time collision avoidance strategies are implemented using artificial potential fields, based on the work of [8], in which positions to be achieved generate virtual attractive forces acting on the manipulator while obstacles generate repulsive forces. Alternative approaches evaluate the distance of the manipulator with respect to the operator to trigger safety stops or to change the state of the manipulator, thus reducing its speed. In [9] photoelectric barriers have been combined with depth cameras devices with body tracking capabilities. In other works, a variety of BVs have been used to better account for measurement errors of human position and to speed up the process of evaluating separation distances. Among them, the procedure used in [10] relies on Sphere Swept Lines (SSLs) whereas the authors in [11] combined SSLs with axis aligned bounding boxes. A more recent approach is proposed in [12], where the controller verifies with multiple safety criteria the execution of a given nominal trajectory piecewise, ensuring safety with formally verified path-consistent fail-safe manoeuvres. The positions of both the manipulator and the operator over time are predicted by computing their reachable occupancies. Our proposed method favours implementation simplicity since we consider only the safety-rated monitored stop criterion, and we perform safety checks on-the-fly supporting stop trajectories which can also be non-consistent with the nominal path. In some instances, worst-case considerations (as suggested in the technical specification [6]) are used to make these checks instead of predicting accurately the occupancies of the bodies.

In the following section we introduce the formula of the PSD along with the definition of both the BVs and the intersection tests used to verify it. Section 3 presents a method for determining the dimensions of these BVs dynamically, considering their actual velocity. A practical application of the described safety control is reported in Sect. 4 and conclusions follow in Sect. 5.

## 2 Preliminaries

The protective separation distance $S_p$ is the minimum distance that a robot must maintain with respect to an operator in order to perform autonomous tasks within a collaborative workspace. It is composed of four contributions accounting for the operator's change in location $S_h$, the space that the robot can cover within the system reaction time $S_r$, the robot system stopping distance $S_s$ and various sources of error $\xi$:

$$S_p = S_h + S_r + S_s + \xi$$

where $\xi$ is defined as the sum of the sensing system measurement tolerance $Z_d$, the position uncertainty of the robot system $Z_r$ and the intrusion distance $C$ (i.e. the portion of the body that can intrude the sensing field before it is detected). The technical specification [6] provides two formulae for each contribution to $S_p$: they can be computed as either a constant value, based on worst-case scenario, or as a function of the actual speed of the operator or the manipulator.

Our safety control algorithm continuously verifies that the manipulator's links and its end effector respect the separation distance requirement. To achieve this, intersection tests between bounding volumes enclosing both the manipulator's links and the operator's upper limbs and torso (Fig. 1a and b) are carried out iteratively. Sphere-Swept-Lines (SSLs) were chosen as BVs. For clarity of subsequent description, we define SSLs, the concept of safety zones and the procedure for intersection tests.

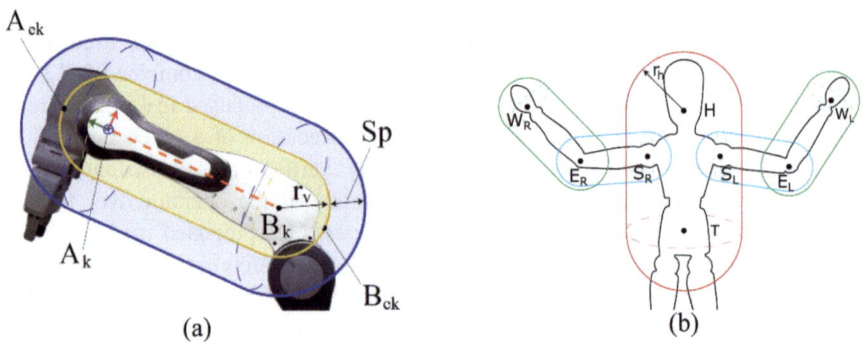


**Fig. 1.** (a) Example of a SSL describing the volumetric occupancy of a link of the manipulator (in yellow) and a Safety Zone related to the same link (in blue). (b) Example of the five SSL enclosing the operator's upper body.

**Definition 1.** SSLs are formally defined as the Minkowski sum $\oplus$ of a sphere $\mathcal{S}$ and a line segment $\mathcal{P} : \mathcal{CW} = \mathcal{S} \oplus \mathcal{P}$. Each one of the $k$ manipulator's SSLs encloses one link of the manipulator. Hereafter, quantities that are referred to the *k-th* link are denoted by the subscript $k$. As shown in Fig. 1a, the $SSL_k$ is defined by the coordinates of the end-points of $\mathcal{P}_k$ (hereafter defined as $A_k$ and $B_k$) and the radius $r_v$ which corresponds to the distance between the furthest point on the surface of any one of the corresponding links and its longitudinal axis (i.e. the minimum radius of $\mathcal{S}$ required to over-approximate the geometry of every link). It is convenient to associate the coordinates of the end-point $A_k$ of each SSL with the origin of a link-fixed frame of reference as shown in Fig. 1a. This gimmick allows us to facilitate the evaluation of the position in space and speed of the SSLs of the manipulator. The end-point $B_k$ is chosen with respect to $A_k$ to tightly over-approximate the geometry of the link.

The position of each operator's SSL segment $\mathcal{P}_h$ is described by the coordinates of two of his skeletal joints (see Fig. 1b). The magnitude of each of the radii $r_h$ were chosen from the 95-th percentile values of suitable anthropometric data of males aged 19–65 [13].

**Definition 2.** Safety zones $\mathcal{SW}$ are SSLs based on the previously defined manipulator's SSLs but additionally accounting for its protective separation distance $S_p$. At any given instant, these safety zones conceptually describe the space that the robot system requires to come to a stop safely from its current condition. The operator's safety from unintended collisions can be guaranteed by triggering a stop of the system as soon as any one of its SSL intersects a safety zone. The radius of a swept sphere $\mathcal{S}_{\mathcal{SW}}$ that identifies a safety zone surrounding the robot is therefore defined as follows:

$$r_{\mathcal{SW}} = r_v + \mathrm{S_p} = r_v + \mathrm{S_h} + \mathrm{S_r} + \mathrm{S_s} + \xi. \qquad (1)$$

**Definition 3.** Intersection tests check each *k-th* robot safety zone $\mathcal{SW}_k$ against the SSLs surrounding the operator iteratively. Intersection tests between two SSLs can be simplified by an intersection test of two spheres centered on the closest points between the corresponding line segments. Thus, the safety conditions required by the speed and separation monitoring collaborative method of operation are respected whenever the robot keeps its safety distance or if it does not move:

$$d_{hr} > r_{\mathcal{SW}} + r_h \vee \dot{\boldsymbol{q}}(\mathrm{t}) = 0 \quad \forall t$$

where $\dot{\boldsymbol{q}}(\mathrm{t}) \in \mathbb{R}^N$ is the speed of the manipulator in joints space, $d_{hr}$ is the Euclidean distance between the two closest points of the segments $\mathcal{P}_k$ and $\mathcal{P}_h$.

## 3 Proposed Method

Our proposed safety control scheme aims at determining the contribution $S_s$ to the size $r_{\mathcal{SW}k}$ of the safety zones dynamically and on-the-fly based on the evolution of the actual speed of the enclosed link over the course of a potential stopping trajectory. $S_r$ and $S_h$ on the other hand are evaluated only once, considering the worst-case scenarios as suggested in [6] (i.e. using the maximum directed speed of the manipulator and a constant speed of 1,6 m/s to account for the speed of the operator respectively). We assume that an arbitrary stopping trajectory can be computed at any time and that the upper bound of the computational time of $r_{\mathcal{SW}_k}(t)$ is known. The latter is to be included in the system reaction time. At a given instant, when the safety control is performed, we assume that the manipulator needs to be stopped. The potential stopping trajectory and the speed profile of the manipulator along it are computed allowing us to estimate the stopping distances for each link. In turn, according to (1), the radii $r_{\mathcal{SW}}(t)$ are obtained and intersection tests between the operator's SSL and the safety zones can then be carried out as defined. The potential stopping trajectory previously evaluated will be implemented only if the safety test detects an intersection. Motion of the system will otherwise continue as planned. Hereafter, whenever clarity of description is not affected significantly, the time dependence of time varying variables is omitted for brevity. The *k-th* link of a manipulator is over-approximated by a SSL having radius $r_{v_k}$ and $A_{e_k}$ and $B_{e_k}$ as extreme points of its spherical end-caps (see Fig. 1a). The distance $S_{s_k}$ covered by a link to come to a stop can be computed as:

$$S_{s_k} = \int_{t_0+t_r}^{t_0+t_r+t_s} v_{s_k}(t)\,dt \tag{2}$$

where $v_{s_k}$ is the speed of the link in the course of stopping. Here, $t_0$ is the instant in which the safety control procedure begins, $t_r$ is the reaction time of the system (including acquisition and processing of data and the issuing of the stop command) and $t_s$ is the time the robot system requires to come to a halt once the stop command has been issued. Following the work in [9], a necessary and sufficient condition to ensure that a link respects a given distance constraint with respect to an obstacle point is to consider the velocity of its extreme points, in our case $\dot{A}_{e_k}(t)$ and $\dot{B}_{e_k}(t)$. Then, the maximum linear speed of a link $v_{s_k}$ defined in Eq. (2) is equal to:

$$v_{s_k}(t) = \max\left(\left|\dot{A}_{e_k}(t)\right|,\ \left|\dot{B}_{e_k}(t)\right|\right) \quad \text{for} \quad t \in [t_0+t_r, t_0+t_r+t_s].$$

To compute $\dot{A}_{e_k}(t)$ and $\dot{B}_{e_k}(t)$, we shall determine beforehand the evolution along the potential stopping trajectory of the velocities of the link-fixed frame to which the $\mathcal{SW}$ is associated. Within the time frame that goes from $t_0$ to $t_r$, we compute the potential stopping trajectory $\boldsymbol{q}(\mathrm{t}) \in \mathbb{R}^N$ and its timing law $\dot{\boldsymbol{q}}(\mathrm{t}) \in \mathbb{R}^N$ in joint space for $t \in [t_0+t_r, t_0+t_r+t_s]$. At this point, the joint space velocity profile over the potential stopping trajectory is known. Differential kinematics allows us to relate these velocity profiles to the Cartesian velocities of the link-fixed frames of reference. Let us define the partial joint-space position $\boldsymbol{q}_{1:k}(t) \in \mathbb{R}^k$ and velocity vector $\dot{\boldsymbol{q}}_{1:k}(t) \in \mathbb{R}^k$ of the manipulator. The partial geometric Jacobian $\boldsymbol{J}_{kP}(\boldsymbol{q}_{1:k}) \in \mathbb{R}^{[3xk]}$ is the matrix that maps the contributions of the joints $m \in [1,\ldots,k]$ to the linear velocity $\dot{\boldsymbol{p}}_k$ of the *k-th* link-fixed frame. The components $\dot{\boldsymbol{p}}_k$ can thus be evaluated using:

$$\dot{\boldsymbol{p}}_k = \sum\nolimits_{m=1}^{k} \frac{\partial \boldsymbol{p}_k}{\partial q_m}\dot{q}_m = \boldsymbol{J}_{kP}(\boldsymbol{q}_{1:k})\,\dot{\boldsymbol{q}}_{1:k}.$$

Similarly, $\boldsymbol{J}_{kO}(\boldsymbol{q}_{1:k}) \in \mathbb{R}^{[3xk]}$ is the matrix that maps the contributions of the joints $m \in [1,\ldots,k]$ to the angular velocity $\boldsymbol{\omega}_k$ of the *k-th* link:

$$\omega_k = \sum\nolimits_{m=1}^{k} \omega_{k-1,k} = \boldsymbol{J}_{kO}(\boldsymbol{q}_{1:k})\,\dot{\boldsymbol{q}}_{1:k}.$$

Since the manipulator's $\mathcal{SW}$s are defined as shown in Fig. 1a, the linear velocity of the point $\dot{A}_k$ corresponds to the velocity of the origin of the *k-th* link-fixed frame. Its magnitude is equal to the Euclidean norm of $\dot{\boldsymbol{p}}_k$. However, to evaluate the velocities of the extreme points of a link's SSL $\dot{A}_{e_k}$ and $\dot{B}_{e_k}$, required to compute $v_{s_k}$, we also account for the contribution of its angular velocity:

$$\dot{A}_{e_k} = \dot{A}_k + \omega_k \times (A_{e_k} - A_k) = \dot{A}_k + \omega_k \times \left(\frac{(A_k - B_k)}{|A_k - B_k|} \cdot r_v\right),$$

$$\dot{B}_{e_k} = \dot{A}_k + \omega_k \times (B_{e_k} - A_k) = \dot{A}_k + \omega_k \times \left(\frac{(B_k - A_k)}{|B_k - A_k|} \cdot (|B_k - A_k| + r_v)\right).$$

A case study in which the proposed collision avoidance method is applied for collaborative sorting operations follows.

## 4 Experimental Evaluation

The collision avoidance strategy presented in the previous section has been tested in a collaborative sorting operation. The robot system used in this application is the 7 degrees of freedom Panda arm by Franka-Emika. A reference frame is fixed to each link of the manipulator using the Denavit-Hartenberg notation [14]. Please note that when two consecutive links are connected by a revolute joint that would only allow rotation about the axis of the two SSLs (as shown in Fig. 1a) these may be enclosed in one SSL since every point on the surface of the links falls inside this SSL independently of the joint configuration. We used this consideration to reduce the number of intersection tests that are performed. A Microsoft Kinect v2 was used as sensing device to acquire data about the skeletal joints position of the operator within the collaborative workspace. The accuracy of the tracking device is taken into account when computing the PSD in the term $\xi$. Data regarding measurement accuracy of the device used were considered for each joint of the human separately, using the results presented in [15].

Aim of the test is measuring and comparing the time required to sort eight objects using two different safety algorithms: one evaluates the PSD following the worst case scenario hypothesis as presented in the technical specification [6] (static safety zones) while the other implements our proposed method (dynamic safety zones). As shown in Fig. 2a, the workspace is divided into two areas: the hatched one is restricted to the operator and what is left, and within the reach of the robot, constitutes the collaborative workspace. The manipulator is placed on a table (labelled M) at the border between the two regions: it has access to two picking points ($G_p$ and $R_p$) located in the collaborative area and can reach two boxes ($B_G$ and $B_R$) that lie in the restricted part.

On the operator's workbench (H) a third box is placed. It is filled by objects of two different colors (i.e. red and green). At the beginning of the task a green object lies in $G_p$ and a red one in $R_p$. The operator is instructed to wait into a designated area by her or his workbench while the manipulator begins the picking operation (Fig. 2b). The robot picks and places one object at the time, alternating between red and green objects picked from the corresponding point. The operator shall supply an object of the correct color every time the robot releases one into a box: as soon as an object is placed into either box, an acoustic signal prompts the operator to reach a drop area (M) by following a labelled path (Fig. 2c), place an object of the corresponding color in the free

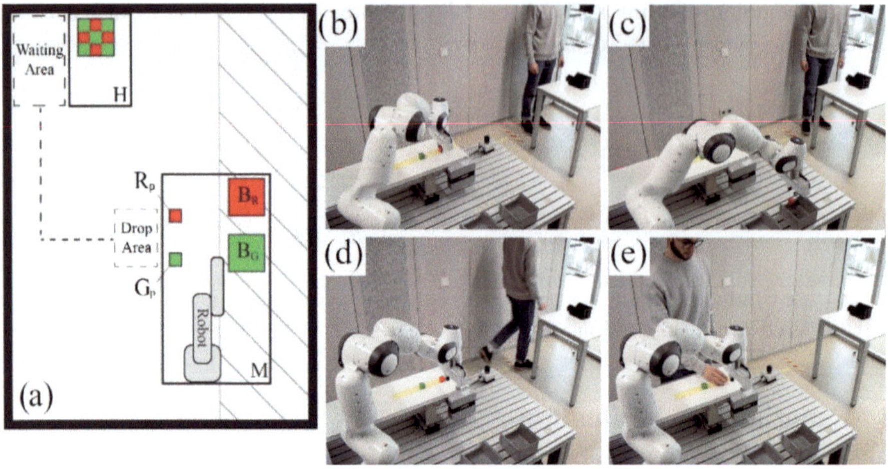


**Fig. 2.** Illustration of the validation test layout and task.

picking point (Fig. 2e) and to go back to the waiting area (Fig. 2d). 10 different operators of mixed gender performed the same task twice using the two safety algorithms under examination. The control routine of the manipulator timed the tasks. Table 1 shows the mean values and standard deviations of the data collected using the two safety control schemes. The results of these tests show a significant difference between the means of task competition time between the two safety control schemes, p-value < 0,0001, with an improvement of about 10% when using our proposed method.

**Table 1.** Standard deviations and mean values of the timed task.

| Static safety zones | Dynamic safety zones |
|---|---|
| $\overline{t_{wc}} = 132{,}14\,\mathrm{s}$ | $\overline{t_{sz}} = 119{,}10\,\mathrm{s}$ |
| $\sigma_{wc} = 3{,}27\,\mathrm{s}$ | $\sigma_{sz} = 1{,}83\,\mathrm{s}$ |
| Sample size = 10 | Sample size = 10 |

## 5 Conclusions

In this paper we presented a novel safety scheme implementing a collision avoidance strategy using safety zones. We introduced the approach used to evaluate online and on-the-fly the contribution to the PSD due to the robot system's stopping distance $S_s$. Lastly, we tested the efficiency improvement of our approach compared to the basic implementation of the ISO technical specification [6] with static safety zones.

Further suggested steps to improve the practical effectiveness of this method are (1) the development of a robust body tracking algorithm employing depth cameras that is also suitable for industrial application and less susceptible to the risk of occlusions and (2) the implementation of a scheme able to compute and verify possible evasive trajectories that would altogether minimize the instances in which a stop of the robot system is required.

**Acknowledgement.** This research has received funding by the European Regional Development Fund ERDF 2014–2020 under the grant agreement within the project DeConPro - code FESR1017 (CUP B56J16001730001).

## References

1. Kagermann, H., Wahlster, W., Helbig, J. (eds.) Recommendations for implementing the strategic initiative Industrie 4.0. Acatech–National Academy of Science and Engineering, Frankfurt (2013)
2. Gualtieri, L., Rauch, E., Rojas, R., Vidoni, R., Matt, D.T.: Application of axiomatic design for the design of a safe collaborative human-robot assembly workplace. In: MATEC Web Conference, vol. 223 (2018)
3. Robla-Gómez, S., Becerra, V.M., Llata, J.R., Gónzales-Sarabia, E., Torre-Ferrero, C., Pérez-Oria, J.: Working together: a review on safe human-robot collaboration in industrial environments. IEEE Access **5**, 26754–26773 (2017)
4. ISO/TC 299: ISO 10218-1:2011 Robots and robotic devices – Safety requirements for industrial robots - Part 1: Robots. International Organisation for Standardisation, Geneva (2011)
5. ISO/TC 299: ISO 10218-2:2011 Robots and robotic devices – Safety requirements for industrial robots - Part 2: Robot systems and integration. International Organisation for Standardisation, Geneva (2011)
6. ISO/TC 299: ISO/TS 15066:2016 Robots and robotic devices – Collaborative robots. International Organisation for Standardisation, Geneva (2016)
7. Ericson, C.: Real-Time Collision Detection, 1st edn. Elsevier Inc., San Francisco (2005)
8. Khatib, O.: Real-time obstacle avoidance for manipulators and mobile robots. In: Proceedings of IEEE International Conference on Robotics and Automation, St. Louis, vol. 2, pp. 500–505 (1985)
9. Zanchettin, A.M., Ceriani, N.M., Rocco, P., Ding, H., Matthias, B.: Safety in human-robot collaborative manufacturing environments: metrics and control. IEEE Trans. Autom. Sci. Eng. **13**(2), 882–893 (2016)
10. Corrales, J.A., Candelas, F.A., Torres, F.: Safe human-robot interaction based on dynamic sphere-swept line bounding volumes. Robot. Comput. Integr. Manuf. **27**(1), 177–185 (2011)
11. Corrales, J.A., Gómez, G.G.J., Torres, F., Perdereau, V.: Cooperative tasks between humans and robots in industrial environments. Int. J. Adv. Robot. Syst. **9**(3), 94–103 (2012)
12. Beckert, D., Pereira, A., Althoff, M.: Online verification of multiple safety criteria for a robot trajectory. In: Proceedings of IEEE 56th Annual Conference on Decision and Control (CDC), Melbourne, pp. 6454–6461 (2017)
13. Pheasant, S.: Bodyspace – Anthropometry, Ergonomics and the Design of Work, 2nd edn. Taylor & Francis Group, London (2003)
14. Uicker, J.J., Denavit, J., Hartenberg, R.S.: An iterative method for the displacement analysis of spatial mechanisms. J. Appl. Mech. **31**(2), 309–314 (1964)
15. Wang, Q., Kurillo, G., Ofli, F., Bajcsy, R.: Evaluation of pose tracking accuracy in the first and second generations of Microsoft Kinect. In: Proceedings of IEEE International Conference on Healthcare Informatics, ICHI, Cancun, pp. 380–389 (2015)

# Versatile Collaborative Robot Applications Through Safety-Rated Modification Limits

Mathias Brandstötter[1(✉)], Titanilla Komenda[2,3], Fabian Ranz[2,3], Philipp Wedenig[4], Hubert Gattringer[5], Lukas Kaiser[1], Guido Breitenhuber[1], Andreas Schlotzhauer[1], Andreas Müller[5], and Michael Hofbaur[1]

[1] Institute for Robotics and Mechatronics, Joanneum Research, Lakeside B08a, 9020 Klagenfurt am Wörthersee, Austria
Mathias.Brandstoetter@joanneum.at

[2] Fraunhofer Austria Research GmbH, Theresianumgasse 7, 1040 Vienna, Austria

[3] Institute of Management Sciences, Vienna University of Technology, Theresianumgasse 27, 1040 Vienna, Austria

[4] Institute for Economic and Innovation Research, Joanneum Research, Leonhardstraße 59, 8010 Graz, Austria

[5] Institute of Robotics, Johannes Kepler University Linz, Altenberger Str. 69, 4040 Linz, Austria

https://www.joanneum.at/robotics

**Abstract.** A robot is a flexible tool and handling device. However, by eliminating safety fences in robotic applications, flexibility is limited as regulations on personal safety must be followed. Any modification on the system or application will require a new risk assessment before the system can be put back into operation. This circumstance costs time and money and stands in contradiction to the nature of a robot as a versatile and adaptable device. By introducing safety-rated modification dimensions and determining admissible variations indicating the limits up to which the system or application can be changed in compliance with safety regulations, a potential solution to overcome this restriction is presented. A model-based strategy for estimating and validating the aforementioned modification limits is proposed, up to which changes on the system can be made without conducting a new risk assessment. The proposed approach gives rise to a novel safety concept for collaborative robotic applications, which ensures flexibility while taking into account safety standards.

**Keywords:** Robot safety · Mobile manipulation · Reconfiguration · Risk assessment


K. Berns and D. Görges (Eds.): RAAD 2019, AISC 980, pp. 438–446, 2020.
https://doi.org/10.1007/978-3-030-19648-6_50

## 1 Introduction

The market for sensitive robot systems is already well established by numerous suppliers like Universal Robots, KUKA, Fanuc, ABB, Franka Emika, F&P Robotics, etc. A collision detection feature enables those so-called collaborative robots to work together with humans. By means of torque measurements in the joints or its estimation from measured actuator currents, collisions with the robot can be detected [2]. The decisive factor for this sensitivity in force and torque control is an accurate dynamic model of the robot that predicts the motor torques in the joints. The comparison of torque values for respective joints allows conclusions about any forces acting on the system from outside [8]. As a result of this functionality, protective fences are not required with sensitive robotic systems, as they can react to occurring forces. In this way, Human-Robot Collaboration (HRC) can be implemented and personal safety is ensured at a first level. However, operating robotic systems by identifying, measuring and validating potential contact situations with limit values stated in ISO/TS 15066 [10] is not trivial, as the real bio-mechanical load for the human during a contact situation depends on various parameters and can deviate from the predefined settings in the robot controller [3]. The high safety requirements, the complexity of the safety-related behaviour of a robotic system, and the lack of engineering tools for analysing those influences still inhibit their extensive application in industrial settings [6].

Especially one specific problem arises in the practical use of "fenceless" robots: Any change to the robot system or the application requires a new extensive risk assessment. By this substantial restriction, requiring the renewal of the risk and measurement assessment of the application after each modification, the essential flexibility of a classical industrial robot – to be adaptable at any time – is actually lost by a collaborative robot. In addition, collaborative robots (stationary and mobile) are preferably used in dynamic environments, which means that the robot must adapt dynamically to its changing requirements. This work focuses on the current need of companies for enhancing the flexibility of industrial HRC applications while at the same time ensuring the mandatory personal safety at any time.

In this paper, a new approach is proposed, where safe limits for modifying a basic application of a robotic system are defined. Thus, changes to the system and to the application can be executed in various dimensions, so-called *modification dimensions*, with limits known to the user. The safety assessment and approval of a desired modification is carried out automatically on the basis of the underlying safety system model. With this approach, we are pursuing two overarching objectives:

**Goal one:** Collaborative robots can be used flexibly and safely as fenced-in industrial robots.
**Goal two:** Sensitive mobile manipulators can be used in an industrial context.

This work is organized as follows. In Sect. 2, the proposed concept for the automated evaluation of modified collaborative robot applications is presented.

The four main phases *Modelling*, *Analysis*, *Identification*, and *Operation* are described in more detail. The essential aspect of *Modification Dimensions* relevant to this work is explained in Sect. 3. The initial ideas for a necessary software framework is presented in Sect. 4 and we give conclusions of this work in Sect. 5.

## 2 Concept Description

In this work, we disengage ourselves from the usual approach of assessing the safety of collaborative robotic applications. We propose a use-centric concept that allows a user to apply a collaborative robot flexibly. Thus, the safety assessment concept (see Fig. 1) is introduced.

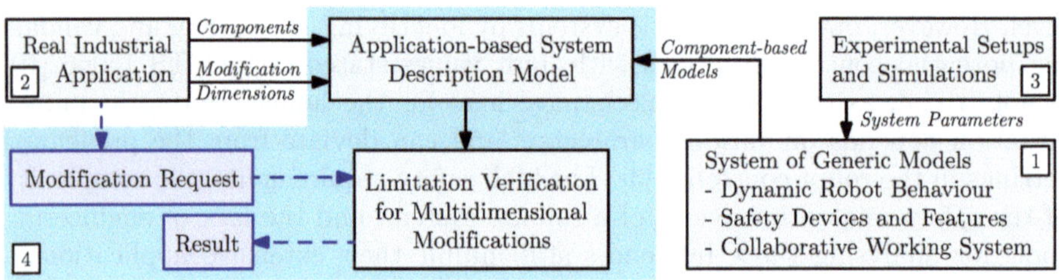


**Fig. 1.** The internal structure of the introduced safety concept for HRC systems. It contains of the four phases defined by (1) Modelling, (2) Analysis, (3) Identification, and (4) Operation.

In the initial phase (Modelling), a set of generic system models is generated from which collaborative applications can be built. This includes the dynamic behaviour of the robot, the safety devices as well as the collaborative working system. The second phase (Analysis) focuses on investigating real industrial applications and identifying required industry-relevant modification dimensions which are integrated in an application-based system description model. Furthermore, the generic system models are optimized with specific parameters in the third phase (Identification), which are identified in experimental tests on individual components or in simulated environments. The final phase (Operation) describes the utilization phase, in which modification limits on the system are identified and validated. As a result of the safety model, a set of *safety-rated modification limits* is derived, describing the admissible changes in variability without the need of an additional safety assessment and measurement.

### 2.1 Modelling

The modelling of the system components is an essential part of the preliminary phase. The larger and more generic the set of predefined models, the faster a given use case can be replicated mathematically.

**Dynamic Robot Model.** The dynamic behavior of a robot is governed by the equations of motion (EoM) in the typical form of $\mathbf{M}(\mathbf{q})\ddot{\mathbf{q}} + \mathbf{g}(\mathbf{q}, \dot{\mathbf{q}}) = \boldsymbol{\tau}_M + \boldsymbol{\tau}_{\text{ext}}$. Therein $\mathbf{q}$ are the minimal coordinates, $\mathbf{M}$ is the mass matrix, $\mathbf{g}$ represents Coriolis, centrifugal, gravity and friction terms and $\boldsymbol{\tau}_M$, $\boldsymbol{\tau}_{\text{ext}}$ are the motor torques and the external torques, respectively. In the presented approach, we focus on generic models that are valid for different robot types.

**Safety Device and Feature Model.** To guarantee human safety for HRC applications, different proprioceptive and exteroceptive sensors are used. In general, proprioceptive sensors measure internal values of the robot like the working speed of the robot's arm. In contrast to proprioceptive sensors, exteroceptive sensors acquire information about the robots environment, like distance measurements from the robot to an object [9]. The provided sensor data is used to statistically model the information density that can be achieved with a certain set of sensors. To develop these statistical models, the quality of the sensor data plays an essential role. To quantify the quality of sensor data, the sensitivity, the error and the precision of every single sensor are of particular importance [9]. A sensor with high sensitivity, high precision and low error should influence the statistical models more than a sensor with the opposite properties. That means, a weighted statistical model needs to be developed for each sensor type. By means of the developed statistical models, an estimation for the human safety will be possible, as the risk on the human changes by adding or removing sensors to or from the HRC system.

**Collaborative Working System Model.** Besides robots and safety devices, the collaborative working system is additionally defined by the physical equipment available, the tasks for execution, and the products that are subject to the task execution (see Fig. 2).

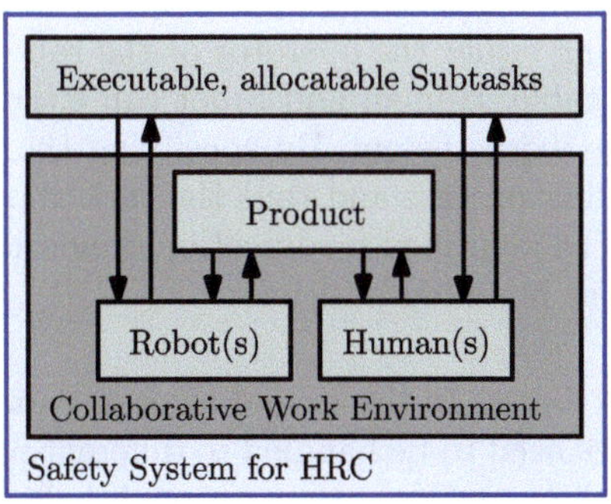


**Fig. 2.** Simplified work system model for HRC safety evaluation

The physical equipment as well as the human model are represented by 3D volume models that use boundary representation. Thus, the model contains directly addressable and manipulable surfaces, edges, radii, etc., which are essential for evaluating safety issues regarding pinching areas (see also ISO 13854). The system-inherent work tasks are generated by a hierarchical task analysis, whereby a decomposition into subtasks takes place on the basis of superordinate objectives of the work system, as for example in [5]. These subtasks are characterized by completed actions, which cannot be divided further meaningfully and serve as allocation objects in the task division between humans and robots. They are modelled in motion-oriented synthetic process languages (MTM-1, MTM-MRK), as originally proposed by [7], and can thus be evaluated temporally as well as with regard to their influence on safety. The integration of the task allocation model into the overall safety-rated model is essential, as a variation in task allocation might lead to a new trajectory or rather workspace of the robot and thus new possible colliding points with the environment but also the human.

## 2.2 Analysis

Every given collaborative robot application consists of a finite number of components and parameters that define the whole system. Thus, the analysis of industrial use cases is essential to tackle the industrial need when it comes to defining an overall safety-rated approach. Based on the analysis, a set of generalized models is derived, which is used to map specific systems with relevant safety-rated aspects. Here, the focus was on processes in the field of manufacturing, assembly and logistics. The main objective was to identify required modification dimensions and their representations that are needed by industrial stakeholders in order to meet their entrepreneurial goals of time, cost and quality.

**Handling Use Case.** In the first use case, products need to be placed in testing units. The products are provided in magazines and need to be placed back in the magazines after testing. Specific types of products go to specific testing units. It is essential that the work system to be designed is applicable for both, human and robot. The idea was to place testing units in individual towers that can be turned facing either the operator or the robot, while the magazines are provided on a turntable. Human and robot can either work simultaneously or consecutively on the work system. By specifying the resource, i.e. robot or operator, for the handling process and thus the task allocation, the robot path changes dynamically. The identified modification dimensions in this use-case are task allocation as well as handling positions.

**Post-processing Use Case.** In the second use case, products need to be post-processed. The products need to be handled to different post-processing stations in order to remove welding spatters, test a through hole, clean the inside of the product and place it in a magazine. Here, products can differ in size. Furthermore, the work system should again be applicable for an operator to work either simultaneously with the robot (e.g. when post-processing pre-serial parts) or consecutively and thus having again an effect on the robot's path. The identified modification dimensions in this use-case are task allocation and product size.

**Machine Loading Use Case.** In the third use case, a robot on a mobile platform should load and unload parts from a milling machine where parts can differ not only in size but also in material and thus affecting the weight of the part tremendously. The identified modification dimensions are product weight and the position of the robot base resulting in changes in the robot path.

**Summarized Lab Use Case.** By defining the lab use case, we wanted to combine the identified modification dimensions in the industrial use cases. In this sense, the lab use case consists of an articulated robot on a mobile platform with the task of loading and unloading a machine and gripping the raw material from a magazine while placing the processed parts on a conveyor belt. Thus, the lab use case covers the industry-relevant modification dimensions, i.e. change of workpiece in size and weight due to changes in product requests, change of pick and place positions on the individual station due to changes in station design and layout, and changes to the robot path due to changes in the robot base position and task allocation. Furthermore, the robot's velocity is subject to change when it comes to modification requests regarding production volume and/or cycle time.

### 2.3 Identification

The identification is used to determine the actual parameters of the real system. This can be done either on the real setup or on simulated laboratory setups. It is well known, that the dynamic robot parameters like link masses, inertia terms, friction coefficients etc. appear linear in the EoM. Therefore they can be rewritten to $\boldsymbol{\Theta}(\mathbf{q}, \dot{\mathbf{q}}, \ddot{\mathbf{q}})\mathbf{p} = \boldsymbol{\tau}_M$, with the regressor matrix $\boldsymbol{\Theta}$ and the parameter vector $\mathbf{p}$. This linear form is employed for the identification using optimized persistent identification trajectories (e.g. Fourier series). These trajectories are used as desired values for the position controlled robot while the necessary motor torques are measured. Therewith an overdetermined system of identification equations is generated that can be solved for the unknown robot parameter $\mathbf{p}$ with a least squares error minimization, see [4] for details.

Additional measurements are done on a simulated laboratory setup to determine the relevant parameters during a collision between a human and the robot while modifying the modification dimensions. Contact forces are recorded in a simulated pick and place task while changing the position of the collision in few centimeter intervals within the entire robot workspace. Also the pose, speeds and velocities of the robot are changed systematically. To measure the contact forces, a special measurement device is used, which allows the biofidel properties to be configured according to ISO/TS 15066 [10] and DGUV information FB HM-080 [1].

### 2.4 Operation

During the use of a robot system, a need for adaptation may arise after a certain time. These changes can take multiple forms, but have a direct impact on the safety risk regardless of their complexity and magnitude. In this important phase, a modification request can be made, which will provide information on the guarantee of human safety in case of multidimensional variations on the system and/or application.

## 3 Modification Dimensions in Collaborative Applications

In order to motivate the need of modification dimensions and limits, they are defined by means of an illustrative example and by the findings of analyzed industrial applications. As analyzed in the previous chapter, the four most relevant identified modification dimensions are: (a) workpiece geometry and mass, (b) process sequence, (c) positions (robot's base, workpieces, workstations, workers), and (d) end effector trajectory regarding path, velocity and acceleration. In addition to those mentioned earlier, there are a number of other modification dimensions, such as robot type and structure, gripper geometry, type of safety function, and position of safety sensors.

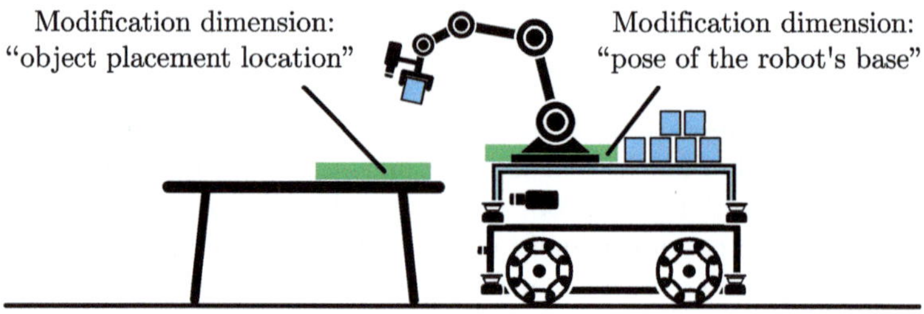


**Fig. 3.** Simplified representation of two exemplary modification dimensions

For a better understanding two modification dimensions are illustrated in a scenario with a mobile manipulator, see in Fig. 3. The use of a mobile manipulator in industry is considered primarily in the logistics sector. For this reason, the focus is on the transport of goods of all kinds and require pick-up and drop-off at designated locations or areas. Because these locations are not always defined accurately and can vary over the lifespan of the application, one approach is to allow the variation of the object placement location. Also the position of the mobile base can be a modification dimension which corresponds to a modification of the pose of the manipulator on top of the mobile platform. The possibility of these changes during run time without the need of re-certifying the safety would enable plenty of applications and reduce the operational costs significantly.

Due to the fact, that the safety of this scenario depends on various parameters, like the environment of the object placement location, the handled objects, the presence of humans or the configuration and speed of the manipulator, the described approach doesn't aim for a universal safety, that can be guaranteed in any situation, but will define limits for the change within the modification dimensions, that are crucial for the application. In the illustrated example these limits could be a bounded surface on the table, where objects can be placed; and a bounded space relative to the mobile platform, where the manipulator can be mounted.

## 4 Proposed Software Framework

A software framework, called ADJUST, will be developed to support stakeholders in decision making during the operation phase. It integrates the dynamic robot model, safety devices and feature models (information density models) as well as a collaborative working system model – each covering different aspects of the collaborative robot application – into one software tool. The primary goal is to determine, if safe operation of a modified application is either possible, not possible or requires an updated risk assessment and re-certification. Based on a given application scenario, it can rate a desired application modification of defined modification dimensions regarding a possible safety violation. Thus, ADJUST supports the operator during design time of modifying an application. In addition to a boolean output, it also gives insight into the quality of a modification. This means that the framework does not only tell the user, if an application is safe or not, but also compares a modification to the status quo and assesses, if safety is improved or degraded by the planned modification. However, it is important to state, that automatic assessable modification limits are very narrow and exceeding these will require an updated risk assessment and re-certification of the application.

## 5 Conclusions

This paper presents a new and comprehensive safety concept for collaborative robotic systems. The proposed framework accounts for the variability of given HRC scenarios in industrial applications. This framework comprises three main models that need to be parameterized for a particular application and robot. Each of them covers a different aspect of the application, in particular (a) the dynamics of the robot, (b) the information density that can be achieved with a certain set of sensors, and (c) the impact of the work-flow and human factors. For a given data set of a particular application, the framework allows the determination of relevant modification limits and thus the boundaries of the workspace, where safe operations of the application are still guaranteed and the desired flexibility of the robotic system is retained without necessarily conducting a new risk assessment.

**Acknowledgments.** The research results of this work originate from the project DR.KORS – Dynamic reconfigurability of collaborative robot systems, funded by the Austrian Research Promotion Agency (FFG) with the project number 864892.

## References

1. DGUV: Collaborative robot systems - Design of systems with "Power and Force Limiting Function" function. Technical Report FB HM-080 Issue 08/2017, Deutsche Gesetzliche Unfallversicherung (2017)
2. Haddadin, S., De Luca, A., Albu-Schäffer, A.: Robot collisions: a survey on detection, isolation, and identification. IEEE Trans. Robot. **33**(6), 1292–1312 (2017)
3. Kirschner, D., Schlotzhauer, A., Brandstötter, M., Hofbaur, M.: Validation of relevant parameters of sensitive manipulators for human-robot collaboration. In: Proceedings of the 26th International Conference on Robotics in Alpe-Adria-Danube Region (RAAD 2017) (2017)
4. Neubauer, M., Gattringer, H., Bremer, H.: A persistent method for parameter identification of a seven-axes manipulator. Robotica **33**, 1099–1112 (2014)
5. Riedelbauch, D., Henrich, D.: Fast graphical task modelling for flexible human-robot teaming. In: 50th International Symposium on Robotics, ISR 2018, pp. 1–6. VDE (2016)
6. Saenz, J., Elkmann, N., Gibaru, O., Neto, P.: Survey of methods for design of collaborative robotics applications-why safety is a barrier to more widespread robotics uptake. In: Proceedings of the 2018 4th International Conference on Mechatronics and Robotics Engineering, pp. 95–101. ACM (2018)
7. Schröter, D.: Entwicklung einer methodik zur planung von arbeitssystemen in mensch-roboter-kooperation. Ph.D. thesis, University of Stuttgart (2018)
8. Siciliano, B., Khatib, O.: Springer Handbook of Robotics. Springer, Cham (2016)
9. Siegwart, R., Nourbakhsh, I.R., Scaramuzza, D.: Introduction to Autonomous Mobile Robots, 2nd edn. The MIT Press, London (2011)
10. Technical Specification: ISO/TS 15066:2016. Robots and robotic devices – Collaborative robots. Technical reports, International Organization for Standardization (2016)

# Kinematic and Force Experiments on Cadavers for the Specification of a Tele-Operated Craniotomy Robot

Terence Essomba[1], Juan Sandoval[2], Med Amine Laribi[2(✉)], Chieh-Tsai Wu[3], Cyril Breque[2,4], Said Zeghloul[2], and Jean-Pierre Richer[4]

[1] Department of Mechanical Engineering, National Central University, No. 300, Zhongda Road, Zhongli City, Taoyuan County 32001, Taiwan

[2] Department of GMSC, Pprime Institute CNRS, ENSMA, University of Poitiers, UPR 3346, Poitiers, France
med.amine.laribi@univ-poitiers.fr

[3] Department of Neurosurgery, Chang Gung Memorial Hospital, No. 5, Fuxing Street, Taoyuan 333, Taiwan

[4] Simulation Laboratory, Faculty of Medicine, University of Poitiers, 6 rue de la Miletrie, Poitiers 86000, France

**Abstract.** This paper presents a preliminary study for the characterization of Burr hole craniotomy procedures, in order to define the specifications for the design of a tele-operated craniotomy robot. Surgeon's gestures are recorded through a motion capture system during the execution of several craniotomy tasks, allowing to characterize the drilling tool range movements. Moreover, a tele-operated platform is used to record the interaction forces produced between the drilling tool and the human skull during the procedure, obtaining an order of magnitude of the force feedback a dedicated tele-operated craniotomy system would need to provide during a craniotomy procedure.

**Keywords:** Craniotomy · Tele-operation · Motion capture · Force interaction · Cadavers

## 1 Introduction

In most neurosurgery operation, the surgeon needs to access the patient's brain. However, it is protected by the skull, which considered ad the most solid bone of human anatomy. To allow a physical access to the brain, it is required to perform a craniotomy. The neurosurgeon used series of surgical instruments to cut and remove a piece of skull, namely "bone flap" [1]. Several types of neurosurgical drills can be uses depending on the type of craniotomy to perform [2]. Today, some surgical tools dedicated to craniotomy are all semi-automatic. This feature can provide a higher level of safety compared to fully classical tool [3, 4]. But there is one type of drill that offers no automatic safety functionality but that allows higher accuracy. It is usually required in some complicated cases which are not suitable with the use of semi-automatic ones. The tip of this drill has the shape of a ball allowing to drill both radially and axially.


K. Berns and D. Görges (Eds.): RAAD 2019, AISC 980, pp. 447–454, 2020.
https://doi.org/10.1007/978-3-030-19648-6_51

It is referred to as "diamond burr" or "rosen burr" depending on the surface geometry of the drill burr cut [5]. Although this instrument provides more freedom of motion and flexibility, it is operated only by highly experienced neurosurgeon due to the higher level of dexterity that its manipulation requires.

Although several robotic systems for craniotomy has been reported in the literature, these cases remain unusual in the practice as only a small number were reported in literatures. The first instance of robotic craniotomy has been reported in 1998, with the use of an industrial manipulator PUMA 260 to perform craniofacial surgery [6]. Another robotic manipulator called "Neurobot" has been developed for craniotomy [7]. An anthropomorphic manipulator has been reprogrammed in 2003 to control the force interaction between the drill and the patient's skull in otoneurosurgery [8]. The RobaCKa system, which is a combination between a CASPAR industrial robot and an optoelectronic navigation system (Polaris) have been used for the first craniotomy trials on real patients [9]. A Steward platform (hexapod) has been equipped with a drill mounted on an active prismatic joint [10]. A mechanism that decouples position and orientation has been used for a series of simulated craniotomy on a plastic skull model [11]. The "CRANIO" robotic system has been used in a series of experiments [12]. A reconfigurable hexapod directly mounted on the patient's head has been introduced for keyhole craniotomy [13].

Only fully automatic procedures have been proposed in literature that require a difficult pre-operative procedure to program a very accurate trajectory. During the post-operative phase, a very accurate registration will be also necessary to insure that the references of both pre-operative and intra-operative trajectories are precisely matched. The solution to this problem is to design a tele-operated robotic system that can allow a neurosurgeon to perform a craniotomy. The use of this procedure will not require any pre-operative trajectory planning as the operator directly control the robot by using a control device. In the present study, a series of motion capture and force experimentation are performed to the kinematic and force interaction characterization of Burr hole craniotomy procedure. The collection of these data will allow the kinematic and force optimization of a mechanism for the design of a tele-operated craniotomy robot.

The paper is organized as follows. The experimental protocol used for the proposed study is presented in Sect. 2. The recorded data are presented and analyzed in Sect. 3. Last section proposes the conclusion of the proposed research work.

## 2 Experimental Protocol

The development of robotic systems for craniotomy needs the identification of the need specifications linked to the application which can be performed in two steps. The first one focuses on the identification of the nature of the movements required to cut and remove a piece of skull. The second one concerns the measures of the interaction effort between the drilling tool and the skull.

One of the main tasks encountered in this procedure is to well manipulate the drill burr cut requesting more freedom of motion and flexibility. An experienced neurosurgeon was asked to perform this task on both, a training station (on 3D printed skull model) and in real situation (on cadaver), as shown in Fig. 1, in order to analyze the motion and the technique he uses.

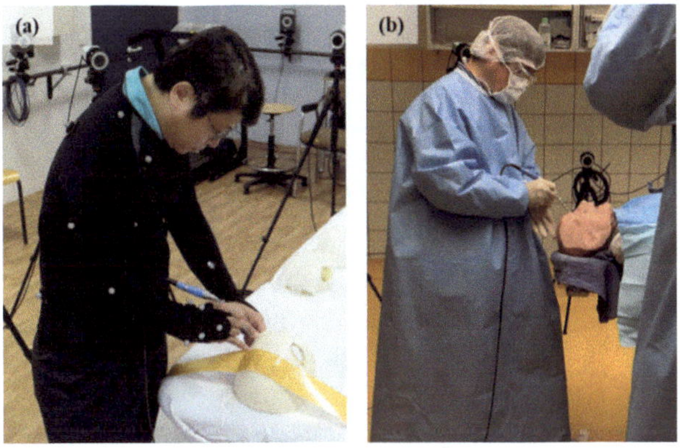


**Fig. 1.** Experimental setup: (a) Training situation in 3D printed skull model - (b) Training in real situation (on cadaver).

The aim is to characterize the motion of the expert hand during a craniotomy and estimate the interaction effort. The motion capture system is used to record the surgeon gestures. The choice of marker sets as well as segment reference definition are based on the experience of the biomechanics community [14].

A set of 10 markers were fixed on the expert and the drill. A motion capture system composed by 8 high speed cameras was used to track and record the position of these markers. A scheme depicting the system is presented in Fig. 2.

Qualisys Tracking Manager (QTM) is the software used to process all the records, to label and to regroup into segments those markers being considered as part of a

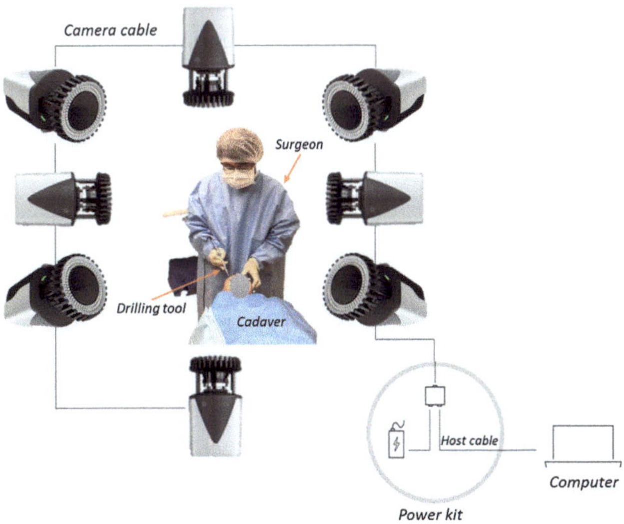


**Fig. 2.** A scheme of the used Qualisys motion capture system, including 8 Miqus cameras, a camera cable, a power unit, and a computer.

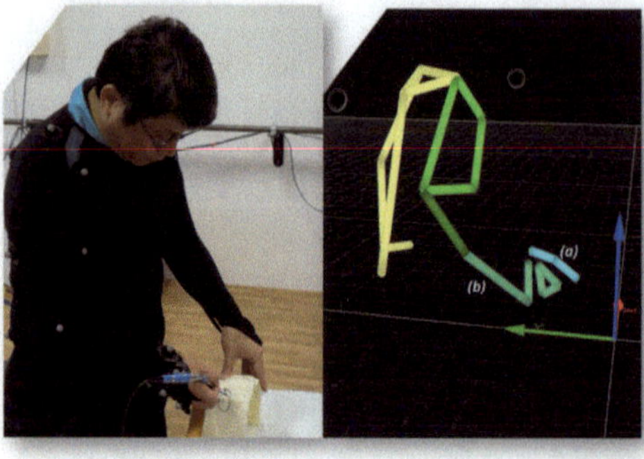


**Fig. 3.** Reconstruction of several detected markers into segments. (a) Drilling tool and (b) Expert's arm.

rigid body. One of these segments is the drilling tool represented in blue in Fig. 3. The 3-D coordinates of each marker is then exported and used to compute the drilling tool orientation.

In order to record the efforts exerced by the drilling tool on the skull, we used a tele-operated robotic platform using a 3-DoF master device (Novint Falcon) and a 7-DoF collaborative robot as the slave device (Franka Emika). The surgeon pilotes the translational movements of the drilling tool held by the robot, by moving the ball grip of the Novint Falcon. The buttons of the ball frip can also be activated by the surgeon to modify the orientation of the drilling tool. Each joint of the Franka robot is provided with a torque sensor, allowing to measure the external torques due to the contact with the skull. These external torques $\tau_{ext} \in \Re^7$ are projected to the cartesian space to obtain the contact forces/torques $F_{ext} \in \Re^6$ between the tool (at the tcp of the robot arm) and the skull, through $F_{ext} = (J^T)^+ \tau_{ext}$, where $J \in \Re^{6\times7}$ is the jacobian matrix which denotes the relationship between the joint space and the cartesian (end-effector) space.

We use a ROS-based framework to allow the data exchanges between the master and the slave system. Figure 4 shows a block diagram of the platform.

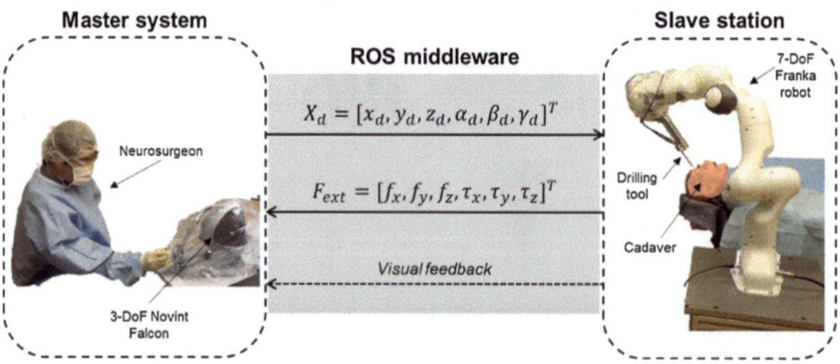


**Fig. 4.** Tele-operation platform with force feedback.

## 3 Data Analysis: Gestures and Effort

The data analysis of the medical gesture is an important step allowing the task characterization on which the robot design is based. The aim is to quantify the gesture of a medical expert. This gesture is to be reproduced by the robotic system to be designed. The results of these studies need be discussed with the medical personnel for a better comprehension of the gesture itself.

The robot system needs 5 Degree of Freedom (DoF), 2 rotations with a remote center of motion (RCM) and 3 translations to exert a force along the tool axis and follows the path on skull. The RCM is located at the contact point between the drilling toll tip and the patient skull. As shown in Fig. 5, the workspace is defined by a cone. The apex angle of the cone depends on the location on the cranial bones (as will be discussed in the next paragraph and shown in Table 1).

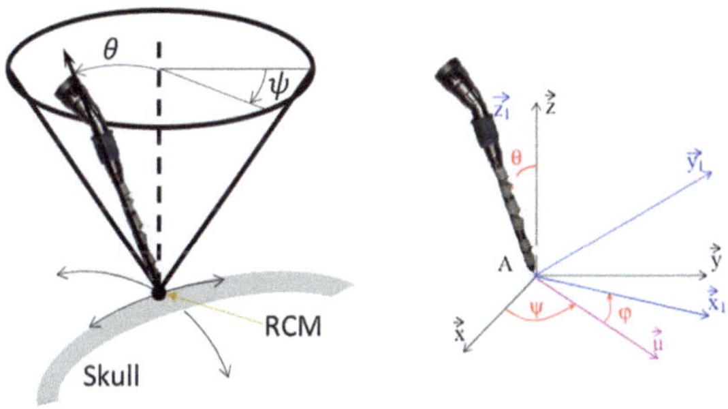


**Fig. 5.** Swept workspace by drilling tool with the 5 DoF.

The orientation of the drilling tool is defined using Euler angles. Only precession and nutation rotations, $\psi$ and $\theta$, are deduced from the recorded gestures. The intrinsic rotation, $\varphi$, can be easily identified but not considered. In fact, this angle does not impact the final results due to symmetry of the drilling tool, spherical ball. The orientation of the drilling tool is linked to the craniotomy locations on skull. Examples of location are depicted on Fig. 6.

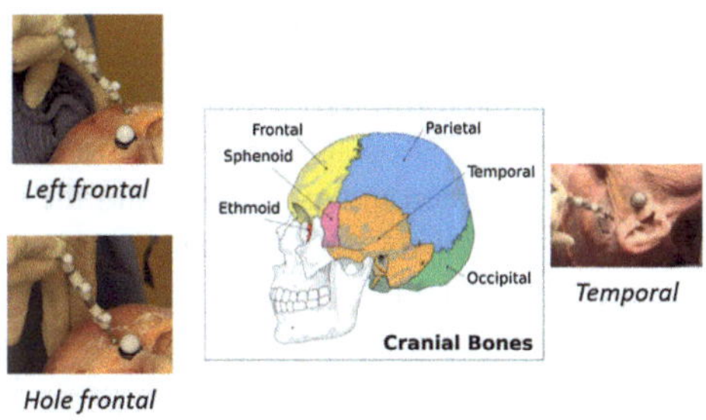


**Fig. 6.** Craniotomy locations on skull and their illustrations during real operations on cadaver.

Only angle boundaries for four locations have been given in this paper, on Table 1, which represent the most explored locations in craniotomy. One observe that the sweeping intervals are not greater than 18° all combined locations (see Fig. 7 for more details).

**Table 1.** Bounding values for orientation angles.

| Orientation angles | | Locations | | | |
|---|---|---|---|---|---|
| | | Hole frontal | Left frontal | Right frontal | Left temporal |
| $\theta$ (°) | Min | 131 | 129 | 132 | 135 |
| | Max | 137 | 134.7 | 140 | 140.5 |
| $\psi$ (°) | Min | −78 | −113 | −77 | −130 |
| | Max | −69 | −95 | −66.3 | −124 |

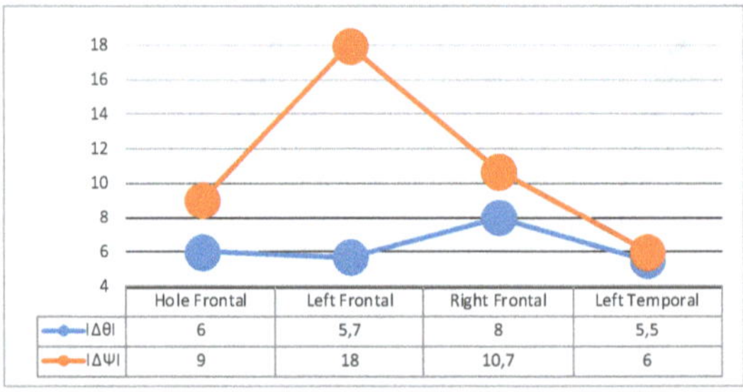


**Fig. 7.** Sweeping intervals of orientation angles (°) for all locations on the skull.

In order to identify the order of magnitude of the interaction forces generated during a craniotomy, Neurosurgeon used the tele-operation platform presented in Fig. 4 to execute different craniotomy tasks on cadavers. Figure 8 shows the recorded forces for a linear milling trajectory, executed on the frontal bone of the skull. Other similar trajectories were executed, obtaining the same order of magnitude for the tangential force, i.e. around 13 N. The identification of this order of magnitude will be usefull for the design of a 6-DoF master device capable to provide the neurosurgeon with a precise force feedback.

The impact of the drilling speed and the robot stiffness parameters on the variation of the interaction forces was evaluated through a simple drilling task performed on the frontal bone, which results are resumed on the Table 2. The obtained results allow to confirm the inverse relationship between the axial force and the drilling speed. Since a collaborative robot has been employed for the experience, a compliant control strategy is used to guarantee a safe physical human-robot interaction. Therefore, the robot stiffness parameter allows to define the level of compliance of the robot, where a higher stiffness produces the increase of the interaction forces, as proved in the last row of the Table 2.

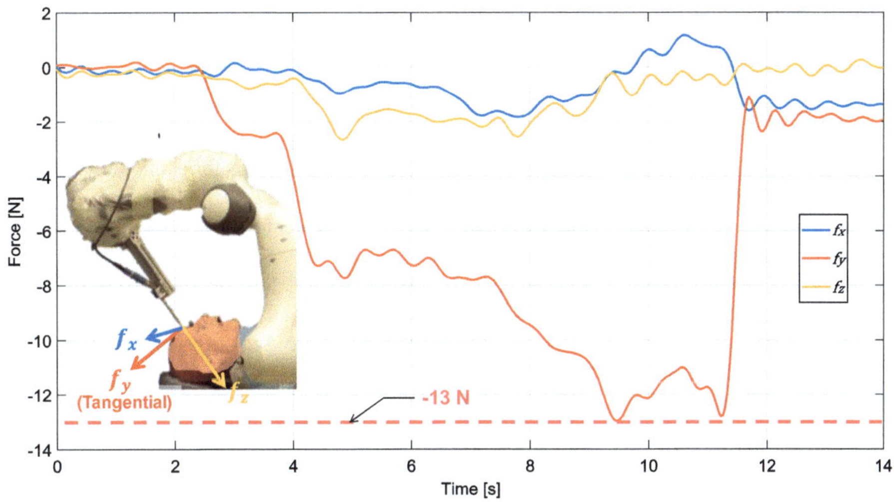


**Fig. 8.** Interaction forces recorded between the drilling tool and the skull during the execution of a tangencial trajectory.

**Table 2.** Maximum axial force for a simple drilling task.

| Drilling speed [RPM] | Robot stiffness [N/m] | Maximum axial force $f_z$ [N] |
|---|---|---|
| 25000 | 600 | 7.4 |
| 50000 | 600 | 5.3 |
| 80000 | 600 | 2.1 |
| 80000 | 750 | 13 |

# 4 Conclusion

A study for the characterization of Burr hole craniotomy procedures has been presented in this paper, allowing the analysis of the surgeon's gestures performed throughout the execution of the surgical task. A motion capture system was employed to identify the kinematics of the surgeon's movements, whereas a tele-operated platform was used to determine the order of magnitude of the interaction forces generated between the drilling tool and the human skull. These experimentations were carried out under real conditions, i.e. on cadavers. The collected data will be exploited in the future work for the design of a tele-operated craniotomy robot.

**Acknowledgements.** The present work has been supported by a collaboration project between Taiwan's MOST (107-2911-I-008-505) and France's CNRS (PRC2142). This work was also sponsored by the French government research program *Investissements d'avenir* through the Robotex Equipment of Excellence (ANR-10-EQPX-44).

## References

1. Bast, P., Popovic, A., Wu, T., Heger, S., Engelhardt, M., Lauer, W., Radermacher, K., Schmieder, K.: Robot-and computer-assisted craniotomy: resection planning, implant modelling and robot safety. Int. J. Med. Robot. Comput. Assist. Surg. **2**, 168–178 (2006)
2. Hsiao, M.-H., Kuo, C.-H.: A review to the powered drilling devices for craniotomy. J. Med. Devices **6**(1), 017557 (2012)
3. Bofinger, G., Wolfle, W.: Skull Trepanation Drill. U.S. Patent No. 4,319,577, U.S. Patent and Trademark Office, Washington, DC (1982)
4. Ahola, J.J., Harris, D. G.: Blade Guard for a Surgical Tool. U.S. Patent No. 6,001,115, U.S. Patent and Trademark Office, Washington, DC (1999)
5. https://www.wfns.org/WFNSData/Document/Aesculap_Basic_Sets_of_Neurosurgical_Instruments.pdf?dl=0
6. Burghart, C., Raczkowsky, J., Rembold, U., Wörn, H.: Robot cell for craniofacial surgery. In: Proceedings of the 24th Annual Conference of the IEEE Industrial Electronics Society, Aachen, Germany, 31 August-4 September, pp. 2506–2511 (1998)
7. Sim, C., Ng, W.S., Teo, M.Y., Loh, Y.C., Yeo, T.T.: Image-guided manipulator compliant surgical planning methodology for robotic skull-base surgery. In: International Workshop on Medical Imaging and Augmented Reality, Shatin, Hong Kong, China, 10–12 June, pp. 26–29 (2001)
8. Federspil, P.A., Geisthoff, U.W., Henrich, D., Plinkert, P.K.: Development of the first force-controlled robot for otoneurosurgery. Laryngoscope **113**(3), 465–471 (2003)
9. Korb, W., Engel, D., Boesecke, R., Eggers, G., Kotrikova, B., Marmulla, R., Raczkowsky, J., Wörn, H., Mühling, J., Hassfeld, S.: Development and first patient trial of a surgical robot for complex trajectory milling. Comput. Aided Surg. **8**(5), 247–256 (2003)
10. Tsai, T.C., Hsu, Y.L.: Development of a parallel surgical robot with automatic bone drilling carriage for stereotactic neurosurgery. Biomed. Eng. Appl. Basis Commun. **19**(4), 269–277 (2007)
11. Matinfar, M., Baird, C., Batouli, A., Clatterbuck, R., Kazanzides, P.: Robot-assisted skull base surgery. In: Proceedings of the IEEE/RSJ International Conference on Intelligent Robots and Systems, San Diego, CA, USA, 29 October-2 November, pp. 865–870 (2007)
12. Cunha-Cruz, V., Follmann, A., Popovic, A., Bast, P., Wu, T., Heger, S., Engelhardt, M., Schmieder, K., Radermacher, K.: Robot- and computer-assisted craniotomy (CRANIO): from active systems to synergistic man-machine interaction. Proc. Inst. Mech. Eng. Part H J. Eng. Med. **224**(3), 441–452 (2010)
13. Kobler, J.P., Kotlarski, J., Öltjen, J., Baron, S., Ortmaier, T.: Design and analysis of a head-mounted parallel kinematic device for skull surgery. Int. J. Comput. Assist. Radiol. Surg. **7**(1), 137–149 (2012)
14. Cavanagh, G.W.: ISB recommendations for standardization in the reporting of kinematic data. J. Biomech. **28**(10), 1257–1261 (1995)

# Efficient, Risk-Encoding Octrees for Path Planning with a Robot Manipulator

Tobias Werner(✉), David Harrer, and Dominik Henrich

Lehrstuhl für Robotik und Eingebettete Systeme,
Universität Bayreuth, 95440 Bayreuth, Germany
tobias.werner@uni-bayreuth.de
http://robotics.uni-bayreuth.de

**Abstract.** Recent research in robotics envisions shared human-robot workspaces to combine individual advantages of humans and robots. As part of this vision, robot manipulators must avoid collisions with humans and other a priori unknown obstacles in the shared workspace. State-of-art approaches (e.g. certainty grids, bounding volume hierarchies) test robot poses for collisions under individual limitations (e.g. memory or processing overhead, assumption of global views or noiseless sensors). In contrast, we contribute a sample-based pose test alongside a hierarchical representation of risk over the shared workspace. Our contribution has low memory and processing overhead, and allows for local, outdated, or noisy sensor views. Experiments with real-world data validate this claim and show advantages and limits of our approach over competing variants. We conclude that our pose test and risk representation enhance real-time path planning for robot manipulators in current and future use cases.

**Keywords:** Path planning · Robot manipulator · Risk-based environment model · Shared human-robot workspaces

## 1 Introduction

Human-robot collaboration promises to combine individual virtues of humans and robot manipulators. Respective research envisions various future use cases for robots, from flexible industrial automation to applications in small businesses and the service sector. However, human-robot collaboration mandates shared human-robot workspaces, which contain a priori unknown obstacles (e.g. humans or human-placed objects). The robot manipulator must thus be able to avoid obstacles in real-time. State-of-art solutions for real-time obstacle avoidance (e.g. [13,14]) propose a two-tiered approach: Real-time workspace monitoring with a multi-camera system finds 3D workspace volumes that are occupied by obstacles. Concurrently, real-time path planning finds collision-free or risk-minimized robot trajectories by testing robot poses against the occupied workspace volumes.

Pose tests are a crucial aspect when developing a joint monitoring and path planning solution. Tests must be efficient to enable real-time robot reactions, but


K. Berns and D. Görges (Eds.): RAAD 2019, AISC 980, pp. 455–462, 2020.
https://doi.org/10.1007/978-3-030-19648-6_52

tests also depend on the results of workspace monitoring (e.g. data structure, available information about obstacles). For instance, representing obstacles as a certainty grid (e.g. [11]) lends itself to sampling-based risk queries per pose (e.g. [1]), while a bounding volume hierarchy over obstacle points (e.g. [12]) enables hierarchical distance queries per pose (e.g. [3]). A choice of obstacle representation and pose test furthermore limits the capabilities of later path planners: Certainty grids, for example, natively support planning risk-minimized robot trajectories, while voxel grids with binary occupation (e.g. [9]) do not. In total, a holistic approach to monitoring and pose tests is necessary.

Formally, state-of-art path planners (e.g. [7]) for a robot manipulator with $d$ joints use a generic pose test $t : \mathbb{R}^d \rightarrow C$. This generic test evaluates a pose $p$ to one value $t(p)$ in regard to some criterion $C$, such as to discard randomly sampled poses or to weight multiple trajectory alternatives. For example, $C = \{\text{true}, \text{ false}\}$ for collision-free trajectory planning (i.e. $t(p)$ indicates collision or no collision at pose $p$), and $C = \mathbb{R}$ for risk-minimized trajectory planning (i.e. $t(p)$ names the risk of a collision at pose $p$ or a suitable risk heuristic).

Pose tests in turn ideally compare the workspace volume $V : \mathbb{R}^d \rightarrow 2^{\mathbb{R}^3}$ that the robot occupies for a pose $p$ to any obstacles. Let $O : 2^{\mathbb{R}^3} \rightarrow C$ be the ground-truth result of this comparison, thus $t(p) = O(V(p))$. However, a ground-truth $O$ is not readily available (e.g. due to a priori unknown obstacles, sensor noise, or inherent discretization). This leads to our problem statement: We seek an approximation $O_H : 2^{\mathbb{R}^3} \rightarrow C$ of $O$. Note the approximation $O_H$ encompasses both a data structure for available information on workspace obstacles and a corresponding per-pose test algorithm. To support real-time path planning, the data structure of $O_H$ must be efficient to build, and the pose test $O_H$ must be fast to compute, which gives two conflicting constraints.

Current state-of-art approaches to an approximation $O_H$ have several limits. We discuss select approaches and their limits in Sect. 2. In contrast to existing approaches, we contribute an alternative data structure and query for $O_H$: We use a hierarchical representation for $O_H$ that is both efficient to build and to test. Additionally, our $O_H$ supports risk queries. We detail our data structure and pose test algorithm in Sect. 3. Experimental results with real-world data and a critical comparison against state-of-art approaches from afore follow in Sect. 4 and substantiate our conclusion in Sect. 5.

## 2 Related Work

Data structures currently in use for $O_H$ include point clouds with a superimposed tree structure (e.g. [12]), voxel grids (e.g. [9]), distance grids (e.g. [10]), distance octrees (e.g. [6]), certainty grids (e.g. [11]), or octrees over binary (e.g. [13]) or probabilistic (e.g. [4]) payload. Robots are commonly represented as exact triangle meshes (e.g. [14]), as convex volume approximations (e.g. [3]), or as sampling-based representations (e.g. here), all optionally with a superimposed bounding volume hierarchy. In general, hierarchical data structures take more time to build, but require less memory and afford more efficient pose tests.

Pose tests, on the other hand, depend on the chosen data structure. Exact tests on primitives (e.g. [15]) give exact results, but are slow if $O_H$ consists of many primitives. Sampling-based pose tests (e.g. here) check individual robot points against $O_H$ and are particularly suited for grid-based data structures. Optimizing variants (e.g. [2]) require a volumetric data structure, but are more efficient than sampling if highly precise results are required. Finally, hierarchical collision or distance tests (e.g. [3]) accelerate all other pose tests, but require superimposed hierarchical data structures and thus increase build times.

Basic holistic variants use voxel grids and sample these over the robot model to check for collisions (e.g. [9]). While intuitive to implement, this approach does not scale well with increasing resolution of $O_H$. Efficient holistic approaches, in contrast, build bounding volume hierarchies over robots and obstacles and evaluate these with exact (e.g. [15]) or optimizing (e.g. [3]) pose tests.

## 3 Our Approach

As a basis for a risk-encoding obstacle approximation $O_H$, we need to establish an understanding of risk in the workspace of a robot: Risk commonly is defined as the probability of an event times the loss resulting from that event (e.g. [5]). The event, in our case, is a collision between the robot manipulator and any obstacle in the work cell. We assume a worst-case, constant loss (e.g. independent of robot speed). We furthermore assume that there is a direct correlation between the distance to indeterminately moving obstacles and collision probability.

Under above assumptions, we propose to use the closest distance to obstacles as a heuristic for risk associated with a single point of the workspace. The risk approximation $O_H$ for a robot pose then is the minimum closest distance to obstacles over all points in the workspace volume that the robot manipulator occupies in that pose. Extended risk options may additionally weight the per-point or per-pose distance (e.g. by sensor noise, data age, or robot speed).

In the following, we discuss a holistic approach to a risk-based $O_H$. To this end, we propose a variant of octree as a data structure to encode distance-based risk over the workspace. Notably, we extend the well-known Brushfire algorithm to efficiently build such an octree from a traditional octree over binary obstacle occupation. Finally, we examine sampling of the risk octree at points on the surface of the robot model as a pose test.

### 3.1 Hierarchical Brushfire Algorithm

The Brushfire algorithm (e.g. [10]) calculates an approximation of the distance to the closest obstacle for all cells in a grid representation of the workspace. To do so, the Brushfire algorithm initially marks distances for cells in free space as unknown, for cells in obstacles as zero. Brushfire then iteratively propagates and increments known closest distances from grid cells to any neighboring cells of unknown distance.

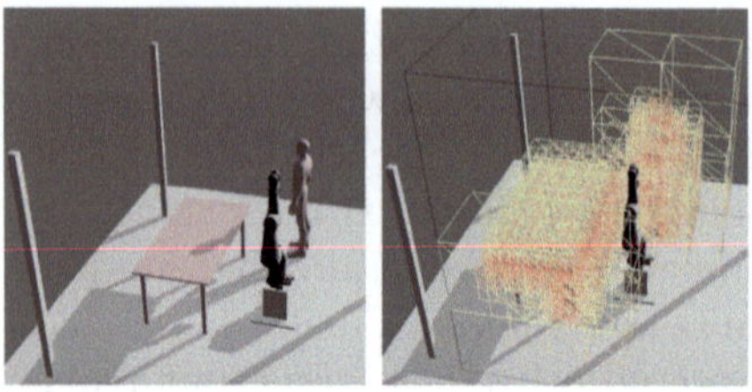

**Fig. 1.** A mock-up workspace (left) and an excerpt of the risk-encoding octree (right) with high-risk leaf nodes close to obstacles (red) and low-risk leaf nodes far from obstacles (orange).

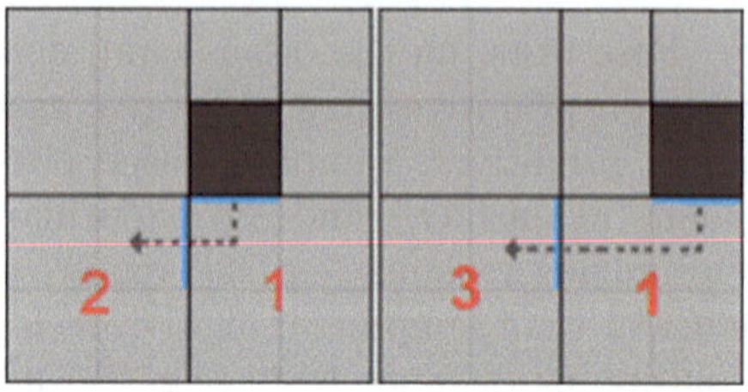


**Fig. 2.** An octree section over free space (light) and obstacles (dark). Any Brushfire distances over node corners depend on the propagation direction (dotted arrow). Marks (blue) keep track of this.

We extend the Brushfire algorithm to work on the nodes of an octree over binary obstacle occupation (e.g. [13]): We initialize closest distances of all nodes in free space to unknown, closest distances of all occupied nodes to zero. Thereafter, we iteratively propagate and increment closest distances from nodes to any larger neighboring nodes. We stop once each node has a closest distance. See Fig. 1 for a workspace and the resulting risk-encoding octree.

In the above outline, actual distance propagation requires further elaboration: Since nodes of an octree in general do not share the same size, we usually increase any propagated distance not by a fixed cell size, but by the size of the originating octree node. However, to get a correct and reasonably exact lower bound on the distance to the closest obstacle, we must consider three special cases: Propagation of distances over corners, propagation to smaller nodes, and refinement of too large nodes. An in-depth discussion of each special case follows. See Algorithm 1 for accompanying pseudo-code.

**Special Case: Propagation over Corners.** In a grid, each Brushfire iteration increases the propagated distance by the size of a cell regardless of the direction to the closest obstacle. In an octree, however, we can only increase the propagated distance by one full node size if the next obstacle lies in the opposing direction of the neighboring node. When propagating around a corner, adding a full node size yields too large distances. Figure 2 illustrates this problem.

We solve above problem by tracking from which direction a closest distance propagated to some octree node. To this end, we introduce node marks: We classify all sides of octree nodes in regard to global coordinates (e.g. positive $X$ side or negative $Z$ side) and assign local 2D coordinates. Marks then consist of an identifier for a cube side and a position in local 2D coordinates.

By placing and evaluating marks, we can propagate correct closest distances over corners: When propagating a closest distance from an originating node to a neighboring target node, we need to check the originating node for marks on sides perpendicular to the shared side between both nodes. Any mark closest to the shared side is the correct mark to propagate distance from. Instead of a full node size, we add only the distance between mark and target node to the closest distance of the originating node to find the closest distance of the target node. Finally, we place a new mark on the shared side of the target node, with local coordinates closest to the selected mark of the originating node.

**Special Case: Propagation to Smaller Nodes.** Our Brushfire variant iterates over the octree from smaller nodes to larger ones. This iteration order is efficient, as it requires no tracking or searching of all (possibly many) smaller neighbors for large nodes. However, our iteration order also implies that marks cannot propagate distances into the opposite direction: from large nodes to neighboring, smaller nodes. In turn, the distances for smaller nodes may be wrong when the direction to the closest obstacle leads through a larger node.

To solve this problem, we separately propagate distances from larger nodes to smaller nodes. To this end, we store distance information not only in leaf nodes, but also in inner nodes of the octree. Each inner node then propagates distance information to its children: If a child has a distance that is larger than or equal to the distance of a parent inner node and if this child shares one or more marks with this parent, we copy the marks of the parent to the child. Note this possibly entails an offset in local coordinates. If the parent additionally has a lower closest distance than the child, then the child discards its marks and assumes the closest distance of its parent. See Fig. 3 for an example.

**Special Case: Refinement of Too Large Nodes.** Depth and thus resolution of the original, binary octree depend on matches in the structure of octree and obstacles. This implies that there may be large octree nodes close to obstacles if the obstacle and node surfaces are similar. In turn, our hierarchical Brushfire may associate a large volume of the workspace with a single closest distance.

Too small distances still give a conservative and thus safe approximation of risk. However, path planners find better (e.g. more reliable) trajectories if the risk approximation $O_H$ is close to the ground-truth obstacle occupation $O$.

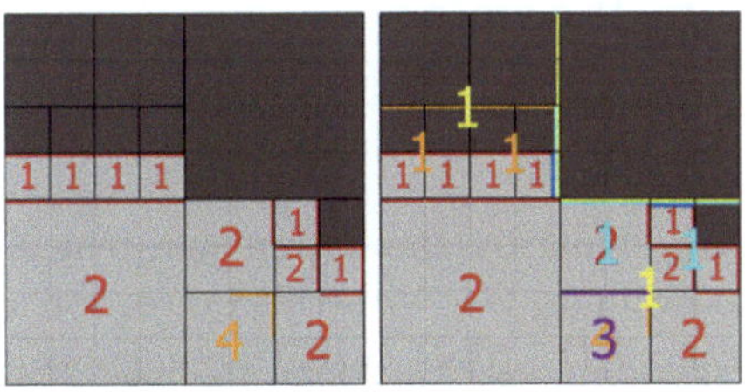

**Fig. 3.** Left: Incorrect closest distances (e.g. orange four), as not propagated from the large, occupied upper-right node. Right: Propagating intermediate distances on inner nodes fixes this issue (additional numbers and marks).

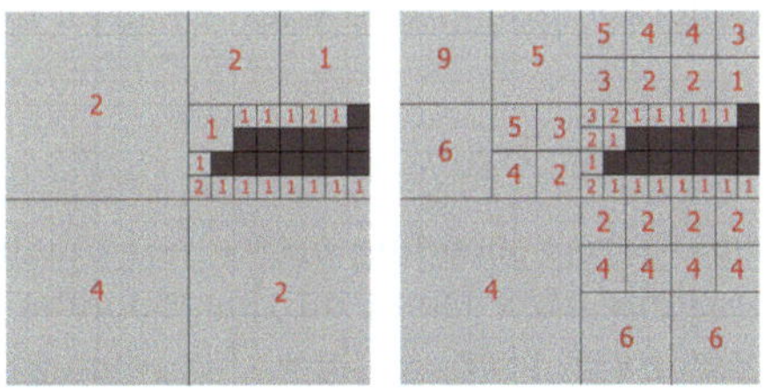

**Fig. 4.** Left: Overlapping structures of octree and obstacles cause large nodes close to the obstacles. Distance errors impede path planning in remote regions. Right: Adaptive subdivision improves distances on remote nodes.

```
Input   : Octree O_H over binary obstacles
Output: Octree O_H with additional risk encoding

heap openlist, topdownlist        // Ordered by increasing node size

Insert all occupied nodes of O_H into openlist

while !openlist.empty() and !topdownlist.empty() do
    foreach node k = openlist.pop() do
        foreach neighbour n of k do
            Calculate distance d for n from k
            if d < n.distance then
                save d as n.distance and copy marks
                if n.isleaf() then
                    openlist.insert(n)
                else
                    topdownlist.insert(n)

    foreach node k = topdownlist.pop() do
        foreach child c of k do
            if !c.occupied() and c.distance > k.distance
                                and k has mark over c then
                copy distance and marks from k to c
                if c.isleaf() then
                    openlist.insert(c)
                else
                    topdownlist.insert(c)
```

**Algorithm 1.** Hierarchical Brushfire Algorithm.

Our octree Brushfire therefore increases the precision of $O_H$ by splitting large nodes close to obstacles: Whenever we store a closest distance in a node, we evaluate a splitting heuristic with thresholds for distance and node size. If the heuristic is met, we split the node and treat any new children as specified under propagation to smaller nodes. This procedure results in an octree with finer resolution around obstacles. Figure 4 illustrates this process.

### 3.2 Pose Tests

Once a risk-encoding octree is available (e.g. built by the above hierarchical Brushfire algorithm), all that remains is to evaluate $O_H$ with this octree for a given robot pose. To do so, we propose to sample the robot surface by regularly spaced points. After rigid transformations to world coordinates for a given robot pose, we can push each point through the octree to determine a matching leaf node. Fusing values of all leaf nodes (e.g. closest obstacle distance) then gives the desired result of $O_H$. The actual fusing method depends on the risk encoded in the octree. For a distance-based risk, we choose the closest distance over all of the samples as the output of $O_H$.

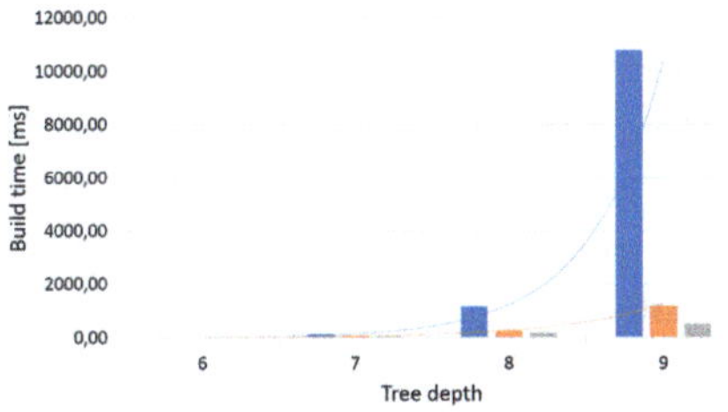


**Fig. 5.** Build times for our hierarchical Octree brushfire without (gray) and with (orange) refinement as opposed to a grid variant (blue) over select resolutions.

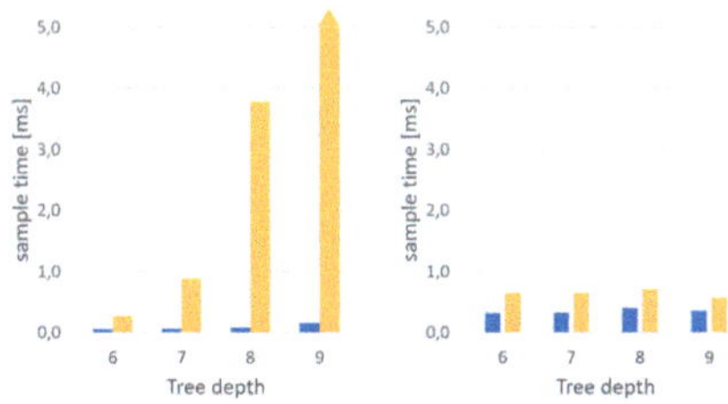


**Fig. 6.** Sampling times for robot poses with points (blue) or triangles (orange) as samples in a grid (left) or our risk-encoding octree (right).

## 4 Evaluation

In the following, we evaluate our hierarchical octree Brushfire and the associated sampling-based pose test. As ground-truth obstacles, our mock-up workspace contains a human and a table. Furthermore, we use a CAD model of a Stäubli TX90-2L robot arm for pose tests. This model consists of 286 triangles, which we approximate by around 3000 points of mostly equal spacing. See Fig. 1 for an image of our mock-up workspace.

We first compare building times of our Brushfire algorithm with and without refinement against a traditional, grid-based variant. To this end, we used grids of various resolutions and octrees of matching lowest-level resolutions. Overall, our octree Brushfire outperforms the grid-based variant, with the benefit growing super-linearly by resolution. Figure 5 shows respective timings.

Next, we evaluate pose test times on our risk-encoding octree against a risk-encoding grid. As a counterpart to our approximate point sampling test, we used a naive triangle-based pose test to determine exact risks. Each experiment involved timing pose tests for 200 random poses. Triangle tests on the grid quickly become infeasible with growing grid resolution due to the number of cells to check for intersection. In contrast, our risk-encoding octree effortlessly handles triangles even at a large resolution. Pose tests with point sampling are substantially faster on the grid, since no octree traversal is necessary. Figure 6 presents timings for each combination of data structure and pose test.

Overall, slower sampling times partially offset the benefit of our efficient octree Brushfire. A reasonable comparison thus needs to amortize the build time over sampling times. Our experiments show that for a grid with $2^9 = 512$ subdivisions and a matching octree of depth 9, our hierarchical Brushfire amortizes building costs only for less than about 100,000 pose tests per build of the data structure. Otherwise, the grid is faster. In practice, reasonable real-time path planners are bound to real-time update rates around or above 10 Hz (e.g. [14]). At the measured 0.5 ms for 200 pose tests, such planners can test at most 10,000 poses per update and thus do not reach the break-even point of grids. We conclude that our octree outperforms grids in practically relevant cases.

## 5 Conclusion

In the preceding, we have contributed a risk-encoding octree and an algorithm that builds this octree from an existing, binary octree over obstacle occupation. The resulting risk-encoding octree balances acceptable build times against fast pose tests and improves timings over naive grid variants for practically relevant numbers of pose tests. Future work examines further effects of weighting risk by robot speed or sensor noise.

**Acknowledgements.** This work has partly been supported by the Deutsche Forschungsgemeinschaft (DFG) under grant agreement He2696/11 SIMERO.

## References

1. Borenstein, J., Koren, Y.: Real-time obstacle avoidance for fast mobile robots. IEEE Trans. Syst. Man Cybern. **19**, 1179–1187 (1989)
2. Cameron, S.: Enhancing GJK: computing minimum and penetration distances between convex polyhedra. In: ICRA (1997)
3. Henrich, D., Cheng, X.: Fast distance computation for on-line collision detection with multi-arm robots. In: ICRA (1992)
4. Hornung, A., et al.: OctoMap: an efficient probabilistic 3D mapping framework based on octrees. Auton. Robots **34**, 189–206 (2013)
5. ISO: ISO31000 Risk management–principles and guidelines. International Organization for Standardization (2009)
6. Jung, D., Gupta, K.K.: Octree-based hierarchical distance maps for collision detection. J. Robot. Syst. **14**, 789–806 (1997)
7. Kavraki, L., Svestka, P., Overmars, M.H.: Probabilistic roadmaps for path planning in high-dimensional configuration spaces (1994)
8. Lacevic, B., Osmankovic, D., Ademovic, A.: Burs of free C-space: a novel structure for path planning. In: ICRA (2016)
9. Ladikos, A., Benhimane, S., Navab, N.: Efficient visual hull computation for real-time 3D reconstruction using CUDA. In: CVPRW (2008)
10. Lau, B., et al.: Efficient grid-based spatial representations for robot navigation in dynamic environments. Robot. Auton. Syst. **61**, 1116–1130 (2013)
11. Moravec, H.P.: Sensor fusion in certainty grids for mobile robots. In: Sensor devices and systems for robotics (1989)
12. Pan, J., Chitta, S., Manocha, D.: FCL: a general purpose library for collision and proximity queries. In: ICRA (2012)
13. Werner, T., Henrich, D.: Efficient and precise multi-camera reconstruction. In: ICDSC (2014)
14. Werner, T., Henrich, D., Riedelbauch, D.: Design and evaluation of a multi-agent software architecture for risk-minimized path planning in human-robot workcells. In: Kongress Montage Handhabung Industrieroboter (2017)
15. Werner; T., Henrich, D., Sand, M.: Sparse and precise reconstruction of static obstacles for real-time path planning in human-robot workspaces. In: ISR (2018)

# Structural Compliance Effects on the Accuracy and Safety of a R-CUBE Haptic Device

Giuseppe Carbone[1(✉)], Antonio Acinapura[1], Domenico Mundo[1], Ibrahimcan Gorgulu[2], and Mehmet Ismet Can Dede[2]

[1] DIMEG, University of Calabria, Rende, Italy
giuseppe.carbone@unical.it
[2] Izmir Institute of Technology, Izmir, Turkey
candede@iyte.edu.tr

**Abstract.** This paper addresses the contribution of structural compliance on stiffness and safety of a R-CUBE Haptic Device. Structural compliance is determined in several poses via FEM analysis and addressed by referring to local and global indices of performance. Results are also compared with evidences from experimental tests. Comparison of numerical and experimental data allows to identify and separate the contributions to the overall compliance that are due to the structural stiffness, and other contributions such as joint clearance, pose and loading conditions.

**Keywords:** Parallel robots · Haptic devices · Stiffness · Safety

## 1 Introduction

Haptic feedback is mostly combining force or contact feedback to the position information so that users can push, pull, feel, and manipulate objects in virtual environments instead of just seeing a video representation, [1, 2]. Over the last decades, a variety of haptic interfaces have been developed ranging from simple single-DOF (Degrees of Freedom) devices to complex, multi-DOF wearable devices often with a trade-off between complexity/workspace and user-friendliness/costs. A haptic device architecture is also affected by the intended application, which can range from augmentation of graphical user interfaces (GUIs) to master interfaces in teleoperation or rehabilitation [1, 2]. The PHANToM desktop haptic interface is probably the most commonly used commercial haptic interface, [3].

In a haptic interaction, the position of the end-effector must be measured to generate an accurate haptic stimulus. However, compliant displacements can significantly affect the positioning accuracy, due to the loading condition/poses. Correspondingly, force/torque feedback to the user is generated inaccurately. These effects might become critical also from a safety viewpoint in applications such as the teleoperation of a robot for surgery tasks [2]. However, safety aspects are still needing in-depth investigation as available literature and standards such as [4–6] still do not provide proper guidelines for these specific applications. In fact, they mostly aim at eliminating avoidable hazards or using adequate protections to prevent risks, which is often not implementable for


K. Berns and D. Görges (Eds.): RAAD 2019, AISC 980, pp. 463–470, 2020.
https://doi.org/10.1007/978-3-030-19648-6_53

haptic devices where a close interaction with human is mandatory. Hence, safety aspects should be here conceived as implementing actions that can reduce and/or mitigate risks, especially from robot mechanical design viewpoint.

In this paper, we limit our attention to the contribution of the structural compliance of a haptic device to both accuracy and safety. Considering rigid joints and no control effects is herewith assumed as the worst-case scenario. We address our attention to the specific case of study of a R-CUBE haptic device prototype. In particular, Sect. 2 describes the R-CUBE architecture; Sect. 3 investigates the structural compliance in several poses via Finite Element Analysis (FEM). Local and global indices of performance are computed and discussed. Results are compared with evidences from experimental tests. The comparison of numerical and experimental results allows identifying and splitting the contributions to the overall compliance that are due to the structural stiffness from the joint clearance or due to the pose and loading conditions.

## 2 R-CUBE Architecture

In the last decades, research on parallel manipulators has been quite rich due to their expected advantages, for example, in terms of stiffness, accuracy, dynamics, as compared with serial manipulators. However, most of these manipulators have coupled motions between the position and orientation of the end-effector as well as they have complex control features also due to singularities. Recent research on parallel manipulators with a reduced number of degrees of freedom (DOFs) has been aiming at possibilities of achieving kinematic architectures with reduction of the number of active DOFs and decoupling of the position and orientation of the end-effector, [7–9].

R-CUBE manipulator is a 3 DOFs parallel manipulator, whose kinematic scheme can be modeled as proposed for example in the scheme of Fig. 1.

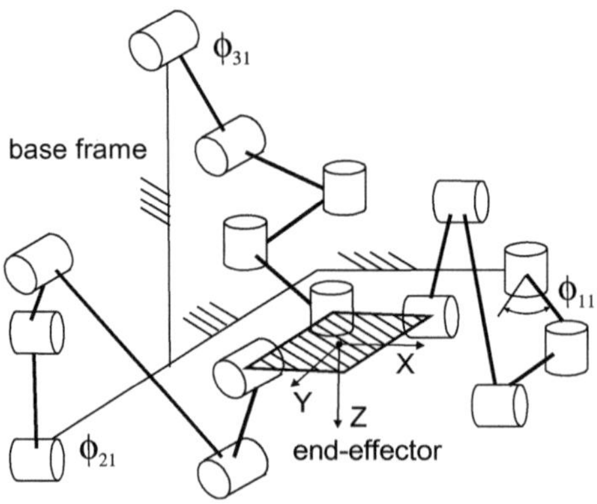


**Fig. 1.** A kinematic sketch of the R-CUBE manipulator.

One of its main features is the possibility to achieve decoupled translational motions along X-, Y- and Z-axes, by rotating the first joint angle of each chain ($\phi_{11}$, $\phi_{21}$ and $\phi_{31}$ in Fig. 1). It is composed of 3 kinematic chains that connect the end-effector to the base frame. Each kinematic chain has 4 links and 5 revolute joints as shown in the scheme of Fig. 1. All the actuators can be conveniently located at the base frame. Singularity analysis shows that the singularities can be easily avoided by means of a careful design [9]. The above features make R-CUBE architecture suitable as haptic device. Further details on R-CUBE can be found at [7–10].

## 3 Structural Performance

Several attempts have been made to properly model stiffness performance such as reported in [10, 11]. A stiffness model may be implemented for predicting and compensating compliant displacements on the end-effector for achieving a more accurate and safe haptic control. However, several aspects contribute to the overall stiffness behavior including structural stiffness, joint clearance, joints/motor stiffness, loading conditions and pose of the manipulator.

In this section, we focus on the structural stiffness. The calculation of the structural stiffness is assessed through linear FE-based static simulations within a commercial package. The FE model of the R-CUBE manipulator is created by discretizing the geometry of links and joints as a connected set of four-sided solid elements with four grid points as shown in Fig. 2. The aim of the static simulations is to calculate the displacement of the center of gravity of the end-effector (mobile platform). The FE mesh density needs to be properly set up for reducing the overall size of the FE model and, thus, the required computational cost. In particular, the dimension of the discrete elements is here set up to decrease its size going from the base to the end-effector. This choice is made since the displacements are expected to grow accordingly.

Auxiliary rigid body elements are used in order to join the mating links of each chain, obtaining a decoupling of the structural and joints stiffness. The associated material is assumed as a Poly-Lactic Acid (PLA) with specifications from a commonly used material for 3D printing. The material properties have been obtained experimentally by means of a bending test on a single 3D printed link. Table 1 shows the relationship between the applied bending force ΔF and the produced compliant displacement Δx that has been experimentally measured. The stiffness of a link can be estimated as

$$k = \frac{\Delta \mathrm{F}}{\Delta \mathrm{x}} \tag{1}$$

On the other hand, the bending stiffness of the link is defined as [11, 12]

$$k = 3EI/L^3 \tag{2}$$

where E is the Young Modulus of the material, L is the length of the tested link and

$$I = 0.15 \frac{\pi D^4}{64} \tag{3}$$

where D is the diameter of the examined link. The second moment of area [11, 12] is considered at 15% as the used infill value for the 3D printed element with PLA. Hence, knowing that D = 6 mm and L = 129.6 mm for the tested link and combining Eqs. (1) and (2), it is possible to compute the Young Modulus as E = 2.7 GPa.

The base of the R-CUBE manipulator is fixed to the ground while the force is applied at the midpoint of end-effector. The direction of the load is along one of the X-, Y- and Z-axes according to the considered analysis case. The aim of the simulations is to compute the compliant displacement of the center of the end-effector along the direction of the applied load. The static analysis tests estimate the compliant displacements ($\Delta$x, $\Delta$y, $\Delta$z) of the center of the end-effector. Two reference configurations (Fig. 2) have been chosen to perform the static simulation tests. The simulation results are shown in Table 2 for two different loading conditions. Namely, it has been considered the case with only gravity and the case gravity plus external load of 1.8 N that is applied along X-, Y- or Z-directions. Results allow to analyze the influence of the own weight of the structure on the overall results. The results show that the effect of the own weight of the structure is not very significant with a maximum contribution of about 18% along Z-axis. Numerical simulations have been also performed to estimate the main structural eigenvalues that have been chosen as a local stiffness index for the structure, such as proposed in [11].

**Table 1.** Results of bending tests on a single link with PLA infill equal to 15%.

| F [N] | 0.34 | 1.32 | 2.30 | 3.29 | 4.27 | 5.25 |
|---|---|---|---|---|---|---|
| x [mm] | 0.29 | 0.72 | 1.25 | 1.79 | 2.35 | 2.91 |

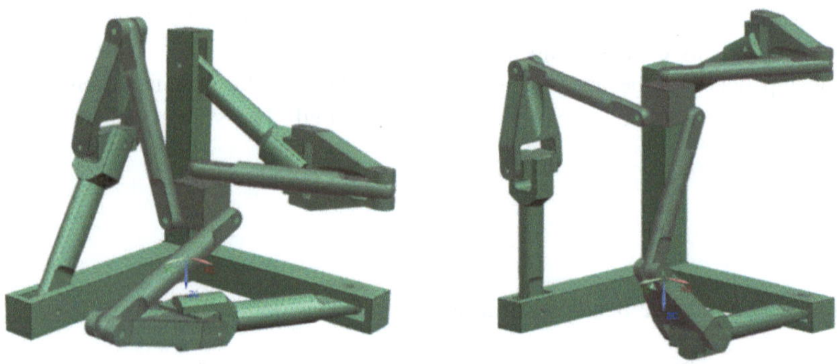

a) b)

**Fig. 2.** FEM model of the considered R-CUBE haptic device: (a) Test Pose 1 (TP1); (b) Test Pose 2 (TP2).

Table 3 shows the values $\lambda_i$ (i = 1, …, 10) of the first ten eigenvalues that are computed for four test poses TP1, TP2, TP3 and TP4 whose joint configuration is described in Table 4. The stiffness performance can be locally analyzed by referring to the maximum and minimum eigenvalues of the stiffness matrix as they provide information on the maximum and minimum deformations at a given pose, [11]. Moreover, the simulation results in Table 3 show that the minimum value of the eigenvalues of each test pose is similar to the others. On the other hand, a difference can be identified in terms of maximum eigenvalues. In fact, TP1 shows a maximum natural frequency of 653.37 Hz that is the smallest one among the four simulated configurations. The FE simulations are also used to address the safety aspect that is due to accidental collisions between the end-effector of the manipulator and an operator.

**Table 2.** FEA computed compliant displacements under only gravity or gravity plus a load of 1.81 N that is applied along X-, Y- or Z-directions.

| | TP1 | | TP2 | |
|---|---|---|---|---|
| | Force + Gravity | Gravity | Force + Gravity | Gravity |
| $\Delta x$ [mm] | 0.054 | 0.001 | 0.164 | 0.006 |
| $\Delta y$ [mm] | 0.057 | 0.004 | 0.148 | 0.009 |
| $\Delta z$ [mm] | 0.071 | 0.013 | 0.114 | 0.017 |

**Table 3.** FEA simulated first 10 natural frequencies at test poses TP1, TP2, TP3 and TP4.

| | TP1 | TP2 | TP3 | TP4 |
|---|---|---|---|---|
| $\lambda_1$ [Hz] | 96.19 | 98.25 | 95.55 | 96.79 |
| $\lambda_2$ [Hz] | 125.12 | 101.75 | 103.23 | 108.25 |
| $\lambda_3$ [Hz] | 131.67 | 135.33 | 144.42 | 142.15 |
| $\lambda_4$ [Hz] | 175.67 | 214.08 | 221.65 | 206.63 |
| $\lambda_5$ [Hz] | 189.29 | 227.04 | 231.59 | 233.25 |
| $\lambda_6$ [Hz] | 285.54 | 248.97 | 248.02 | 248.08 |
| $\lambda_7$ [Hz] | 343.71 | 356.16 | 349.15 | 342.72 |
| $\lambda_8$ [Hz] | 439.15 | 419.22 | 414.35 | 409.63 |
| $\lambda_9$ [Hz] | 503.22 | 698.75 | 682.64 | 616.14 |
| $\lambda_{10}$ [Hz] | 653.37 | 725.43 | 716.71 | 658.26 |

**Table 4.** Joint configuration of the selected testing poses.

| Joint Configuration [deg] | TP1 | TP2 | TP3 | TP4 |
|---|---|---|---|---|
| $\phi_{11}$ | −30 | 0 | 0 | 0 |
| $\phi_{21}$ | −30 | 0 | +30 | −30 |
| $\phi_{31}$ | −30 | +30 | +30 | +30 |

It is to note that the main mechanical risk cases for a robot can be identified as a direct collision with humans or indirect collisions where a human can be hit by elements that a robot hurls or drops. The latter case is not being expected for a haptic device. A possible way to describe the safety reliability of a service robot can be based on [13]. In particular, the safety performance of the R-CUBE can be assessed by computing an index that considers the acceleration of the impact between the device and the head of an operator. The Safety Index for Robots (SIR), [13], relies on the definition of safety used in automotive industry such as the widely used HIC (Head Injury Criterion) index, [13]. Considering the case of the R-CUBE end-effector moving at uniform velocity $v$, it is possible to assume the system robot-operator as a mass-spring-mass model, with K being the combined stiffness of the R-CUBE and the human head, $M_r$ the mass of R-CUBE and $M_{oper}$ the impacted operator-head mass, [13]. Thus, the SIR index can be computed as

$$SIR = 2560 \times (2/\pi)^{\frac{3}{2}}(1/9.81)^{\frac{5}{2}}\left(K/M_{oper}\right)^{\frac{3}{4}}\left(M_r/\left(M_r + M_{oper}\right)\right)^{\frac{7}{4}}v^{\frac{5}{2}} \tag{4}$$

Considering that the stiffness of the manipulator changes with the pose, it is convenient to analyze the variability of the SIR values among the different configurations as well as different directions of impact. Table 5 shows the SIR values associated to four different testing poses along X-, Y- and Z-directions. A constant speed of 2 m/s has been here assumed by referring to common operating conditions. Analyzing the obtained results, one can note that the SIR values associated to TP1 are larger than the others. Accordingly the first tested pose represents a safer configuration. On the other hand, the poses TP2, TP3 and TP4 have similar safety behaviors.

**Table 5.** Computed SIR values along three directions, for a constant speed of 2 m/s.

| SIR | TP1 | TP2 | TP3 | TP4 |
|---|---|---|---|---|
| $SIR_X$ | 0.28 | 0.12 | 0.12 | 0.15 |
| $SIR_Y$ | 0.27 | 0.13 | 0.14 | 0.12 |
| $SIR_Z$ | 0.23 | 0.16 | 0.13 | 0.19 |

## 4 Experimental Validation

A specific test rig has been developed for measuring the compliant displacements of a R-CUBE prototype, also by considering previous experiences as reported in [14–17]. The proposed test rig includes a 3D printed R-CUBE prototype, mechanic stops to fix the structure in the desired pose, a laser sensor for measuring the compliant displacements, a set of pulleys and weights to apply a desired loading condition at a desired pose, a NI6009 USB National Instruments data acquisition board, and a dedicated LabView Virtual Instrument for data acquisition and processing. It is worth noting that the model of R-CUBE haptic device has been built with the same material that has been considered in the FEM analyses. Single components have been experimentally checked to verify that the material properties match with FEM. The proposed test rig can apply a specific load aligned along X-, Y- and Z-directions. The used laser sensor measures the corresponding compliant displacement along the direction of the applied load. The applied load is obtained by using cables and pulleys with calibrated weights of known mass. Namely, a mass of 185 g is used to apply a force of 1.8 N. Results are reported in Table 6.

The comparison of experimental and numerical results shows that the stiffness performance is significantly affected by the pose and loading conditions. Among the analyzed poses, TP1 shows the stiffest behavior. Also the safety performance (based on SIR index) shows TP1 as the most performing pose. On the other hand, TP4 shows the best performance in terms of modal shapes, since it has the biggest value of maximum natural frequency. The effect of the structural weight is negligible along X- and Y-directions while it shows a maximum contribution of 18% along Z-direction as compared with the effect of a load of 1.81 N. Comparing numerical and experimental results in Tables 2 and 6, one notes that the structural stiffness is not the most significant source of compliance with compliant displacements due to structural stiffness being even more than 95% lower than overall measured compliant displacements. The above effects prove that joint stiffness and clearance, as well as loading conditions and pose play the key role in the overall stiffness and safety performance.

**Table 6.** Experimental results of compliant displacements at the selected testing poses.

| | TP1 | TP2 |
|---|---|---|
| | Force + Gravity | Force + Gravity |
| Δx [mm] | 0.340 | 1.020 |
| Δy [mm] | 0.450 | 0.930 |
| Δz [mm] | 0.230 | 1.700 |

## 5 Conclusions

This paper addresses the significance of structural properties on stiffness and safety of a R-CUBE Haptic Device. Namely, FEM analysis is used to compute indices of the performance of a R-CUBE prototype such as maximum compliant displacements, the minimum eigenvalues, and safety index SIR at four test poses. Results are also compared with evidences from experimental tests. This gives the possibility to highlight the contribution of structural properties on overall stiffness and safety performance as well as the significance of loading conditions, pose, and joint stiffness/clearance.

**Acknowledgments.** The paper presents results from the research activities of the project ID 37_215, MySMIS code 103415 "Innovative approaches regarding the rehabilitation and assistive robotics for healthy ageing" co-financed by the European Regional Development Fund through the Competitiveness Operational Programme 2014–2020, Priority Axis 1, Action 1.1.4, through the financing contract 20/01.09.2016, between the Technical University of Cluj-Napoca and ANCSI as Intermediary Organism in the name and for the Ministry of European Funds.

## References


1. Hayward, V., Astley, O.R., et al.: Haptic interfaces and devices. Sens. Rev. **24**(1), 16–29 (2004)
2. Okamura, A.M.: Haptic feedback in robot-assisted minimally invasive surgery. Curr. Opin. Urol. **19**(1), 102 (2009)
3. Salisbury, J.K., Srinivasan, A.M.: Projects in VR. Phantom - based haptic interaction with virtual objects. IEEE Comput. Graph. Appl. **17**, 6–10 (1997)
4. ISO 14121–1: Safety of machinery. Risk assessment. Part 1: principles (2007)
5. EN ISO 13855: Safety of machinery – the positioning of protective equipment in respect of approach speeds of parts of the human body (2010)
6. ISO/DIS 13482: Robots and robotic devices – safety requirements for non-industrial robots – non-medical personal care robot (2011)
7. Li, W., Gao, F., Zhang, J.: R-CUBE, a decoupled parallel manipulator only with revolute joints. Mech. Mach. Theory **40**(4), 467–473 (2005)
8. Takeda, Y., Kamiyama, K., Maki, Y., Higuchi, M., Sugimoto, K.: Development of position-orientation decoupled spatial in-parallel actuated mechanisms with six degrees of freedom. J. Robot. Mechatron. **17**(1), 59–62 (2005)
9. Biligincan, T., Can Dede, M.I.: Development of a R-CUBE-based general purpose haptic device system. In: 10th Biennial Conference on Engineering Systems Design and Analysis, ASME 2010, Istanbul, pp. 675–682 (2010). https://doi.org/10.1115/esda2010-24596
10. Gorgulu, I., Dede, M.I.: Computation time efficient stiffness analysis of the modified R-CUBE mechanism. In: International Conference of IFToMM ITALY, pp. 231–239. Springer (2018)
11. Carbone, G.: Stiffness analysis and experimental validation of robotic systems. Front. Mech. Eng. **6**(2), 182–196 (2011)
12. Hibbeler, R.C.: Mechanics of Materials. Pearson, London (2014)
13. Cordero, C.A., Carbone, G., Ceccarelli, M., Echavarri, J., Munoz, J.L.: Experimental tests in human-robot collision evaluation and characterization of a new safety index for robot operation. Mech. Mach. Theory **80**, 184–199 (2014)
14. Carbone, G., Ceccarelli, M.: Experimental tests on feasible operation of a finger mechanism in the LARM hand. Mech. Based Des. Struct. Mach. **36**, 1–13 (2008)
15. Carbone, G., Iannone, S., Ceccarelli, M.: Regulation and control of LARM hand III. Robot. Comput. Integr. Manuf. **26**, 202–211 (2010)
16. Hernández-Martínez, E.E., Ceccarelli, M., Carbone, G., López-Cajún, C.S., Jáuregui-Correa, J.C.: Characterization of a cable-based parallel mechanism for measurement purposes. Mech. Based Des. Struct. Mach. **38**(1), 25–49 (2010)
17. Major, K.A., Major, Z.Z., Carbone, G., Pisla, A., Vaida, C., Gherman, B., Pisla, D.L.: Ranges of motion as basis for robot-assisted poststroke. Hum. Vet. Med. **8**(4), 192–196 (2016)

# Robot Vision

# Adaptive Recognition for Tracking of Moving Objects

Karol Dobrovodský(✉) and Pavel Andris

Institute of Informatics, Slovak Academy of Sciences,
Dúbravská cesta 9, 842 37 Bratislava, Slovakia
{utrrdobr, utrrandr}@savba.sk

**Abstract.** An adaptive multilocator for online recognition and tracking of moving objects is considered. The vision system is based on the fast real time sub image localization module aimed to recognize patterns in order to track the motion of the selected object. In order to reduce mistakes concerned with the identity, the system is able to recognize several similar objects simultaneously. Nevertheless, the operator can point out an object on the screen to correct the tracking process. An adaptation of the actual patterns according to the lastly localized objects in the area of interest is performed in each step of the real time process. The aim of the paper is to present an effective adaptation and correction procedure based on constant primary patterns defined by the operator and varying secondary patterns defined by the automatic adaptation.

**Keywords:** Vision system · Online recognition · Pattern adaptation · Real time localization

## 1 Introduction

The second generation vision systems for online recognition and tracking of moving objects offer to the operator an active image processing with the recognition of patterns, [1, 2]. Nevertheless the vision system is not able to distinguish between the object of interest and its background. The fast real time sub image localization module aimed to recognize patterns in order to track the motion of the selected object is working with the initial pattern without the ability to recognize the object of interest in its new positions. An operator must perform the first recognition of the new object of interest and define the pattern sub image which should be recognized. The vision system is continuously updating the pattern according to the successfully recognized sub image. Nevertheless this adaptation process may not be stable, [3]. The updated pattern may slide from the tracked object to the background and thus it is desirable to work out a concept of primary and secondary patterns in order to correct the adaptation process from time to time.

The third generation systems offer to the operator automated active image processing with the recognition of new objects of interest. The disadvantage of this approach is the great number of dummy objects especially in complex surroundings.

In order to avoid fictitious objects, we are expanding the second-generation system with intensive pattern recognition by introducing primary and secondary patterns.


K. Berns and D. Görges (Eds.): RAAD 2019, AISC 980, pp. 473–478, 2020.
https://doi.org/10.1007/978-3-030-19648-6_54

Primary patterns are defined by the operator and thus being considered more reliable. The identity of constant primary patterns is guaranteed. On the other hand, varying secondary patterns adapting to new positions of the tracked object may lose their belonging to the object. The proposed concept of primary and secondary patterns is a continuation of the basic fast real time pattern localization module concerned in [3] towards a more reliable adaptive vision system for recognition and tracking of moving objects as compared with vision systems considered in [4–7].

## 2 Primary Patterns

Primary patterns are defined by the operator simply by pointing out on the screen. The system takes a sample of the nearby area where the operator indicated. Sample size is always the same and maximum regardless of the actual value that is used for recognition. The object of recognition is assumed to be in the center of the defined primary pattern. This may not be easy to realize especially when a moving object has to be pointed out for recognition. To facilitate the definition process, the two step procedure is offered to the operator. The first step is to magnify the image as a magnifying glass. The second step is pointing out the center of the object.

The current pattern size used for localization in the area of interest may change at any time during the tracking process. The optimal pattern size is the size that frames the silhouette of the object as tightly as possible.

In order to define the primary pattern precisely, the operator may recall an enlarged vicinity of the object pointed out and point out the center more precisely so that the object of interest gets more accurately into the center of the pattern. In addition, it is possible to adjust the frame size so that the pattern will contain less background.

There are two basically different situations when tracking one moving object. The first situation is tracking one of more very similar objects moving in a group. We have to distinguish between similar objects having different identities. The second case is tracking one object that looks differently from different points of view. We have to distinguish between different forms of that particular object with one identity. In general, primary patterns should guarantee the identity of the tracked object.

## 3 Secondary Patterns

Each step of the real time recognition process begins with using primary patterns for localization in the area of interest. For each primary pattern, a localization module is looking for all similar places with the similarity level over the specified value. All of these locations are findings that are subject to further processing.

We must be aware of the fact that number of such findings may be large, but there may also be cases where there are no findings. In such cases the secondary pattern from the previous step is used. Further processing reduces the number of all findings by selected criteria to roughly the number of primary patterns. The resulting findings become secondary patterns for the next step. The secondary patterns are taken from the localized places of the corresponding findings. Thus the level of similarity in the next step may potentially rise with the risk of identity loss. The success of the concept depends on the experimentally achieved balance between the required level of similarity and the loss of identity.

# 4 Reduction of Findings

All findings above the specified level of similarity should be reduced in a preprocessing phase. There are several criteria for reduction:

- critical mutual distance
- distance from the central point of the area of interest
- level of similarity

All findings under the critical mutual distance are joined into one object. The critical distance is approximately half the size of the current pattern. The resulting objects are then sorted by the distance from the center of the area of interest and finally by the level of similarity. The objects are formally numbered after the reduction.

At this point, it should be emphasized that after the connection of critically close objects the correspondence between the patterns and the corresponding findings is lost.

# 5 Identity Estimation

The last phase of the recognition process is the identity estimation. Formally numbered objects should be paired with objects having been identified in the previous step of the repeated real time process. We must have in mind that to all objects may be given already existing identities as well as any pair may not be identified. It depends on criteria fulfilled.

Every identity has a short history of the previous positions. Based on this history, a likely position is predicted in the next step. Predictions of positioning of objects with identity are similar to new findings, which are only formally numbered and identity is unavailable. The pairing runs freely according to the criteria. The arising couples define couples of the same identity. The scoring criterion is not only the proximity of the positions but also the sufficient alignment of the movement direction. New objects obtain the new identity.

Identities that have not been found will become a ghost. The ghost is a fictitious object with an identity that has not been recognized. The visual system thus prevents the object from being lost. Only one object remains in the form of a ghost until the operator does not cancel it. All other objects can be ghost one step only. If a new object appears near the ghost, the identity of the ghost is given to the object.

## 6 Search Mode

The localization process starts running in the Search mode. A small database of primary patterns may be available for this phase. Primary patterns localized in the area of interest are offered to the operator as recognized objects. The area of interest on the screen is typically 128 × 128, 256 × 256, 512 × 512 pixels, but any other dimensions are possible. It depends on frequency of image processing and the maximal number of patterns. The system is able to work with several primary patterns simultaneously (Fig. 1).

**Fig. 1.** Primary patterns.

Nevertheless, the operator can point out an object on the screen to be recognized. From this moment a new primary pattern is defined and a search for similar sub-images is repeatedly performed in real time.

All recognized places within the area of interest with the degree of similarity higher than the selected level (typically 70%, 80%, 90%) are reduced according to the selected criteria. The reduced number of findings will become secondary patterns defined by recognized places. Thus secondary patterns reflect possible changes of objects. In the next step, the secondary patterns will be used optionally in case of primary patterns failure.

## 7 Track Mode

The operator can make a choice from possibly more objects offered for tracking simply by pushing knobs on the joystick. Having selected one, the Track mode is started. The tracked object is only one, but in the case of several similar objects in a group, it is necessary to identify and associate the identity with the accompanying objects too. Otherwise, there is an undesired jump on a companion object probable.

In the event of the object recognition failure, the system predicts the motion based on the history of the determined positions.

It is clear that the normalized cross correlation is not an ideal approach to feature tracking since it is not invariant in respect of imaging scale, rotation, and perspective distortions. We eliminate these limitations, addressed in [8, 9], by continuously updating the pattern every 40 ms. Updated patterns are called secondary patterns (Fig. 2).

**Fig. 2.** Two from secondary patterns.

The prediction of the future position of the tracked object is calculated in the Track mode. The output of the recognition process is an object that meets the maximum allowable distance criterion from its predicted position. If it is not the case, i.e. there is no such object recognized, the output is the predicted position and the system signals blind monitoring (a ghost). The operator can stop the process in case of a dummy object tracking. The maximum allowable distance of the object from its predicted position is determined experimentally with respect to the size of pattern.

## 8 Conclusion

We presented a concept of primary and secondary patterns to eliminate limitations of the Pearson coefficient application in real time recognition of moving objects. Pointing out an object of interest on the screen is more universal way how to start tracking. The operator defines primary patterns, which are constant. On the other hand, the system is able to update the secondary pattern automatically in each step of the tracking process and thus to avoid difficulties with scale, rotation, and perspective distortions. Nevertheless, some generally oriented basic pattern images may be stored in a small database to reduce the time needed to define new primary patterns.

**Acknowledgment.** We thank the support of the Scientific Grant Agency of the Ministry of Education of the Slovak Republic and the Slovak Academy of Sciences project number 2/0155/19.

## References

1. Rodgers, J.L., Nicewander, W.A.: Thirteen ways to look at the correlation coefficient. Am. Stat. **42**(1), 59–66 (1988)
2. Gonzales, R.C., Woods, R.E.: Digital Image Processing, 3rd edn. Adison-Wesley, Reading (1992)
3. Dobrovodsky, K., Andris, P.: Real time sub image localization for Tracking. In: Proceedings of the 27th Conference on Robotics in Alpe-Adria-Danube Region, RAAD 2018. Mechanisms and Machine Science, vol. 67, pp. 588–596. Springer, Switzerland (2019)
4. Andris, P., Dobrovodsky, K.: Developing an embedded system based on a real-time version of Linux. In: Proceedings of the 23rd International Conference on Robotics in Alpe-Adria-Danube Region, CD, IEEE catalog no. 34043, 3–5 September, Smolenice, Slovakia (2014)
5. Dobrovodsky, K., Andris, P.: Real time recognition and tracking of moving objects. Comput. Inf. **33**(6), 1213–1236 (2014)
6. Dobrovodsky, K., Andris, P.: Aiming procedure for the tracking system. In: Advances in Robot Design and Intelligent Control: Proceedings of the 24th International Conference on Robotics in Alpe-Adria-Danube Region (RAAD). Advances in Intelligent Systems and Computing, vol. 371, pp. 369–378. Springer, Switzerland (2016)
7. Dobrovodsky, K., Andris, P.: Cooperative distance measurement for an anti-aircraft battery. In: Advances in Service and Industrial Robotics: Proceedings of the 26th Conference on Robotics in Alpe-Adria-Danube Region, RAAD 2017. Mechanisms and Machine Science, vol. 49, pp. 95–101. Springer, Switzerland (2018)
8. Lewis, J.P.: Fast template matching. In: Vision Interface 1995, pp. 120–123. Canadian Image Processing and Pattern Recognition Society, Quebec City, Canada, 15–19 May 1995
9. Lewis, J.P.: Fast normalized cross-correlation. In: Proceedings of Vision Interface, pp. 120–123 (1995)

# Probabilistic Orientation Resolution for Near Symmetrical Objects Using Depth Images

Nawal Hafez[1(✉)], Vincent Dietrich[2], and Michael Zwick[1]

[1] Technical University of Munich, Arcisstrasse 21, 80333 Munich, Germany
nawal.hafez@tum.de
[2] Siemens Corporate Technology, Otto-Hahn Ring 6, 81739 Munich, Germany

**Abstract.** Estimating object poses accurately is crucial in various applications. Especially in industrial applications, pose estimation can be challenging when objects and work pieces are (almost) symmetrical and without texture. Common pose estimation methods cannot easily distinguish between several nearly symmetrical configurations of the same object.

In this paper, we propose an approach to improve pose estimation of nearly symmetric objects. Our algorithm estimates a distribution of all possible orientations based on an initial pose, a CAD model of the object, and a depth image. We evaluated our approach on simulated as well as real sensor data.

**Keywords:** Computer vision · Pose estimation · Symmetry

## 1 Introduction

Estimating the correct orientation of objects is important in many applications. For example, let's consider the assembly of a mechanical shaft as depicted in Fig. 1. This object is symmetrical up to a small detail, the key seat. However, this subtle difference is important for the correct assembly of the shaft. In industrial applications, there are numerous of such nearly symmetrical objects and in many cases these objects are texture-less. We define nearly symmetrical objects as objects displaying partial symmetry evident to the human eye as the shaft or a cup with handles.

The contribution of this paper is a novel approach to improve pose estimation of objects using prior knowledge, in particular, information on object symmetry. Our approach can be applied as a post-processing step to a traditional object recognition pipeline. We assume that objects have already been classified and localized, and that this localization needs to be refined or verified. An overview of our approach is given in Fig. 1.

Relevant work is presented in Sect. 2. An algorithm is introduced in Sect. 3 and its characteristics are analyzed with respect to simulated and real experiments in Sect. 4. Within these experiments, we analyze noise robustness, usability with real data, and capability of dealing with near circular symmetry.


K. Berns and D. Görges (Eds.): RAAD 2019, AISC 980, pp. 479–487, 2020.
https://doi.org/10.1007/978-3-030-19648-6_55

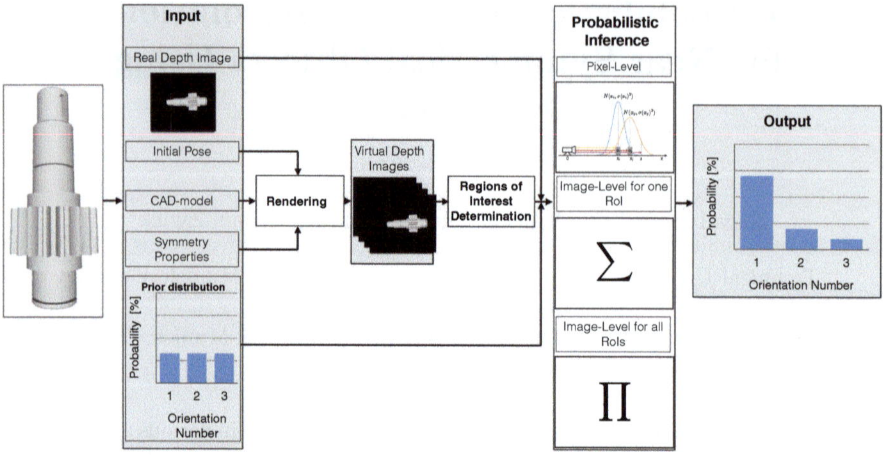


**Fig. 1.** Overview of the approach: An initial pose, the symmetry properties and also a CAD model of the object are known beforehand. Based on this initial pose and its properties, depth images are rendered. The differences in these images are examined to discern the correct orientation. Because of uncertainty of pose and camera noise, this task is dealt with probabilistically.

## 2 Related Work

Object recognition is a widely researched topic. Existing approaches are capable of providing a good estimate for the initial pose. We use such an algorithm as a first step. Popular strategies used for object recognition include feature-based, template matching and learning-based methods [1].

Feature-based methods rely on finding features that can distinguish objects from each other. Other methods rely on template matching, such as LINEMOD [2] which is the approach we use to conduct this work, as it performs sufficiently well on texture-less objects. Learning based-approaches [3,4] generally rely on training in order to be able to recognize objects. An example for a learning-based approach which specifically targets nearly symmetrical object is BB8 [3]. However, to the best of our knowledge most methods do not provide a probabilistically sound distribution.

In our work, we rely on data taken from the T-LESS dataset [5]. It contains industry related objects that are texture-less and display symmetry characteristics which underline the challenges such objects place on pose estimation algorithms [5,6]. Other relevant work includes evaluation methods that take into account the object's symmetry [7].

We use depth images in this paper to determine the orientation. Targeted objects are mostly texture-less and do not display a lot of features, which is why we opt for depth images.

A probabilistic visual verification system dealing with pose uncertainty and sensor errors is developed in [8]. However, it doesn't focus on nearly-symmetric objects.

## 3 Approach

The approach presented in Fig. 1 is divided into three parts: 3.1 the generation of the necessary prior information 3.2 the determination of the regions of interest 3.3 the probabilistic inference to estimate the distribution over all possible orientations.

### 3.1 Prior Information Generation

We assume a pose estimate is given as input, which is prone to small pose and symmetry errors. The object's symmetry properties are currently manually annotated but could be determined in an automatic manner [9]. The cardinality of the symmetry group of object X, $|S(X)|$, corresponds to the number of possible orientations. The symmetry group only includes rotational symmetries.

### 3.2 Region of Interest Determination

We define regions of interest (RoIs) as the areas in images where we suspect there is information about the most probable orientation. To find these regions we calculate the absolute depth difference between each pair of possible orientations given by S(X). A threshold is needed to neglect differences in the depth values that cannot be detected by the depth sensor. Difference that are bigger than the threshold value are regarded as part of the region of interest between two possible orientations. An example is presented in our experiments in Fig 4

### 3.3 Probabilistic Inference

In this section, we introduce our approach to calculate the probability distribution in three steps and use the notation in Table 1.

**Table 1.** Table of notations

| | | | |
|---|---|---|---|
| $x_{i,n}$ | depth value for pixel No. n belonging to ground truth depth image No. i | $z_n$ | depth value for pixel No. n from the image generated by the Camera |
| Z | depth image generated by the camera | $\phi_{i,j}$ | RoI between images i & j |
| $X_i$ | class i | M | set with all possible orientations |
| k | index k for the most likely orientation | p(x) | probability mass function |
| f(x) | probability density function | P(A) | probability function |

**Pixel Level**

We compare orientations pairwise. Figure 2 shows the probability of the depth values, $x_i$ and $x_j$, given the measurement $z$ is sought after. We assume that the likelihood of the measurement given the value x, can be modeled as a Gaussian, $N(\mu, \sigma^2)$. Applying Bayes theorem and accounting for priors with equal distributions results in the following equation:

$$p(x_i|z) = \frac{f(z|x_i)}{f(z|x_i) + f(z|x_j)}. \tag{1}$$

**Image Level for One RoI**

We use a linear opinion pool [10] for the pixel-wise probability fusion:

$$P(X = X_i|Z, \phi_{i,j}) = \frac{\sum_n^{|\phi_{i,j}|} p(x_{i,n}|z_n)}{|\phi_{i,j}|}. \tag{2}$$

The probability $P(X = X_i|Z, \phi_{i,j})$ is the probability of belonging to class $i$ given one specific RoI. Because we treat orientations pairwise, the probability that it belongs to the other orientation can be calculated with the help of the $2^{nd}$ axiom of probability.

**Image Level for All RoIs**

In the final step, we combine the probabilities of all regions. We regard the probabilities calculated for each region as classifiers and combine them with the product rule [11] due to their statistical conditional independence:

$$P(X_i|Z, \phi_{i,j}, \phi_{i,j+1}, ...,) = \frac{\prod_{j \neq i}^{|M|} P(X_i|Z, \phi_{i,j})}{\sum_n^{|M|} \prod_{j \neq i}^{|M|} P(X_i|Z, \phi_{i,n})}. \tag{3}$$

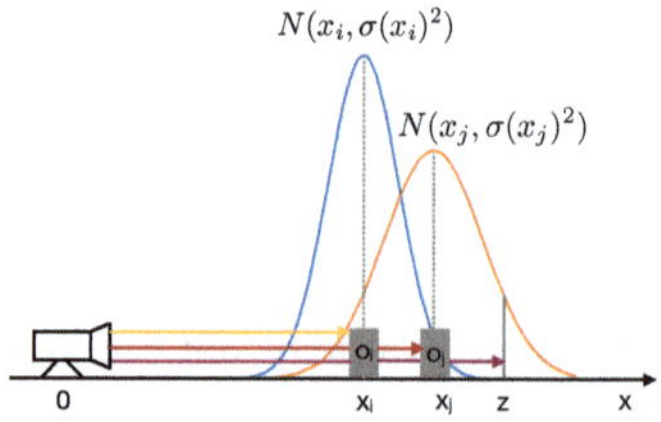


**Fig. 2.** Lateral noise model

## 4 Evaluation

The sensitivity and applicability of the proposed algorithm is evaluated based on different experiments with simulated and real data in order to understand its characteristics. We start with introducing the noise models used in our four experiments. Sensitivity to the initial pose error is tested in simulation to show tolerance towards errors. For real data experiments we use a perception pipeline consisting of LINEMOD as an exemplary pose estimation method and refine it with ICP [12]. We conduct experiments to examine the viewpoint effect on the confidence of the result and the post-processing step effect on the perception pipeline to show its relevance. Experiments also include a runtime analysis.

### 4.1 Noise Models

**Camera Noise Models**

It is beneficial to add noise to the synthetic depth image to bridge the simulation reality gap and define the model for the probability inference. We use a one dimensional per pixel Gaussian noise model [8]. A simple experiment is designed similar to the one in [13] to discern the model parameters of the Intel® RealSense$^{TM}$ SR300, leading to:

$$\sigma(d) = 0.0772(d - 0.0179)^2 + 0.0056, \tag{4}$$

with $\sigma$ being the standard deviation and $d$ the distance. All experiments use the SR300 as a camera or its noise model to create synthetic images with a threshold of 3–5 mm to determine the RoIs.

**Simulated Pose Noise**

We add noise to the ground truth pose so that it is comparable to the perception pipeline. The initial pose error is modeled by adding Gaussian noise to the 6D-pose. We assume that each element is independent. This noise model is only required for the simulated experiments.

### 4.2 Sensitivity Analysis on Simulated Data

*Experimental Setup:* Noise is applied to the ground truth position based on the noise model in 4.1 and we use the noisy position and a synthetic depth image as input to the post-processing step. The noise effects vary because of the object's distance to the sensors. We therefore define the *relative pixel error metric*, $r$, which is an error transformation from pose to pixel space. The goal is to compare contrasting objects at different distances. It is calculated by:

$$r = \frac{f_x \left| \frac{o_x}{o_z} - \frac{n_x}{n_z} \right|}{w} + \frac{f_y \left| \frac{o_y}{o_z} - \frac{n_y}{n_z} \right|}{h}, \tag{5}$$

where $w$, $h$ are width and height of the bounding box, f is the focal length, the $o$ is the original position vector and the $n$ indicates the new position vector. The RoIs are typically only small areas of the overall object which is why relative pixel errors have a very strong effect on the result quality. We test objects in distances between 0.3 and 1.0 m.

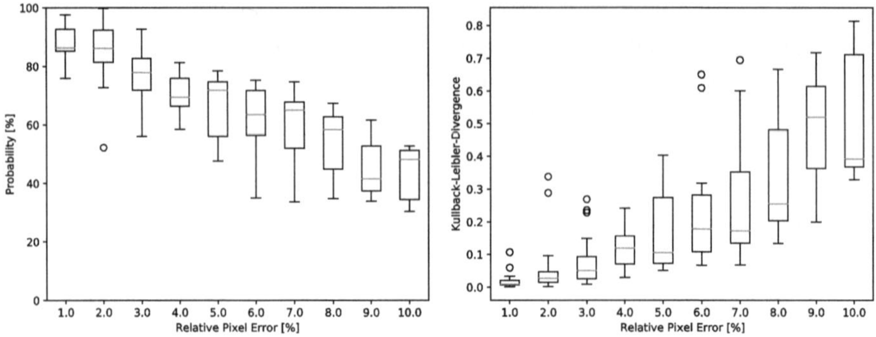


(a) Effect of position noise (b) Distribution divergence

**Fig. 3.** The effect of position noise displayed with the help of box plots, where circles are outliers and areas between the whiskers are the four quartile groups.

*Results:* One expected outcome is that the perception result quality decreases with increasing noise as visualized in Fig. 3a. In the range between 1% and 6% the probabilities are higher than 50% ensuring that it is most likely that the correct orientation is predicted. Probabilities may be as low due to the inaccuracy of the initial pose which leads to adjacent orientations becoming similarly probable. Another way of examining the effect of position noise is the Kullback-Leibler Divergence (KL) [14]. For small relative changes the divergence is close to 0 which means little to no information loss.

### 4.3 Viewpoint Analysis

Objects with circular symmetry are especially challenging due to high number of possible orientations. The mechanical shaft is a suitable example for such an object, see Fig. 1 with $|S(X)| = 19$ equal to the number of shaft teeth.

*Experimental Setup:* Experiments are conducted on real data in a table top experiment. We examine a group of poses with the keyseat:
1. in the middle region, 2. around the edges of the visible area and 3. invisible.

*Results:* The distribution visualization in Table 2 gives a better sense of how reliable the result is. For pose one and two the predicted orientations are very likely, while for pose three multiple orientations have similar probabilities. These examples show the importance of the viewpoint but also underline the importance of the distribution which can be used for decision making in robotic applications.

**Table 2.** Probability of correct orientation for different poses of the shaft

| Pose | 1: visible feature | 2: feature on edge | 3: invisible feature |
|---|---|---|---|
| Images | 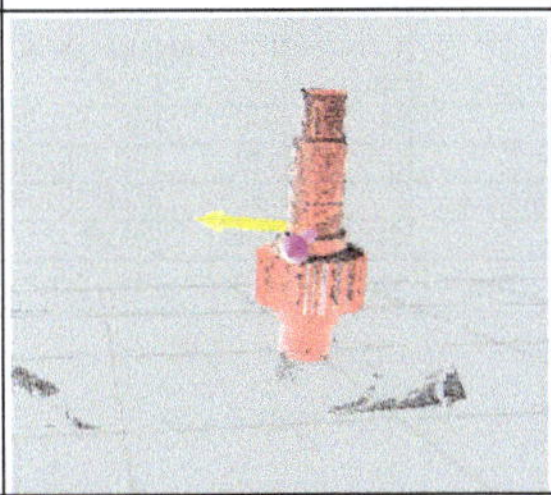 | 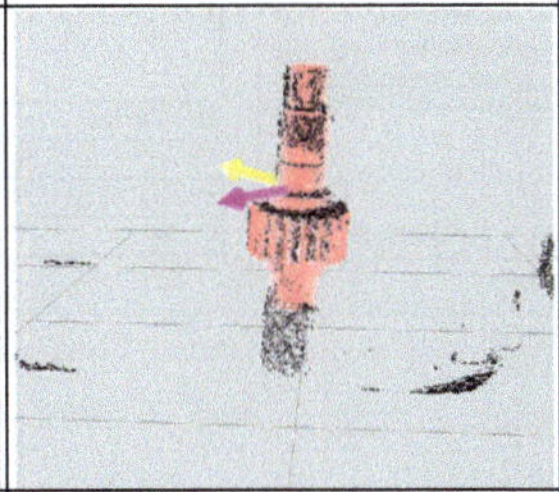 |  |
| Distribution | 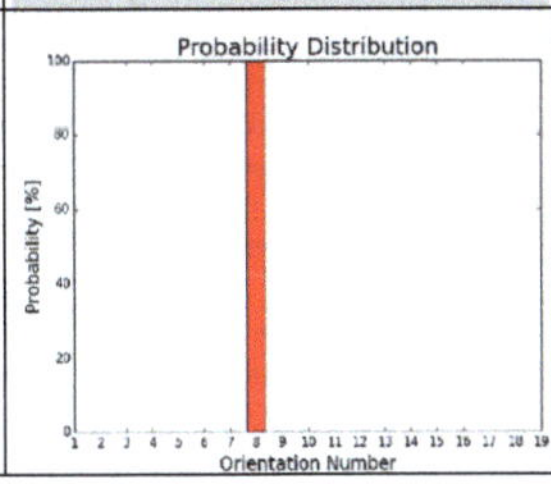 | 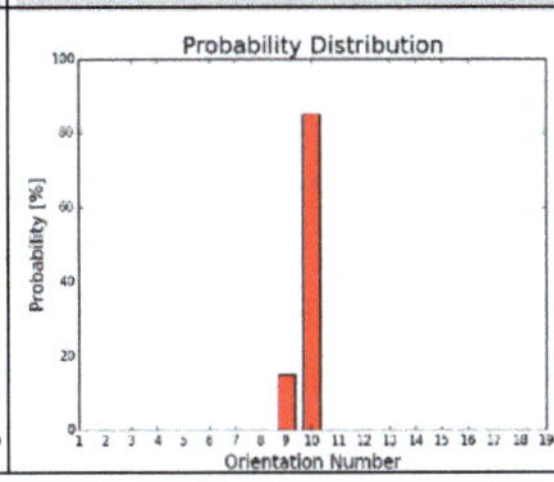 | 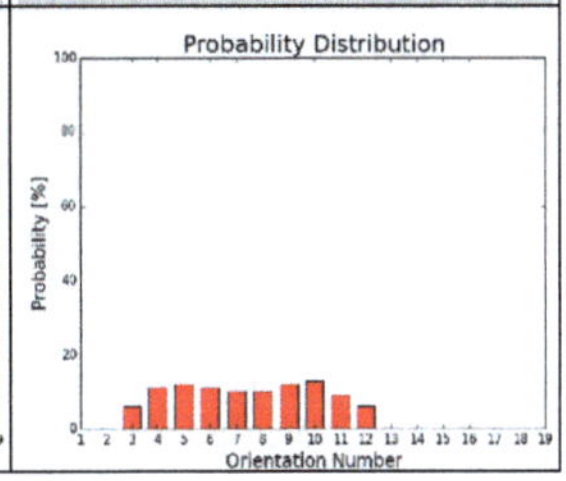 |

### 4.4 Effect of Post-processing Step on Perception Pipeline

*Experimental Setup:* In the following, we conduct a qualitative table top experiment using the SR300 in the range 30–50 cm commands to localize object number 6 from the T-LESS dataset [5]. We choose this object because it displays the targeted properties and four scenes in particular to analyze, see Fig. 4. For each scene, we localize the object 20 times. The aim of these experiments is show that our approach can improve an existing perception pipeline by adding it as a post-processing step.

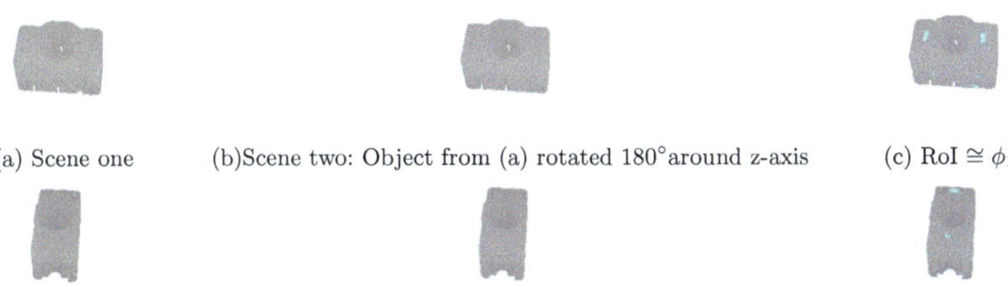

(a) Scene one (b)Scene two: Object from (a) rotated 180°around z-axis (c) RoI $\cong \phi_{1,2}$

(d) Scene three (e) Scene four: Object from (d) rotated 180°around z-axis (f) RoI $\cong \phi_{3,4}$

**Fig. 4.** Poses and their corresponding RoIs

*Results:* The results are shown in Table 4. The success rates of LINEMOD + ICP vary and are not reliable. On the other hand, the proposed algorithm is capable of finding the right orientation each time, even though the RoIs are small. This shows that it is beneficial to use the proposed algorithm as an extension to usual object detection pipelines.

**Table 3.** Post-processing step analysis

| Pose | pipeline: LINEMOD + ICP $[\frac{\#successful\ tests}{\#all\ tests}]$ | piepline: LINEMOD + ICP + Post-processing step (our approach) $[\frac{\#successful\ tests}{\#all\ tests}]$ |
|---|---|---|
| 1 | 13/20 | 20/20 |
| 2 | 8/20 | 20/20 |
| 3 | 20/20 | 20/20 |
| 4 | 0/20 | 20/20 |

**Table 4.** Runtime analysis

| Part | Obj_06 Runtime [ms] | Shaft Runtime [ms] | Average of Atomic Operation [ms] |
|---|---|---|---|
| Rendering | 61 | 290 | 18 / orientation |
| Determining RoI | 8 | 179 | 1.2 / comparison |
| Probability Calculation | 5 | 15 | 0.09 / comparison |
| Total | 74 | 484 | 19.29 |

### 4.5 Runtime Analysis

Runtime is crucial in time critical applications, e.g. industrial assembly. The runtime overview is presented in Table 4 on an Intel Core i5-3470 3.2 GHz CPU with a GeForce GTX 750 Ti graphic card. We measure it for the object from T-LESS dataset and the mechanical shaft. Rendering is the most time consuming part of the algorithm. It is possible to further cut the runtime significantly by using parallelization and GPU based processing.

## 5 Conclusion

Object recognition is a widely researched topic because of the abundance of its applications. Texture-less objects displaying a near symmetry present a challenge even for state-of-the-art methods. While the position can generally be precisely estimated, the similarity of several orientations can lead to a wrong orientation estimate. A possible solution is presented in this paper. It entails using existing pose estimation methods and improving the estimated pose via a post processing step. The proposed algorithm exploits symmetry properties as prior knowledge to improve the localization of nearly symmetrical objects. It is deduced from basic probability theory and yields a distribution of the pose given a depth image of the setting.

In conclusion, the proposed approach has a number of advantages. Being a model-based approach it is easily reusable for different objects and settings as well. The approach delivers a sound probabilistic distribution, which is why it stands out in comparison to other methods. It is tailored for industrially relevant objects that are texture-less and display near-symmetry. The approach is also capable of dealing with objects with many possible orientations. In experiments we could show its robustness to pose uncertainty and camera noise and demonstrate the applicability to real settings.

# References


1. Hodaň, T., Michel, F., Brachmann, E., Kehl, W., Buch, A.G., Kraft, D., Drost, B., Vidal, J., Ihrke, S., Zabulis, X., Sahin, C., Manhardt, F., Tombari, F., Kim, T.K., Matas, J., Rother, C.: BOP: Benchmark for 6D object pose estimation. In: ECCV, pp. 19–35 (2018). https://doi.org/10.1007/978-3-030-01249-6_2
2. Hinterstoisser, S., Lepetit, V., Ilic, S., Holzer, S., Bradski, G., Konolige, K., Navab, N.: Model based training, detection and pose estimation of texture-less 3D objects in heavily cluttered scenes. In: ACCV, pp. 548–562 (2012). https://doi.org/10.1007/978-3-642-37331-2_42
3. Rad, M., Lepetit, V.: BB8: A scalable, accurate, robust to partial occlusion method for predicting the 3D poses of challenging objects without using depth. In: ICCV, pp. 3848–3856 (2017). https://doi.org/10.1109/ICCV.2017.413
4. Tekin, B., Sinha, S.N., Fua, P.: Real-time seamless single shot 6D object pose prediction. In: CVPR (2018)
5. Hodan, T., Haluza, P., Obdržálek, Š., Matas, J., Lourakis, M., Zabulis, X.: T-LESS: an RGB-D dataset for 6D pose estimation of texture-less objects. In: WACV, pp. 880–888 (2017). https://doi.org/10.1109/WACV.2017.103
6. Drost, B., Ulrich, M., Bergmann, P., Härtinger, P., Steger, C.: Introducing MVTec ITODD - a dataset for 3D object recognition in industry. In: ICCVW, pp. 2200–2208 (2017). https://doi.org/10.1109/ICCVW.2017.257
7. Brégier, R., Devernay, F., Leyrit, L., Crowley, J.L.: Symmetry aware evaluation of 3D object detection and pose estimation in scenes of many parts in bulk. In: ICCV, pp. 2209–2218 (2017). https://doi.org/10.1109/ICCVW.2017.258
8. Choi, C., Rus, D.: Probabilistic visual verification for robotic assembly manipulation. In: ICRA, pp. 5656–5663 (2016). https://doi.org/10.1109/ICRA.2016.7487786
9. Mitra, N.J., Guibas, L.J., Pauly, M.: Partial and approximate symmetry detection for 3D geometry. In: TOG, pp. 560–568 (2006). https://doi.org/10.1145/1141911.1141924
10. Bibby, C., Reid, I.: Robust real-time visual tracking using pixel-wise posteriors. In: ECCV, pp. 831–844 (2008). https://doi.org/10.1007/978-3-540-88688-4_61
11. Kittler, J., Hatef, M., Duin, R.P., Matas, J.: On combining classifiers. In: TPAMI, pp. 226–239 (1998). https://doi.org/10.1109/34.667881
12. Besl, P.J., McKay, N.D.: A method for registration of 3-D shapes. In: TPAMI, pp. 586–607 (1992). https://doi.org/10.1109/34.121791
13. Khoshelham, K., Elberink, S.O.: Accuracy and resolution of kinect depth data for indoor mapping applications. Sensors, 1437–1454 (2012). https://doi.org/10.3390/s120201437
14. Joyce, J.M.: Kullback-Leibler divergence. In: International Encyclopedia of Statistical Science, pp. 720–722 (2011). https://doi.org/10.1007/978-3-642-04898-2_327

# Generation of Elevation Maps for Planning and Navigation of Vehicles in Rough Natural Terrain

Hannan Ejaz Keen(✉) and Karsten Berns

Technische Universität Kaiserslautern, 67663 Kaiserslautern, Germany
{keen,berns}@cs.uni-kl.de

**Abstract.** We propose a system, which generates a precise elevation map of natural rough terrain. It uses basic sensors such as LiDAR and Stereo camera to generate point clouds. Based on the requirements of high precision, specific techniques are integrated to receive adequate maps in real-time. The presented methodology allows an easy extension of existing maps.

**Keywords:** Rough terrain · Digital Elevation Maps · Terrain mapping · Sensor data fusion · Point cloud stitching

## 1 Introduction

A precise map of the environment is necessary for an efficient path planning and navigation in off-road vehicles for rough terrain. The needed data collection can be achieved by ground vehicles (GVs) or Aerial Vehicles (AVs) on which LiDAR sensor and Stereo camera systems are installed. Nonetheless the transformation of data and stitching of map from GVs is comparatively easy because in case of AVs the localization specially to the height axis is imprecise due to disturbances. On the other hand AVs can fly high normally over every type of terrain with less effort which is of course not possible for GVs. However the major drawback in AVs is precise land mapping with flying perspective.

Although there exist some of the elevation map plugins which use Google map to create elevation, the map is not much precise for navigation and planning due to trees/forests. Moreover, Digital Elevation Maps (DEM) of certain areas are available for public use nonetheless the best resolution one can get from those maps is of 1 m where resolution is the distance of each point with its neighboring points. This resolution is not feasible for precise navigation of autonomous vehicles such as tractors on the field or specific GVs for rough natural terrain.

In order to address the above stated shortcomings, a system has been proposed that generates the elevation maps with accuracy of less than 50 cm and 15 cm of resolution. The accuracy is measured by computing the root mean squared error (RMSE) between the resultant map and ground truth. For the


K. Berns and D. Görges (Eds.): RAAD 2019, AISC 980, pp. 488–495, 2020.
https://doi.org/10.1007/978-3-030-19648-6_56

experimental scenario, we have used vehicle sensor data navigating over agriculture field. The approach works on real-time constraints as it keeps unique points of the data while discarding redundancy between consecutive frames. Because of the huge amount of raw data and changes in environment, it is not possible to store and perform an off-road mapping.

The paper is organized as follows: First, it discusses the state-of-the-art related to localization, mapping and reconstruction. Thereafter it explains the approach to generate the elevation maps of rough terrain. Lastly, a detailed explanation of experimentation and evaluation studies are discussed.

## 2 Related Work

Precise Mapping with real-time constraints often requires good localization to decrease the transformation time at each time step. Individually the inertial sensors such that Inertial Measurement Unit (IMU) etc. have translational or rotational errors which can be reduced by fusion of multiple sensors. Extended Kalman Filter [1] is widely used for the integration of Global Positioning System (GPS) and Inertial Navigation System (INS) where GPS is used to aid INS which helps in estimating the errors in the INS. Nemra and Nabil [2] proposed State-Dependent Riccati Equation (SDRE) non filtering to fuse GPS and INS. In this paper, they have solved issues related to linearization but the algorithm does not guarantee the real-time constraints. Researches [3] and [4] have also been presented another form of estimation algorithms i.e. "square-root" family algorithms. Another technique has been proposed by Hinzmann et al. [5] where monocular visual-inertial camera rig as well as GPS are fused to estimate the precise pose.

Mapping and Reconstruction of environment has been looked over numerous times by the researchers since last decade. A work conducted by Hoppe et al. [6] provides the algorithm for surface reconstruction from unorganized points where boundaries and geometry of manifolds are unknown and automatically inferred from provided data. Due to the complexity of this approach, it cannot be processed in real-time. Moreover, an array of dense point cloud and its alignment to construct a map of surrounding is a challenging task. Turan et al. [7] have worked towards this challenge and proposed a comprehensive medical 3D reconstruction method for endoscopic capsule robots. Other methods for reconstruction include the stereo matching algorithms where many improvements have been made. Hamzah et al. [8] have used stereo matching to generate disparity map which provides the information in form of point cloud. Moreover, another technique such that Spatial Pyramid Matching (SPM) [9] divides the image in multiple sub frame and computes histograms of local features. According to their results, the algorithm finds useful discriminative information even in highly variable data sets. Another research focuses towards the use of disparity map estimation [10] for road surface reconstruction. This paper proposes a novel disparity refinement approach to improve the accuracy of estimated disparity map.

# 3 Methodology

The proposed system generates the precise elevation map of natural rough terrains using standard localization techniques (RTK-GPS, IMU, Visual Odometry), as well as LiDAR and Stereo Camera for point cloud generation. Moreover, it follows [5] for precise localization integrated with visual odometry from ORB SLAM [11] algorithm which uses bag of words approach [12] for place recognition and loop closure. The localization subsystem is not discussed in detail. The paper only focuses on the concept of surface mapping. The methodology presented below is developed for the mapping of agricultural fields where there are less obstacles, a precise map is obtained. The overall <60 cm of accuracy and real-time constraints of the system are enough for the kind of application under consideration. The precision of final map depends on the filtering and transformation of sensors raw data. The system is divided in three essential subsystems to avoid unnecessary data provided by LiDAR and Stereo camera. First step is the filtration of outliers and extraction of plane from terrain data. This helps the system in achieving real-time constraints to discard unwanted points and to make computation faster. The second subsystem performs transformation of data from local frame to the global frame. This is significant for the precision of final map. The third subsystem is stitching of the cloud and performing interpolation to fill out the gaps between points. This subsystem provides the complete elevation map of terrain by concatenating the data of each frame. The overall methodology is shown in Fig. 1.

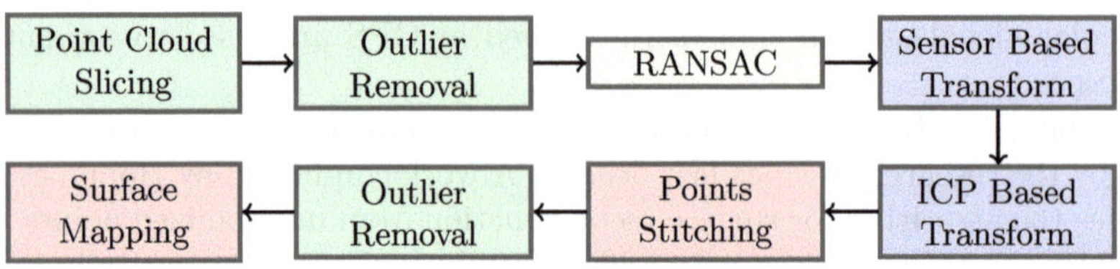


**Fig. 1.** Proposed methodology

## 3.1 Filtering and Plane Extraction

The high data frequency of LiDAR results in huge redundancy in consecutive frames' points. Therefore, the back half of the LiDAR data is sliced and discarded using PassThrough filter. This slicing helps in fast computation during stitching stage and filters a lot of redundant points at each time step. Moreover, the sensor data often has many outliers which should be removed for precise mapping (as outliers degrade the quality of final map). Radius based outliers removal technique has been used to discard all the irrelevant points that do not possess specific number of neighbors within a certain range. The careful values of radius and number of neighbors have been selected. These parameters control the number of outliers (noise) aimed to be removed from data. Consequently, this preprocessed data has lesser points and noise as compared to raw data. Moreover, the data still contains a lot of unwanted information such as trees, poles and other vertical objects which need to be further discarded. The Random Sample Consensus - RANSAC has been implemented with parallel plane model to extract the plane of terrain. The green blocks in Fig. 1 show the steps of filtration whereas yellow block shows the plane extraction.

### 3.2 Point Cloud Transformation

This subsystem is responsible for the transformation of LiDAR and Stereo data from local frame to the global frame of reference. It is divided in two stages. First, the sensor based transformation block fuses IMU, GPS and Visual Odometry to get pose on which LiDAR and stereo data are mapped. This step is mandatory for the real-time working of the system as it decreases the computation required by Iterative Closest Point (ICP) stage. In second stage of transformation, ICP [13] has been implemented to compute the transformation between two point clouds. Sensor based transformation in addition to ICP takes less time to compute transformation as compared to ICP alone. The key parameter of this algorithm is the Maximum Correspondence Distance which is maximum distance between a point and its corresponding nearest neighbor in order to be considered in the alignment process.

### 3.3 Stitching and Mapping

There are two essential modules of this subsystem: Point Cloud Stitching and Surface Mapping. The first module stitches the transformed point cloud of each frame. As the frame rates of sensors are quite high (max. approx. 20 ms), each point cloud constitutes more than 25% of redundant points from previous clouds. If this redundancy is not removed, the size of total map increases exponentially which results in abundant redundant points. Standard stitching algorithms do not eliminate the redundant points in real-time. Therefore, this module saves complete cloud as Octomaps. The second module uses bi-cubic spline interpolation to reconstruct surface from the point cloud. This interpolation estimates the best fit, for it identifies all points likely to be part of the terrain, including potentially identifying outliers that are supposedly part of the terrain, but are not used in the best-fit solution. The pcl ON_NURBS Library is used to generate smooth and parametric surface representation of map's point cloud.

## 4 Experimentation and Evaluation

A tractor mounted with Starfire 6000 RTK-GPS receiver, microstrain's 3DM-GX3-25 IMU, Stereo Labs ZED camera and HDL-32E LiDAR sensor above the wind screen is used (as shown in Fig. 2). The GPS receiver has an accuracy of ~2 cm in open fields and ~25 cm or more in shaded areas whereas LiDAR sensor

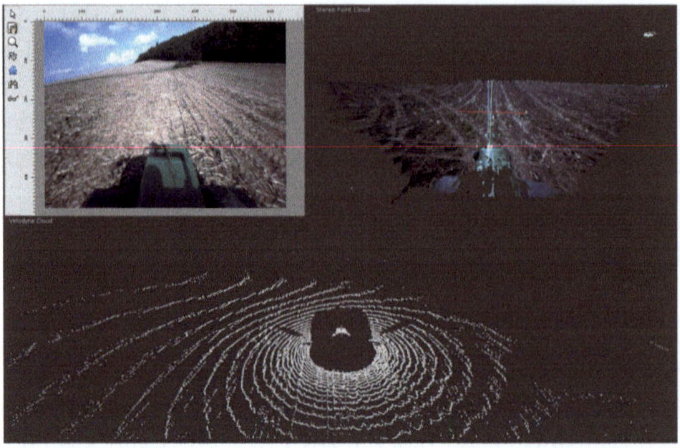

**Fig. 2.** LiDAR and Stereo camera based point cloud

has accuracy of ~2 cm. The whole system runs over core i7 processor with 16 GB RAM and NVIDIA 1050 GPU. During the locomotion, the speed of the tractor is kept at 18 m/h. Sensors' data is collected at a frequency of approximately 38 frames per second.

There are a lot of choices of parameters in the system which can be tuned according to specific application requirements such as high precision and low computation cost. Figure 3 shows the results of first three stages proposed in the methodology section. The data refinement and outliers removal can easily be visualized at each stage. Since the LiDAR has 360° field of view, on every forward movement of the tractor, the 3D points repeat between consecutive frames. Hence, it becomes justifiable to perform slicing in the initial stage.

**Fig. 3.** -From Left- (a) Raw LiDAR data (b) Slicing stage (c) Outliers removal stage (d) RANSAC stage

Slicing (Fig. 3b) is basically a pass through filter which is performed on specific axis. In our case it is performed from the middle of the point cloud perpendicular to the vehicle direction axis up till 50 m front view. After slicing the point cloud is visualized as 180° wide in front direction of the vehicle. This is a major step in reduction of point cloud to at least one half. The point cloud still contains outliers, so radius based outlier is performed with radius of 25 cm and threshold of 10 minimum neighboring points. This stage leads to a decrease in more than 23k points and increases the computation by a factor of 2.5. Finally, the extraction of the rough terrain is done by RANSAC. A plane fitting model is used with 38° angular deviation and 1.3 m distance threshold. This results in removal of vertical geometrical objects such as trees and poles but the overall elevation of the ground remains intact as can be seen in Fig. 3(d).

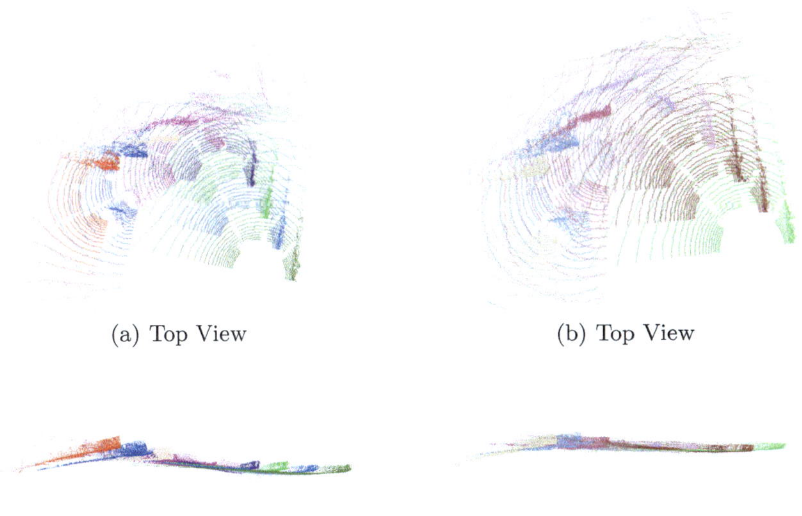

(a) Top View (b) Top View

(c) Side View (d) Side View

**Fig. 4.** (a) and (c) Sensor based transformation (b) and (d) ICP

During the alignment process as seen in Fig. 4, every next frame is aligned according to the initial frame. At the time of turning the vehicle, the inertial sensor shows considerable drift in its values. This drift is eliminated by ICP algorithm. Due to alignment process ICP performs transformation with comparably higher speed. Figures 4a and c show the results of sensor based transformation at every 5th frame. The remaining minor drift accumulated from the previous stage is also discarded by the ICP. This process increases the accuracy of the terrain map. Each scan is mapped with different color so that mapping can be easily visualized.

In Fig. 5, Point cloud is mapped on the trajectory of vehicle pose to visualize the correspondence. Green point cloud is depicted over the pose with blue dotted line. Overall results show that average time to complete the whole process for finding the terrain takes 37 ms, where more time is spent at corners of the field because rotation makes ICP to take more time for better convergence. The results are compared with the careful elevation measurement of 1200 points using Laser Level. The system provides ~42 cm accuracy and ~18 cm resolution which also degrades at corners as rotation produces more errors in sensor values.

For general comparison, Fig. 6a is image of the same field taken from google map and the path followed by tractor during data gathering. At the last stage the

(a) Top View (b) Side View

**Fig. 5.** Point-cloud stitching using octomap

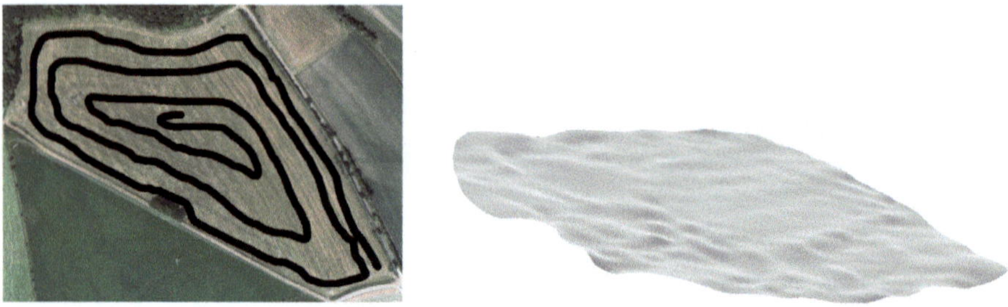

(a) (b)

**Fig. 6.** (a) Google map of the field (b) Surface reconstruction of whole area using cubic spline

point cloud of the generated map is visualized in the form of surface. The surface reconstruction stage fills out the gaps between the points of map using bi-cubic spline interpolation technique as shown in Fig. 6b. It can easily be concluded that the system, not only computes the elevation of terrain but is also capable of modeling its roughness.

## 5 Conclusion and Future Work

Elevation maps are required by autonomous vehicles to perform multiple navigation and path planning tasks on irregular or rough terrains. The DEMs are seemingly apt representations for research but can never be used in real life applications because of their low (1 m) resolution. It can be because of two important reasons: the 1 m resolution is not adequate to perform precise path planning and the environment is dynamic which can be changed as compared to the available map. Our system follows the methodology proposed in Fig. 1 and generates real-time precise elevation maps with ~42 cm average accuracy and ~18 cm resolution. In future, the generated maps will be used for path planning and navigation of off-road autonomous vehicles.

## References


1. Kumar, N.V.: Integration of inertial navigation system and global positioning system using Kalman filtering. Aerospace Engineering, Department of Aerospace Engineering Indian Institute of Technology, Bombay, Mumbai (2004)
2. Nemra, A., Aouf, N.: Robust INS/GPS sensor fusion for UAV localization using SDRE nonlinear filtering. IEEE Sens. J. **10**(4), 789–798 (2010). https://doi.org/10.1109/JSEN.2009.2034730
3. Morf, M., Kailath, T.: Square-root algorithms for least-squares estimation. IEEE Trans. Autom. Control **20**(4), 487–497 (1975). https://doi.org/10.1109/TAC.1975.1100994
4. Van Der Merwe, R., Wan, E.A.: The square-root unscented Kalman filter for state and parameter-estimation. In: 2001 Proceedings of the IEEE International Conference on Acoustics, Speech, and Signal Processing (ICASSP 2001), vol. 6. IEEE (2001)
5. Hinzmann, T., Schönberger, J.L., Pollefeys, M., Siegwart, R.: Mapping on the fly: real-time 3D dense reconstruction, digital surface map and incremental orthomosaic generation for unmanned aerial vehicles. In: Hutter, M., Siegwart, R. (eds.) Field and Service Robotics. Springer Proceedings in Advanced Robotics, vol. 5. Springer, Cham (2018)
6. Hoppe, H., DeRose, T., Duchamp, T., McDonald, J., Stuetzle, W.: Surface reconstruction from unorganized points. SIGGRAPH Comput. Graph. **26**(2), 71–78 (1992). https://doi.org/10.1145/142920.134011
7. Turan, M., Pilavci, Y.Y., Ganiyusufoglu, I., et al.: Machine Vision and Applications, vol. 29, p. 345 (2018). https://doi.org/10.1007/s00138-017-0905-8
8. Hamzah, R.A., et al.: Improvement of stereo matching algorithm for 3D surface reconstruction. Signal Process. Image Commun. **65**, 165–172 (2018)
9. Lazebnik, S., Schmid, C., Ponce, J.: Beyond bags of features: spatial pyramid matching for recognizing natural scene categories. IEEE (2006)
10. Fan, R., Ai, X., Dahnoun, N.: Road surface 3D reconstruction basedon dense subpixel disparity map estimation. IEEE Trans. Image Process. **27**(6), 3025–3035 (2018). https://doi.org/10.1109/TIP.2018.2808770
11. Mur-Artal, R., Tardós, J.D.: ORB-SLAM2: an open-source SLAM system for monocular, stereo and RGB-D cameras. IEEE Trans. Robot. **33**(5), 1255–1262 (2017)
12. Gálvez-López, D., Tardós, J.D.: Bags of binary words for fast place recognition in image sequences. IEEE Trans. Robot. **28**(5), 1188–1197 (2012)
13. Besl, P.J., McKay, N.D.: A method for registration of 3-D shapes. IEEE Trans. Pattern Anal. Mach. Intell. **14**(2), 239–256 (1992). https://doi.org/10.1109/34.121791

# Learning to Predict 2D Object Instances by Applying Model-Based 6D Pose Estimation

Hannes Kisner(✉), Tim Schreiter, and Ulrike Thomas

Lab of Robotics and Human-Machine-Interaction,
Chemnitz University of Technology, 09126 Chemnitz, SN, Germany
{hannes.kisner,tim.schreiter,ulrike.thomas}@etit.tu-chemnitz.de

**Abstract.** Object detection and pose estimation still are very challenging tasks for robots. One common problem for many processing pipelines is the big amount of object data, e.g. often it is not known beforehand how many objects and which object classes can occur in the surrounding environment of a robot. Especially available model-based object detection pipelines often focus on a few different object classes. However, new deep learning algorithms have been developed in the last years. They are able to handle a big amount of data and can easily distinguish between different object classes. The drawback is the high amount of training data needed. In general, both approaches have different advantages and disadvantages. Thus, this paper presents a new way to combine model-based 6D pose estimation with deep learning to reduce time for training and to improve the 6D pose estimation pipeline.

## 1 Introduction

The applications for robots range from industrial environments to service robots. These scenarios lead to interactions between robots and their environments. To this end, they are equipped with sensors to gather information about their surroundings. In many cases, object detection is one approach to find object instances, i.e. which objects are there, how many instances and how does the 6D pose of each object look like. Many approaches exist to solve this problem. Usually RGB-D cameras are used providing 3D point clouds with depth and texture information. Processing pipelines often sart with filtering the scene into regions of interest. This leads to restricted areas and a lowered amount of information for the following detection steps. Object detection algorithms evaluate the received data and try to find correspondences between the scene data and the data in a beforehand provided object database. Model-based approaches calculate feature vectors that include texture and shape information for each object [1]. Because objects are modeled by dense point clouds, they have to calculate a high amount of feature vectors. For correct registration of known objects feature vectors in the scene have to be compared with many vectors in the database, to find the correct object classes and poses. This leads to a time-consuming search procedure, if the database contains many different objects. However, model-based


K. Berns and D. Görges (Eds.): RAAD 2019, AISC 980, pp. 496–504, 2020.
https://doi.org/10.1007/978-3-030-19648-6_57

approaches show high accuracies in pose estimation [2], if they are known a priori. To overcome the computational drawbacks of model-based fitting, deep learning algorithms are chosen by a lot of researches. Especially Convolutional Neural Networks (CNNs) outperform object detection in scenes with various objects [3]. They outperform model-based approaches if a big amount of training data is provided. Well trained CNNs are even able to detect highly occluded objects, e.g. in bin-picking scenes. However, a major drawback of the training process is that it needs a lot of data to get a sufficient recognition rate. In our paper we combine both approaches. The model-based estimaton supports to label images which can be feed into CNNs to be trained automatically. In Sects. 2.1 and 3 we describe the pipeline in detail. Section 4 evaluates the presented methods. The final section includes a summary and concludes the paper.

## 2 Vision Pipeline

### 2.1 Basic Pipeline

Our new object pose estimation pipeline is based on a model-based approach [4,5]. The baseline of the new approach is shown in Fig. 1. It requires two inputs. One is the scene itself and the other one is a database which contains the CAD-model of each object and a feature map (to identify the object itself). With respect to the feature vectors from [4], the database is evaluated before processing the scene. Furthermore, we assume the scene is captured by an RGB-D camera or another system that can provide depth and texture information.

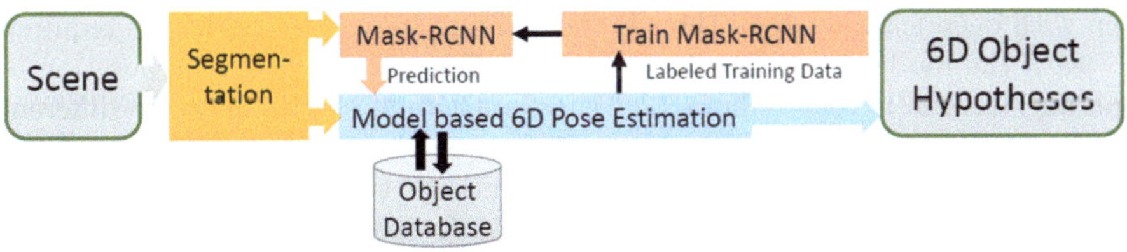


**Fig. 1.** Steps of the new introduced object detection pipeline.

The basis for our vision pipeline is the segmentation of the scene. As we have an unordered set of information, we have to apply a process that finds different groups of points that belong to the same region. Therefore, a spectral clustering method, as proposed in [5], choosen because of its efficiency. This algorithm segments groups of points that are near together and have nearly the same properties (normal orientation and texture information). The main advantage of this process is, that we do not need the exact number of regions in the scene beforehand. This satisfies our demands well because in the real world a segmentation process is influenced by noise and an unknown number of segments. Additionally, the Spectral Clustering does not need any assumptions, thus it can be applied to simple and complex scenes. We use the object pose estimation pipeline from [4], as it provides accurate 6D poses. The main target of an object detection process is to find the same regions in the scene and the matching object. Therefore, we store the properties of the objects as feature vectors in a database, similar as in [4]. During the detection process, we find correspondences between the database and the scene. This results in hypotheses and their confidence values. These values represent how well the shape and texture fit into the corresponding regions. A drawback of this algorithm is the exhaustive search for feature vectors. With a higher amount of objects, the amount of information increases and the search becomes more time-consuming.

### 2.2 Learning by Automatic Labeling

To overcome this problem a deep learning approach may help. A high amount of data is necessary to achieve good results (see [6]). However, it can occur, that CNNs fail to predict correctly in uncertain scenarios. Thus, a verification step would be necessary. This leads to our approach, which combines a deep learning method with a model-based pipeline. Therefore, two possibilities exist. The first is a prediction of objects in the whole scene, without a segmentation process. A CNN would be trained, that detects bounding boxes for each object. The determined boxes are not restricted to the exact object pixels and may include background pixels, e.g. during noise. The second possibility is shown in Fig. 1. There, we add a prediction step between the segmentation and the search for hypotheses. Leading to different advantages, the most important is that we get assumptions about possible object classes. Moreover, we are able to detect a possible object with specific parameters, e.g. different object sizes. The problem with deep learning algorithms is the training step. A lot of data is required to learn the model to distinguish between different objects. The generation of training data is a very tedious and time-consuming step. There are different approaches to overcome this problem. But, often the transfer to a new robotic environment is difficult. The generation of synthetic training data is a promising idea, but not yet appropriate enough. Even different illuminations in the environment or different camera models (unique hardware noise) lead to different results for CNNs. Therefore, Sect. 3.3 explains our approach to overcome this problem by generating scene specific training data by the object pose estimation pipeline from Sect. 2.1.

## 3 New Pose Estimation Approach

The following section describes how we combine deep learning with a model-based detection pipeline and how we automatically acquire the needed training data. Providing a sufficient amount of training data for neural networks is essential to achieve high accuracies. Moreover, the model structure itself is important for pose estimations, e.g. which type of model can be assumed.

### 3.1 Object Instance Segmentation

In recent years the central topic of object detection was the usage of CNNs. There is a huge amount of models and architectures, that compet with one another in object detection challenges like the Pascal VOC [7]. Most of them are improved over the years, like the YOLO series [3] or the RCNN series [8]. There has been a recent publication [6] in form of the MaskRCNN. It seems to be the current state-of-the-art method, when it comes down to object instance segmentation. Since the network assigns each object a class- and each pixel a masklabel, it results in a decent representation of the boundaries of the detected object, thus, becoming independent from the background of images. This is a major advantage in comparison to the standard method of using bounding boxes. For the approach proposed in this work, we use the implementation of the MaskRCNN from [9].

### 3.2 Transfer Learning

Most of CNNs implementations achieve object detection with training of large data sets, e.g. "MS COCO" [10]. Henceforth, these implementations have the problem to generalize on more specific tasks. Transfer learning offers the possibility to transfer the learned knowledge to new domains. Therefore, it is possible to train with small data sets and thus, being considered in our approach.

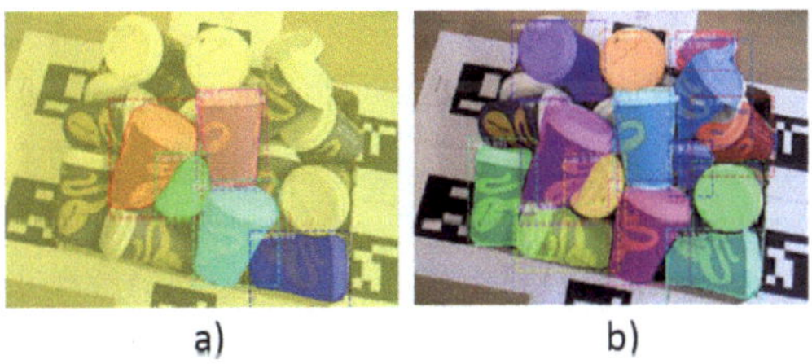

**Fig. 2.** Results of the MaskRCNN model (a) before and (b) after using transfer learning.

Figure 2(a) visualizes a result for a bin-picking scene, if we will use the MaskRCNN model out of the box (trained on the MS COCO dataset). The model fails to recognize even half of the cups. One of the main reasons is that there is an object detected that stretches over the entire image. The object "dining table" dominates and lowers the attention for other objects. However, if we retrain the model using transfer learning and an appropriate training dataset, we can achieve better results. We achieve this more accurate results by retrain the network using our automatically labeled training data.

### 3.3 Dataset Generation

Having a sufficient amount of training data available is a key for any deep learning algorithm. Nowadays it is easy to acquire datasets. But there is a gap in finding datasets for any scenario. This is caused by several circumstances. For instance, in real-world scenarios a high amount of different object classes exist. Furthermore, any workspace differs from others, even hardware components are not always the same. For instance Fig. 3 visualizes a possible environment. There are two cups, one socket, and one blue mug. A socket is initially a textureless object. But with the surrounding light, e.g. shading of texture appears. However, the resulting bright and dark regions are more complicated to evaluate. Therefore, to improve accuracies when working with neural networks, one has to generate his own data. This is a time-consuming process and led us to automate the generation of training data. As mentioned, there are different ways to generate training data, e.g. synthetic data [11]. But the estimation of the illumination or the camera noise model is still very difficult. Thus, we train on various environmentally specific data. Therefore, our approach makes use of a model-based 6D pose estimation pipeline and benefit from its advantages. Firstly, the object detection pipeline from Sect. 2.1 provides very accurate 6D poses. Furthermore, the segmentation is independent of the object itself. Thus, the approach is able to filter objects from their surroundings. Moreover, during the generation of training data, the scene is somehow structured. We assume that we are able to control the scene and know which objects are there. Thereby, we are able to adjust the parameters of the detection pipeline with nearly no false predictions. This is important to generate exact training data.

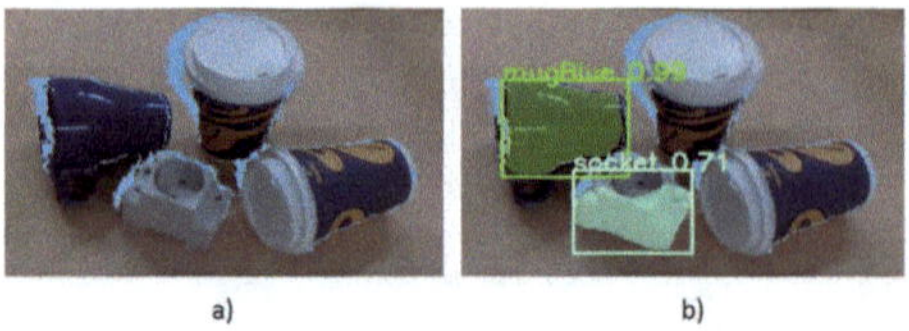


**Fig. 3.** (a) Example of a real data image in an example workspace. (b) A resulting prediction of the CNN. For a more focused visual presentation, we ignored the prediction/detection of the cups. The light blue pixels represents unknown depth values.

**Pose Estimation for 2D Images.** The process of automatically generating training images is visualized in Fig. 4. As shown, the pipeline for generating data and during the online execution just differs in the prediction step. This has to be skipped, because we first have to train the model with training data. The input data is a point cloud. With the known camera calibration, we are able to calculate for each pixel in a 2D image the depth information. If the pixel coordinates are saved, we are able to switch between the 3D point cloud and a 2D image, like in Fig. 3. With this image and the related 3D point cloud segmentation, we can extract binary masks for each segmented region. We then process each mask with morphology algorithms to reduce noise. Now, if the object pose estimation found a good hypothesis for the specified object and region, we save the contour of the mask in our training dataset. Thus, we automate the creation of training images.

Moreover, the same procedure is executed during the process from Sect. 2.1, but with a prediction step for each region. This is the initial guess for the model-based pose estimation pipeline and thus reducing the time to look-up for the correct object id.

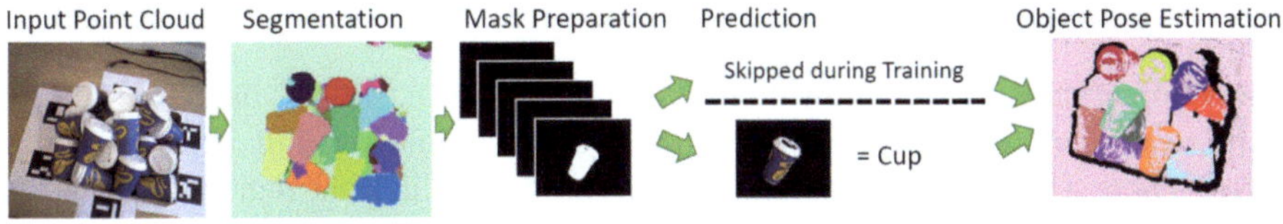


**Fig. 4.** Generation of training instances

**Interaction to Increase Training Data.** With the proposed approach it is possible to generate images with marked object instances. Most deep learning approaches need hundreds of images to achieve accurate results. Thus, we have to execute the approach several times. To this end, we have to consider that only with an appropriate amount of different viewpoints of the objects the training process is successful. In our scenarios, we assume the camera to be fixed. Hence, we have to interact with the scene. Furthermore, we assume to be in a robotic workspace and the robot is able to pick and place the object. With this assumption and the pose estimation pipeline, we can further automate our data generation. As shown in Fig. 2 we also need to get training data of many object instances that are occluded. However, the proposed approach is more related to a proof of concept. Thus, we did not yet implement such a difficult interaction process between the scene and the robot. Therefore, we placed manually additional objects in different locations into the scene.

# 4 Evaluation

## 4.1 Dataset

We apply the object detection pipeline to complex bin-picking scenes. In such scenes, multiple object instances lay near together, thus most of them are occluded. Therefore, we choose for evaluation purposes the dataset from [12].

**Fig. 5.** Results of the proposed pipeline for the dataset. The first column displays the original point cloud. The second column shows the segmentation result. The third and fourth columns visualize the results of the prediction and pose estimation.

Figure 5 visualizes example scenes. The cup dataset consists of 55 point clouds of bin-picking scenes. In each scene, multiple cups are placed and the number of visible objects is unknown beforehand. With our process, we evaluated the bounding boxes of the objects and the detected 6D poses.

### 4.2 Segmentation and Pose Estimation

The segmentation process is evaluated in [5]. Therefore, we just have to evaluate if the results are appropriate for our needs. An exemplary result is visualized in Fig. 5. There, the occluded objects are separated into different regions. Very large regions, e.g. the desks, are separated well and also noisy regions are separated, thus not influencing the segmentation process. Moreover, we do not have to make any assumptions about the scenes, e.g. knowing the number of objects, which relates to the advantage of the segmentation. However, the segmentation fails in some regions, because the similarity calculation has its disadvantages. But, this is not significant for proofing the whole concept and can be ignored. The pose estimation pipeline is also evaluated in [4]. As the accuracy is high, it mostly depends on the segmentation process. But, with the proposed regions, the process estimates correct poses, as shown in Fig. 5. The disadvantage is the process tries to fit any object in any region. Therefore, we have to evaluate each region with each object in our database. Hence, if we have to detect multiple objects classes, it would increase the execution time significantly.

**Table 1.** Mean Average Precision of the used models.

| | coco | cup_30 | cup_250 | cup_300 |
|---|---|---|---|---|
| Approach I (mAP) | 0.07 | 0.09 | 0.32 | 0.39 |
| Approach II (mAP) | 0.07 | 0.33 | 0.41 | 0.50 |

### 4.3 Training Data and Model Evaluation

During the training process, we generated few amounts of scenes, similar to those visualized in Fig. 3. As we do not have ground truth data, we are not able to evaluate these images. Thus, we increased the needed confidence for an object pose to a high value. Thereby, we did not receive false positives, which is important for the training data. We trained the model, as explained in Sect. 3, with 30 images (30 object instances) and with 200 images (250 object instances). Table 1 visualizes the evaluation results on the used dataset. Therefore, we evaluated four different models and two different approaches. Model "coco" represents the original MaskRCNN model, trained on the coco dataset. Model "cup_30" and "cup_250" represents the generated models through our training process. Furthermore, we trained a model on 80 images (300 object instances), that we labeled manually. We evaluated all models with two approaches. In approach one, with the segmentation process from [5] as a preprocess to filter the point-cloud into meaningful reagions. We skipped the segmentation in approach two and thus predicting the objects on the complete point cloud. The evaluation of the Mean Average Precision (mAP) calculates over all scenes the intersection of union error measurement. We used the visible bounding boxes of each object as ground truth data. The evaluation shows, that using transfer learning on real data improves the accuracy much. Furthermore, the mAP increases with an increasing amount of object instances. Moreover, there is a difference between the two approaches, which results from the segmentation process. This changes the scene a lot (black masks) and thus making it harder for the prediction of object instances. But, the segmentation is beneficial for the following pose estimation, as it removes background pixels. All in all, it is possible to increase the amount of training data, especially with more complex scenarios while increasing the overall accuracy. Furthermore, it has to be mention, that we did not receive any false positive during the evaluation of the best cup models. This fact and that we have an initial guess about the object class, proof our concept. Additionally, we tested on real data in our environment, e.g. Fig. 3. The blue mug has similar color and shape as the cup from the dataset. Due to sensor noise or different shading, a standard CNN predicts the false object (false positive). However, with the environment specific training dataset we can overcome this problem. Thus, the CNN predicts the mug and socket correct and does not predict false positives.

## 5 Conclusion

In this work, we propose a new object detection and pose estimation pipeline by combining the advantages of deep neural networks and a model-based 6D pose estimation. Furthermore, we use the concept of object instance detection to overcome the problem of background pixels. Thus, we are able to detect multiple occluded objects in very complex bin-picking scenes. Additionally, we propose an approach to automate the generation of the necessary training data while using the advantage of an accurate model-based pose estimation in the combination with an interaction of a robot. This leads to the possibility to automatically generate and label environment specific training data to adjust the detection network. The evaluation shows sufficient results for first usages. Next, we aim to able to produce more complex training scenes with the robot, e.g. stacking of objects. Thus reducing the manual interaction completely.

**Acknowledgements.** This work is supported by the European Social Fund (ESF) and the Free State of Saxony.

## References

1. Salti, S., Tombari, F., Di Stefano, L.: Shot: unique signatures of histograms for surface and texture description. Comput. Vis. Image Underst. **125**, 251–264 (2014)
2. Hinterstoisser, S., Lepetit, V., Rajkumar, N., Konolige, K.: Going further with point pair features, CoRR (2017)
3. Redmon, J., Divvala, S.K., Girshick, R.B., Farhadi, A.: You only look once: Unified, real-time object detection, CoRR (2015)
4. Kisner, H., Thomas, U.: Efficient object pose estimation in 3D point clouds using sparse Hash-Maps and Point-Pair features. In: 50th International Symposium on Robotics (ISR 2018), Munich, Germany, June 2018
5. Kisner, H., Thomas, U.: Segmentation of 3D point clouds using a new spectral clustering algorithm without a-priori knowledge. In: Proceedings of the 13th International Joint Conference on Computer Vision, Imaging and Computer Graphics Theory and Applications (2018)
6. He, K., Gkioxari, G., Dollár, P., Girshick, R.: Mask R-CNN. In: Proceedings of the IEEE International Conference on Computer Vision (ICCV) (2017)
7. Everingham, M., Eslami, S.M.A., Van Gool, L., Williams, C.K.I., Winn, J., Zisserman, A.: The PASCAL visual object classes challenge: a retrospective. Int. J. Comput. Vis. **111**, 98–136 (2015)
8. Girshick, R.B., Donahue, J., Darrell, T., Malik, J.: Rich feature hierarchies for accurate object detection and semantic segmentation, CoRR (2013)
9. Abdulla, W.: Mask R-CNN for object detection and instance segmentation on keras and tensorflow (2017). https://github.com/matterport/Mask_RCNN
10. Lin, T.-Y., Maire, M., Belongie, S., Hays, J., Perona, P., Ramanan, D., Dollár, P., Zitnick, C.L.: Microsoft COCO: common objects in context. In: Computer Vision – ECCV 2014 (2014)
11. Hinterstoisser, S., Lepetit, V., Wohlhart, P., Konolige, K.: On pre-trained image features and synthetic images for deep learning, CoRR (2017)
12. Doumanoglou, A., Kouskouridas, R., Malassiotis, S., Kim, T.: 6D object detection and next-best-view prediction in the crowd, CoRR (2015)

# Dataset Generation Using a Simulated World

Axel Vierling(✉), Tanittha Sutjaritvorakul, and Karsten Berns

Technische Universität Kaiserslautern, 67663 Kaiserslautern, Germany
{vierling,tanittha,berns}@cs.uni-kl.de

**Abstract.** In this paper we focus on a missing part of solving the detection of human workers around the load and the truck of a crane. Current solution attempts use Convolutional Neural Networks to detect these workers in images. Especially the detection around the load can not be solved with the use of public datasets as the viewing angles differ to much. We therefore propose and evaluate an approach which uses a simulation engine to create automatically labeled images to train a network for this use case.

**Keywords:** Safety · Crane · Simulation · Dataset Generation · CNN

## 1 Introduction

In recent times robotics and automation is an ever growing field of interest. Automation is omnipresent in defined areas such as production halls. More complex areas e.g., construction sites, still need a lot of research. In order to fully automatize a construction site many tasks need to be considered. These tasks can be divided into earthmoving tasks, see [1], and high-rise or low-rise construction, [2]. Especially in the high-rise construction the automation of cranes or other lifting mechanisms is of particular interest. There has already been some work in this direction, see [3] and [4], but safety aspects were mostly disregarded. The reliable detection of human workers, and other objects, is of utmost priority. A safety concern unique to cranes is that potentially hazards can occur near the crane truck itself, but also around the load, see Fig. 1. A step towards solving this problem will be the main focus of this paper.

In [5] the authors propose a camera setup and algorithms, which focus on the detection of workers around the load. The proposed algorithms are mainly based on Convolutional Neural Networks (CNNs), among others Faster R-CNN [6]. However one of the main drawbacks of using CNNs is the need for huge datasets. One of the most renown benchmarks for CNNs, the ILSVRC challenge [7], provided 456,567 labeled images for the training of object detection networks. Since the detection of workers around the load is a unique challenge, primarily because the workers are seen from above and also because helmets and reflective vests are not comparable to normal clothes, there exists no publicly available


K. Berns and D. Görges (Eds.): RAAD 2019, AISC 980, pp. 505–513, 2020.
https://doi.org/10.1007/978-3-030-19648-6_58

**Fig. 1.** Grove Manitowoc GMK 3060 crane

dataset satisfying these specific requirements. Manually labeling such a vast amount of images is not feasible. Therefore a possibility of creating labeled data tailored to these specific requirements is needed. Recently a lot of work in the field of using simulations to generate datasets has been done. In [8] the authors use the game *Grand Theft Auto 5* and extract images and annotations in order to have a dataset featuring urban environments and humans closely mimicking real-world scenarios. In [9] the author use the KITTI dataset, see [10], to create a clone of the environment inside the *Unity* game engine. These created scenes are then used in to train Convolutional and Reinforcement Networks. In [11] the *Unreal Engine 4* (UE4) is used and each texture in a scene is adapted by changing the vertex color. A server query is used to get the needed ground truth (GT) information out of the simulation. This gives a possibility to extract the GT for specific scenes, but a drawback is that adapting the vertex color removes a modeling parameter. None of the described approaches fulfills our needs completely. Therefore in this paper first the capabilities of simulation engines are discussed and methods to extract needed information are described. The generated data is compared to a real world dataset and the performance of a Faster R-CNN architecture trained with different quantities of real and simulated data is evaluated.

## 2 Generating Simulated Datasets

In order to generate a simulated dataset the main requirements of a suitable simulation engine are the possibility to generate realistic looking images, this includes realistic rendering, appropriate environments, arbitrary viewing angles, and of course the ability to automatically create annotations. The UE4 seems to be the most suitable engine. In the following section it will be shown, that in fact UE4 fulfills all our requirements.

### 2.1 Visual Capabilities of Simulation Engines

The first mentioned requirement of a simulation engine are realistic looking images. In Fig. 2 one can see three nearly real looking images from the UE4 engine. Figure 2a shows a forest floor with leaves and critters. Figure 2b shows an indoor environment with furniture rendered in UE4. Figure 2c shows a car on

(a) Forest environment (b) Indoor environment (c) Street environment

**Fig. 2.** Photorealism in Unreal Engine 4

an avenue. So the UE4 is able to generate realistic looking images in different settings.

However not only different settings are interesting, also different scenarios in such settings are of interest. Also different weather conditions e.g., rain [12] and lightning can heavily effect the performance of object detection algorithms and it is therefore desirable to include such circumstances in a dataset. In Fig. 3 one can see the simulation of different weather and lightning conditions in UE4. In Fig. 3a snowflakes were added to simulation. Figure 3b and c show a similar scene once in night time and once in bright daylight with a prominent overexposure. This in facts highlights an advantage of simulated data compared to real datasets, as such effects can be added at any time in simulation while in real datasets one needs to wait till a certain weather forms.

(a) Snow (b) Darkness/Night (c) Glare/Daylight

**Fig. 3.** Different weather and lightning conditions in the simulation environment

One other requirement of a suitable simulation engine is an application specific environment i.e., a construction site, and an user-defined choice of viewing angles. In Fig. 4 one can see different images of a road construction scenario. Where in Fig. 4a and b a similar scene is displayed from different viewing angles. Comparing Fig. 4b, c and d one can also see different actions which are performed by the persons and different attires i.e., a worker with a helmet and a human in everyday clothing. Different actions imply different postures which are resulting in a different shape of a human in an image. These shapes are again interesting for training and evaluating an object detection algorithm.

With the described abilities of the UE4 it is shown, that all the demanded visualization requirements are fulfilled by the UE4. The automatic extraction of annotations is still missing and will be described now.

(a) Top view (b) Walking (c) Kneeing (d) Talking

**Fig. 4.** Different actions and viewing angles of simulated workers

### 2.2 Generating Ground Truth Data

To generate ground truth data one needs to extract the position and bounds of each visible object. The UE4 engine has two different pipelines. The render pipeline, where only some information is available for the end user, and the physic pipeline which is fully available to the end user. Therefore it is easy to extract the collision bodies of the physic pipeline. This body is, in its most simple form, a box, which is bigger than the actual object. This collision box is then the initial guess for the bounding box of a human, where the bounding box of an object is defined as the smallest rectangle containing all pixels belonging to this object.

In order to decide if a collision box is in the view of the camera, one first needs the frustum of the camera. Let $T, R$ be the translation and the rotation of the camera in world coordinates. Let $\alpha_{fov}$ be the field of view of the camera and $w$ and $h$ be the width and height corresponding to the resolution and the distance $z$. Then if the $z$-axis is normal to the image plane the frustum can be computed with the following equation:

$$V = T^{-1}R^{-1}\begin{pmatrix} 0&0&1&0\\ 1&0&0&0\\ 0&1&0&0\\ 0&0&0&1 \end{pmatrix}\begin{pmatrix} c&0&0&0\\ 0&\frac{w}{h}c&0&0\\ 0&0&1&1\\ 0&0&-z&0 \end{pmatrix}, \text{ where } c = \frac{1}{\tan(\frac{\alpha_{fov}}{2})} \tag{1}$$

Since the collision box is not tight a second camera is used to get the bounding box. The first camera renders the original RGB-Image, $Im_1$ of the scene. The second camera uses a *Custom Depth Buffer*, which assigns a custom distance value, encoded in a color, to each object in the scene. This distance value is chosen unique for each object and called the $ID$ of the object. In order to easily distinguish each objects ID the following color is assigned to corresponding pixels: $(H, S, V) = (\frac{360ID}{256}, \frac{1}{2} + \frac{ID}{512}, 1 - \frac{ID}{512})$. This pixel-wise segmented image is called $Im_2$. Afterwards the bounding boxes can be calculated with Algorithm 1.

So far the focus was on the detection of persons with 2D-Image data. However one might want to include other sensors e.g., a laserscanner for the automation of a crane, since having 3D perception information of surrounding objects for robotic applications is very helpful. This 3D information, can be used further, among others, for navigation, interaction, control, decision making and of course detection. Likewise 2D-image data, also 3D information can be extracted from UE4. From the physic pipeline of the engine one can directly extract the position, $(x, y, z)$ coordinates, as well as the orientation, roll, pitch, yaw or $(\phi, \theta, \psi)$ coordinates i.e., the pose of an object can be extracted. Then the pose of an object in UE4 can be converted into a pose in robot coordinates as follows.

The UE4 coordinate system is left-handed while the robotic coordinate system is commonly right-handed. The unit of measurement in UE4 is 1 uu (Unreal Engine unit) which equals to 1 cm. Conversion from UE4 poses $P_{ue}$ to real world robot poses $P_{rh}$ is then done by Eq. 2:

$$P_{rh} = \left[0.01x_{ue}, -0.01y_{ue}, 0.01z_{ue}, -\phi_{ue}, \theta_{ue}, -\psi_{ue}\right]^T \tag{2}$$

Now all the requirements which were needed to generate suitable datasets are met. However it is not clear how these simulated datasets compare to real datasets.

**Data:** $Im_1$ and $Im_2$, list of objects with unique IDs
**Result:** Bounding boxes for each visible object in the scene
**for** *Each object* $O_i$ *in object list* **do**
  **if** $O_i$ *is visible in* $Image_1$ **then**
    Extract collision boxes $CB_i$ of $O_i$ from physic pipeline;
**end**
**for** *each* $CB_i$ **do**
  Transform $CB_i$ to image coordinate system of $Im_1$;
  **for** *each* $CB_i = \left[x_i^{max}, x_i^{min}\right] \times \left[y_i^{max}, y_i^{min}\right]$ **do**
    $x_{min} := x_i^{max}, x_{max} := x_i^{min}, y_{min} := y_i^{max}, y_{max} := y_i^{min}$;
    **for** $k \in \left[x_i^{max}, x_i^{min}\right]$ **do**
      **for** $j \in \left[y_i^{max}, y_i^{min}\right]$ **do**
        **if** *Pixel of* $(k, j)$ *in* $Im_2$ *corresponds to class of* $CB_i$ **then**
          **if** $k < x_{min}$ **then**
            $x_{min} := k$;
          **if** $k > x_{max}$ **then**
            $x_{max} := k$;
          **if** $j < y_{min}$ **then**
            $y_{min} := j$;
          **if** $j > y_{max}$ **then**
            $y_{max} := j$;
      **end**
    **end**
  **end**
**end**

**Algorithm 1.** Generation of bounding boxes

## 3 Simulated Versus Real Datasets

In order to compare a simulated dataset to a real one a comparable setting is needed. Therefore in this section the use-case of detecting workers around a crane is not considered but the focus is shifted to an application which is similar to an already existing dataset. As real life dataset the Caltech dataset [13] is chosen, i.e. a dataset recorded from a car in order to detect pedestrians in a city environment. In the Caltech Dataset around 132,000 frames with a labeled pedestrians are included. Persons shown in an image are only labeled as pedestrians if they are nearer than 60 m and are at least 20% visible. The same principle for labeling is used in the simulation. Around 22,000 labeled images for the datasets were created from simulation. The generated dataset is from now on referenced as UE4_dataset. Exemplary images from the Caltech dataset and the UE4_dataset can be seen in Fig. 5. While Fig. 5a and b are distinguishable they show similar characteristics. From a qualitative evaluation they show similar settings and similar lightning conditions as well as similar scales. However if a more in depth comparison is used they show qualitative differences.

One can e.g., compare the mean image of the UE4_dataset, Fig. 6a, and the mean image of the Caltech Dataset, Fig. 6b. In both mean images a certain standard case is still visible. A road in the middle and some vegetation next to

(a) Image from the Caltech Dataset (b) Image from the Unreal Dataset

**Fig. 5.** Comparison of an image from the Caltech and the simulated dataset

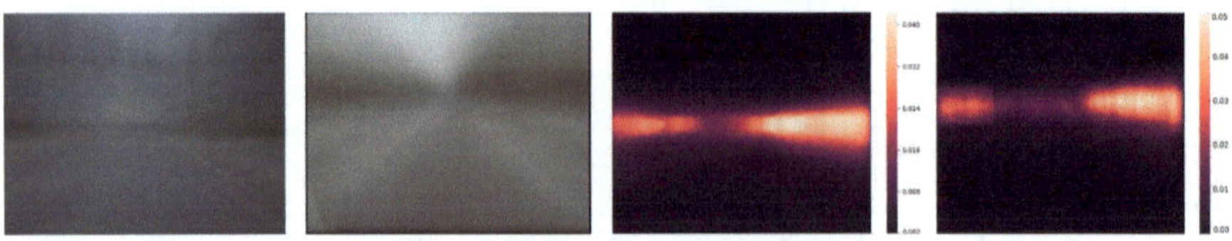

(a) Unreal (b) Caltech (c) Unreal (d) Caltech

**Fig. 6.** Mean images (a and b) and heatmaps of ground truth (GT) bounding boxes (BB) (c and d) this road is showing in both mean images. A quite visible difference is that in the mean UE4_dataset image the road curb and the columnar structure of the buildings are clearly visible. One can also compare the heatmaps of the bounding boxes in the UE4 and in the Caltech case, Figs. 6c and d. In the heatmap of the bounding boxes in simulated images the $y$-position is higher. Also the variance in bounding box size and position is less.

All in all one can say that even with the invested effort the variance in the real dataset is inherently higher. From this fact alone one can not conclude that this lack of variance has a negative effect on training a detection network, which will be considered next.

## 4 Training with Simulated Data

In this section the UE4_dataset is evaluated by comparing it to the Caltech dataset with regard to performance of a CNN trained on each dataset. As a basis the Faster R-CNN architecture [6] pretrained on the ImageNet dataset with ResNet101 [14] as a feature extractor is used. The Tensorflow API is chosen as a learning framework and the default hyperparameters are not adapted for training. Therefore biased optimization of hyperparameters for any dataset is impossible. The number of training iterations is 200,000 for each dataset. Three different cases are compared. In the first case the network is only trained on the UE4_dataset, i.e. roughly 20,000 images. In the second case the network is trained purely on the Caltech dataset. For the fixed number of iterations the best performance was achieved when only roughly 7000 images were used as training set. For the third case the network was trained with the whole Unreal dataset for 200,000 steps and then fined tuned with around 2000 images from the Caltech dataset for 100,000 steps. The results can be seen in Table 1. Evaluation for all networks is done on Caltech data which is not part of any training dataset. Examples of the evaluated data can be seen in Figs. 7a and c. To better understand the different performance one can further investigate the Faster R-CNN architecture. In Faster R-CNN first a set of potential objects with a certain objectiveness score is detected, afterwards these object candidates are classified to find the relevant class, also a class "background" is possible. A heatmap of the detected object candidates can be seen in Figs. 7b and d.

**Table 1.** Results of training for different cases. Precision is calculated for a detection with an Intersection over Union (IoU) > 50% with the ground truth

| Training set | Average precision | # Images | # Iterations |
|---|---|---|---|
| Pure Unreal (PU) | 0.101 | around 20k | 200k |
| Pure Caltech (PC) | 0.44 | around 7k | 200k |
| Fine-tuned (FT) | 0.83 | around 22k | 300k |

(a) PU network (b) PU network (c) FT network (d) FT network

**Fig. 7.** Evaluation of object detections (a and c) and object candidate heatmaps (b and d)

## 5 Conclusion and Outlook

As a conclusion one can say that a *Dataset Generation Using a Simulated World* is not enough to train a network. Training solely on simulated images does not yield good results, see Table 1. But network first trained on simulated data and then fine-tuned on real data seems to perform better in a quantitative evaluation, Table 1, and in a qualitative evaluation, Fig. 7. So one can conclude that with a current simulation engine the need for real datasets is not removed. But it is possible to reduce the needed amount tremendously. Coming back to the initial use-case of detecting workers around the load and the truck of a crane, only around 1000 to 2000 hand labeled images are needed. Additionally using simulated images first and then fine-tuning with real images increases the accuracy of the detection network. Further work could be invested in using generative adversarial networks in order to close the gap between the simulated and the real world. Also the quality of the presented simulated images is not yet perfect, compared to e.g., Fig. 2.

## References

1. Ha, Q., Yen, L., Balaguer, C.: Earthmoving construction automation with military applications: past, present and future. In: Proceedings of the 35th ISARC, Berlin, Germany, pp. 1180–1190 (2018)
2. Cai, S., Ma, Z., Skibniewski, M., Guo, J., Yun, L.: Application of automation and robotics technology in high-rise building construction: an overview. In: Proceedings of the 35th ISARC, Berlin, Germany, pp. 309–316 (2018)
3. Lee, G., Kim, H.-H., Lee, C.-J., Ham, S.-I., Yun, S.-H., Cho, H., Kim, B.K., Kim, G.T., Kim, K.: A laser-technology-based lifting-path tracking system for a robotic tower crane. Autom. Constr. **18**(7), 865–874 (2009)
4. Kim, D.-W., An, S.-H., Cho, H., Jeong, J.-W., Lee, B.-H., Doh, N.L., Kang, K.-I.: Development of conceptual model of construction factory for automated construction. Build. Environ. **44**(8), 1634–1642 (2009)
5. Vierling, A., Sutjaritvorakul, T., Berns, K.: Crane safety system with monocular and controlled zoom cameras. In: Proceedings of the 35th ISARC, Berlin, Germany, pp. 873–879 (2018)
6. Ren, S., He, K., Girshick, R.B. Sun, J.: Faster r-cnn: Towards real-time object detection with region proposal networks. CoRR, vol. abs/1506.01497 (2015)

7. Russakovsky, O., Deng, J., Su, H., Krause, J., Satheesh, S., Ma, S., Huang, Z., Karpathy, A., Khosla, A., Bernstein, M., Berg, A.C., Fei-Fei, L.: Imagenet large scale visual recognition challenge. Int. J. Comput. Vision (IJCV) **115**(3), 211–252 (2015)
8. Richter, S., Vineet, V., Roth, S., Koltun, V.: Playing for data: ground truth from computer games. In: Computer Vision - ECCV 2016, pp. 102–118 (2016)
9. Gaidon, A., Wang, Q., Cabon, Y., Vig, E.: Virtual worlds as proxy for multi-object tracking analysis. CoRR, vol. abs/1605.06457 (2016)
10. Geiger, A., Lenz, P., Urtasun, R.: Are we ready for autonomous driving? the KITTI vision benchmark suite. In: Conference on Computer Vision and Pattern Recognition (CVPR) (2012)
11. Qiu, W., Yuille, A.: UnrealCV: connecting computer vision to unreal engine. In: Hua, G., Jégou, H. (eds.) Computer Vision - ECCV 2016 Workshops, pp. 909–916. Springer International Publishing, Cham (2016)
12. Fu, X., Huang, J., Zeng, D., Huang, Y., Ding, X., Paisley, J.: Removing rain from single images via a deep detail network. In: 2017 IEEE Conference on Computer Vision and Pattern Recognition (CVPR), pp. 1715–1723, July 2017
13. Dollar, P., Wojek, C., Schiele, B., Perona, P.: Pedestrian detection: an evaluation of the state of the art. IEEE Trans. Pattern Anal. Mach. Intell. **34**(4), 743–761 (2012)
14. He, K., Zhang, X., Ren, S., Sun, J.: Deep residual learning for image recognition. CoRR, vol. abs/1512.03385 (2015)

# Convolutional Encoder-Decoder Networks for Robust Image-to-Motion Prediction

Barry Ridge[1,2(✉)], Rok Pahič[2], Aleš Ude[2], and Jun Morimoto[1]

[1] ATR Computational Neuroscience Laboratories, Kyoto, Japan
{barry.ridge,xmorimo}@atr.jp
[2] Department of Automatics, Biocybernetics, and Robotics, Jožef Stefan Institute, Ljubljana, Slovenia
{rok.pahic,ales.ude}@ijs.si

**Abstract.** A deep encoder-decoder network was previously proposed for learning a mapping from raw images to dynamic movement primitives in order to enable a robot to draw sketches of numeric digits when shown images of same. In this paper, the network architecture, which was previously constructed entirely with fully-connected linear layers, is modified to include convolutional layers in order to improve the image encoder component and make the network more robust to noise. The convolutional layers are pre-trained as part of an MNIST digit classifier and adapted for use in the encoder-decoder network, before the network is trained using a dataset composed of digit images and corresponding writing trajectories. This architecture was tested on several challenging noisy digit datasets and the use of convolutional layers is shown to provide a robust improvement in results.

**Keywords:** Deep neural networks · Dynamic movement primitives

## 1 Introduction

Effectively learning to predict action mappings directly from perceptual input is a highly challenging problem in robotics research that has seen a broad variety of approaches attempting to solve it in different settings. One example is that of object-action complexes (OACs) [8], which is a grounded representation that binds objects, actions, and attributes in a causal model. The particular setting under consideration in this work is depicted in Fig. 1, in which a robot must learn direct mappings between handwritten characters in input images and the motion trajectories needed to draw them. Our previously proposed fully-connected encoder-decoder network architecture [9] used dynamic movement primitives (DMPs) [6] for movement representation and this proved to be an effective choice both for representation and learning with the neural network and ultimately for control of the robot when drawing the actual digits. The fully-connected architecture, however, was not ideal for image representation.


K. Berns and D. Görges (Eds.): RAAD 2019, AISC 980, pp. 514–523, 2020.
https://doi.org/10.1007/978-3-030-19648-6_59

In this paper, we investigate a different architecture that combines the benefits of convolutional layers for image encoding with those of a fully-connected encoder-decoder architecture for DMP parameter prediction and image-to-motion representation in a low-dimensional latent space. This combination allows for relatively robust prediction compared to the previously proposed architecture, even when the input images are heavily corrupted by noise. The use of convolutional layers has the added benefit of significantly reducing the number of network parameters and by pre-training these layers on images from a similar image domain, the learning process is further improved.

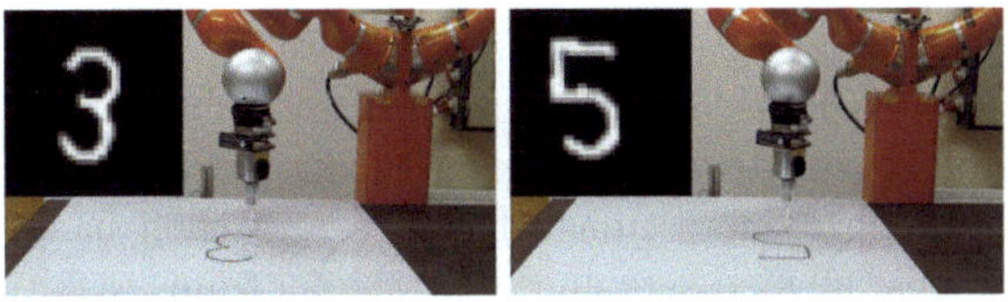


**Fig. 1.** Writing digits with a robot using image-to-motion encoder-decoder network prediction. The movements are generated using DMPs predicted by the network from the images shown in upper left corners.

Autoencoders [4], as well as variational autoencoders [5], have been demonstrated to be quite effective when it comes to calculating DMP-based representations of human motion. Since our focus is on learning direct mappings between images and actions, instead of using such autoencoder networks in which the DMP encoding occurs in the latent space, we use an encoder-decoder architecture in which the image is encoded from the input layer, the DMP parameters are predicted at the output layer and the transformation and generalization of the image-to-motion representation occurs in the low-dimensional latent space. Encoder-decoder networks in combination with convolutional layers have proven to be useful in computer vision. A well-known example is SegNet [2], in which pre-trained convolutional layers from a convolutional neural network (CNN) were adapted to form a fully-convolutional encoder-decoder architecture for semantic pixel-wise segmentation. We also use pre-trained convolutional layers from a CNN in this work, although the architecture differs in that the convolutional layers are only used to form the encoder part of the network. The CNN that we use here is also comparatively basic, but the use case is a very different one in which a more limited image recognition approach is sufficient and the architecture would nonetheless be extensible for more advanced scenarios.

Prior to the adoption of deep learning methods for such tasks, an effort was made by Ali [1] at modeling individual brush strokes of calligraphic characters using Gaussian Mixture Models, combining the brush strokes using Gaussian Mixture Regression and reproducing brush stroke trajectories using DMPs. The reproduced stroke trajectories were iteratively refined using reinforcement learning for learning examples in the database, but each reproduction started from scratch with no generalization between different examples or to new unseen cases. Usually when CNNs are used for supervised learning of perception-action

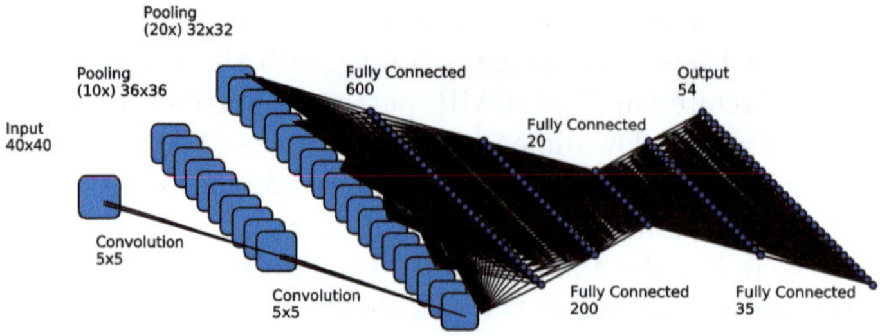


**Fig. 2.** The CIMEDNet architecture.

couplings, they are used in combination with another neural network in two separately trainable parts. In [13], Yang et al. first used a deep convolutional autoencoder for finding camera image features and then in combination with recorded robot angles, formed sequences for the learning task dynamics with a time delay neural network. Pervez et al. [11] used a pre-trained CNN for finding task parameters from input images, while using a another fully-connected neural network to learn to generate forcing terms from the clock signal and task parameters, before combining both networks in an end-to-end training scheme. Both of these two examples produce the next step from the image of the current step while working in online loop, whereas our method, by contrast, uses just single images for generating entire trajectories.

## 2 Convolutional Image-to-Motion Encoder-Decoder Networks

The structure of the data under consideration in this work is the same as in [9] where the input and output data pairs take the form $\mathbf{D} = \{\mathbf{C}_j, \mathbf{M}_j\}_{j=1}^{M}$ where $M$ is the number of input and output training pairs, $\mathbf{C}_j \in \mathbb{R}^{H\times W}$ are the input images of width $W$ and height $H$, and $\mathbf{M}_j$ the corresponding movements associated with each image, i. e. $\mathbf{M}_j = \{\boldsymbol{y}_{i,j}, t_{i,j}\}_{i=1}^{T_j}$. Here $\boldsymbol{y}_{i,j} \in \mathbb{R}^{\mathrm{d}}$ are the vectors describing the movement's degrees of freedom, e. g. Cartesian positions or joint angles, $t_{i,j} \in \mathbb{R}$ the measurement times for the $j$-th movement, and $d$ is the number of degrees of freedom. However, it should be noted that in this paper, we convert the movements $\mathbf{M}_j$ to DMPs and construct all of the datasets used to train the network models as follows: $\mathbf{D}' = \{\mathbf{C}_j, \mathbf{k}_j\}_{j=1}^{M}$, where $\mathbf{k}_j$ are the DMP parameters calculated for each movement $\mathbf{M}_j$ and are represented as

$$\mathbf{k}_j = \left\{\{\boldsymbol{w}_k\}_{k=1}^{N},\ \tau,\ \boldsymbol{g},\ \boldsymbol{y}_0\right\}. \tag{1}$$

The construction of DMPs and the nature of the parameters $\{\boldsymbol{w}_k\}_{k=1}^{N}$, $\tau$, $\boldsymbol{g}$ and $\boldsymbol{y}_0$ are explained in detail in the following subsection.

### 2.1 Motion Representation with DMPs

Letting a time-dependent movement trajectory be denoted as $\boldsymbol{y}(t) \in \mathbb{R}^d$, a DMP specifying this trajectory is given by the following system of differential equations

$$\tau\dot{\boldsymbol{z}} = \alpha_z(\beta_z(\boldsymbol{g} - \boldsymbol{y}) - \boldsymbol{z}) + \mathrm{diag}(\boldsymbol{g} - \boldsymbol{y}_0)\mathbf{F}(x), \tag{2}$$

$$\tau\dot{\boldsymbol{y}} = \boldsymbol{z}, \tag{3}$$

where $\boldsymbol{y}_0 \in \mathbb{R}^d$ is the initial position on the trajectory, $\boldsymbol{g} \in \mathbb{R}^d$ the final position on the trajectory, $\mathrm{diag}(\boldsymbol{g} - \boldsymbol{y}_0) \in \mathbb{R}^{d\times d}$ a diagonal matrix with components of vector $\boldsymbol{g} - \boldsymbol{y}_0$ on the diagonal, $\mathbf{F}(x) \in \mathbb{R}^d$ a nonlinear forcing term, $\boldsymbol{z} \in \mathbb{R}^d$ a scaled velocity of motion, and $x \subset \mathbb{R}$ the phase defined by the following equation

$$\tau\dot{x} = -\alpha_x x. \tag{4}$$

The phase $x$ is used instead of time to avoid explicit time dependency. It is fully defined by setting its initial value to $x(0) = 1$. Equation system (2)–(4) constitutes a *dynamic movement primitive* (DMP). If the parameters $\tau, \alpha_x, \alpha_z, \beta_z \in \mathbb{R}$ are defined appropriately, e. g. $\tau, \alpha_x > 0$ and $\alpha_z = 4\beta_z > 0$, then the linear part of equation system (2)–(3) becomes critically damped and $\boldsymbol{y}$, $\boldsymbol{z}$ monotonically converge to a unique attractor point at $\boldsymbol{y} = \boldsymbol{g}$, $\boldsymbol{z} = 0$. The forcing term $\mathbf{F}(x)$ is usually defined as a linear combination of radial basis functions

$$\mathbf{F}(x) = \frac{\sum_{k=1}^{N} \boldsymbol{w}_k \Psi_k(x)}{\sum_{k=1}^{N} \Psi_k(x)} x, \tag{5}$$

$$\Psi_k(x) = \exp\left(-h_k\,(x - c_k)^2\right), \tag{6}$$

where $c_k$ are the centers of Gaussians distributed along the phase of the trajectory, and $h_k$ their widths. The role of $\mathbf{F}$ is to adapt the dynamics of (2)–(3) to the desired trajectory $\boldsymbol{y}(t)$, thus enabling the system to reproduce any smooth movement from the initial position $\boldsymbol{y}_0$ to the final configuration $\boldsymbol{g}$. This can be accomplished by computing the free parameters $\boldsymbol{w}_k \in \mathbb{R}^d$ using regression techniques. See [12] for more details.

$\alpha_z$, $\beta_z$, and $\alpha_x$ are usually constants that do not change between movements. Thus the neural network needs to learn the other parameters of differential equation system (2)–(4) to fully specify a DMP as defined in Eq. (1).

### 2.2 Network Architecture

In our improved architecture, images are encoded via convolutional layers that are pre-trained as part of a basic CNN classifier that was trained on the original MNIST dataset. The input is a $40 \times 40 \times 1$ grayscale pixel image, followed by a convolutional layer with $5 \times 5$ kernel size and 10 feature maps, a convolutional layer with $5 \times 5$ kernel size and 20 feature maps, a 0.5 dropout layer, a fully-connected layer of size 320, a fully-connected layer of size 50 and the output layer of size 10 matching the number of digits. After training the classifier, the fully-connected layers are removed and the convolutional layers are retained and are used to form the first layers of the encoder in our proposed architecture. These convolutional layers are illustrated on the left side of Fig. 2. In different variations of the architecture, the encoder is either composed entirely by the convolutional layers or the convolutional layers are followed by additional fully-connected layers before reaching the decoder. The latter case is illustrated in Fig. 2 which shows two fully-connected layers with sizes of 600 neurons and 200 neurons respectively in the encoder following the two convolutional layers. These two variations of the encoder structure as well as their differences in terms of performance are described further in Sect. 3.

Following the final convolutional layers or fully-connected layers of the encoder depending on the precise network architecture, at the bottleneck of the network that forms the latent space representation, a decoder is formed via a number of fully-connected layers that gradually expand the number of units in each layer until the final output layer which has a size set to 55 units in order to match the DMP parameters $\{\boldsymbol{w}_k\}_{k=1}^{N}$, $\tau$, $\boldsymbol{g}$ and $\boldsymbol{y}_0$. The layers of the decoder are illustrated on the right side of Fig. 2 starting with the bottleneck of size 20, followed by a layer of size 35 and finishing with the output layer. This is the same decoder structure as used [9] and we retain it here as-is, having found it to be effective throughout our experiments for this particular use case. The cost function used to evaluate the output of the network is the same as that of Equation (9) in [9], which is defined for the $j$-th DMP as follows:

$$\mathrm{E}_p(j) = \frac{1}{2}\left(\sum_{k=1}^{N} \|\boldsymbol{w}_k - \boldsymbol{w}_{k,j}\|^2 + (\tau - \tau_j)^2 + \|\boldsymbol{g} - \boldsymbol{g}_j\|^2 + \|\boldsymbol{y}_0 - \boldsymbol{y}_{0,j}\|^2\right), \quad (7)$$

where $\{\{\boldsymbol{w}_k\}_{k=1}^{N},\ \tau,\ \boldsymbol{g},\ \boldsymbol{y}_0\}$ denotes the output of the neural network and $\{\{\boldsymbol{w}_{k,j}\}_{k=1}^{N},\ \tau_j,\ \boldsymbol{g}_j,\ \boldsymbol{y}_{0,j}\}$ the DMP parameters from the training data $\mathbf{k}_j \in \mathbf{D}'$. For further details on the gradient calculations required for minimizing the cost function via backpropagation we refer the reader to [9].

## 3 Experiments

In our experiments, we trained both the fully-connected image-to-motion encoder-decoder architecture (IMEDNet) and the convolutional architecture (CIMEDNet) on various digit image and motion trajectory datasets. We experimented with two different versions of the new architecture, one of which used exclusively pre-trained convolutional layers in the encoder CIMEDNet used pre-trained convolutional layers followed by some additional fully-connected layers in the encoder with the decoder constructed entirely with fully-connected layers. The IMEDNet architecture was the same as described in [9] with fully-connected hidden layer sizes of 1500, 1300, 1000, 600, 200, 20, and 35 neurons, respectively. The CIMEDNet architecture was as described in Sect. 2.2 and as illustrated in Fig. 2 where the two additional fully-connected layers in the encoder are shown to have sizes of 600 and 200 respectively. We used PyTorch [10] in order to implement all networks and trained our models on NIVIDIA GTX 1080 and 1080Ti GPUs. When pre-training the MNIST CNN classifier, we used a stochastic gradient descent optimizer, a negative log-likelihood loss, a batch size of 64, a learning rate of 0.01 with momentum 0.5, and trained for 10 epochs. This achieved a 98% accuracy which we deemed sufficient for our purposes in extracting the trained convolutional layers for use in the encoder-decoder networks.

When training the encoder-decoder networks, we used the Adam optimizer [7] with a learning rate of 0.0005 and the epsilon parameter for numerical stability set to 0.001. In order to avoid learning plateaus, the optimizer parameters were periodically reset to initial values every 500 epochs. A mean squared error loss was used to evaluate the cost function in Eq. (7) and the batch size was set to 128 for weight updates. As a stopping criterion, if the best validation loss was unchanged after 60 epochs, training was halted. The above training procedure was used for both the IMEDNet and CIMEDNet architectures. However, in the case of CIMEDNet, we also experimented with either freezing the convolutional layer weights or training the entire network end-to-end. The results for these different training regimes are cataloged in Table 1.

### 3.1 Datasets

In order to construct **D**, we employed the same scheme described in [9] to generate $40 \times 40$ images of synthetically written digits and associated two-dimensional artificial writing trajectory movements. Briefly, the synthetic trajectory data was generated using a combination of straight lines and elliptic arcs. These geometric elements were used to generate grayscale digit images and their parameters were varied according to a uniform distribution. The resulting images were processed with a Gaussian filter and some moderate salt-and-pepper noise was added to the foreground pixels. Finally, both the generated trajectories and the resulting images were transformed using affine transformations composed of translation, rotation, scaling, and shearing. These parameters were again taken from a uniform distribution. For the DMP representation of the trajectories, 25 radial-basis functions were selected for every dimension. The weights of these basis functions form together with the common time constant (1 parameter) and the start and the goal values of a planar movement ($2 \times 2$ parameters), the full set of 55 DMP parameters that represent the motion. Using this procedure, several datasets were generated both with and without similar noise as used in the noisy MNIST (n-MNIST) datasets [3] as follows:

- **s-MNIST:** 2000 pairs of images and trajectories without any added noise were generated for each digit, for a total of 20000 samples that were split in a 70%/15%/15% ratio between training/validation/test data,
- **s-MNIST-AWGN-19.0:** 300 samples per digit/3000 total samples, using additive white gaussian noise with a signal-to-noise ratio of 19.0,
- **s-MNIST-AWGN-9.5:** 300 samples per digit/3000 total samples, using additive white gaussian noise with a signal-to-noise ratio of 9.5,

- **s-MNIST-MB:** 300 samples per digit/3000 total samples, using a motion blur filter emulating a linear motion of the camera of 5 pixels and a 15° motion in the counterclockwise direction,
- **s-MNIST-RC-AWGN:** 300 samples per digit/3000 total samples, using a contrast range scaled down to half as well as additive white gaussian noise with a signal-to-noise ratio of 9.5.

It should be emphasized that in the results that follow, only the s-MNIST dataset was used for training the presented models.

**Table 1.** DMP reconstruction statistics. The results are in pixels. The best result for each dataset is highlighted in boldface.

| | s-MNIST | s-MNIST-AWGN-19.0 | s-MNIST-AWGN-9.5 | s-MNIST-MB | s-MNIST-RC-AWGN |
|---|---|---|---|---|---|
| IMEDNet (End-to-End) | $0.22 \pm 0.08$ | $0.56 \pm 0.20$ | $1.66 \pm 0.60$ | $\mathbf{0.35 \pm 0.15}$ | $2.32 \pm 0.77$ |
| CIMEDNet (Frozen Conv.) | $0.26 \pm 0.10$ | $0.54 \pm 0.20$ | $1.48 \pm 0.55$ | $0.47 \pm 0.25$ | $2.19 \pm 0.76$ |
| CIMEDNet (End-to-End) | $\mathbf{0.19 \pm 0.08}$ | $\mathbf{0.36 \pm 0.14}$ | $\mathbf{1.02 \pm 0.45}$ | $0.36 \pm 0.12$ | $\mathbf{1.93 \pm 0.66}$ |

### 3.2 Results

The main quantitative results are presented in Table 1 while qualitative results for selected samples are presented in Fig. 3. After training on the noiseless s-MNIST dataset each of the models were tested on all five of the noiseless and noisy s-MNIST datasets described in the previous section. The CIMEDNet architecture was trained with two separate training regimes in which the convolutional layer weights were frozen and the models were trained end-to-end respectively. For the quantitative evaluation, dynamic time warping was used to measure the mean pointwise pixel distance between the trajectories generated by the DMPs predicted by the networks from the digit images and the actual digit trajectories.

As can be seen in Table 1, the CIMEDNet model that is trained end-to-end significantly out-performs the IMEDNet model on both the noiseless s-MNIST dataset and on most of the noisy s-MNIST datasets, apart from the dataset featuring motion blur noise. We reason that this may be due to the fact that motion blur can significantly distort overall object shape and edge profiles and given that convolutional neural networks function the basis of exploiting hierarchies of image filters often heavily represented by edge detectors, this may impact on their effectiveness in such circumstances. The CIMEDNet that was trained with frozen convolutional layers also fared well, beating the IMEDNet model on the same noisy datasets despite not scoring as well on the noiseless dataset. This indicates that the feature detectors in the convolutional layers allow for more robust generalization whereas fully-connected layers are more inclined to overfit.

The qualitative result samples in Fig. 3 are also interesting. Results using the s-MNIST-RC-AWGN dataset are omitted as the noise levels are so pathologically difficult that the qualitative results are comparatively worthless. However, the CIMEDNet model often performs surprisingly well given that it was not trained or fine-tuned on the noisy data. Both models appear to produce highly legible writing trajectories that closely match the actual trajectories in the case of the s-MNIST-MB dataset, but the CIMEDNet model is demonstrably superior to IMEDNet in many cases with the s-MNIST-AWGN-19.0 and s-MNIST-AWGN-9.5 data, producing much more legible results and demonstrating the robustness of the convolutional layers in dealing with even high noise levels.

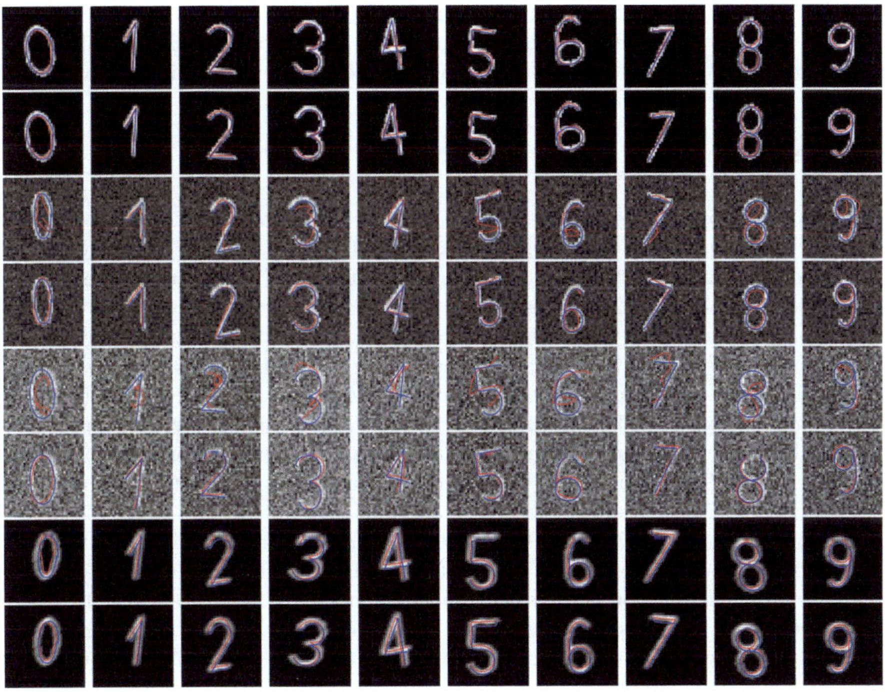

**Fig. 3.** Example results for IMEDNet (rows 1, 3, 5 & 7) & CIMEDNet trained end-to-end (rows 2, 4, 6 & 8). Rows 1 & 2: s-MNIST, rows 3 & 4: s-MNIST-AWGN-19.0, rows 5 & 6: s-MNIST-AWGN-9.5 and rows 7 & 8: s-MNIST-MB. Original trajectories are shown in blue, trajectories calculated by the neural networks are shown in red. Samples in matching dataset rows are identical.

## 4 Conclusions and Future Work

We have presented an extended form of an encoder-decoder neural network for image-to-motion prediction that employs convolutional layers in the encoder in order to make the image recognition component more robust to noisy input.

We have demonstrated that this architecture outperforms its predecessor on a variety of different kinds of noise. Regarding future work, we intend to further expand the capabilities of this model by incorporating layers from more powerful pre-trained CNN models into the encoder and training the network on more challenging image sets. One challenge here lies in either finding suitable image datasets that include trajectory information in their target outputs or in finding other means of producing images with corresponding motion trajectories, e.g. by gathering both in a robot simulation environment.

**Acknowledgement.** This work has received funding from the EU's Horizon 2020 RIA AUTOWARE (GA no. 723909); the Slovenian Research Agency under GA no. J2-7360; JSPS KAKENHI JP16H06565; NEDO; the Commissioned Research of NICT; the NICT Japan Trust (International research cooperation program); and JST-Mirai Program Grant Number JPMJMI18B8, Japan. The authors also wish to thank Marcel Salmič for his significant contribution to the PyTorch network implementations.

## References

1. Ali, O.: Robotic calligraphy: learning from character images. Ph.D. thesis (2015)
2. Badrinarayanan, V., Kendall, A., Cipolla, R.: SegNet: a deep convolutional encoder-decoder architecture for image segmentation. IEEE Trans. Pattern Anal. Mach. Intell. **39**(12), 2481–2495 (2017)
3. Basu, S., Karki, M., Ganguly, S., DiBiano, R., Mukhopadhyay, S., Gayaka, S., Kannan, R., Nemani, R.: Learning sparse feature representations using probabilistic quadtrees and deep belief nets. Neural Process. Lett. **45**(3), 855–867 (2017)
4. Chen, N., Bayer, J., Urban, S., van der Smagt, P.: Efficient movement representation by embedding dynamic movement primitives in deep autoencoders. In: 2015 IEEE-RAS 15th International Conference on Humanoid Robots (Humanoids), pp. 434–440 (2015)
5. Chen, N., Karl, M., van der Smagt, P.: Dynamic movement primitives in latent space of time-dependent variational autoencoders. In: 2016 IEEE-RAS 16th International Conference on Humanoid Robots (Humanoids), pp. 629–636 (2016)
6. Ijspeert, A.J., Nakanishi, J., Hoffmann, H., Pastor, P., Schaal, S.: Dynamical movement primitives: learning attractor models for motor behaviors. Neural Comput. **25**(2), 328–373 (2013)
7. Kingma, D.P., Ba, J.: Adam: a method for stochastic optimization. In: 3rd International Conference for Learning Representations (ICLR), San Diego (2015)
8. Krüger, N., Geib, C., Piater, J., Petrick, R., Steedman, M., Wörgötter, F., Ude, A., Asfour, T., Kraft, D., Omrčen, D., Agostini, A., Dillmann, R.: Object-action complexes: grounded abstractions of sensory-motor processes. Robot. Auton. Syst. **59**(10), 740–757 (2011)
9. Pahič, R., Gams, A., Ude, A., Morimoto, J.: Deep encoder-decoder networks for mapping raw images to dynamic movement primitives. In: 2018 IEEE International Conference on Robotics and Automation (ICRA), Brisbane, Australia, pp. 5863–5868 (2018)
10. Paszke, A., Gross, S., Chintala, S., Chanan, G., Yang, E., DeVito, Z., Lin, Z., Desmaison, A., Antiga, L., Lerer, A.: Automatic differentiation in PyTorch. In: NIPS 2017 Autodiff Workshop (2017)

11. Pervez, A., Mao, Y., Lee, D.: Learning deep movement primitives using convolutional neural networks. In: 2017 IEEE-RAS 17th International Conference on Humanoid Robotics (Humanoids), pp. 191–197 (2017)
12. Ude, A., Gams, A., Asfour, T., Morimoto, J.: Task-specific generalization of discrete and periodic dynamic movement primitives. IEEE Trans. Robot. **26**(5), 800–815 (2010)
13. Yang, P.C., Sasaki, K., Suzuki, K., Kase, K., Sugano, S., Ogata, T.: Repeatable folding task by humanoid robot worker using deep learning. IEEE Robot. Autom. Lett. **2**(2), 397–403 (2017)

# Multi-robot and Communication Systems

# A Swarm Algorithm Inspired by Tree-Dwelling Bats. Experiments and Evaluations

Ján Zelenka(✉), Tomáš Kasanický, and Ivana Budinská

Institute of Informatics, Slovak Academy of Sciences, Bratislava, Slovakia
{zelenka,kasanicky,budinska}@savba.sk
http://www.ui.sav.sk/w/en/dep/mcdp/

**Abstract.** The paper describes a new swarm based algorithm inspired by the behaviour of tree bats. The algorithm is designed to search through unknown environment and look for objects of interest. Individual robotic swarm agents are equipped with memory to make a more efficient search of the space. The proposed algorithm has been evaluated in comparison with a uniform search. Changing a parameter "shout" changes the behaviour mode of the algorithm. By decreasing the parameter, parallel scan by smaller groups of agents is achieved. A higher value of the parameter leads to a sequential scanning with a large group of agents.

**Keywords:** Swarm intelligence · Swarm robotics · Bio-inspired algorithms

## 1 Introduction and Related Works

Swarm robotics is an approach, which use of numerous, simple and autonomous robots to accomplish complex tasks. Robots in a swarm coordinate their behaviours in a decentralised manner. An inspiration for creation of such systems can be often found in nature. Scientists study species, which exhibit emergent behaviour. For example social insect, flocking of birds or schooling of fish. Swarm robotics was also inspired by bats - the only flying mammals on the Earth. In 2010, Xi-She Yang presented a formal description of a behavioural algorithm, where he idealized some of the echolocation characteristics of microbats [1]. Since the time of publication of this work, the Bat algorithm found application in various fields, including optimization, classification, image processing, task scheduling, data mining and many others. Properties of echolocation and communication among bat individuals might, as well, find application in the field of robotics, particularly to solve problems related to collision avoidance of multiple robots [2]. The majority of bat-inspired algorithms developed until now are based on certain echolocation characteristics of microbats, however behavioural algorithm, which try to mimic the group movement of bats, are more rare. To our knowledge, the first attempt to simulate movement of tree-dwelling bats was


K. Berns and D. Görges (Eds.): RAAD 2019, AISC 980, pp. 527–534, 2020.
https://doi.org/10.1007/978-3-030-19648-6_60

published by Ruczyński and Bartoń [3]. In their agent-based model, they aimed to explore the benefits of tree selection, memory and eavesdropping (compensation behaviours) during search for tree cavities by bats with short and long perception range. In this paper we present a second version of a more advanced agent-based model, which is able to fully resolve coverage problem. The model is based parallel with swarming behaviour of tree-dwelling bats revealed by recent studies [4,5].

In the article [6], we described a unique behavioural model of a group management of tree-dwelling bats. The model is a result of research that integrates behavioural biology with applied informatics and robotics. In the research we are focusing on three direction of developing the model:

- simulation of biological behaviour that responds the real observation of bats swarming,
- development of computing algorithms based on biological parallels,
- application of the algorithms on scanning and searching tasks in swarm robotics.

The SkyBat model had several shortcomings that we removed by adding additional parameters. The most significant changes were made by adding the parameter "memory", causing the algorithm not to converge to one location, but is attracted to locations that have not yet been visited. Another added parameter is the scream, which models the distance of attracting other agents.

The presented work focusses on the parametrization of the swarm based algorithm motivated by the bats behaviour. The algorithm is intended for application on coverage problems. We suppose an environment with several objects of interests with different values of quality. The basic goal of a group of agents (bats) is to find one or more objects of interests with the highest quality within a limited time interval. The agents gather in the objects of interests and jointly perform the required tasks (e.g. if the object of interest is a source of pollution, they jointly clean the place). In the article we show that by choosing the appropriate parameters, it is possible to switch between two different modes:

- the objects are sought and served sequentially by the larger groups, and
- the objects are sought and served in parallel by small groups of agents.

It is supposed that each agent judge the quality of the object and on the basis of this judgement it starts attracting other individuals to perform the required task jointly. The algorithm shows the signs of swarm behaviour. Individuals are able to cooperate relatively primitive rules to achieve common goals, individuals can move from one location to another without a group leader, the group can perform a flexible non-centralized decision-making in a dynamically changing environment. Such an approach can be used to efficiently locate pollution sites and then dispose of pollution using a group of mobile robots. We assume that there are places with different degrees of pollution. If an agent finds a pollution and does not have enough capacity to eliminate it, it will begin to attract other agents that they work together to clean up the site until the pollution is completely eliminated. Agent activities are not centralized, and multiple site cleanups are executed. Changing the "shout" parameter changes the behavior of the entire group. In areas where more minor pollution is expected, it is advisable to reduce the size of this parameter. At a high parameter value, larger groups of agents are created that can eliminate large pollution quicker, but the group's work parallelization is reduced.

The paper is organized as follows: Sect. 2 describes the algorithm, simulation experiments and results are described in Sect. 3 and discussion and future work ideas are presented in Conclusions.

## 2 Algorithm Description

For the testing of the algorithm we chose objects of interest (OoI) to be found and served. If an individual finds an OoI, it judges its quality. If the quality of the OoI is over a set threshold $\boldsymbol{T}$, the individual starts to perform a required task and in the same time to attract other individuals.

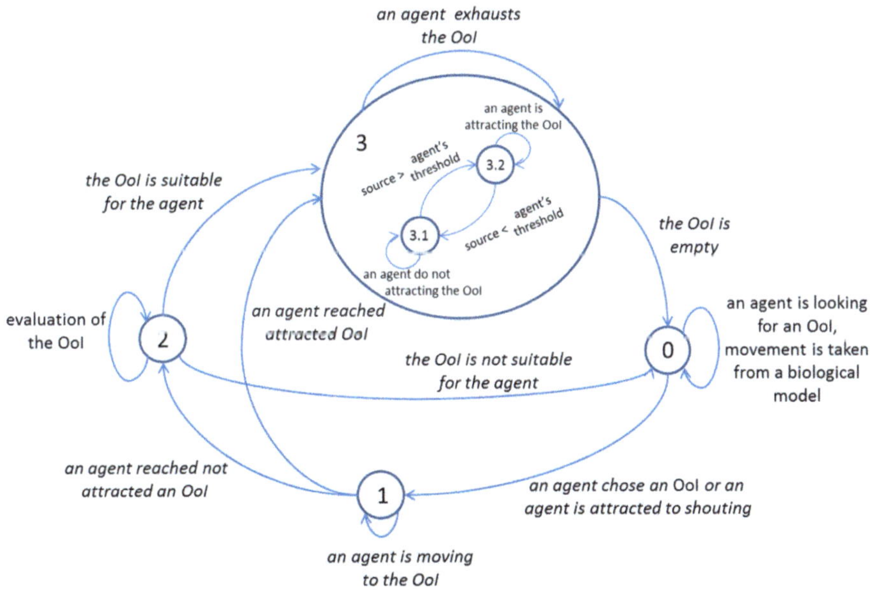


**Fig. 1.** The agent's state diagram.

Every agent has an ability to:

- move through the space (movement algorithm is taken from the biological model SkyBat [6]),
- recognize objects of interest in his perimeter (a scanning range is defined by a distance $\boldsymbol{R}$ and angle),

- perform the required task (e.g. clean pollution, consume resources),
- attract other agents by shouting (the audibility radius $\boldsymbol{D}$),
- remember the visited places. The memory is in the form of a stack (FIFO) of defined size $\boldsymbol{M}$.

An individual may occur in the following states (Fig. 1):

- ***0*** it passes through the space and looking for the OoI,
- ***1*** it is attracted by the shout of another agent or he chooses an unattracted OoI and heading towards the source of the shout,
- ***2*** it judges the source,
- ***3*** (***3.1***) it is performing the task and (***3.2***) starts attracting other individuals by shout if the source exceeds the threshold $\boldsymbol{T}$.

Figure 1 shows the agent's state diagram. The agent in state ***0*** moves through the space and searches for suitable sources. If a source appears, at a distance $\boldsymbol{d} < \boldsymbol{R}$ where ($\boldsymbol{R}$ is a scanning range of the agent), the agent switches to the state ***1*** and moves towards the source. Once a target has been reached, the agents is in the state ***2***. The agent checks for the suitability of the object. If the quality of the object is bigger than threshold $\boldsymbol{T}$, the agent changes to state ***3***. However, if the quality of the object is low, the agent continues in searching mission (state ***0***). The agent remembers the position of the object. Whenever a signal from another agent is encountered, it is switched to state ***1*** and moves towards a new target.

Figure 2 shows agents in different states in the space. The source is represented by a black square. The agent represented by the yellow square is in the

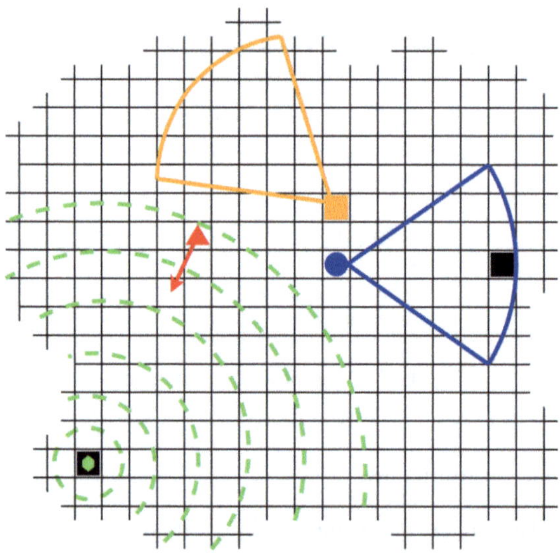

**Fig. 2.** An illustration of different agent states.

state ***0***. Agents represented by a red triangle or a blue circle are in a state ***1*** and going to the source. Finally the agent represented by a green hexagon is consuming the source, and attracting others agents in the range of its shout.

## 3 Simulation Experiments and Results

We have prepared the following environment to test the parameters and the overall concept of the algorithm. Let a space be divided by the uniform grid of $n$ x $n$ size. In this area, there are ***X*** objects of interest. These objects are randomly distributed within the space. Each object has a value that represents its quality ***E***. The main parameter we evaluated was the time spent (exhaustion) of all objects with the quality value higher than ***T***. The space changes dynamically. Agents do not have general information about the space, the position and quality of objects. Communication between agents is only through the attraction signal transmitted after finding a suitable object (similarly to a fission-fussion dynamics in bats swarming).

There are several parameters that affect the experiment result:

- a ratio of the number of agents to the number of cells and the number of object of interest in the area,
- a threshold ***T*** to evaluate the quality of the object and then it decides to start or not attracting,
- a distance ***D*** to which the shout of agent is coming that appropriate object has been found,
- a range ***R*** of the sensor for recognition of objects,
- an agent's memory size ***M***. ***M*** represents the number of objects visited by the agent(FIFO stack),
- a population size ***P***,
- motion rules (keep the direction and velocity until an OoI or an obstacle occurs, then change the direction.

We set the following hypotheses to be confirmed by experiments:

1. Hypothesis 1: It is assumed that the large value of parameter ***D*** (reaching the size of the space) has a strong affluent on the size of group.
2. Hypothesis 2: The level to which the robot attracts ***D*** other individuals to the point of attraction affect the quality of the searching process.

Experiments Evaluation:

Hypothesis 1:

If the parameter ***D*** is set to a size near to a size of exploring space, a large group of agents will be created in the space, including all individuals. The algorithm with parameter ***D*** set to a bigger value performs a sequential search, where sources are visited one by one. The results show that the size of the memory ***M*** determines that the agent is forced to move into the non visited area of the space. The time between the growth number of agents in each group corresponds with the parameter ***M***, after the grouping is more promoted sequential

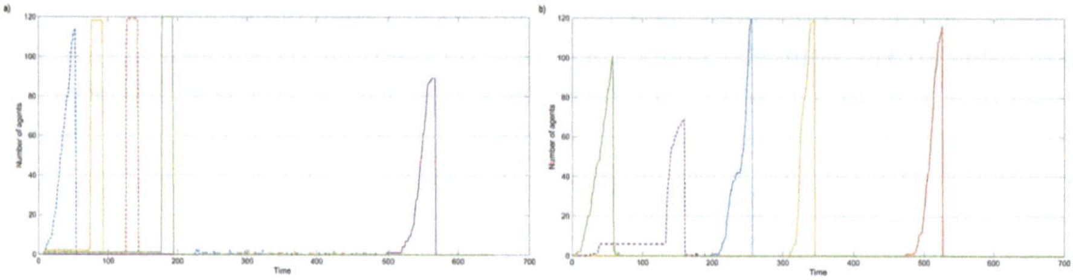


**Fig. 3.** Simulation for 120 agents in the area of 1000×1000 [m], memory set to $\boldsymbol{M} = 1$(a) object ($\boldsymbol{M} = 10$(b)), attraction range $\boldsymbol{D} = 1000$ m. The 5 of OoI with quality higher than $\boldsymbol{E}$ is located in space and his number of attending agent in time.

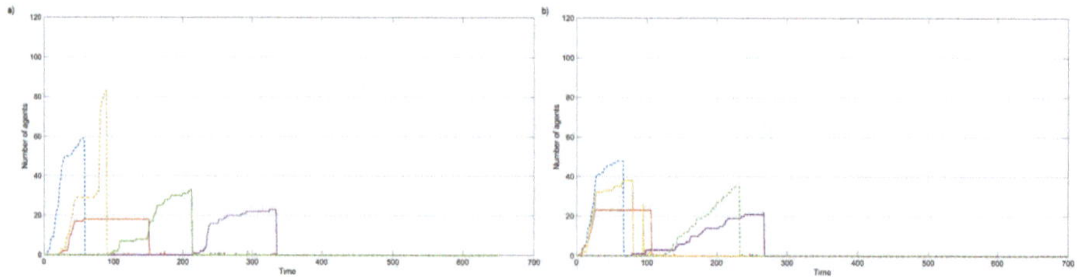


**Fig. 4.** Simulation for 120 agents in the area of 1000×1000 [m], memory set to $\boldsymbol{M} = 1$(a) object ($\boldsymbol{M} = 10$(b)), attraction range 350 m. The 5 of OoI with quality higher than $\boldsymbol{E}$ is located in space and his number of attending agent in time.

reconsideration (Fig. 3a). Otherwise (Fig. 3b), when the $\boldsymbol{M}$ parameter has lower value (eg. 10) supports parallel searching (the number of agents interested by individual OoI increases gradually, which is caused by a greater dispersion agents in the area).

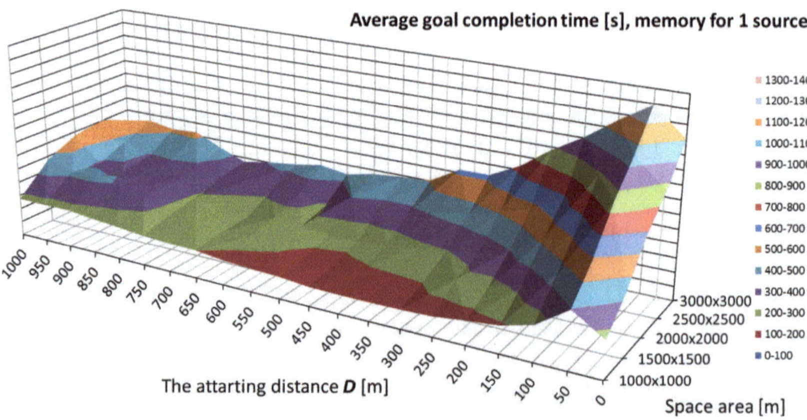


**Fig. 5.** Comparison of the algorithm with an uniformed search with a memory set to $\boldsymbol{M} = 1$ visited object.

Hypothesis 2:

The introduction of the process, in which an agent starts attracting other agents, leads to significantly better result in comparison with uninformed search (blind search). Uninformed search is represented in the charts on Figs. 5, 6 and 7 by the point, where $\boldsymbol{D}$ is set to 0. Graphs also showed the effect of the memory on the process of searching the space. Bigger values of parameter $\boldsymbol{D}$ and increasing the memory size $\boldsymbol{M}$ cause a more even distribution of agents throughout the space. This fact can be seen in the chart on Fig. 7, a gradual increase of number of agents. Graph on the Fig. 4 shows a step change of the agent population in individual sources, which represents the fact that agents are not spread into the space.

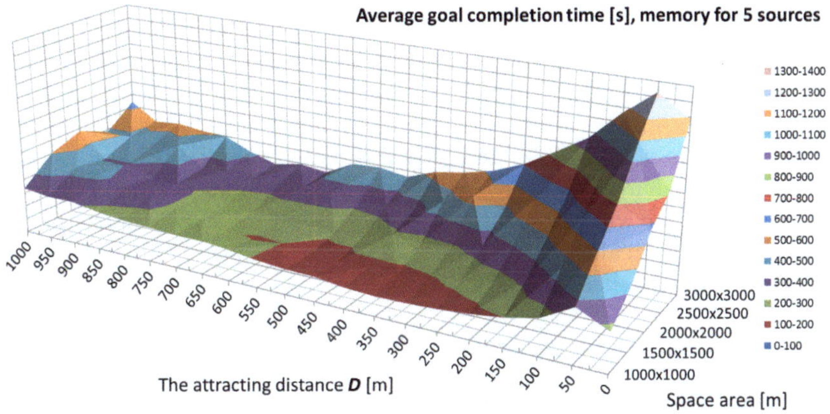


**Fig. 6.** Comparison of the algorithm with an uniformed search with a memory set to $\boldsymbol{M} = 5$ visited object.

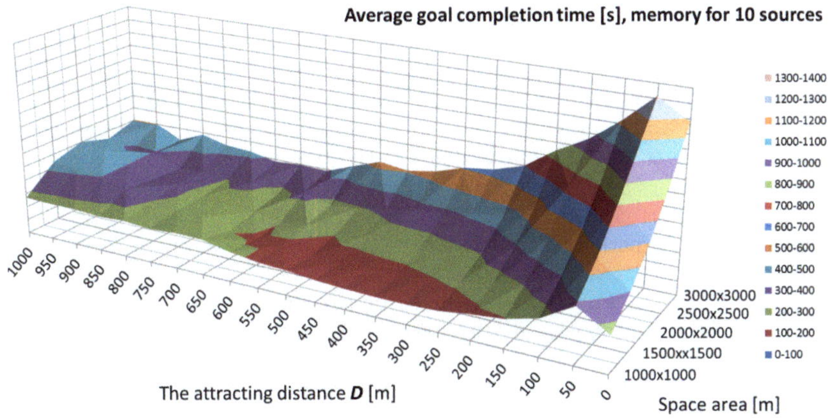


**Fig. 7.** Comparison of the algorithm with an uniformed search with a memory set to $\boldsymbol{M} = 10$ visited object.

The simulations were performed using MDCS (Matlab Distributed Computing Server - 96 workers server) at CVTI SR. At least 21 repetitions were performed for each setting (in graphs on Figs. 5, 6 and 7). The average simulation time of one setting lasted approx. 16 s. The simulation of one graph lasted approx. 6.5 h. The total number of tasks for this paper is over 75,680 simulations to date.

## 4 Conclusions

The presented algorithm in experiments demonstrated its ability to find optimal solution from a space of solutions. It is clear from the results that the performance of the presented algorithm is better than the uninformed search. By selecting the parameters you can change the searching behaviour of the presented algorithm.We encourage the search will be done by one large group of agents in sequential manner, by increasing the agent's shout range. On the other hand, the lower range of agent's shout support parallel scans with multiple minor groups. In addition to polluting elimination, this approach can be used in areas where communication is not possible, and therefore not a coordinated control of the robots. Such an environment may be e.g. in space or in underground mining.

In the future, we will use a learning method to set up and optimize the shout parameter, which will allow the parameter size to be adaptively altered according to the goal's effective performance.

**Acknowledgment.** This work has been supported by the Slovak Research and Development Agency SRDA (grant No. APVV-17-0116) and the Slovak Scientific Grant Agency VEGA (grant No. 2/0155/19). The simulations were performed using MDCS (Matlab Distributed Computing Server - 96 workers server) at CVTI SK.

## References

1. Yang, X.-S.: A new metaheuristic bat-inspired algorithm. In: Gonzales, J.R., et al. (eds.) Nature Inspired Cooperative Strategies for Optimization. Studies in Computational Intelligence, vol. 284, pp. 65–74. Springer, Heidelberg (2010)
2. Hossain, Q.D., Uddin, M.N., Hasan, M.M.: Collision avoidance technique using biomimic feedback control. In: Proceedings of the International Conference on Informatics, Electronics and Vision (ICIEV), pp. 1–6. IEEE (2014)
3. Ruczyński, I., Bartoń, K.A.: Modelling sensory limitation: the role of tree selection, memory and information transfer in bats' roost searching strategies. PloS one **7**(9), e44897 (2012)
4. Naďo, L., Kaňuch, P.: Dawn swarming in tree-dwelling bats – an unexplored behaviour. Acta Chiropterologica **15**, 387–392 (2013)
5. Naďo, L., Kaňuch, P.: Swarming behaviour associated with group cohesion in tree-dwelling bats. Behav. Process. **120**, 80–86 (2015)
6. Zelenka, J., Kasanický, T., Budinská, I., Naďo, L., Kaňuch, P.: SkyBat: a swarm robotic model inspired by fission-fusion behaviour of bats. In: Mechanisms and Machine Science: Advances in Service and Industrial Robotics. RAAD 2018, vol. 67, pp. 521–528. Springer, Cham (2019). ISSN 2211-0984

# Secure Sharing of Robot and Manufacturing Resources in the Cloud for Research and Development

Florin Anton(✉), Theodor Borangiu, Silviu Răileanu, Silvia Anton, Nick Ivănescu, and Iulia Iacob

Department of Automation and Applied Informatics, University Politehnica of Bucharest, Bucharest, Romania
{florin.anton, theodor.borangiu, silviu.raileanu, silvia.anton, nick.ivanescu, iulia.iacob}@cimr.pub.ro

**Abstract.** Cloud systems are nowadays more and more present in industry, being part of Industry 4.0 solutions; this can be seen at production level, in manufacturing systems and in robotics. Cloud services are used to improve and interconnect the manufacturing shop-floor processes with the higher-level enterprise components (enterprise resource planning systems, logistics, supply chains and other operational systems). The paper presents a solution for cloud – manufacturing system integration where the cloud system is used to remotely access and control a manufacturing system for research, development and training purpose. The access to the manufacturing system is configured in the cloud as a service, the connection to the manufacturing system being implemented through a set of virtual machines which are deployed and custom configured in cloud. The paper describes the architecture of the system, the deployment scenarios and presents the limitations and performances of the system obtained during the tests.

**Keywords:** Cloud manufacturing · Cloud services · Remote control · Cloud robotics · Manufacturing system

## 1 Introduction

Cloud computing has been part of the computer science domain and has been evolved from technologies like grid computing and virtualization. Cloud services have been adopted in many fields including Industry 4.0, where cloud services represent an important part of the high level control and management of industrial systems.

In the manufacturing field, cloud is used to create services accessed by the manufacturing processes to store and exchange data in order to process large volumes of data or to execute computing intensive tasks.

Analyzing the recent scientific literature, it can be observed that cloud systems are used to create ubiquitous manufacturing control systems [1] which establish an integrated manufacturing environment across multiple levels; cloud allows for efficient resource sharing and is also used to offer low cost, enhanced performance services in Manufacturing Execution Systems (MES) [2]. Production planning, operations


K. Berns and D. Görges (Eds.): RAAD 2019, AISC 980, pp. 535–543, 2020.
https://doi.org/10.1007/978-3-030-19648-6_61

scheduling and resource allocation are also tasks which are optimally solved using cloud systems [3–6] and autonomous agents running in cloud [7].

The cloud adoption in manufacturing and robotics brings not only benefits but also raise question marks regarding the privacy, data exposure and other issues mainly if the cloud infrastructure used is a public one [8].

Robots are an important part in manufacturing and they are also integrated in the cloud manufacturing concept. The studies cover mobile robotics but also industrial robotics, both being used in the manufacturing process [9].

The integration of cloud systems with robots is done using cloud services [10] in order to: enhance the performance of robots by exporting the processing intensive tasks in cloud [11]; reduce the power consumption [12]; improve the interaction and negotiation of industrial robots [13]; or bring powerful cognitive services in robotics [14].

The paper presents a solution for remote controlling a manufacturing line equipped with multiple robots, for training and research purposes, using a cloud infrastructure. The paper is structured on five sections as follows: an introduction; a second section presenting the concept and the architecture of the solution; the third section presents the cloud system architecture and service request mechanism; the fourth section presents the provisioning processes; the paper ends with a fifth section presenting the experimental results and the conclusion.

## 2 Remote Control of Robotic Manufacturing Systems Using Cloud

The paper's objective is to present a cloud-based system capable to offer remote control facilities for a manufacturing cell in order to allow users to access, configure and interact with a cloud manufacturing system for research and education activities.

The manufacturing cell is presented in Fig. 1; it is composed by a closed loop conveyor system controlled by a Bosch-Rexroth PLC which routes pallets to six robot systems used to assembly and process products. Products are assembled on pallets.

The pallets are introduced and removed from the line by a Cartesian robot; the processing tasks are executed by four robots (two horizontally articulated and two vertically articulated robots, the latter two having also the task to feed two CNC machines).

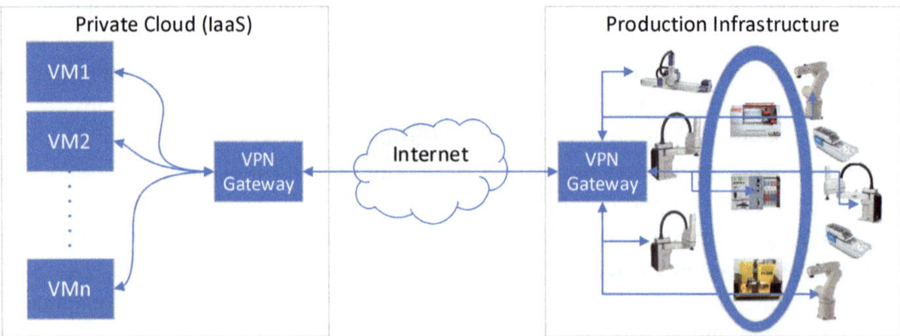


**Fig. 1.** Cloud – manufacturing cell integration.

A 5th robot (also horizontally articulated) is using two flexible feeders to feed the four processing robots with necessary parts if their associated part stacks are depleted in the assembly process. Each robot has one or multiple video cameras fixed or arm mounted; on the ceiling of the manufacturing shop floor there are also three Axis PTZ cameras which can be used to monitor the entire manufacturing process.

The cloud system used for this multiple robot cell and described in the next section is a private cloud system of IaaS type (Infrastructure as a Service), which is connected with the manufacturing line server through a Virtual Private Network (VPN).

In order to work with the system a user must be registered in the cloud self-service interface and must create a request for a new project containing the needed virtual machines. The virtual machines (VM) which can be used are created based on a template which offers different functionalities:

- PLC Control VM: this VM allows to program and control the manufacturing central PLC that is responsible for pallet driving on the shop-floor conveyor.
- Robot Control VM: this VM allows programming the robots in standalone or collaborative applications, with or without artificial vision.
- Observation Cameras Control VM: this VM is used to control the Axis PTZ observation cameras orientation and settings; the VM is also offering a web service for other users in order to visualize the images from the cameras (the users being unable to directly control the orientation and settings). This VM is required only in one copy for each project.
- Basic VM: this VM is based on a template which contains the operating system (Windows or Linux) and development tools. The VM is used in order to create new services or applications which can be executed in cloud and integrated with the manufacturing cell.

When the user requests a project, he is specifying the VM's required, the hardware configuration of each VM and the time range when the project must be available (the start and end date of the project).

After the project is created the user is notified about how to connect to VM's. The connection is made using digital certificates, the connection being encrypted; the communication between the cloud and the manufacturing cell is also encrypted the data being transferred through a VPN.

## 3 The Cloud System Architecture and the Service Request

The architecture of the cloud system uses an Infrastructure as a Service (IaaS) cloud, based on IBM Service Delivery Manager (ISDM). Regarding the hardware structure, the cloud shares the same components as the IBM CloudBurst 2.1 medium size platform, being composed by 14 HS22 Blade servers (12 cores per server) and 72 GB of RAM. The storage space is provided by an IBM DS3400 device having three expansions, the total raw space being 28.8 GB. The management server has 16 CPU cores and 24 GB of RAM. The proper system shutdown in a case of a power failure is assured by 2 APC 10KVA ups systems [15].

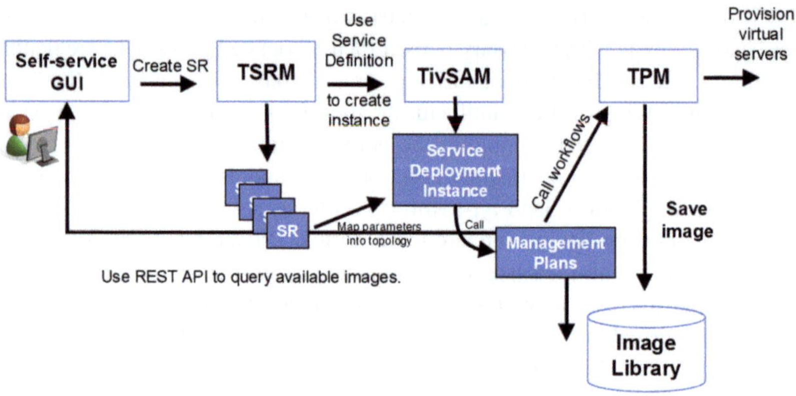


**Fig. 2.** The request and handling of the service

The cloud management machines are used to process the service requests and to administer the process planning for all services offered by the infrastructure. A Self-service GUI is accessed using a web browser and allows to place one or more service requests in the system. The user (researcher or teacher) is using the GUI to authenticate and to place service requests in the system; the service requests are generated by creating a project which includes the requested services (Fig. 2).

The information required to create a project includes: the name of the project, the lifetime of the project (start and end dates), the pool of resources required to create the VMs (which offers also the information regarding the network connectivity of the VMs: number of network adapters, Virtual LANs and network configuration parameters (IP, Netmask, Gateway and DNS)), the template used to deploy the VMs (containing the operating system and additional software, the list of templates being queried by using REST API), the agents and other software components which can be installed after the provisioning of the VMs, and the hardware resources (HDD size, RAM and virtual memory and CPU).

After the project request is finalized and submitted, the request is received by the system as a service request (SR) and sent to Tivoli Service Request Manager (TSRM) which verifies the available resources; if the resources are available, TSRM will allocate them to the SR after it is approved by the cloud administrator. The SR is then placed in a queue until the start date.

Right before the start date, TSRM splits the project into individual SRs for each virtual machine contained in the project and will send for each SR a service definition (SD) to the Tivoli Service Automation Manager (TivSAM). This SD reflects the topology of the service (network parameters for each virtual machine and the management plans for administering the VMs). The management plans are formed by procedures (step by step actions) to execute VM actions like: resetting a password, installing new software, removing or adding a new VM, etc.

When TivSAM receives the service requests from TSRM it will create a service deployment instance the function of which is to forward the management plans to Tivoli Provisioning Manager (TPM). The management plans are sent as workflows (or procedures for management plans execution). For each operation, it must execute (creation of a VM, saving the image of a VM, removing a VM, etc.) the TPM will execute a workflow associated with that operation.

## 4 Provisioning the VMs

From the point of view of the workflow required to provision a project request, a number of steps must be executed: first the client will use the Self-service GUI to submit a project request; this request is then sent to the administrator of the cloud system who must approve or reject the project request.

If the project is approved, TSAM will allocate the requested resources for the project in the timeframe of the project (resources which will be used during the lifetime of the project). When the start date of the project occurs TPM deploys and configures the VMs. After this step the project is operational and the user who requested the project and his team members are notified about how to access the VMs. Later the project can be modified to add or remove VMs.

Figure 3 shows the lifetime of the project; in the first step the team administrator (user or client) creates a project request; after the request is placed in the system it will be sent to the administrator to be approved and receives the status "Waiting for approval". The cloud administrator is notified about the project request and can reject the project, in which case the project request will have the status "Rejected". If the request is approved the project's status will be "Queued". In both cases the user who requested the project will be notified about the decision of the cloud administrator.

The project is placed in a queue until its start date; then TPM is deploying and configures each VM from the project. In this stage the project status is "In progress". In the situation when from different reasons the VM creation of configuration fails, then the status is changed in "Failed" and the cloned VMs are erased, the resources freed and the user is notified.

If the VM's are successfully deployed the project status becomes "Resolved", the user who requested the project is notified and will be able to access the VMs created for the project, in this stage the project status will be "Operational".

During the exploitation phase of the project when the VMs are used, other requests can be made in order to change the configuration of a specific VM (hardware or software configuration, for example changing the hardware resources of a VM or installing additional software) or to add/remove VMs to/from the project. These requests can be automatically approved if the request is not exceeding a certain threshold (for example if the RAM of a VM is changed but is not exceeding 4 GB of RAM) or can be sent to the administration for approval; if they are approved then the cloud system is executing the operation.

Two days before the expiration date, the system notifies the user who requested the project and informs him that the project is about to expire so the user can eventually make a request to extend the lifetime of the project if needed. If the project is cancelled by the user or if the end date occurs, the status of the project is changed to "Decommissioning" and TPM is shutting down the VMs, saves an image for each VM if it was requested by the user and after that operation is easing the VMs. After this operation

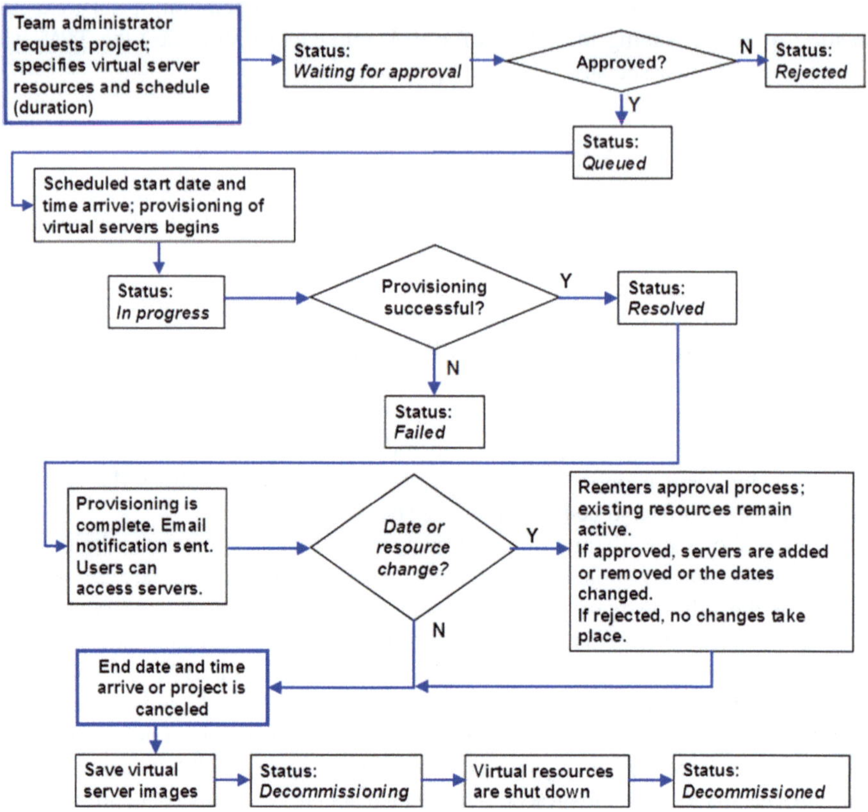


**Fig. 3.** The lifecycle of the service

TSAM release the resources allocated to the project, the project status becomes "Decommissioned" and the team administrator of the project is notified about the deletion of the project.

## 5 Experimental Results and Conclusion

In order to test the cloud service solution an infrastructure of six Adept/Omron robots was used to access services offered by the cloud infrastructure. The conducted tests were used to evaluate the project provisioning speed for the cloud infrastructure, each project having a number of VMs offering the services.

The requests for the projects specified an immediate deployment; in this case the following stages were executed:

- In order to provision a project the system was queried, and the infrastructure was prepared for the VM deployment. This stage was executed in about 10 min and did not heavily depend on the number of VMs which had to be provisioned.

- The deployment of VMs in the cloning stage was done using 8 threads (one file was copied in 8 different locations in parallel); this process used the storage system (a DS3400) and the blade servers.
- After the template of the VM was cloned, the VMs were reconfigured at the hardware level.
- Another stage was related to the software reconfiguration of the VM; in this stage additional software was installed (this adds more time depending on the size of the software and the installation procedure). For example in our case a small package of 50 MB was installed in about 5 min because of the installation procedure which modified the operating system configuration, registers and needed to restart the VM two times.

Figure 4 presents the results obtained. Multiple projects have been created starting with a project with 10VMs and then by adding another 10VMs until we deployed a project with 100VMs. The template used was installed with a Windows 7 32bit operating system and a HDD of 10 GB extended to 60 GB in the hardware reconfiguration stage.

The results (time for deployment) are presented in minutes for the following actions: system query, VM cloning, hardware and software reconfiguration of VMs, total deployment time without additional software installation (THW), and with additional software installation.

From the point of view of system usage, the part which consumes the most of the bandwidth is the process visualization which is based on video streaming from the three PTZ cameras which are mounted on the ceiling of the manufacturing cell and are down looking; their image fields cover all working spaces of the shop-floor.

From the tests we concluded that in order to have a good image quality (using MPEG4 image format and 512 kbps constant bit rate) and a maximum number of frames per second the user needs a connection which has a bandwidth of at least 1 Mbps.

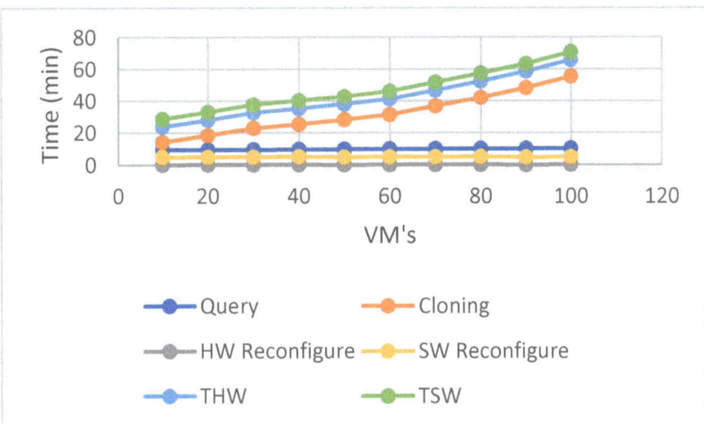


**Fig. 4.** Time for deployment

The VM mechanism above exposed is able to offer cloud manufacturing (robot) services CMfg (CRob) that can be used in a dual industrial control and computing model that:

- Transforms a pool of manufacturing resources (robots, their controllers and capabilities), products (recipes, customer requirements) and orders (producing modes, schedules, business specifications) into on-demand manufacturing services.
- Enables pervasive, on-demand network access to a shared pool of configurable high-performance computing (HPC) resources (servers, storage networks, applications) that can be rapidly provisioned and released as services to various end-users e.g., shop-floor devices, robot-vision systems, system scheduler, etc. with minimal management effort or service provider interaction. This means that CRob uses Cloud Computing (CC) facilities for high performance purpose - such as event-triggered robot motion control, intelligent robot actions, prediction-based planning and scheduling of robot tasks, etc.

From the point of view of security the system offers the level of security given by the lowest security control level imposed by the controls which have been used for the implementation.

The security is assured by using a virtual private network (VPN) which connects the robot infrastructure with the cloud system, the cloud system being a centralized cloud system he can be isolated from the Internet.

The connectivity between the robot systems and the cloud infrastructure is based on encrypting the data between two virtual private network servers as follows: The robot systems are isolated from the internet inside a private local area network (LAN) using a special allocated virtual LAN (VLAN) for the robots and other devices (like the PLC which controls the conveyor belt which connects the robots). The private lan is then connected to a VPN gateway which runs a single service which is the VPN service and is connected with another VPN gateway which is in the LAN of the cloud system.

The selected security solution is based on isolating the infrastructure, principle which has been used in many industrial systems, unfortunately this solution will be less and less implemented due to the changes in Industry 4.0 where the interconnection between the low layers of production (production execution systems) and the upper levels (Enterprise Resource Planning, SCADA systems, production monitoring and management) is required and more requested.

The security solution used here can be much easily implemented in current manufacturing structures where the principle of isolation is still in use.

We can conclude that the presented solution is viable and efficient for training and collaborative research purpose and allows outsourcing high performance computational services for industrial robots with secure integration in applications developed for cloud manufacturing structures.

## References

1. Wang, X.V., Wang, L., Mohammed, A., Givehchi, M.: Ubiquitous manufacturing system based on cloud: a robotics application. Robot. Comput.-Integr. Manuf. **45**, 116–125 (2017)
2. Cheng, Y., Zhao, D., Hu, A.R., Luo, Y.L., Tao, F., Zhang, L.: Multi-view models for cost constitution of cloud service in cloud manufacturing system. In: Zhou, M., Tan, H. (eds.) Advances in Computer Science and Education Applications. Communications in Computer and Information Science, vol. 202. Springer, Heidelberg (2011)
3. Wang, X.V., Givehchi, M., Wang, L.: Manufacturing system on the cloud: a case study on cloud-based process planning. Procedia CIRP **63**, 39–45 (2017)
4. Xu, X.: From cloud computing to cloud manufacturing. Robot. Comput.-Integr. Manuf. **28**(1), 75–86 (2012)
5. Giriraj, M., Muthu, S.: From cloud computing to cloud manufacturing execution assembly system. In: Ponnambalam, S.G., Parkkinen, J., Ramanathan, K.C. (eds.) Trends in Intelligent Robotics, Automation, and Manufacturing, IRAM 2012. Communications in Computer and Information Science, vol. 330. Springer, Heidelberg (2012)
6. Helo, P., Phuong, D., Hao, Y.: Cloud manufacturing – scheduling as a service for sheet metal manufacturing. Comput. Oper. Res. available online 27 June 2018
7. Liu, Y., Wang, L., Wang, Y., Wang, X.V., Zhang, L.: Multi-agent-based scheduling in cloud manufacturing with dynamic task arrivals. Procedia CIRP **72**, 953–960 (2018)
8. Henzel, R., Herzwurm, G.: Cloud manufacturing: a state-of-the-art survey of current issues. Procedia CIRP **72**, 947–952 (2018)
9. Nascimento Jr., A., Cardozo, E., Souza, R.S., Guimarães, E.G.: A platform for cloud robotics. IFAC-PapersOnLine **49**(30), 48–53 (2016)
10. Xia, C., Zhang, Y., Wang, L., Coleman, S., Liu, Y.: Microservice-based cloud robotics system for intelligent space. Robot. Auton. Syst. **110**, 139–150 (2018)
11. Qureshi, B., Koubâa, A.: Five traits of performance enhancement using cloud robotics: a survey. Procedia Comput. Sci. **37**, 220–227 (2014)
12. Hou, G., Zhou, K., Qiu, T., Cao, X., Li, M., Wang, J.: A novel green software evaluation model for cloud robotics. Comput. Electr. Eng. **63**, 139–156 (2017)
13. Wang, S., Zhang, C., Liu, C., Li, D., Tang, H.: Cloud-assisted interaction and negotiation of industrial robots for the smart factory. Comput. Electr. Eng. **63**, 66–78 (2017)
14. Krueger, V., Rovida, F., Grossmann, B., Petrick, R., Crosby, M., Charzoule, A., Garcia, G. M., Behnke, S., Toscano, C., Veiga, G.: Testing the vertical and cyber-physical integration of cognitive robots in manufacturing. Robot. Comput.-Integr. Manuf. **57**, 213–229 (2019)
15. Kannadasan, R., Prabakaran, N., Boominathan, P., Krishnamoorthy, A., Naresh, K., Sivashanmugam, G.: High performance parallel computing with cloud technologies. Procedia Comput. Sci. **132**, 518–524 (2018)

# Communication Between Robots over Intelligent Objects Realized by RFID Tags

Christian Thormann(✉) and Alexander Winkler

Hochschule Mittweida, University of Applied Sciences, Mittweida, Germany
{thormann,alexander.winkler}@hs-mittweida.de
https://www.inw.hs-mittweida.de/webs/winkler3

**Abstract.** This article presents an approach to communication between robot controllers via RFID transponders, where the partners in communication read and write data into a joint transponder. In some scenarios, cable-based communication, e.g. using connections of digital inputs and outputs or fieldbuses, can be substitute. Using the new approach, it is not necessary to integrate the robot controller into the network of the factory. Since at all times no more than one robot can communicate with the transponder, an effective algorithm of collision avoidance has been implemented. We propose an approach that uses random waiting times for avoiding collisions. The approach is verified by a practical experiment, where two robots hand over a workpiece. Additionally, further scenarios of communication over RFID tags in intelligent workpieces are mentioned.

**Keywords:** Robot control · Robot-robot cooperation · Communication · Radio frequency identification · Collision avoidance · Intelligent workpieces

## 1 Introduction

In cases where multiple robots work in a shared workspace, the communication between the robots is a very important issue. Possible scenarios are, e.g., the handing over of workpieces, the cooperative handling of objects or collision avoidance between the robot manipulators. One method of communication is bidirectional connection between the digital inputs and the digital outputs of the robot controllers. However, this approach is not appropriate for flexible scenarios, where more than two robots are involved. Another common possibility is communication via fieldbuses such as *PROFINET*, *EtherCAT* or *DeviceNet* [3]. If mobile robots are involved in a multiple-robot scenario, e.g., automated guided vehicle systems moving on a factory floor towards stationary robots, cable-based communication is not applicable. Wireless data exchange is also conceivable, e.g., using *Wi-Fi* or *Bluetooth*. Considering *Ethernet*-based communication or *Wi-Fi*,


K. Berns and D. Görges (Eds.): RAAD 2019, AISC 980, pp. 544–552, 2020.
https://doi.org/10.1007/978-3-030-19648-6_62

in some cases, it may be undesirable to integrate all communication partners into the factory-floor network. Reasons can be, e.g., the security of the network with respect to viruses and Trojans, or the availability of the network.

An important technology in industrial manufacturing and logistics is radio-frequency identification (RFID) [1]. This is used to identify objects which have been equipped with RFID tags (RFID transponders). These can be divided into active and passive transponders. Active transponders contain a battery as a power supply. These types are not considered in this contribution. If a passive transponder is located in the electromagnetic field of an RFID antenna, communication between the transponder and the RFID reader is possible. The power for the transponder is provided by the electromagnetic field. Using RFID tags with read/write functionality, it is also possible to store and modify information in the tags.

In this paper, an approach will be developed and investigated to realize communication between robots using RFID technology. For this purpose, workpieces are equipped with read/write transponders. All communication partners contain an RFID antenna and an RFID reader. The communication should occur over the workpiece. The advantage of this approach is as follows: It is not necessary to build a communication network between all partners, which is an advantage with respect to flexibility, security and availability. The communication is decentralized, without risk of viruses and Trojans.

Decentralization is also important in the context of the term Industry 4.0, which refers to the computerization of industrial manufacturing. The design principles of Industry 4.0 are, e.g., the interconnectedness of different devices in manufacturing and the decentralized decisions of the devices [2]. Therefore, decentralized communication will support decentralized decisions.

As well as allowing communication over the transponders in the workpieces, they should also be used for local storage of object-specific data. With such information, attempts can be made to improve the robot task with respect to cycle time, energy consumption, etc. A workpiece which controls the robot may be also denoted an intelligent workpiece. Some proposals are mentioned in this contribution.

This paper is further organized as follows. In the next section, the approach of communication over RFID transponders will be presented. Thereafter, an example is shown where this approach is used for handing over workpieces between two robot manipulators. In Sect. 4, additional scenarios for robot work cells in which this method of communication may be used are mentioned, together with a discussion how intelligent workpieces can improve robot tasks. In this context, proposals for further research in this field are also provided. Finally, in the last section, the conclusion is given.

## 2 Communication over RFID Tags

As already mentioned in the previous section, there are various possibilities for realizing communication between several robot controllers. The principles

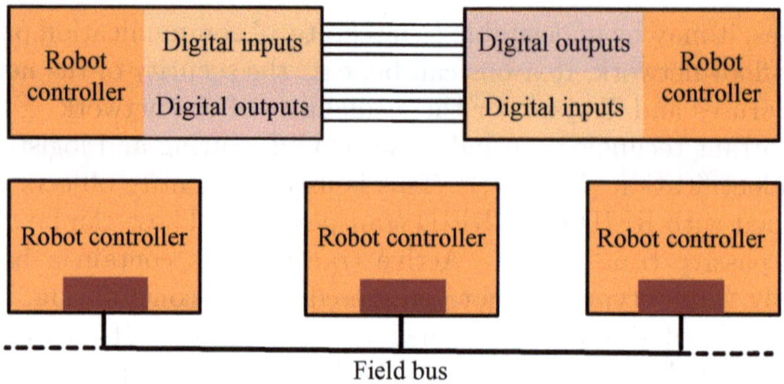


**Fig. 1.** Classical ways of communication between robot controllers.

of communication via digital inputs/outputs and via fieldbuses can be seen in Fig. 1. IO-based communication is inflexible and is only convenient for a small number of robot controllers. Fieldbus-based communication is also inflexible and may be expensive in cases where different kinds of fieldbuses are involved, e.g., when different robot controllers are used. IP-based fieldbuses may carry a risk of attacks to the factory network.

For these reasons, we propose that communication over RFID tags may be practical in some scenarios. The communication partners are equipped with an RFID reader and an antenna. The read/write transponder can be seen as a part of the data transmission medium. This idea is illustrated in Fig. 2. Because communication is impossible for certain orientations of the antenna and workpiece, it may be expedient to apply more than one transponder on the workpiece.

**Fig. 2.** Novel approach of communication over RFID tags.

Usually, access to the memory of the transponder is performed byte-by-byte. There are two different possibilities for communication. A shared memory area can be used, where all partners have read and write access. Alternatively, certain memory areas can be written by only one participant, similarly to the bidirectional input/output connection. It is very important that at most one antenna is active at all times. This means that the fields of all the other antennas located within the range of the transponder have been deactivated. Otherwise, the communication will be disturbed. For this reason, an algorithm has been developed and implemented for collision avoidance.

This algorithm will be described subsequently for the case of simple synchronization of two robots. *Robot I* should wait until the flag in the transponder is set to 1 by *Robot II*. For this purpose, *Robot I* tries to read the value of the flag. If the read operation is unsuccessful, e.g., due to a collision, *Robot I* must wait for a random time until it tries again. If the value of the flag is read successfully, it is used to test whether the condition is fulfilled. If so, the program of *Robot I* can be continued. Otherwise, it waits for a random time again, until *Robot I* resumes reading from the transponder. If *Robot II* is ready to set the flag to 1, it attempts to write this value into the transponder. If the write operation fails, *Robot II* now waits for a random time. Otherwise, *Robot II* resumes its own program. In some cases, it may be appropriate to wait a given time before continuing. The random waiting time should be a discrete function. The flow chart for communication between two robots, including collision avoidance, can be seen in Fig. 3. This approach should be next verified using a real scenario.

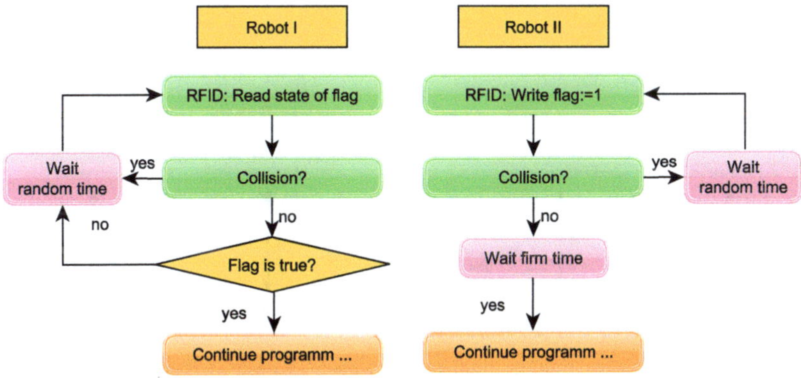


**Fig. 3.** Flow chart of communication between two robots.

## 3 Handing over of Workpieces

In this section, an example of communication over RFID tags will be presented. The experimental setup consists of two low-pay-load six-axis articulated robots. One is a classical industrial robot of type *ABB IRB 120* (*Robot I*). The second is the lightweight robot *UR3* from *Universal Robots* (*Robot II*). The robots are equipped with servo-electric grippers of type *WSG50* and *WSG32*, respectively. On one finger of each gripper, an RFID antenna of type *ISC.ANT40/30* with side lengths of 30 mm and 40 mm is mounted. These are connected to RFID readers of the *OBID ID ISC.LRM1002-E* type. Both devices are from *FEIG ELECTRONIC*. The antennas allow a transmission power of 1 W, which leads to a maximum communication range of approximately 7 cm. The workpieces are equipped with read/write transponders of type *TW-R20-B128* from *Turck*. They contain a memory (EEPROM) for local data storage with a size of 128 bytes. The experimental setup is shown in Fig. 4.

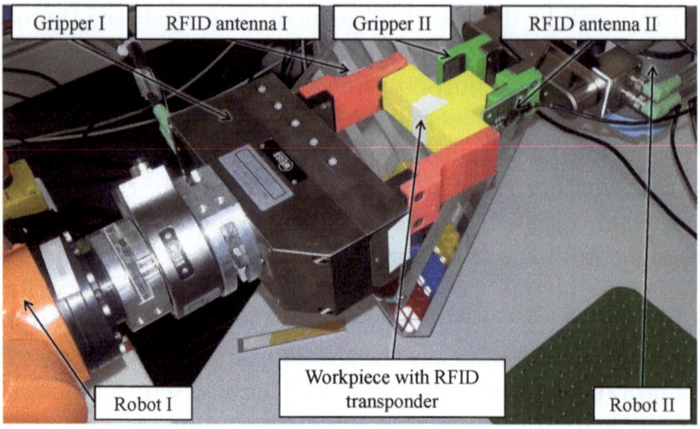


**Fig. 4.** Experimental setup for handling over workpieces.

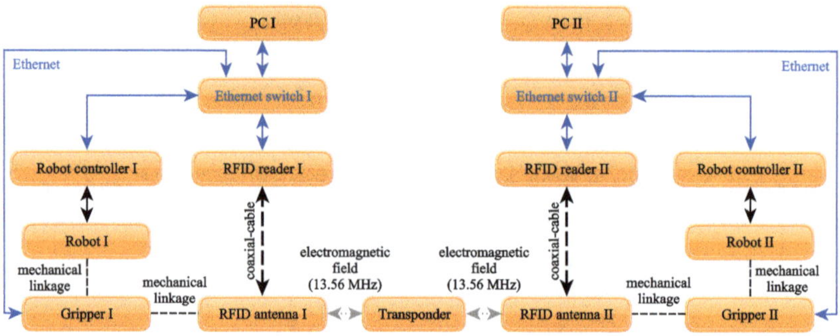


**Fig. 5.** Communication structure of the experimental setup.

The communication structure associated with the experimental setup is presented in Fig. 5. As can be seen, both robot systems are particularly isolated. In each system, a PC communicates with the robot controller, the RFID reader and the gripper, via an *Ethernet* connection. The PC can be understood as the master. It sends commands to the slaves and receives answers. For this purpose, in every robot controller, a program of the interpreter type is implemented. It would be also possible for the robot controller to take over the function of the master and to communicate directly with the gripper and the RFID reader. However, due to its universality, the former concept is chosen. If another type of robot is used, only a new interpreter program needs to be written.

The task is handing over the workpiece from *Robot I* to *Robot II*. It is clear that for successful handing over, the grippers must move in a coordinated manner, avoiding dropping the workpiece or jamming the robots. For this purpose, in the transponder, two flags $R1$ and $R2$ are defined and initialized with the value $false$. First, *Robot I* takes the workpiece, sets *R1* to *true* and brings the workpiece to the transfer position, where *Robot II* is waiting. *Robot II* closes its gripper, after the transponder is detected. Then, flag $R2$ is set to *true*. Simultaneously, *Robot I* reads the value of $R2$, in order to know when the workpiece is held by *Robot II*. Thereafter, *Robot I* opens its gripper and informs *Robot II* by writing *false* to flag $R1$. *Robot II* waits until it receives the value *false* when reading $R1$. After the workpiece is released by *Robot I*, both robots can move away. During the communication over the transponder, the proposed algorithm for collision avoidance works. An example of such a sequence can be seen in Fig. 6.

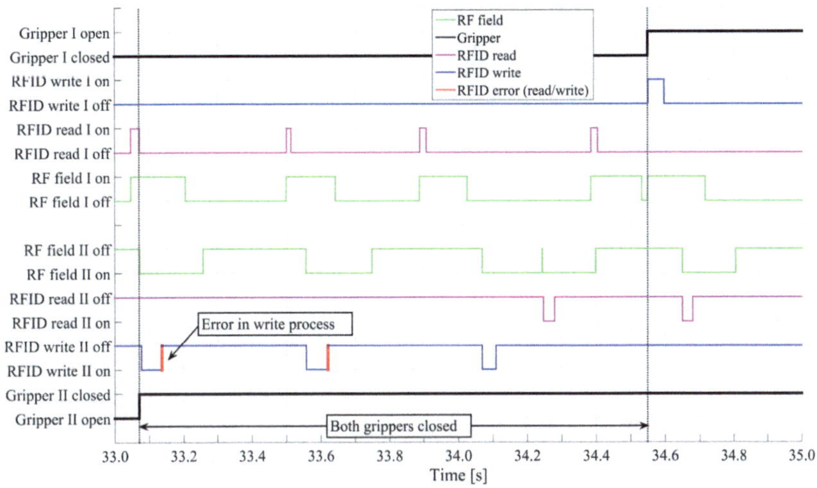


**Fig. 6.** Sequence of communication during handing over of a workpiece.

We have successfully show that it is possible to use the internal memory of a transponder to realize the communication between the robots. However, the possibility of local data storage on the workpiece gives room for additional improvement. For the present task of handing over objects using servo-electric grippers the following idea seems to be interesting. After the first grasping of the workpiece by *Robot I*, its width can be determined from the controller of the gripper. This value can then be written to the transponder. Before handing over, *Robot II* reads out the width and uses this value to pre-position its fingers. Such a feature may be useful for grasping sensitive workpieces quickly and carefully [8].

## 4 Further Scenarios

Some further examples of scenarios where transponder-based communication between robots over intelligent objects may be useful will be presented in the following paragraphs. Further possibilities for research in this field will be discussed.

### 4.1 Force-Based Cooperation Between Robots

In many applications, multiple robots tool a workpiece together, e.g., in welding processes. For this purpose, the robots work in the mode of position control. Another application of cooperation between robots is the cooperative handling of large and heavy workpieces with the objective of sharing the load. Here, it is necessary to avoid undesired high contact forces and torques, which may occur due to small position differences between the robots. Such forces and torques can be reduced by passive compliances, e.g., so-called RCC elements (remote center compliances) [4]. Another possibility is active force/torque control (FTC) of the cooperating robots [5]. For this purpose, at least one of the robots is equipped with a force/torque sensor (FTS) for measuring contact forces and torques. The measured values are then used to modify the robot's motion. A good overview of classical approaches to FTC is given in [10]. In [6], FTC is used during the cooperative motion of two robots.

As well as synchronized starting of the motions of the involved robots, communication during the motion may improve the quality of the FTC. In a scenario where a force-controlled slave robot follows a position-controlled master robot, it can be pre-controlled, e.g., by the current velocity of the master. This value may be sent to the slave over the transponder of the handled workpiece. In cases where the contact forces exceed certain limits, the master should reduce its velocity to give the slave a better chance of eliminating the control deviation. This information could also be provided via the RFID tag. A similar approach, denoted velocity control, is presented in [9] for force-controlled contour-following of a single robot.

Due to the influences of gravitational forces and torques, for FTC it is necessary to know the mass the and center of mass of the robot load. For intelligent workpieces, these values can be stored in their transponders and can be used for compensation of gravitational influences in the measured values of the FTS.

### 4.2 Localization Before Grasping

As we have already shown in [8], RFID technology can also be used for localization of objects. With respect to the handing over of workpieces between two robots, it cannot be assumed that their coordinate frames are synchronized every time, especially when considering flexible work cells. However, it is not necessary here to calibrate the tool frames of the robots with high precision. Displacements between the grippers within a certain range may corrected by localization of the transponder in the workpiece. This idea seems to provide an interesting avenue for further investigations. In this context, automated guided vehicles could also be included.

### 4.3 Collision Avoidance

In some processes, multiple robots work in a shared workspace. A very important issue is to prevent robot collisions. For this purpose, the idea of communication over RFID tags may also be used. For example, transponders can be placed at the entrance of the shared workspace. The information inside the transponder locks or unlocks the workspace. Such a solution may be especially suitable if mobile robots are involved. As well as this simple semaphore function of the transponder, more information may be provided by the transponder. It may contain information about the positions and dimensions of obstacles in a certain workspace. The data can be read by a robot entering the workspace and can be used for collision avoidance. Thus, the robot can plan its motion in a dynamic environment without the use of special sensors, such as cameras [7], laser scanners or ultrasonic detectors. In cases where the environment changes, the information must updated.

## 5 Conclusion

In this article, an approach was presented which realizes communication between multiple robot controllers without any direct connections between inputs and outputs, cable-bases network or wireless network. Communication is performed over RFID tags, which can be understood as part of the transmission medium. For successful data exchange, an algorithm for collision avoidance was developed, which controls the electromagnetic fields of the RFID antennas involved. The proposed approach was verified by practical experiments. For this purpose, the task of the handing over of workpieces between two robots was chosen. In addition, further possibilities for RFID-based communication between robots are presented and the potential of intelligent objects is discussed.

## References


1. Finkenzeller, K.: RFID Handbook: Fundamentals and Applications in Contactless Smart Cards and Identification, 2nd edn. Wiley, London (2003)
2. Hermann, M., Pentek, T., Otto, B.: Design principles for industrie 4.0 scenarios. In: Proceedings of 49th Hawaii International Conference on System Sciences (HICSS), pp. 3928–3937 (2016)
3. Klasen, F., Oestreich, V., Volz, M.: Industrial Communication with Fieldbus and Ethernet. VDE Verlag, Frankfurt (2011)
4. McKerrow, P.J.: Introduction to Robotics. Addison Wesley, Boston (1995)
5. Reinhart, G., Zaidan, S., Hubele, J.: Is force monitoring in cooperating industrial robots necessary? In: Proceedings of Joint Conference on Robotics - 41th International Symposium on Robotics and 6th German Conference on Robotics, pp. 523–530 (2010)
6. Spiller, A., Verl, A.: Superimposed force/torque-control of cooperating robots. In: Proceedings of Joint Conference on Robotics - 41th International Symposium on Robotics and 6th German Conference on Robotics, pp. 531–537 (2010)
7. Thormann, C., Winkler, A.: Collision avoidance of robots by artificial force field around geometric primitives using two range image sensors. In: Advances in Service and Industrial Robotics, pp. 102–112. Springer (2017)

8. Thormann, C., Winkler, A.: Localization and efficient grasping of objects by a manipulator using RFID technique. In: Proceedings of 22nd International Conference on Methods and Models in Automation and Robotics (MMAR), pp. 670–675 (2017)
9. Winkler, A., Suchý, J.: Force controlled contour following on unknown objects with an industrial robot. In: IEEE International Symposium on Robotic and Sensors Environments, pp. 208–2013 (2013)
10. Zeng, G., Hemami, A.: An overview of robot force control. Robotica **15**(5), 473–482 (1997)

# Average Consensus with Bounded Execution Under Quantization Noise

Martin Kenyeres[1(✉)] and Jozef Kenyeres[2]

[1] Institute of Informatics, Slovak Academy of Sciences, Dubravska cesta 9, Bratislava, Slovakia
martin.kenyeres@savba.sk

[2] Sipwise GmbH, Europaring F15, 2345 Brunn am Gebirge, Austria
jkenyeres@sipwise.com
http://www.ui.sav.sk/w/en/dep/mcdp/

**Abstract.** Quantization in the digital signal processing provides the representation of transmitted information with a finite number of the bits. Therefore, this process is crucial for the communication models with bounded bandwidth channels. This paper is focused on the average consensus algorithm with a finite-time execution and a uniform quantization update rule over high-scale wireless sensor networks modeled as random geometric graphs. The experimental part is concerned with a comparison of various initial configurations of the average consensus algorithm with a uniform quantization scheme with a modified number of bits allocated for quantization and different parameters of the implemented stopping criterion. Our goal is to find the most appropriate configuration in terms of the estimation precision quantified by the mean square error.

**Keywords:** Distributed computing · Average consensus · Quantization noise · Stopping criterion

## 1 Introduction

In the modern telecommunication systems, the data transmission and processing are primarily digital even though the most signals are in an analog form [1]. Thus, the entities forming these systems have to be equipped with an analog-to-digital converter for signal digitalization. Its main component is a quantizer, which affects the quality of the digitized signals and so, its design is a very important topic. As shown in the literature, a properly configured quantizer is a crucial design issue of many real-life applications, among others, WSN[1]-based ones. It can significantly optimize the communication process among the sensor nodes and reduce the memory requirements, thereby extending the lifetime of WSN applications. Therefore, many papers deal with this topic [2,3].

WSNs consist of autonomous entities (referred to as sensor nodes) organized into a cooperative network in order to sense various physical quantities [4]. WSNs

[1] Wireless sensor networks.


K. Berns and D. Görges (Eds.): RAAD 2019, AISC 980, pp. 553–560, 2020.
https://doi.org/10.1007/978-3-030-19648-6_63

can be formed by hundreds or even thousands of the sensor nodes, which allows sensing large-scale areas with high precision. Every sensor node is equipped with an independent microcontroller, a local memory, an energy source, sensors, and actuators [4]. The sensor nodes, often deployed in an ad hoc fashion and working in a self-organizing manner, communicate with each other over limited geographical areas via a wireless medium in order to provide high-quality information about the sensed physical quantity [4]. WSNs find a wide application in robotics (e.g. robot location, robotic exploration, beacons for robots). Due to technological advances, it is technically feasible to manufacture small sensors that are financially affordable but at a cost of communication, computation, memory and energy constraints [5]. Therefore, QoS of WSN-based applications can be significantly reduced since WSNs are not as robust to potential threats as the traditional IT systems. Thus, many of WSN-based applications are equipped with mechanisms for data aggregation to overcome (or at least minimize) the impact of negatives factors on QoS [6]. In this paper, our attention is focused on the average consensus algorithm (abbreviated as AC), a consensus-based algorithm for distributed averaging, with a fully-distributed stopping criterion proposed for WSNs. Its application ensures an efficiently bounded execution of the algorithm and therefore, its implementation is crucial in the real-life applications [7]. Many papers prove the applicability of AC in WSNs due to its lower computational/memory requirements and further advantages such as the absence of any fusion center, routing algorithms etc. [8].

Several papers are concerned with AC under quantization noise [9,10] but no one deals with AC bounded by the fully-distributed stopping criterion from [7].

## 2 Theoretical Background

### 2.1 Model of Average Consensus Algorithm with Bounded Execution

We consider WSNs formed by $n$ sensor nodes that are modeled as indirect, finite graphs $G$ determined by two sets $\mathbf{V}$ and $\mathbf{E}$ as follows: $G = (\mathbf{V}, \mathbf{E})$ [8]. Here, $\mathbf{V}$ is the vertex set gathering all the vertices representing the sensor nodes in a network. The edge set $\mathbf{E}$ is formed by all the edges, which represent the direct connection between two sensor nodes $v_i$ and $v_j$ (labeled as $e_{ij}$). As mentioned above, our attention is focused on distributed AC, which is based on neighbor-to-neighbor communication and an asymptotic convergence of the sensor inner states to the arithmetic mean. The sensor nodes iteratively update their inner states according to the inner states collected from the adjacent area and the inner state from the previous iteration [8]. Thus, AC can be modeled as a difference equation as follows [8]:

$$\mathbf{x}(k+1) = \mathbf{W} \times \mathbf{x}(k) \tag{1}$$

Here, $\mathbf{x}(k)$ poses a column vector of the inner states that is variant over the iterations and whose size is determined by $n$. We assume that $k = 0$ is the beginning of the algorithm and the initial inner state of each sensor node is a scalar value depend on a local measurement. Furthermore, $\mathbf{W}$ is the weight matrix, which is determined by the chosen weight design [8]. Their elements affect several aspects such as the convergence/the divergence, the convergence rate, the robustness etc. In this paper, we use the Constant weight design, whose weight matrix $\mathbf{W}$ is defined as follow [7]:

$$[W]_{ij} = \begin{cases} \epsilon, & \text{if } e_{ij} \in \mathbf{E} \\ 1 - d_i.\epsilon, & \text{if } i = j \\ 0, & \text{otherwise} \end{cases} \tag{2}$$

Here, $\epsilon$ is the mixing parameter, whose higher values result in a higher convergence rate, but too high values cause the divergence of the algorithm. $d_i$ represents the degree of the corresponding vertex/node $v_i$ and so the number of its neighbors. As mentioned earlier, AC asymptotically converges to the arithmetic mean (calculated from the initial inner states of all the sensor nodes in a network) and so, we can write the following [8]:

$$\lim_{k \to \infty} \mathbf{x}(k) = \lim_{k \to \infty} \mathbf{W}^k \times \mathbf{x}(0) = \frac{1}{n}.\mathbf{1} \times \mathbf{1}^{\mathrm{T}} \times \mathbf{x}(0) \tag{3}$$

Here, $\mathbf{1}$ is an all-ones column vector and $\mathbf{1}^{\mathrm{T}}$ is its transposed variant. The existence of the limit from (3) is essential for the proper functionality of AC and is ensured when these three convergence conditions are met [8]:

$$\mathbf{1}^{\mathrm{T}} \times \mathbf{W} = \mathbf{1}^{\mathrm{T}}, \mathbf{W} \times \mathbf{1} = \mathbf{1}, \rho(\mathbf{W} - \frac{1}{n} \cdot \mathbf{1} \times \mathbf{1}^{\mathrm{T}}) < 1 \tag{4}$$

Here, $\rho(\cdot)$ is the spectral radius of the corresponding matrix. As discussed in [8], the lower and the upper bounds adduced in (5) are a sufficient condition for the convergence (unless the underlying graph is bipartite and regular). Ignoring this interval may cause the divergence of the algorithm.

$$\epsilon \in (0, \frac{1}{\max_l\{d_l\}} > \tag{5}$$

As mentioned above, we assume the application of the stopping criterion, ensuring a finite time execution of the algorithm. We implement the one from [7], which is determined by two pre-set constants (*accuracy* and *counter threshold*) and one variable *counter*, independent for each sensor node.

## 2.2 Uniform Quantization Update Rule

In the real-life applications, the sensor nodes can exchange and store the information only with limited precision, i.e. the transmitted and the stored data have to be quantized [11]. Thus, all the sensor nodes transform their inner states into a binary code (we assume 2/4/8/16-bit codes) due to their bandwidth and power constraints [11]. In this paper, we assume a uniform quantization update scheme

**Algorithm 1.** Average consensus with stopping criterion and uniform quantization update rule

In the beginning, each sensor node $v_i \in \mathbf{V}$ sets its initial inner state (i.e. $x_i(0)$) according to e.g. a local measurement and quantizes it (i.e. $Q(x_i(0))$). Subsequently, each sensor node sets its *counter* to zero and become active until it locally completes AC.

At each iteration, each active node $v_i \in \mathbf{V}$ (i.e. the one that has not completed AC) repeates the following:

1. sends a broadcast message with its current inner state to its neighbors.
2. collects the inner states from its neighbors.
3. updates its inner state according to (1).
4. quantizes the new inner state according to (6).
5. verifies whether $|\Delta x_i(k)| < accuracy$. If so, it increment its *counter* by one, otherwise, sets it to zero.
6. verifies whether *counter* = *counter threshold.* If so, it considers AC to be locally completed and becomes inactive. Otherwise, it furthermore participates on the estimation process.

[9], i.e. the range[2] of the initial inner states $(\eta_l, \eta_h)$ is divided into $2^b - 1$ intervals of the same length (referred to as the quantization steps) [9]. Thus, the inner states within the range of the initial states $(\eta_l, \eta_h)$ can take $2^b$ values, which is equal to the number of the quantization levels. Here, $b$ is the number of the bits allocated for quantization. The quantization step can be then defined as $\Delta_{step} = (\eta_h - \eta_l)/(2^b - 1)$. Subsequently, the formula (1) can be reformulated as follows [9]:

$$\mathbf{x}(k+1) = Q(\mathbf{W} \times \mathbf{x}(k)) \tag{6}$$

Here, the operator $Q$ represents the implementation of a quantization scheme (element-wise quantization is assumed). As mentioned above, we focus on a uniform quantization scheme, i.e. the quantization step is constant between each two quantization levels and the inner states are rounded to the nearest quantization level as follows [9]:

$$Q(x_i(k)) = \begin{cases} \lfloor x_i(k) \rfloor, & if\ x_i(k) < \frac{\lfloor x_i(k) \rfloor + \lceil x_i(k) \rceil}{2} \\ \lceil x_i(k) \rceil, & if\ x_i(k) \geq \frac{\lfloor x_i(k) \rfloor + \lceil x_i(k) \rceil}{2} \end{cases} \tag{7}$$

Here, $\lfloor x_i(k) \rfloor$ and $\lceil x_i(k) \rceil$ represents the nearest lowest and highest quantization level. A formalization of AC under quantization noise and with the stopping criterion from [7] is adduced in Algorithm 1.

## 3 Experiments and Discussion

As mentioned above, we examine AC with various mixing parameters and bounded execution by the stopping criterion from [7] under quantization noise. In our experiments, we use 30 random geometric graphs (RGGs) formed by 200 vertices. The initial states of the sensor nodes are random i.i.d. values of the

[2] the difference between the maximal and the minimal initial inner state.

Bernoulli distribution defined as:

$$\begin{aligned} Pr(x_i(0) = 1) = 0.5 \\ Pr(x_i(0) = 0) = 0.5 \end{aligned} \tag{8}$$

The mixing parameter takes the following four values: $0.25.\frac{1}{\max\limits_{l}\{d_l\}}$, $0.5.\frac{1}{\max\limits_{l}\{d_l\}}$, $0.75.\frac{1}{\max\limits_{l}\{d_l\}}$, $\frac{1}{\max\limits_{l}\{d_l\}}$ (abbreviated as $\epsilon = 0.25$, $\epsilon = 0.5$, $\epsilon = 0.75$, $\epsilon = 1$), the number of the bits allocated for quantization takes the following values: 2, 4, 8, 16, and the parameters of the stopping criterion are varied as follows:

- *accuracy* $= 10^{-1}$, $10^{-4}$, $10^{-6}$
- *counter threshold* = 3–100 (with the step equal to one)

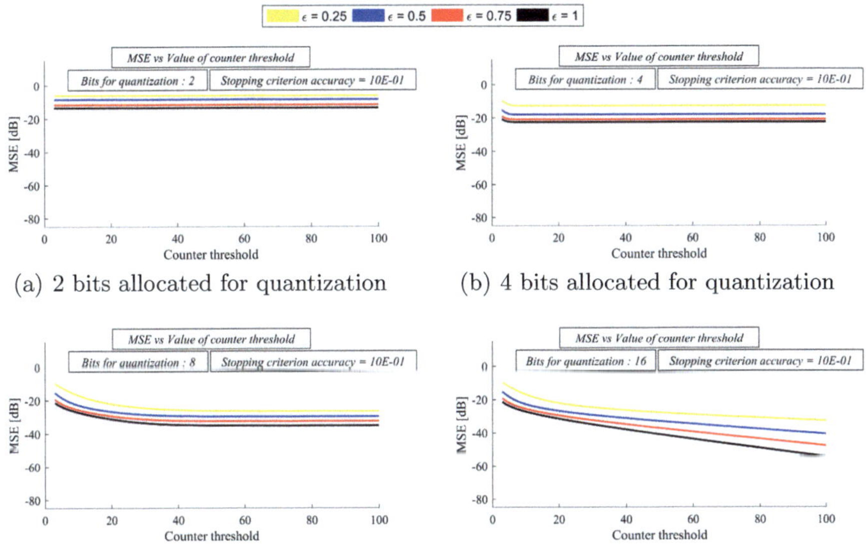


(a) 2 bits allocated for quantization (b) 4 bits allocated for quantization

(c) 8 bits allocated for quantization (d) 16 bits allocated for quantization

**Fig. 1.** MSE [dB] averaged over 30 random geometric graphs as function of *counter threshold* for *accuracy* $= 10^{-1}$

We use the mean square error (MSE) as a metric to quantify the performance of AC. We compute MSE at the iteration when the last sensor node completes AC and so, can write the followings:

$$\text{MSE} = \frac{1}{n}.\sum_{i=1}^{n}\left(x_i(k_l) - \mathbf{1}^{\mathrm{T}} \times \frac{\mathbf{x}(0)}{n}\right)^2 \tag{9}$$

Here, $k_l$ is the label of the first iteration at which AC is completed at all the sensor nodes. We show only MSE over all 30 RGGs.

In Fig. 1(a–d), we show the results obtained when *accuracy* takes $10^{-1}$. We can see that the highest performance is achieved by AC with the mixing parameter $\epsilon = \frac{1}{\max\{d_l\}}$ and the performance decreases as the value of $\epsilon$ is decreased regardless of the number of the bits allocated for quantization. Moreover, it is seen that more allocated bits result in a lower error and so, higher precision of the final estimates. Furthermore, it can be seen that the value of *counter threshold* does not affect MSE when two bits are allocated for quantization. If we use four quantization bits, a higher value of *counter threshold* ensures higher precision. However, from approximately *counter threshold* = 10, MSE does not change for any $\epsilon$ in spite of an increase in *counter threshold.* For eight allocated bits for quantization, the character is the same as in the previous case with the difference that MSE is almost not decreased from approximately *counter threshold* = 65 for each $\epsilon$. For sixteen allocated bits, the value of *counter threshold* affects MSE over the whole examined interval. In Figs. 2(a–d) and 3(a–d), we show the results for *accuracy* = $10^{-4}$ and *accuracy* = $10^{-6}$, respectively. From the results, we can see that the character of MSE is the same as for *accuracy* = $10^{-1}$. Only differences are that the values of *counter threshold* for which MSE is not decreased are lower and for *accuracy* = $10^{-6}$, higher values of *counter threshold* do not affect MSE at all in contrast to two other lower values of *accuracy*.

In the next paragraph, we focus our attention on the impact of *accuracy* on MSE. At first, let us compare Figs. 1(a), 2(a), and 3(a), where two bits are allocated for quantization - we can see that the value of *accuracy* has no impact

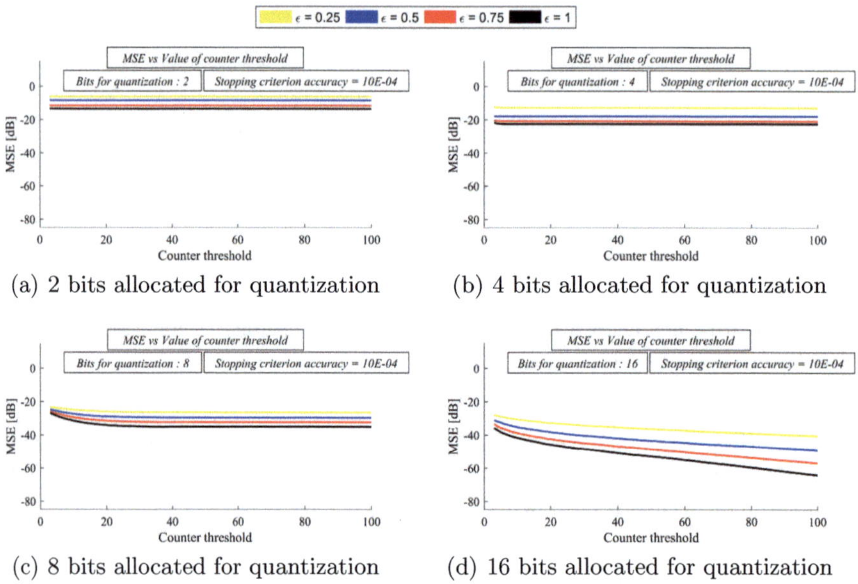


(a) 2 bits allocated for quantization

(b) 4 bits allocated for quantization

(c) 8 bits allocated for quantization

(d) 16 bits allocated for quantization

**Fig. 2.** MSE [dB] averaged over 30 random geometric graphs as function of *counter threshold* for *accuracy* = $10^{-4}$

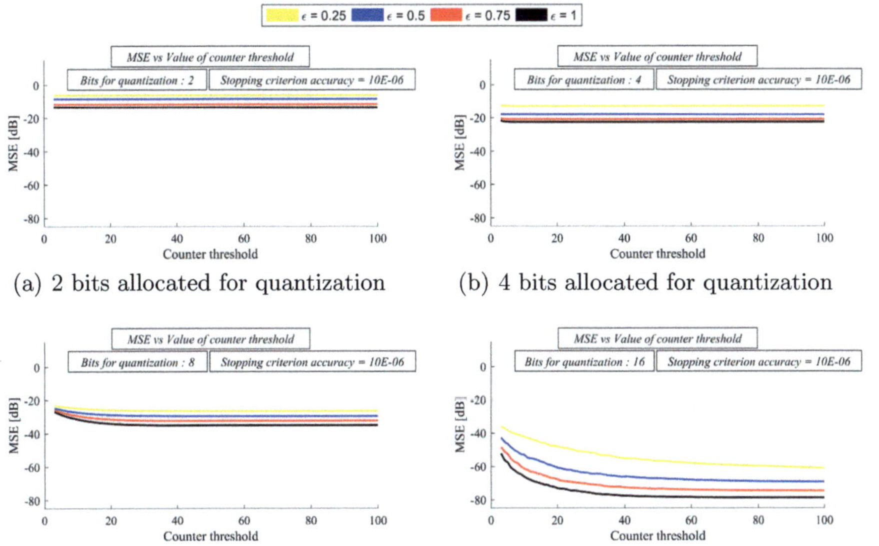


(a) 2 bits allocated for quantization

(b) 4 bits allocated for quantization

(c) 8 bits allocated for quantization

(d) 16 bits allocated for quantization

**Fig. 3.** MSE [dB] averaged over 30 random geometric graphs as function of *counter threshold* for $accuracy = 10^{-6}$

on MSE for this quantizer configuration. When four quantization bits are used, we can see from Figs. 1(b), 2(b), and 3(b) that the configurations with $accuracy = 10^{-4}$, $10^{-6}$ achieve the same performance and outperform the configuration $accuracy = 10^{-1}$. The same character is observed also when eight bits are allocated for quantization - see Figs. 1(c), 2(c), and 3(c). For the configuration with sixteen bits (Figs. 1(d), 2(d), and 3(d)), we can see that a higher value of *accuracy* results in a higher precision of the final estimates.

## 4 Conclusion

In this paper, we analyze AC bounded by a stopping criterion proposed for WSNs with four various mixing parameters and the initial states of a Bernoulli distribution under quantization noise (a uniform quantization scheme is assumed) over 30 RGGs. We vary the configuration of the implemented stopping criterion and examine MSE for a different number of the bits allocated for quantization. From the experimental results, we can see that a higher value of the mixing parameter $\epsilon$ results in lower MSE and so, higher precision of the final estimates. Moreover, an increase in the number of the quantization bits ensures lower MSE regardless of the value of the mixing parameter and *accuracy*, *counter threshold* of the analyzed stopping criterion. Furthermore, the value of *counter threshold* has either no or just a marginal impact on MSE in the scenarios with a lower number of the bits for quantization. In the other scenarios, its impact depends on *accuracy* - a higher value of *counter threshold* ensures lower MSE. However, from the exact iteration whose value depends on *accuracy* (its higher value causes that this occurs for lower values of *counter threshold*), an increase in *counter threshold* does not affect the value of MSE. Furthermore, it depends on the number of the bits allocated for quantization how the value of *accuracy* affects MSE - for 2 bits, its value has no impact on MSE, for 4 bits and 8 bits, it affects MSE only slightly, and for 16 bits, the impact is the most notable.

Thus, the final conclusion of this paper is that a higher value of the mixing parameter and the number of the bits allocated for quantization always ensures higher precision of the final estimates. The configuration of the stopping criterion (i.e. the values of *accuracy* and *counter threshold*) either increases the precision or has no impact on MSE.

**Acknowledgement.** This work was supported by the VEGA agency under the contract No. 2/0155/19 and by the Slovak Research and Development Agency under the contract No. APVV-17-0116.

## References

1. Dincic, M.R., Peric, Z.H.: Multiproduct uniform polar quantizer. Radioengineering **24**(1), 233–239 (2015). https://doi.org/10.13164/re.2015.0233
2. Ciuonzo, D., Rossi, P.S.: Quantizer design for generalized locally optimum detectors in wireless sensor networks. IEEE Wirel. Commun. Lett. **7**(2), 162–165 (2018). https://doi.org/10.1109/LWC.2017.2762305
3. Bai, X., Wang, Z., Zou, L., Cheng, C.: Target tracking for wireless localization systems with degraded measurements and quantization effects. IEEE Trans. Ind. Electron. **65**(12), 9687–9697 (2018). https://doi.org/10.1109/TIE.2018.2813982
4. Stankovic, J.A.: Wireless sensor networks. Computer **41**(10), 92–95 (2008). https://doi.org/10.1109/MC.2008.441
5. Al-Karaki, J.N., Kamal, A.E.: Routing techniques in wireless sensor networks: a survey. IEEE Wirel. Commun. **11**(6), 6–28 (2004). https://doi.org/10.1109/MWC.2004.1368893
6. Izadi, D., Abawajy, J.H., Ghanavati, S., Herawan, T.: A data fusion method in wireless sensor networks. Sensors **15**(2), 2964–2979 (2015). https://doi.org/10.3390/s150202964
7. Kenyeres, J., Kenyeres, M., Rupp, M., Farkas, P.: WSN implementation of the average consensus algorithm. In: Proceeding of the 17th European Wireless 2011 - Sustainable Wireless Technologies, VDE, Germany, pp. 1–8 (2011)
8. Kenyeres, M., Kenyeres, J., Skorpil, V., Burget, R.: Distributed aggregate function estimation by biphasically configured Metropolis-Hasting weight model. Radioengineering **26**(2), 479–495 (2017). https://doi.org/10.13164/re.2017.0479
9. Jafarizadeh, S., Jamalipour, A.: Weight optimization for distributed average consensus algorithm in symmetric, CCS and KCS star networks. Submitted to IEEE Trans. Automat. Contr.
10. Huaqing, L., et al.: Event-triggered distributed average consensus over directed digital networks with limited communication bandwidth. IEEE Trans. Cybern. **46**(12), 3098–3110 (2016). https://doi.org/10.1109/TCYB.2015.2496977
11. Hongbin, L., Fang, J.: Distributed adaptive quantization and estimation for wireless sensor networks. IEEE Signal Process. Lett. **14**(10), 669–672 (2007). https://doi.org/10.1109/LSP.2007.896390

# Author Index





K. Berns and D. Görges (Eds.): RAAD 2019, AISC 980, pp. 561–563, 2020.
https://doi.org/10.1007/978-3-030-19648-6

MIX
Papier aus verantwortungsvollen Quellen
Paper from responsible sources
FSC® C105338
FSC
www.fsc.org

If you have any concerns about our products,
you can contact us on
**ProductSafety@springernature.com**

In case Publisher is established outside the EU,
the EU authorized representative is:
**Springer Nature Customer Service Center GmbH**
**Europaplatz 3, 69115 Heidelberg, Germany**

Printed by Libri Plureos GmbH
in Hamburg, Germany